0-07-049309-X	Pelton	*Voice Processing*
0-07-707778-4	Perley	*Migrating to Open Systems: Taming the Tiger*
0-07-049663-3	Peterson	*TCP/IP Networking: A Guide to the IBM Environment*
0-07-051143-8	Ranade/Sackett	*Advanced SNA Networking: A Professional's Guide to VTAM/NCP*
0-07-054418-2	Sackett	*IBM's Token-Ring Networking Handbook*
0-07-057628-9	Simon	*Workgroup Computing: Workflow, Groupware, and Messaging*
0-07-057442-1	Simonds	*McGraw-Hill LAN Communications Handbook*
0-07-060360-X	Spohn	*Data Network Design*
0-07-063636-2	Terplan	*Effective Management of Local Area Networks: Functions, Instruments and People*
0-07-067375-6	Vaughn	*Client/Server System Design and Implementation*

The McGraw-Hill
Internetworking
Handbook

The McGraw-Hill Internetworking Handbook

Ed Taylor

McGraw-Hill, Inc.

New York   San Francisco   Washington, D.C.   Auckland   Bogotá
Caracas   Lisbon   London   Madrid   Mexico City   Milan
Montreal   New Delhi   San Juan   Singapore
Sydney   Tokyo   Toronto

Library of Congress Cataloging-in-Publication Data

Taylor, Ed, (date.)
   The McGraw-Hill internetworking handbook / Ed Taylor.
      p.   cm.
   Includes index.
   ISBN 0-07-063263-4
   1. Internetworking (Telecommunication)—Handbooks, manuals, etc.
I. Title.
TK5105.5.T392 1994
004.6—dc20   94-24060
                                                          CIP

Trademarks are listed on pages 745 through 750.

   3 4 5 6 7 8 9 0 DOC/DOC 9 0 0 9 8 7 6

ISBN 0-07-063263-4

*The sponsoring editor for this book was Jerry Papke, the editing super-
visor was Peggy Lamb, and the production supervisor was Donald F.
Schmidt. It was set in Century Schoolbook by Cynthia L. Lewis of
McGraw-Hill's Professional Book Group composition unit.*

Printed and bound by R. R. Donnelley & Sons Company.

McGraw-Hill books are available at special quantity discounts to use as
premiums and sales promotions, or for use in corporate training programs. For more information, please write to the Director of Special
Sales, McGraw-Hill, Inc., 11 West 19th Street, New York, NY 10011. Or
contact your local bookstore.

Completed in memory of my Dad

This book is dedicated to:
My wife Lisa
Luther and Kitty Adkins
John and Zelma Gandy
Isaac May
Ted Taylor
All Residents in Florence, MS, U.S.A.
From:
IBONE

Contents

Part 2   Lower-Layer Protocols

Preface

In Retrospect

En route to San Francisco, CA August, 1993 for the Interop convention, I contemplated the history of internetworking as I ate lunch during the flight. After lunch, I wrote my thoughts on a napkin. They were:

> First came something, next came a variety, chaos followed. Pain brought forth integration; ignorance en masse followed. An awareness of gridlock appeared. Education culminated in a glut of information. An attempt to harness this information followed. Mankind awaited its destiny.

8-24-93
Ed Taylor

Purpose of This Book

This book is a definitive reference manual for topics listed in the table of contents. Its scope is limited to selected information about these topics. It is not totally exhaustive about each topic, for to attempt such in one book would be an insurmountable task. From years of experience, I realized select information would create a definitive reference text. The topics and concepts presented here are prevelant in the marketplace.

It is specifically designed for:

Network Integrators

Network Managers/Administrators

Network Designers

Network Consultants

Systems Programmers

Systems Engineers

Corporations with multiple networks

Businesses in general worldwide

Educational institutions that require a definitive reference on internetworking topics

Individuals who desire to know more about topics in the Contents

Others will find it invaluable as well.

Reader.   You can contact me.

Internet: Edtaylor@aol.com

Compuserve: 72714,1417

America Online: Edtaylor

Acknowledgments

I want to thank the following for contributions made that made parts of this book possible.

Ann and Jim Brown, for re-creating figures from my artwork; Bedford, TX

Bob Thomas; Dallas, TX

IBM
Greg Cooley; Dallas, TX
Tom Trotter; Dallas, TX
James S. Miller, Jr.; Dallas, TX
Jan L. Goldberg; IBM Education, Atlanta, GA
Brent Boisvert; Glendale Programming Lab
Sally Walsh; Glendale Progamming Lab
Curtis Hughes; Boulder, CO
Krista Judson; Boulder, CO
Patty Rustin; Boulder, CO
Keith Sarntee; Boulder, CO
Mike Bowers; Research Triangle Park, NC
Donna Mueller; Chicago, IL
Rosalyn Wilson; Dallas, TX

Skill Dynamics
Karen Murphy; Chicago, IL

Digital Equipment Corporation
John Adams
Larry Walker
Stacy Maitland

Cardinal Business Media
Ceil Perry; Dallas, TX
Denny Yost; Dallas, TX

George Dishman; Dallas, TX
Mary B. Thomas; Dallas, TX

Hewlett-Packard

Dave Richardson; Ft. Collins, CO
Tim Jones; Ft. Collins, CO

Dell Computer   Jim Woodward; Austin, TX

Novell Corporation   Linda for documentation

Convex Computer Corporation   Don Davis; Richardson, TX

Apple Computer Corporation   Cupertino, CA

Microsoft Systems Journal   San Francisco, CA

Bernie (B. C.) Hogan   Dallas, TX

McGraw-Hill, Inc.

Jerry Papke; New York, NY
Peggy Lamb; New York, NY
Cynthia Lewis; New York, NY
Don Schmidt; New York, NY
The rest of the NY team

Trademarks are listed on pages 745 through 750.

The McGraw-Hill
Internetworking
Handbook

Networking Fundamentals

1

Commonalities
among Networks

As the title implies, networks have commonalities. Even proprietary vendor networks have commonalities with nonproprietary networks. Network commonalities can be reduced to a single aspect: resource sharing. Networks can be characterized in a number of ways. Some of the networking categories discussed in this chapter employ terms commonly used today, but uncommon insight is provided here.

1.1 Perspective

A look at the past

The origins of computing are based on a *centralized* approach, and a brief look at the past may clarify or broaden the reader's perspective of technology today. Approximately one century ago (actually in 1890) a *tabulating device,* as it was called at that time, computed the U.S. census in record speed. Created by Herman Hollerith, this device used punched cards and calculated the U.S. census in record time.

Prior to Hollerith's tabulating machine was Charles Babbage's *analytical engine,* which—although it was never actually built—Babbage planned to develop to replace his earlier model, the *differential engine,* which he had toiled over for years.

Both Babbage and Hollerith, along with others during the nineteenth century, contributed greatly to the mechanical devices that were the forerunners of the computers we use today. Another major contributor to nineteenth-century computing technology was George Boole. He invented what we know today as *symbolic logic* or *boolean algebra.* His works brought us the basics of the algebraic operations known as AND, NOT, and OR, collectively referred to as *logic gates.*

These operands, and Boole's basic approach to logic, provided a structure for arriving at swift conclusions without verbiage. This, coupled with other inventions during the nineteenth century, laid the framework for shaping the direction of early twentieth-century technical accomplishments.

By the 1930s the scientific community had developed vacuum-tube technology, which became prevalent in the 1940s, and in 1947 the transistor was invented. These technological advancements, coupled with the contributions of the mathematician John Von Neumann, began to solidify what is now considered, in retrospect, to be the foundation of computing technology from the 1950s to the present.

John Von Neumann contributed the fundamentals of computers that apply even today. He described the following as key components for computer architecture:

- Arithmetic logic unit
- Memory
- Input
- Output
- Central control unit (which became the CPU)

These elements became the basis for computers decades ago and are the fundamental components of computers today. A major issue today concerning technological evolution is how these and other components are implemented.

Centralized computing

From a critical analysis it is certainly more reasonable to believe that computing was centralized before it was decentralized. The brief historical perspective just provided coupled with an understanding of the basic history of technology over the past 50 years provide sufficient evidence that centralized computing came first.

Centralized computing could be used to categorically define the 1960s technology to a considerable degree. It is the technology of the 1960s and 1970s that brought the fundamentals together to begin to make *decentralized computing,* as we have come to know it today, possible. Consider the origins of the first viable networks in the late 1960s and early 1970s. At that time these networks were considered state-of-the-art technology, but by today's standards they are rudimentary. Much of the computing in the 1970s would be considered centralized if it were defined by the current standards for centralized computing. It was the technological advances of the mid- to late 1970s and the 1980s that began to make decentralized computing possible.

Decentralized computing

In the 1980s decentralized computing began to come into fruition. Technology, market forces of supply and demand, and a different philosophical approach to computing seemed to usher in what is now considered decentralized computing. A tremendous amount of technology was conceived and developed in the 1980s; much of this technology provided the ability to utilize computers in ways not previously possible.

A few concise examples pave the way for exploration into this matter. From 1980 to 1985 the following technology was brought to market:

- WordPerfect Version 1
- IBM Personal Computer (PC)
- Hayes 300 Smartmodem
- Compaq Computer (introduced an IBM-compatible PC)
- Lotus 1-2-3 (released)
- Intel 286 microchip (revealed by Intel)
- Apple Computer IIe
- Apple Macintosh
- Novell NetWare file server
- Microsoft Word
- Disk operating system (DOS)
- IBM token ring
- Intel 386 microchip
- Microsoft Windows Version 1.0

These examples, which are purposely PC-oriented, began to revolutionize people's perceptions of computing. By 1989, a glut of PC hardware and software was the order of the day. If the PC revolution was not enough to alter general perceptions and use of computers, consider other aspects of the technology.

Local area networks (LANs) were first launched in the 1980s. Hardware and software making a variety of computing functions possible were brought to market. Devices that made it possible to connect heterogeneous networks sprouted on the market. All sorts of network-specific devices were introduced. Routers, bridges, servers, repeaters, and gateways were brought to market and made possible scenarios that are still being contemplated and implemented.

Distributed programming became a reality in the 1980s. Mixing computers and networks from different vendors became the trend, and the notion of distributed heterogeneous networking became a reality. A distinct aspect of the 1980s can be summarized by a philosophical shift in the *way* many thought about *how* computing, networking, and ultimately *work* would be performed.

Decentralized computing can be defined many ways. Today it means for some *where* one does one's work is not very important anymore. In a sense using computers today is similar to using a telephone; the idea of having to go to a particular *place* to use either is becoming a matter of history. In other words, because of the glut of portable PCs and cellular telephones on the market today, physical location of the workstation or the person initiating or receiving phone calls is no longer relevant.

Decentralized computing for some means that a corporation is no longer restricted to having just one physical location and/or duplicating equipment in another location to perform the same task. For others decentralized computing means that a central repository for data can be maintained and used at will from practically anywhere. Decentralized computing also means that printing may occur in one facility (or city) and data entry in another, while actual processing occurs in a third location.

Defining decentralized computing is difficult at best because the term is open to a wide variety of definitions. In fact, centralized computing actually occurs when decentralized computing is implemented. The nature of implementing a decentralized computing environment means that *some* functions (computing) will be performed locally— wherever "locally" is.

Network backbone

The term *network backbone* is confusing unless those participating in the conversation understand what they mean to convey and others understand what they believe another person is communicating. Actually it is best defined as a loose term generally considered slang. Consider the following examples.

A network backbone can be perceived as the upper-layer protocols that make up the bulk of the network or the lower-layer protocols within the network, or the prevalent protocol(s). Still another example could refer to the dominant media used throughout the network. Other uses probably exist, but these are the dominant uses of the phrase.

Physical and logical characteristics

Two terms are used to explain various aspects of networks and sometimes the network itself. Generally the term *physical* refers to something tangible; however, aberrations of this usage exist, as this text points out in later chapters.

In reference to the term *logical*, it is generally used to define a function rather than to imply sound reasoning. For example, a network protocol that is explained in later chapters uses the terms *physical* and *logical* frequently.

In the case of this protocol [system network architecture (SNA)] multiple transmission lines may exist. These are literally physical, tangible objects. Put together, they can connect two devices. These transmission lines that form a transmission group may have data moving through a particular one because of a better throughput. Consider Fig. 1.1.

Figure 1.1 depicts two hosts, two devices, three physical lines (A, B, and C) connecting devices 1 and 2, and a circle around all three lines reflecting that collectively they are considered a group, and it shows data flow over link C.

Reference to these three lines can be thus: Three physical lines exist between devices 1 and 2, and the logical path where data flows in this example is through line C. This is an example and theoretically data could pass over line A or B. Since data could potentially flow through either A, B, or C, reference to the selected path is considered logical. This may not necessarily be the *best* route for data flow, but it is by definition in this example the logical path. In fact it is not only logical but is also the physical path.

In most cases something referred to as *logical* in a network usually maps to something physical. Exceptions may exist, but in most cases this rule applies. Therefore, understanding that logical does not necessarily mean reasonable, or the best way for something to occur, the term *logical* does refer to a characteristic or aspect of an occurrence— or the potential thereof.

Conversely, *physical* does not always mean physical in the sense of something tangible. In some uses of the term, *physical* refers to a function, service, or capability of a device. However, reality does set in, and

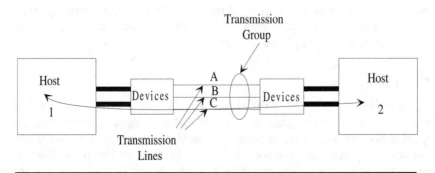

Figure 1.1 Physical and logical characteristics.

if it is logical, it must be based on something physical, but the fact that it is physical does not necessarily mean that it is logical.

Network categories

Various terms can be used to describe a network. Some of the popular terms include

- Local area network (LAN)
- Wide area network (WAN)
- Metropolitan area network (MAN)
- Car area network (CAN)
- House area network (HAN)

Acronyms seem to dominate the computer and networking terminology; the previous list exemplifies this. LANs are generally considered as being in a specific physical locale. Consider the origins of LANs. Some LANs had distance restrictions to a single facility, and if a facility were large, a LAN might be restricted to a particular area within the facility. Today LANs have a somewhat nebulous meaning, but they are still considered physically local.

Wide area networks (WANs) have existed for at least two decades. Today WANs span the globe. Commercial, governmental, and other organizations or systems have existing WANs today. The meaning of the term *wide area network* is fairly standard throughout the computer literature industry in general; most agree that WANs cover large geographic distances.

Metropolitan area networks (MANs) have come of age. These networks are generally agreed to cover a large metropolitan area. One example is the Dallas–Fort Worth, Texas, metropolitan area. One university has a MAN that connects multiple sites located throughout the metroplex, which is some 30 to 40 mi (48 to 64 km). As a general rule, MANs are physically more dispersed than LANs, but not to the point of spanning large geographic distances.

1.2 Topologies

The term *topology* refers to the structure of a network. This means that most networks can be described by physical and logical characteristics. As this section explains, some examples that represent a network are not the same as the network when seen in reality. For a network to exist, something physical must be present, but a logical network (or multiple logical networks) may be mapped to the physical network.

Bus

One aspect of a bus network is the common medium that connects multiple devices together. Physically, bus networks are considered to exist in one central location. Consider Fig. 1.2.

This figure depicts a logical view of a bus; granted, it may be more or less detailed than you may be used to viewing, but it nevertheless conveys the concept of a bus network. Fundamental to the bus network is a single medium. This may be coaxial or fiber-optic cable. Notice that each host connects to the network via a transceiver and each end of the medium is terminated. To a certain degree Fig. 1.2 is a popular representation of bus networks. Figure 1.2 is best described as a logical view of a bus topology.

In reality, when a bus topology is implemented it rarely, if ever, appears as depicted in Fig. 1.2. Figure 1.3 represents a bus topology as it is normally implemented. This figure shows the same components, but the line depicting the medium is not straight. When bus topology is used, the physical cable may be placed between walls separating rooms, in the attic, or between floors in multilevel office buildings.

Star

A star network is best represented by Fig. 1.4.

A star implementation uses a central processor to which multiple hosts are connected. It may appear similar to a ring topology but it differs because no internal ring exists. Basic to the operation of this topology is that data is routed through the central processor.

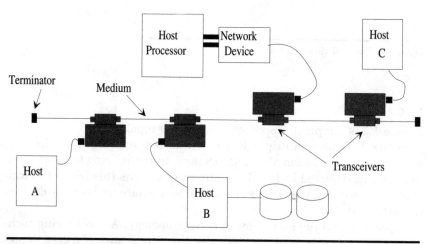

Figure 1.2 Logical view of a bus.

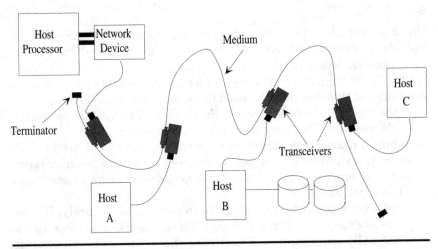

Figure 1.3 Physical view of a bus network.

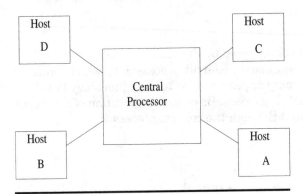

Figure 1.4 Physical view of a star network.

Ring

Figure 1.5 is a typical representation of ring topology.

Figure 1.5 shows multiple devices connected to a ring. This figure is a logical representation of a ring topology. In reality ring topology does not exist as depicted in Fig. 1.5; a drawing such as this is used to conceptually convey a ring implementation and whatever hosts or devices are attached.

Figure 1.6 is a physical view of a ring topology. Actually, ring technology is physically implemented with the appearance of a star implementation. Ring topology is implemented through what is called a

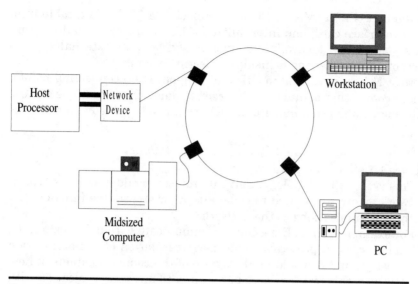

Figure 1.5 Logical view of a token ring.

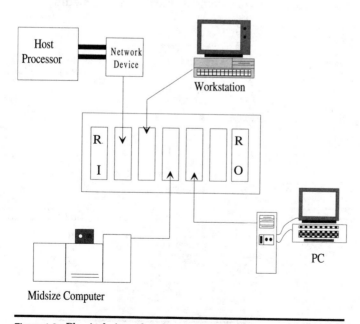

Figure 1.6 Physical view of medium access controller.

media access unit (MAU). Devices connect to a MAU via a cable from their interface card, and internally the MAU is a ring (actually a dual ring). MAUs are generally rectangular and have no external appearance of a ring. The ring is inside connecting each port.

Most MAUs have ring-in (RI) and ring-out (RO) connecting points. These connecting points exist in order to daisy-chain multiple MAUs, thus increasing the number of devices that can be added to a ring.

Tree

Figure 1.7 shows a typical tree topology.

This is a flexible topology. Certain protocols could make this type of network easily segmented to isolate users or streamline functionality of the communications in the network.

For example, if host E needed to communicate with only hosts D, G, F, and C, certain protocols would permit isolation of communication and hosts B and A would not be aware of the communications of host E. However, this topology does permit any-to-any communication with attached hosts.

1.3 Transmission Media

All networks use transmission media to communicate. Two types of media are used to connect network devices: hard and soft media.

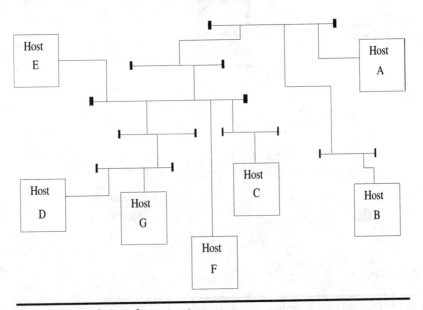

Figure 1.7 Logical view of tree structure.

Hard media

Twisted-pair cable. *Hard media* include various distinctly different types of media. For example, the simplest of the hard media is the twisted-pair cable (see Fig. 1.8).

Twisted-pair cabling is normally copper-stranded cable. In some instances this type of cable has shielding on each stranded group; this is the norm. However, some twisted-pair cables are collectively compounded together and benefit from a shield and an outer jacket housing all the individual shields.

Twisted-pair cable is measured by gauge. The lower the gauge number, the larger the cable physically and the more current it can support. Interpreting cable gauge is not arbitrary; formal definitions apply. For example, ≥20-gauge cabling is typically found in networking scenarios, whereas <20-gauge cable is larger in size. For instance, a number 10- or 8-gauge twisted-pair cable is common in electrical wiring; each strand of this twisted-pair cable is about the diameter of an ordinary pencil and can accommodate considerable electrical voltage and current.

All twisted-pair cabling, regardless of size, has one fundamental characteristic with respect to electric current: resistance. Twisted-pair cable electrical capacity (whether alternating or direct current) is directly proportional to the cable length. The greater the cable length, the greater the resistance and hence the signal loss.

The varying degrees of resistance incurred with difference gauges of cable have been tabulated elsewhere in the literature, but I have witnessed two scenarios: (1) those cases in which signal loss was so great, because of cable length, that devices at the far end of the cable were not functional, and (2) ironically, those showing the converse.

I have participated in wiring a terminal with twisted-pair cable that far exceeded any specifications for the cable, and the terminal worked. The only answer to this paradox must be that, in fact, the maximum distance was not exceeded. The precise point (length) where a cable begins to undergo significant signal loss is contingent on at least two factors: (1) the device driving the cable and cable size and (2) the environment in which the cable is placed.

Coaxial cable. Coaxial cable is prevalent in many networks today and is widely used in cable television installations. Figure 1.9 depicts a coaxial cable.

Figure 1.8 Twisted-pair cabling.

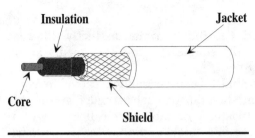

Figure 1.9   Coaxial cable.

Coaxial cable has an outside covering called a *jacket*. Immediately inside the jacket is a material called a *shield*—generally fine wire wrapped around its inner component—which typically serves as a ground. Just inside the shield is a plastic material that serves as insulation. Inside the insulated material is the core (the actual cable). The core is usually solid copper cable.

Coaxial cable also has a means of measurement. Its rating differs from that of twisted-pair cable, but the ideas between the two are similar. The larger the core, the greater the length that a cable can be implemented. However, coaxial cable differs from twisted-pair cable.

Coaxial cable has good resistance to outside interference because of cable design. The cable core is insulated, shielded, and jacketed.

Fiber-optic cable.   Fiber-optic cable appears as shown in Fig. 1.10.

Fiber-optic cable has an outer covering called a *sheath*. Inside some cables have strands of string or strengthening component. Next, as Fig. 1.10 shows, is the cladding. *Cladding* is glass or other transparent material surrounding the optical fibers. Inside the cladding is the *core*; this part of the cable carries the signal.

Fiber-optic cable is distinctly different from any form of copper cable. Fiber-optic cabling does not emit any electromagnetic properties; nor is it sensitive to these properties. Fiber can span distances much farther

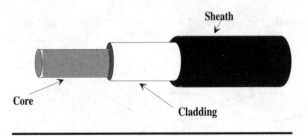

Figure 1.10   Fiber-optic cable.

than copper-based cable before a signal loss factor is significant. However, length is not the only factor in the computation of signal loss. In fiber-optic cable the refractive index is important also. The *refractive index* is a ratio of the speed of light in a vacuum compared to the speed of light in a material such as optical fibers. Pulse dispersion is another factor in the equation used to calculate bandwidth and hence data loss. Simply defined, *pulse dispersion* is the spreading of pulses as they traverse fiber-optic cable.

There is a standard rule for measuring fiber-optic cable. Numbers used to indicate the size of fiber reflect the core size and the cladding size. Some examples of sizes are a 50-μm core and 125-μm cladding (measured from side to side); this type of rating would appear as 50/125. Another example of fiber is a core size of 62.5 μm and 125-μm cladding. Other sizes and types of fiber exist; however, fiber is used to move photons, not electrons.

Soft media

Soft media in fact uses hard tangible components, but it moves data through means other than a cable of some sort.

Satellite communication.   Satellite communications utilize satellites in orbit around the earth. These satellites are *geosynchronous*; this means that they are orbiting the earth at approximately 36,000 km above the equator and remain stationary. Hence, they are synchronous with the earth's revolutions. These satellites remain in a fixed position, permitting signals to be transmitted to and from earth stations. Figure 1.11 illustrates this interesting concept.

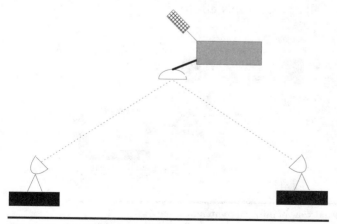

Figure 1.11   Satellite communications.

Three satellites positioned 120° apart can cover the entire earth, from a communications standpoint. In satellite communications a transmitter is used at the point of origin and a receiver is used at the destination point. The term *satellite communications* implies that transmission will be delayed. This delay may be measured in seconds or fractions of seconds, but it nevertheless occurs.

Infrared communication. This type of communication uses different hardware and frequency bands. Infrared communications is a *line-of-sight* method of communication, meaning that if anything interferes with it (physically), the signal is impaired. At the time of this writing infrared LANs are beginning to be introduced into the marketplace. Figure 1.12 conveys the concept of infrared communication.

Figure 1.12 shows two devices are communicating. This could be implemented in physical plants where distances are long and are clear between two points where communication needs to occur.

Microwave communication. This is similar to satellite communication in that transmitters and receivers must be located at the origin and destination points, respectively. Microwave communications, however, do not use orbiting devices in communications; they use line-of-sight communication. Figure 1.13 is an example of this idea.

Microwave communication is possible between sites approximately 25 to 30 km apart. Distance of microwave communication is restricted primarily because of the curvature of the earth. However, the transmitter and receiver tower heights and the terrain also affect this equation. Microwave communications are popular in metropolitan areas and provide an effective means for signal transmission to various locales outside metropolitan areas. Microwave communications are cost-effective as well as technically effective.

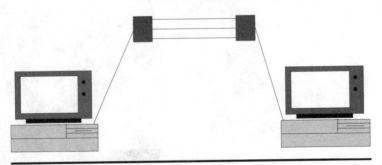

Figure 1.12   Conceptual view of infrared communication.

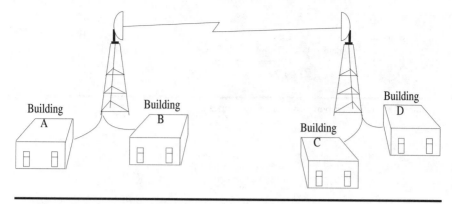

Figure 1.13   Conceptual view of microwave communication.

1.4   Physical Communication Link Configurations

Physical links used in data communications can assume a variety of configurations. Certain technologies described in this book use different technological implementations to achieve their purposes. This section explores some of the most popular implementations.

Consider Fig. 1.14, which is an example of a point-to-point connection.

A *point-to-point connection* is just that: linking one place to another. This type of connection is either switched or dedicated. Figure 1.15 depicts a switched connection.

A *switched communication* line is best described as being in use when needed and not in use when not needed; hence the term *switched*. An excellent example of this is the ordinary telephone. It is used only when necessary, i.e., switched.

A dedicated point-to-point communication line may appear as either scenario shown in Fig. 1.16.

Figure 1.16 illustrates two examples of point-to-point connection. Hosts A and B could represent two hosts in one physical location with

Figure 1.14   Point-to-point communication.

Figure 1.15 Switched point-to-point communication.

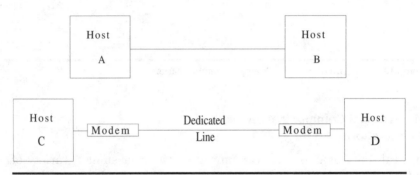

Figure 1.16 Dedicated point-to-point communication.

a line *dedicated* between them. Hosts C and D are linked using a modem and what would be regarded as a *switched* connection. The difference between a point-to-point connection and a switched one is that the former is normally dedicated. Analogously, the modems do not release the line; rather, the signal is maintained. This type of connection is typically referred to as a *leased line* because a circuit must be used to accommodate this service. Leased-line connections are commonly used in large commercial corporations in the United States.

Figure 1.17 depicts what is called a *multipoint* or *multidrop line.*

In this example one host exists with a communication line attached and multiple devices connected to the line. This configuration, which is similar to a street with houses located on it, is easy to understand. At any given time the host may communicate with a device *downstream,* and vice versa. The idea behind this type of communication is one single path for data flow and multiple device accessibility. Scenarios similar to that shown in Fig. 1.17 are generally nonswitched.

Figure 1.18 depicts a ring or loop communication line configuration.

These devices normally operate in what is considered *nonswitched* lines. In some implementations using nonswitched lines devices can be removed from and inserted into the line without disrupting other devices on the line.

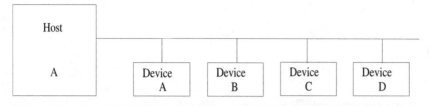

Figure 1.17 Multipoint communication line.

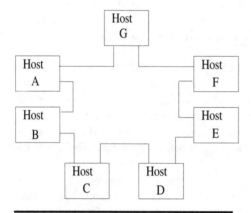

Figure 1.18 Ring or Loop Communication Line.

1.5 Summary

Herman Hollerith, George Boole, Charles Babbage, John Von Neumann, and many others were the forerunners who advanced computer and network technology to where it is today. Each in his own way made significant contributions.

Herman Hollerith formed a company that later became what has been known for decades as International Business Machines (IBM). IBM's growth leaped during the 1940s with World War II in progress. The contributions of this company cannot go without recognition. IBM has become one of the largest companies ever known from various perspectives. Its contribution in the technical arena has left an imprint on the technical community that is destined to last well into the twenty-first century.

George Boole's approach to reasoning differed from that of Aristotle. Boole's contribution has literally impacted technology everywhere. As a product of his work we have the following equation:

$$X = Y$$

$$Y = Z$$

Therefore

$$X = Z$$

This equation, known as the *hypothetical syllogism,* is the foundation of the vast number of programs written in today's computers and networking technology.

Charles Babbage's contribution to the technical community includes his papers and ideas on machines that could compute and print large volumes of numbers. Some of Babbage's ideas were implemented decades later in various ways. Thus, indirectly, his ideas were eventually achieved.

John Von Neumann conceived the fundamental computer architecture existing in the vast majority of personal computers today. In their own way, the components he defined as what a computer required for fundamental operation shape the present computer industry.

Many other individuals contributed ideas, products, and pieces to the puzzle that has become known as *computers.* Their contributions are nonetheless important.

Centralized and decentralized computing exist today. In a sense decentralized computing is actually centralized computing. The notion that a singular task may be performed in a given location with other tasks performed in other locations is at the heart of decentralized computing.

Networks can be categorized by the geographic distance they occupy. The idea of physical and logical phenomena presented the idea that all things logical are mapped to something physical; however, not all things that are physical are logical.

Topologies reflect the actual implementation of a network and also the way in which the human mind perceives it. Media of some sort are a common ground among all networks. Hard media include different types of cabling. Soft media include satellite, microwave, and infrared transmission. Communication lines can be categorized by the nature of link operations. Multiple examples were presented in this chapter.

Conclusions about computers and networks can be drawn. Technology has come a long way in a short time. In a sense a time existed when people seemed to drive the technological engine, but it is clear that today technology is used to drive technology, and the result is exponential to design, manufacturing, and product delivery.

In 1890 when the punched card was used to tabulate the U.S. census people were competing in time, but now the tools have changed and the apparent percentage of time available for competition continually decreases when in reality it is the devices used that have changed. Computers and networking of all sorts have ushered in a generational era that presupposes that a considerable amount is constant as certain elements when in fact it is not.

Computers and technology today have brought us where no one has been before in the history of the world. In a sense technology has changed how business is conducted and lives are lived, and has put us in a position to achieve what has heretofore been unthinkable to some. In less than a century we have gone from horse and carriage for transportation to putting people in space on a regular basis. In fact, in less than 15 years technology has rendered devices that make computing and communication possible from having to be in a physical location to mobility practically anywhere.

2

Data Communication Considerations

Networking, internetworking, and computer communication, by necessity, are involved in what is called *data communication*. I suppose a technical definition exists for the term somewhere, but it is generally used to refer to computer communications involving all layers within a network. A similar word, *telecommunication,* seems to connote more involvement with the lower layers in a network. Both terms reflect fundamental characteristics required for successful communications to occur in a network, among computers, or within a diverse internetworked environment.

This chapter focuses on some topics that arise in work with networks or internetworked environments. Understanding some topics in this section is crucial to troubleshooting problems at lower levels within a network. The better understanding readers have of the abstract concepts of network or communication operations, the more effective they will be in troubleshooting a particular problem.

2.1 Signal Characteristics

Communication between entities is achieved through signals of some sort. This is true with humans or machines. Humans normally use speech, whereas networks, computers, and internetworking devices use electrical or optic signals. These signals have many characteristics, and the signal type (*electrical* or *optic*) determines (to a degree) the characteristics of that signal.

This section explores signals and characteristics related to them. The details presented here are intended as a reference source for readers working at fundamental layers within a network.

Signal types

A *signal* can be defined or characterized in many ways; however, for our purposes here the difference between analog and digital signals is explained.

Analog   An *analog* signal can be described by what it is not. It is neither on or off, positive or negative, nor some other diametrically opposed position. An example would be a dimmer switch used in electrical lighting. The function of the dimmer switch is to vary light intensity without full intensity or zero intensity (light turned off) (unless those two states of intensity are desired).

Digital   A *digital signal* is best defined as being in either an on or off state and with no in-between point. In data communication a digital signal is a binary 1 or 0.

Signaling methods

Two signaling methods exist: baseband and broadband. *Baseband signaling* uses digital signal techniques for transmission, and *broadband signaling* uses analog signal techniques for transmission. Baseband signaling has a limited bandwidth in general, whereas broadband signaling has a large bandwidth potential.

Evaluating signals

Signals, either analog or digital, are based on fundamental trigonometric functions. As a result, understanding some fundamental principles of signaling facilitates the evaluation of waveforms.

A baseline for evaluating waveforms is rooted in the cartesian coordinate system. This system of measurement is mathematically unique for defining or locating a point on a line or plane or in space. Fundamental to this coordinate system is numbered lines that intersect at right angles (see Fig. 2.1).

This coordinate system and other aspects of trigonometry are used in signaling. Here the focus is on certain characteristics of signals.

Signals in computers, networking, or data communication can be categorized as either analog of digital. Consequently, with the aid of this coordinate system along with other tools, it is possible to explain signal patterns.

Analog and digital commonalities

Both analog and digital signals have certain common characteristics. Each can be evaluated for amplitude, frequency, and phase.

Amplitude.   The *amplitude* of a signal refers to height of the signal with respect to a baseline. Height may be a positive or negative voltage. The baseline is a zero-voltage reference point. This amplitude value is proportional to the movement of the curve about the X (horizontal) axis (abscissa) of the coordinate system as shown in Fig. 2.1.

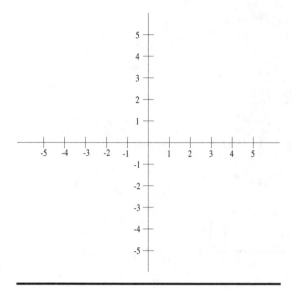

Figure 2.1 Cartesian coordinate system.

Frequency. *Frequency* is the number of cycles a wave makes per second. Specifically, frequency is measured in *hertz* (Hz), which is also known as a unit of frequency. A *cycle* is a complete signal revolution from zero to the maximum positive voltage past zero to the maximum negative voltage, then back to zero. In the cartesian coordinate system this represents a complete revolution from 0 to 360°. Figure 2.2 shows the signal characteristics for one cycle.

Figure 2.2 shows one cycle and the signal with respect to time, amplitude, and frequency.

Phase. *Phase* is normally measured in degrees that represent the location of a waveform. Another way of thinking about phase is that it is the relative time of a signal with respect to another signal. A change of phase without the change of frequency or amplitude results in the scenario depicted in Fig. 2.3.

Figure 2.3 shows three waveforms: *A, B,* and *C.* In this example waveform *A* has a 20° phase angle or leads waveform *B* by 20°. Determination of the leading waveform is derived by visually ascertaining which waveform crosses the *X* axis (abscissa) first; in this case it is waveform *A.* Waveform *C* on the other hand is lagging waveform *A* by 20°.

Figure 2.4 is an example of two signal waveforms (better known as *sine waves*) that have the same frequency but are out of phase with respect to each other.

In general, signals transmitted over a medium are subject to varying frequencies. Phase can be thought of as the distance a waveform is from

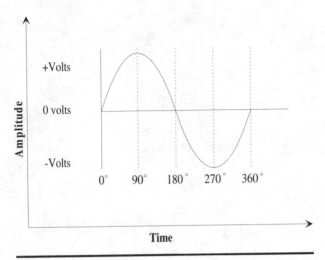

Figure 2.2 Signal characteristics.

its point of origin (which is zero degrees). This is particularly important in examination of transmission characteristics of encoded signals. We will discuss encoded signaling characteristics later; however, the significance of this information becomes real when one uses different measuring scopes to troubleshoot a line with varying frequencies.

Period. A *period* is best described as the length of a cycle. It is defined as the time required for signal transmission of a wavelength.

Waveforms

Waveforms come in many forms. Two are discussed here: the sine wave and the square wave.

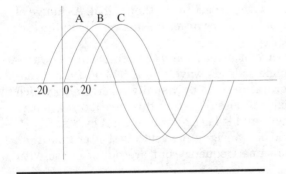

Figure 2.3 Example of a phase angle.

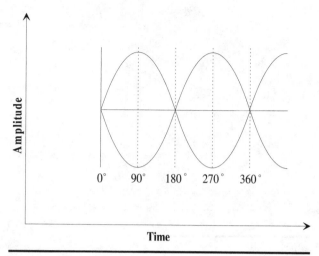

Figure 2.4   Phase differentiation.

Sine wave.   A *sine wave* can be defined as a periodic wave. Characteristically, this is a wave's amplitude based on the sine of its linear quantity of phase or time (see Fig. 2.5).

Square wave.   A *square wave* is a wave with a square shape (see Fig. 2.6).

The square wave has characteristics similar to those of the sine wave except that the *form* has a square versus a *wave* appearance.

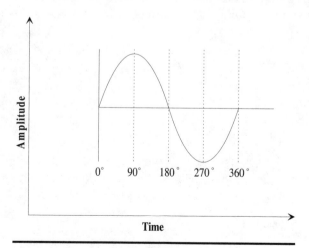

Figure 2.5   Example of a sine wave.

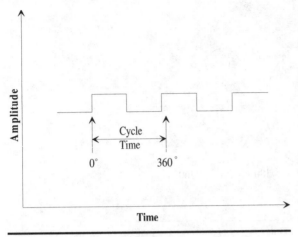

Figure 2.6   Example of a square wave.

2.2 Data Representation

Before discussing characteristics of transmission, it is important to understand how the data is represented. Two methods of data representation are presented here: binary and hexadecimal.

Binary

Binary data representation uses ones and zeros (1s and 0s) to represent alphanumeric characters within computer systems. A *bit* is a single digit, either one or zero. A *byte* is 8 bits.

The binary numbering system is based on powers of 2 and progresses from right to left. Consider the following example:

Powers of 2	Value
2×0	1
2×1	2
2×2	4
2×3	8
2×4	16
2×5	32
2×6	64
2×7	128
2×8	256
2×9	512
2×10	1024
2×11	2048
2×12	4096

One byte is 2×0, which is eight binary digits (either 1s or 0s). Another way of approaching binary representation is by way of the

American Standard Code for Information Interchange (ASCII). ASCII is a method of data representation that uses 128 permutations of arrangements of ones and zeros.

This means that with a computer using an ASCII character set letters, numbers, control codes, and other keyboard symbols have a specific binary relationship. Consider the following examples which show the correlation of a letter or number in the ASCII character set and its binary equivalent.

Letter or number	Binary value
A	01000001
a	01100001
E	01000101
2	00110010
3	00110011
7	00110111
T	01010100
t	01110100
"	00100010
-	00101101

This example means that each time a key is pressed on a keyboard the equivalent binary value is generated. This value is represented inside the computer as a voltage.

Computers for the most part are digital; that is, either a 1 or a 0. Hence, converting letters, numbers, or control codes to a numeric value is straightforward.

Hexadecimal

Hexadecimal or *hex,* as the term is used, refers to a numbering scheme that uses a base 16 for counting. Consider the following example:

Hexadecimal value	Decimal value
0	0
1	1
2	2
3	3
4	4
5	5
6	6
7	7
8	8
9	9
A	10
B	11
C	12
D	13
E	14
F	15

The hexadecimal system of numeric representation is a shorthand, if you will, representing binary values. Consider the following example:

Letter/number	Binary	Hex value
A	01000001	41
a	01100001	61
E	01000101	45
2	00110010	02
3	00110011	03
7	00110111	07
T	01010100	54
k	01101011	6B
m	01101101	6D
z	01111010	7A

As this example indicates, it is easier to represent the letter z by hex value 7A than its binary representation.

Other methods of representation exist, such as Octal code, which uses base 8 for a numbering system and a decimal. Of these, binary and hex are prevalent, and understanding this binary representation is helpful with other data communication concepts.

A final word about data representation. IBM uses its Extended Binary Coded Decimal Interchange Code (EBCDIC) as a prevalent method for representing data, alphanumerics, and control codes. EBCDIC uses an arrangement of 256 ones and zeros (1s and 0s) to make this possible. It should also be noted that ASCII and EBCDIC are not one-for-one interchangeable.

2.3 Transmission Characteristics

Other terms are used to describe concepts and functions of data communication. These are fundamental to networking and interconnectivity of all sorts.

Channel

The term *channel* refers to a concept and component and is best understood in the context of usage. For example, use of channel to refer to a concept could be: "The current system supports multiple channels." Here the term loosely refers to a path into a given system.

Use of the term *channel* to refer to a component could be: "A byte channel supports such-and-such throughput." However, generally the term refers to a data path along which data moves.

Different types of channels exist. Some are described by a grade such as voice, narrowband, wideband, or broadband channels. A wide variety of channels exists today. With increasing movement toward fiber channels and encoding and compression techniques a channel can be exploited in ways not thought possible only a few decades ago.

Bandwidth

A channel's capacity is termed *bandwidth,* which is simply the difference between the highest frequency and the lowest frequency with which signals can be sent simultaneously across the channel. Bandwidth directly reflects data transfer rate of the channel. Obviously the higher the bandwidth, the higher the data rate.

Asynchronous transmission

Asynchronous transmission, also termed "start/stop transmission," is characterized by character-oriented protocols. The reason for this is that data is transmitted asynchronously and timed by the start and stop bits of the frame, primarily the start bit. Consider Fig. 2.7.

Figure 2.7 depicts a start bit, data bits, and a stop bit. In asynchronous transmission, the start bit notifies the receiving entity that data bits follow. Likewise, the stop bit signifies the end of data bits.

A problem with this method of communication exists if the last data bit and the stop bit are the same. If this occurs, the receiving entity is confused. Otherwise, whether the last data bit and the stop bit are the same is relatively unimportant because this is overcome with parity.

Parity is achieved by the originating entity counting the number of bits and appending the outgoing character as necessary to achieve an even or odd parity. The receiver, on the other hand, calculates for parity against 7 data bits and compares it to the parity bit received (the parity bit is the eighth bit transmitted). If the parity sent and the computation on behalf of the receiver do not match, an error has occurred.

Terms normally used with parity are *odd* and *even;* they reflect an accurate representation of the transmission. However, the terms MARK and SPACE are sometimes used as well. When they are, they refer to parity as the bit settings of 1 and 0, respectively.

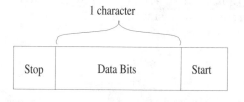

Figure 2.7 Example of an asynchronous frame.

Ironically, *asynchronous serial communication* is a misnomer. In fact, the start bit actually synchronizes the following bits, whereas synchronous serial communication is synchronized by byte. Both are synchronized; asynchronous synchronization is performed on bits, while synchronous synchronization is typically performed at the byte level, as is explained below. The bottom line is that theoretically more overhead occurs with asynchronous communication than with synchronous communication.

Synchronous transmission

Basic to synchronous transmission is the intent to reduce the overhead inherent in *asynchronous* transmission and to provide more efficient error detection and correction. Perspectives of two categories of synchronous protocols are explained here: byte-oriented and bit-oriented protocols.

Byte-oriented.   An example of a byte-oriented protocol used in synchronous transmission is IBM's binary synchronous (BISYNC) protocol. Introduced by IBM in 1967, it appears as shown in Fig. 2.8.

Figure 2.8 shows the beginning field as the synchronization (SYN) character. This precedes all data, and a synchronization (SYN) control character may even be inserted in the middle of a long message to ensure synchronization. Data codes supported in BISYNC are ASCII, EBCDIC, and a 6-bit transparent code.

The start-of-text (STX) character indicates that data immediately follows. The end-of-text character (ETX) follows data. If a BISYNC transmission is lengthy and divided into segments, only the last segment will have an ETX indicator.

The *block-check character* (BCC) can be either a *longitudinal redundancy check* (LRC) or a *cyclic redundancy check* (CRC).

The byte-oriented protocol BISYNC is not as dominant now as it was in the 1970s. Its code dependence and transparency implementation are not sufficiently flexible to support the current popular needs.

Bit-oriented.   Two examples of bit-oriented protocols transmitted synchronously are High-Level Data-Link Control (HDLC) and Synchronous Data-Link Control (SDLC). Figure 2.9 is an example of a SDLC frame.

S Y N	S T X	DATA	E T X	B C C

Figure 2.8 Byte-oriented protocol for synchronous transmission.

F l a g	Address	Control	Data	Frame Check Sequence	F l a g

Figure 2.9 Bit-oriented protocol used in synchronous transmission.

The beginning and ending flag has a reserved value and is always 01111110 (7E). The address field in the SDLC frame contains addresses. The control field (CF) in the frame indicates the type of frame: control, information, or supervisory. The data field contains the data being transmitted.

The frame-check sequence (FCS) is implemented for determining whether errors in transmission have occurred. The last field is the ending flag.

SDLC supports code transparency because it was designed into the protocol. The result of this architecture is good performance with low overhead.

Synchronization of these byte- and bit-oriented protocols is achieved by performing error checks on larger blocks of data, and the result is less overhead.

Serial transmission

Another transmission characteristic is how data is moved from one entity to another. *Serial communication* occurs bit by bit. An example of this is fiber-optic-based data transfer. In this example photons are moved in serial fashion through the medium. Figure 2.10 is an example of serial transmission of a T through a medium.

Parallel transmission

Parallel transmission is movement of data along a channel path in byte form. An example of this is IBM's parallel channels. More information is provided about parallel channels in Part 2, but the essence of trans-

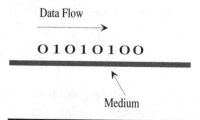

Figure 2.10   Serial transmission.

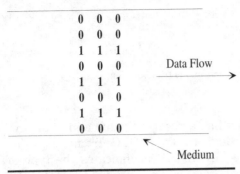

Figure 2.11 Example of parallel transmission.

mission is moving data in bytes rather than sequential bits. Figure 2.11 is an example of parallel data transfer.

Simplex transmission

Simplex transmission, which is a reference to the direction that data can move at any given instance, is best exemplified by analogy to radio-station broadcasting and multiple receivers detecting the signal. Hence, the direction of data flow is one-way.

Half-duplex transmission

In *half-duplex transmission,* data can flow in either of two directions, but not simultaneously; data flows in only one direction at a time. An analogy of this is courteous communication between individuals. Normally, one speaks while another listens; then the reverse happens. Unfortunately, this does not have to be the case with humans, but it does with technology.

Full-duplex transmission

In *full-duplex transmission* data flows in both directions at the same time. The implication here is that simultaneous data transfer can occur and be interpreted by both parties. Each entity can send and receive at the same time.

2.4 Multiplexing

Multiplexing, which involves maximum utilization of a channel, can be accomplished in a variety of ways; however, two are popular and are explained here.

Frequency-division multiplexing (FDM)

FDM is as its name implies: the multiplexing of frequencies. This type of multiplexing can be readily used with analog transmission because frequencies are divided and then are multiplexed onto the medium. Consider Fig. 2.12.

Figure 2.12 shows one medium and three devices connected to the medium. Each device transmits on a different frequency; in this hypothetical example the frequencies are 1, 2, and 3. These hypothetical frequencies could realistically be 10 to 14 kHz, 5 to 9 kHz, and 0 Hz to 4000 kHz. The premise of a FDM multiplexer is that each device uses a range of frequencies and that they stay within that range. The point is that bandwidth of the medium is utilized effectively to accommodate multiple users communicating on different frequencies.

Time-division multiplexing (TDM)

TDM is, as its name implies, multiplexing data via time. Figure 2.13 illustrates this idea.

Time-division multiplexing is the utilization of the medium by time-slicing devices attached to the multiplexer. Its premise for operation is based on this concept of time-based data transfer. In Fig. 2.13 three devices are attached to the multiplexer and the multiplexer itself multiplexes signals from devices 1, 2, and 3 to maximize the channel. Although not shown, on the receiving end of the data path is another multiplexer which demultiplexes the data to its destination point.

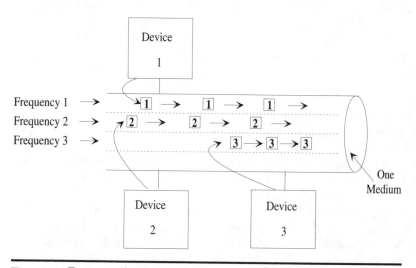

Figure 2.12 Frequency-division multiplexing.

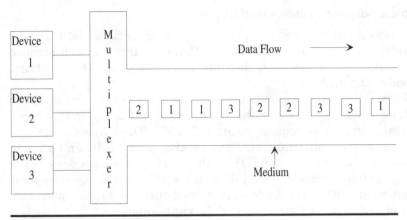

Figure 2.13 Time-division multiplexing.

Types of multiplexers

There are many types of multiplexers. For example, a network-based multiplexer may provide service for a variety of devices, each utilizing different transfer speeds. T1 multiplexers operate at T1 speeds (1.54 Mbits/s). Another type of T1 multiplexer is the fractional T1, which supports fractions of the T1 speeds that are incremented by 56 or 64 kbits/s. Normally, 1 to 23 circuits can be derived from a fractional T1.

2.5 Physical Interface Considerations

At the physical layer (OSI level 1) within a network, numerous topics must be considered. A number of well-written books focus on only the physical layer within a network. In those books the information presented is focused and concise and some reference books applicable are listed in the bibliography.

What is an interface?

If defined in the context of networking and computers, an *interface* can be considered a point for common ground between independent systems. In the networking community two terms apply to this issue regarding the interface: data terminal equipment (DTE) and data circuit-terminating equipment (DCE).

A DTE is generally thought of as user-oriented, whereas a DCE is communication-oriented. Many conversations, whether verbal or printed, tend to separate the DTE and DCE. The DTE is generally referred to as a *terminal* or *personal computer* and the DCE, as the *modem* or *communication device*. This separation or distinction is not always the case, however, as shown later in this text with an example of the DTE and DCE located in the same physical device in large systems.

For the sake of clarification, the DTE is assumed here to be user-oriented and concerned with communications higher up in a given system. The DCE is communication-oriented and concerned with signal transfer to and from the DTE and to and from another DCE. The International Telegraph and Telephone Consultative Committee (CCITT), for instance, has numerous specifications that cover DTE and DCE operations. Some of them are more common than others and are presented in this section, but before examining these details, we will examine other information that may be beneficial.

An interface can also be thought of as a piece of hardware that is inserted into a personal computer, mainframe, or midrange computer. *Interface boards,* as they are called, have not only hardware logic but generally firmware as well. *Firmware* comprises software instructions, stored in a microchip, that perform specific functions. Other characteristics of interface boards are their connecting point (or connector). These connectors have specifications for not only the number of pin-outs but also the protocol for communication through them.

Universal synchronous/asynchronous receiver/transmitter

This term, *universal synchronous/asynchronous receiver/transmitter* (USART), is responsible for numerous functions in communications with both DTEs and DCEs. Before exploring the basics of this component, we will discuss other related terms.

Basic to the USART is the universal asynchronous receiver/transmitter (UART). The UART differs from the USART in that the UART handles only asynchronous communications. UARTs are also known by the terms *asynchronous communication element* (ACE) and *asynchronous communication interface adapter* (ACIA).

The USART or UART performs the function of assembling and reassembling bytes of data. It also handles timing. Specifically, a UART handles both internal clocking of its operations and clocking to handle the receiver and transmitter sections.

The UART is also responsible for framing the serial data unit which is transmitted over a medium. It is responsible for parity, stop bits, start bits, and some error detection. Conversely, if the UART is on the receiving end, it is responsible for unframing (disassembling) the serial data unit.

Another characteristic about the UART is that it is interrupt-driven. No polling occurs. Consequently, basic conditions that exist when an interrupt is generated are transmitter-, receiver-, or break-related. An interrupt can also be generated by a state change in an RS-232 [Electronic Industries Association (EIA) standard] input line.

This one chip on communication interface boards makes communications much easier. In addition, it takes some load off the processor. An additional note about UARTs is that they generally have some common input and output functions native to RS-232 built into them.

Understanding communication terms

Multiple terms are used to convey information about the communication process. Some terms can be categorized and therefore are easier to understand. Some of these terms are discussed in the following paragraphs.

Bit rate.   The term *bit rate,* meaning the number of bits transmitted per second, is generally used in reference to modems.

Baud rate.   The *baud rate* is a measurement of the number of times per second a change occurs in the amplitude, frequency, or phase of a wave. One baud is a change in one of the aforementioned parameters. To calculate the number of bits per second, you can use the following equation.

Number of bauds per second × 1 bit per baud (or appropriate amount

according to specifications) = number of bits per second

Determine the number of bits that equal one baud. This information is usually ascertainable through documentation sources from modem suppliers. It can also be obtained through knowledge of the specifications for the modem and referencing this specification. Next, multiply the number of bauds per second a modem can perform; this is equal to the number of bits per second for that modem.

Modulation techniques.   Three modulation techniques are popular: *amplitude modulation* (AM), which varies the amplitude of a signal without changing its frequency or phase; *frequency modulation* (FM), which changes the frequency to reflect the change in the binary state but maintains the amplitude; and *phase modulation* (PM), which varies the phase of the wave to reflect the binary value.

Encoding techniques.   The term *encoding* refers to how signals are introduced onto the medium and how they appear on the medium when examined. This idea is reflected in Fig. 2.14.

Figure 2.14 is an example of what is called *non-return-to-zero encoding.* This encoding scheme uses each signal change to represent one bit time.

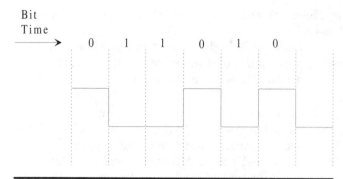

Figure 2.14   Non-return-to-zero encoding.

Figure 2.15 represents a type of encoding known as *Manchester encoding.* This encoding scheme changes the polarity each bit time.

Notice in Fig. 2.15 that each bit time experiences a change in polarity. This method of encoding results in good clocking performance and is widespread throughout LAN technology.

Other encoding schemes include *differential Manchester encoding,* which is a form of Manchester encoding in which the previous bit time is used as a base reference point for interpretation of the signal. *Return to zero* is another scheme that utilizes two signals to represent one bit change. It is similar to Manchester in that polarity is changed each bit time.

2.6   Interface Standards

Interface standards include popular terms such as RS-232, V.35, T1, and X.21. These interfaces and many others are prevalent throughout

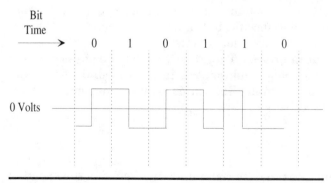

Figure 2.15   Manchester encoding.

the marketplace. There are entire books devoted to explaining these and other interfaces.

Physical-layer interfaces, as do higher layers within a network, have protocols to transfer data. A particular interface specification identifies the protocols of its operation. Some examples are given below.

RS-232. RS-232-D is the follow-on to RS-232-C. The fundamental difference between the two is that the RS-232-D is parallel with the V.24, V.28, and ISO 2110 specifications. The RS-232 standard specifies the pin-outs of a 25-pin cable used for serial communications. Although most of them are not used for typical modem installations used with PCs, the 25 pins are nevertheless assigned.

V.35. This specification comes from the CCITT. It specifies modem operation of 48 kbits/s; however, it is typically implemented at 56 kbits/s.

T1. T1 interfaces have the capability to move data up to 1.54 Mbits/s. T1 lines comprise 24 channels, with each channel using 8 bits. The result is that one T1 line uses a twisted pair for 24 voice signals (i.e., at a 24:1 ratio).

X.21. This CCITT specification is flexible in that different signaling rates are supported. For example, a given DTE and DCE may differ with respect to adherence to specifications. X.21 calls for synchronous operations with public data networks. An example of this scenario is X.25 using X.21 as the interface.

Many other interfaces exist. To list them here would require the remainder of the book; but the purpose is to get oriented into what happens at the physical layer.

Regardless of the vendor, interfaces exist to provide a link between two systems or between a system and the medium. The interface standard used may be vendor-specific, but in most cases vendors adhere to the guidelines the service providers offer.

A final note about interfaces. They do not have to be physical in the sense of being used to bridge cable or some tangible medium. For example, wireless networks require interfaces between network devices, transmitters, and receivers.

2.7 Modems

The function of modems in data communication remains somewhat of a mystery. Modems convert signals from digital to analog and from analog to digital. According to the information provided here, it is clear that analog and digital data representation are distinctly different. A greater mystery than the signal conversion process is the speed at which these tasks are executed.

Modems can be categorized in a number of ways, and the typical categories are presented here.

Asynchronous versus synchronous modems

A popular way to categorize modems is by the method with which they transfer data: asynchronously or synchronously. Asynchronous modems are typically used with personal computers, notebook computers, and subnotebook computers. Synchronous modems are dominant in commercial implementations. With large computer systems synchronous modems can deliver the high speed and reliability required.

Asynchronous modems can also deliver high speeds and good throughput, but high speed is relative. Asynchronous modems are also used in commercial applications, but their role is generally different from that of synchronous modems. For example, synchronous modems can be connected into communication processors which utilize synchronous transmission. Asynchronous modems can also be used in a similar fashion, but the implementation is different.

Asynchronous modems such as those provided by the Hayes Corporation have become the de facto standard for personal computers, notebooks, and subnotebook computers. Hayes has synchronous modems and provides a broad range of modems that support other functions. A commonality of Hayes and Hayes-compatible modems is the AT (IBM PC Advanced Technology) command set.

This command set is used to communicate with nonvolatile random-access memory (RAM), issue commands to change capabilities such as the speaker volume, store numbers, change values to perform specific functions such as putting the modem in automatic answer mode, and enable or disable the carrier signal.

Typical modem standards

The CCITT has set some standards for modem operation. One such standard is the V.32, which supports transmission speeds of up to 14,400 bit/s. This standard also defines how modems should communicate with each other with full-duplex signaling at 4800 and 9600 bit/s over normal switched telephone lines. A drawback to this specification is the lack of errors-handling ability.

Another standard is CCITT V.42. This standard called for an error-correction scheme to be built into the hardware and for use with the link-access procedure for modems (LAPM).

A V.42bis standard is also available and focuses on data compression in hardware. For example, a 9600-bit/s modem could achieve a 38,400-bit/s throughput using compression.

V.32bis is a revision of V.32 which specifies 7200-, 12,000-, and 14,400-bit/s transfer rates along with a negotiating protocol that operates faster than its predecessor.

Data compression

Data compression is another way of characterizing modems. Some implement it, and some do not. An example of data compression is the Microcom networking protocol (MNP). MNP, created by Microcom, was designed to increase data throughput. If MNP is used, both sending and receiving entities must implement it. MNP can be implemented in hardware or software.

Data compression works because algorithms are used to examine data being transmitted and change the data into a compressed version based on a sampling of the data. In turn, the receiving entity must implement data compression techniques to reverse the process.

Wireless modems

Modems are now available that utilize cellular and radio frequency for transmission. Implementation of a modem where used in a wireless environment means the modem connects to a wireless phone (cellular or RF). Radio-based modems use radio-frequency (RF) technology. A network based on such technology is called *ARDIS*. ARDIS, created by IBM and Motorola, is a public data network that covers a considerable geographic distance.

Fax/modems

Fax/modems are just that: They have the capability to send faxes (facsimiles) or to use the modem—in fact, two devices in one. There is a fax chip that drives the fax part and a modem chip that drives the modem part; hence, the difference between speeds for most fax/modem combinations. For example, a typical expression of a fax/modem speed advertisement is 2400-bit/s modem and 9600-bit/s fax. That is what it means; as of this writing, there is no crossover or attempt to increase speed capability. There may be in the future, but now there is a clear delineation between modem operation and fax operation with respect to speed.

Other characteristics may be used to describe modems; however, those described above are the most dominant ones used in the marketplace today, and, if understood, they may provide insight as to how the different characteristics fit together to meet a variety of requirements.

2.8 Summary

As a topic, data communication covers a vast amount of material. This chapter presented the core of data communication topics to build a foundation for the remainder of the text. To have an edge in troubleshooting, an understanding of signal types, signaling methods, and evaluating signals themselves is important. Binary and hexadecimal data representations are at the heart of computer and data communication device internals for converting characters, numbers, and control codes into meaningful machine language. Channels, bandwidth, and asynchronous versus synchronous transmission were explained.

Serial and parallel transmissions were explained, as well as simplex and half- and full-duplex transmission. Frequency-division and time-division multiplexing were highlighted with examples of how their functions delineate their differences.

A topical approach to physical interface considerations covered the DCE, DTE, and USART and UART microchip. Bit and baud rates were explained in light of their differences. Modulation and encoding schemes were also presented. A brief section on interface standards oriented the reader to where and what some examples of interfaces fit into the data communication picture.

Modems were discussed in light of categories based on their functions. Differences between asynchronous and synchronous modems, along with examples of modem standards, were explained. Data compression, wireless modems, and an explanation of how fax/modems operate concluded the last section.

3

Protocol
Fundamentals

Protocols are relatively easy to understand. Consider the analogy of human social behavior compared to a protocol used in computers and networking devices.

Regardless of the culture or location, people have always had traditions and mores. Consider the marriage ceremony. Depending on the culture, people might *perform* ceremonies differently but they serve the purpose of acknowledging the marriage.

In some cultures people who attend a marriage ceremony wear nice clothes, sit in preassigned places, and bring gifts to those getting married; other actions may also be appropriate. For example, in some cases a given culture implicitly dictates that nice clothes be worn to a marriage ceremony. How do the attendees *know* to do this? There are two possible ways.

First, if those attending a marriage ceremony are from the same culture, specifically the same region, and if they interact with others in the community, the acquisition of knowledge of *what* to wear to a marriage ceremony may be considered general knowledge; in other words, one learns this along life's way (i.e., by cultural conditioning). The second method of communication could occur regarding *what* should be worn by those responsible for the occasion informing the prospective attendees explicitly what to wear. This could be done by mail or human communication (regardless of the medium) and other ways. The point is that one entity explicitly communicates to the other entity its own expectations of the other's members. This is a social *protocol*. Many would call this common sense. In fact, considering the different cultural backgrounds that represent the human species, there is little common about it.

Like human culture, computers and networks have protocols, also. The remainder of this section explores network protocols from a variety of different angles.

3.1 Perspective

To define protocols from a nontechnical standpoint, this section provides insight into protocols.

Why protocols are required

Protocols are required because the idea of no protocols does not and cannot exist. If the idea of no protocols exists, then that idea is a protocol itself, logically. Do not be confused. This is a brief lesson in logic! The notion that nothing exists in fact means that something exists, even if only as an idea. Put another way, it is a logical impossibility for nothing to exist. Hence, the lack of a protocol is in fact a protocol.

Protocols are required in the operation of computers, networks, and internetworking devices of all sorts. In computers and networks each function performed, regardless how small, is a quantifiable entity. The entity may be an abstract function, but it is definable and traceable to some tangible line of definition in software or a piece of hardware.

Computer and network protocols are the contrivance of humans. Computers and networks use protocols to define aspects such as physical connections, how data is passed across a link between two or more devices, and other examples. Virtually all aspects of computer network functions can be categorically placed into a protocol category. The issue now becomes what category and how protocols are made.

How standards are made

Standards are practically synonymous with protocols. Some split technical hairs about this matter, but if it is implemented, accepted, and used, it seems that this would qualify it to be called a protocol. Standards come about in different ways and typically reflect the philosophical approaches to a specific technology.

De facto standards. We live in a de facto world; we do something, define what we did, then record the method into some sort of standard, and proclaim those who follow in this manner will do thus and so. Is this not how we live?

An example of something de facto is the DOS operating system used on personal computers. DOS commands the lion's share of the PC operating system market today. This was not planned.

The point is that no group or individual *planned* strategically that DOS would have the popularity and market share it has today. If anyone could have predicted this, it would be equivalent to foretelling user desires and technical developments that would occur. DOS simply *happened* as a result of the user community, and it exists.

Although Microsoft, IBM, and Digital Equipment Corporation (DEC) have versions of it and have made changes to their respective products, it is the user community that has driven DOS with a tried-and-true method called *demand*. Users liked it, developers created popular programs for it, additional hardware was designed and sold to operate with it, and so the spiral continued.

De jure standards. A *de jure standard* is derived by groups. The following are examples of groups that concentrate on standards:

- Institute of Electrical and Electronics Engineers (IEEE)
- American National Standards Institute (ANSI)
- International Standards Organization (ISO)

These groups and others that will be discussed later in this chapter focus on different standards and protocols. Standards and protocols are different. A general definition from the marketplace is that standards are what *should* be the case, while working protocols are what *get* implemented—and they may or may not be a "standard."

The main distinction between de jure and de facto is that de jure standards are composed by groups of people from different backgrounds who contribute ideas, resources, and other items to aid in the development of a standard. De jure groups could be considered steering committees. The questions are: "How powerful is the steering?" and "Does a de jure standard reflect what the marketplace wants?"

De jure standards are prevalent in the networking arena. Some standards are more prevalent than others. However, the intent of de jure groups seems to be for promotion and betterment of a technology.

Proprietary standards. *Proprietary standards* are owned outright by a corporation or other entity. Most corporations in the networking business today are embracing multiple networking solutions. This appears to indicate that corporations of all sizes and focuses are attempting to broaden their focus, so it is not perceived as proprietary. Ironically, not too long ago proprietary networking, or computing, was the norm; today most companies shun the use of the term *proprietary* anything.

Proprietary products are not necessarily more restrictive than nonproprietary products. In fact, the term *open* is now frequently used to describe support for multiple standards or protocols. Actually, some proprietary protocols are more robust than those which have not been refined.

Interpreting which standard or protocol is best should be user-driven. In most cases this is the inevitable result.

Examples of standards-making bodies

This section lists and briefly describes the intent behind a variety of groups involved in standards making. Some of the names included in this group may be categorized as forums, consortiums, organizations, and other entities. The list is not exhaustive, but it includes the dominant and most popular groups.

American National Standards Institute (ANSI). ANSI is the agency that published the American Standard Code for Information Interchange (ASCII). This is a method used to represent alphanumeric characters and control keys. ANSI coordinates standards and is parallel in many respects to other standards-making bodies.

CCITT. CCITT is the acronym for the International Telegraph and Telephone Consultative Committee. This entity, like ANSI, has been in existence for decades. The CCITT has a number of specifications. Some examples are the X and V series of recommendations. Actually, some recommendations are more than that—they are protocols that have widespread implementation. Two examples of CCITT recommendations suffice to convey the impact of the organization on the technical community. The X.400 electronic mail (E-mail) system is a product of the CCITT as well as V.35. The latter is a specification for duplex data transfer over leased lines.

Corporation for Open Systems (COS). This organization is nonprofit and focuses on the promotion of ISO standards. Specifically, it is involved in testing products for ISO compliance. Its thrust is providing information about ISO guidelines and helping corporations meet these guidelines.

European Computer Manufacturers Association (ECMA). This group now has representatives from companies located around the world. The focus of the group is on the European region of telecommunication and related standards. Companies such as IBM, DEC, and AT&T are members.

National Institute for Science and Technology (NIST). NIST was known previously as the National Bureau of Standards (NBS). It is a U.S. government organization that provided workshops that resulted in implementor agreements. These workshops have become formally organized and internationally known. NIST is one of the major contributors to the International Standards Profiles as part of the ISO.

Institute of Electrical and Electronics Engineers (IEEE). The IEEE is a professional organization of individuals in multiple professions. I am a member of the IEEE. This organization is active in standards that will be explored later in this book. For our purposes it is sufficient to understand that this organization focuses efforts on lower-layer protocols such as the 802 series. This topic is explored in greater detail in Chap. 4.

International Standards Organization (ISO). The ISO is a dominant organization whose focus is widespread. Many other standards-making body work is accepted and incorporated in ISO standards. One example of an ISO standard is the open-systems interconnection (OSI) model. This model is a seven-layer reference reflecting what should occur at each layer within a network.

X/Open. This organization focuses on specifying product requirements for vendors creating product specifications. These products include a variety of offerings. One example of their focus is with POSIX. Another technology group functioning in similar areas is Unix International (UI). Actually UI works with this organization in market specifications for applications.

GOSIP. Government open-systems interconnection profile (GOSIP) is a protocol that specifies standards to be adhered to by suppliers of products to the U.S. government. Some of the specifications included in various versions of GOSIP include some CCITT specifications, OSI, and other standards and protocols. Its thrust is support for OSI protocols supported in equipment supplied to the U.S. government.

Open Software Foundation (OSF). OSF is a nonprofit organization research-and-development (R&D) group consisting of vendors, users, educational institutions, and other entities. It began in 1988 and participates with other standards-making groups such as the IEEE, X/Open, and ANSI. The focus of OSF is distributed computing, distributed management, and a graphical user interface (GUI) (Motif). UNIX is considered to be a part of open systems, but not a total solution. It does include representatives from vendors such as DEC, IBM, Hewlett-Packard, Convex Computer, MIT, Standford, and a few hundred other entities that represent a variety of concerns.

Unix International (UI). This organization focuses on the AT&T UNIX operating system. Like OSF, it has hundreds of members. However, UI is oriented toward the success of the operating system as a standard and its continual refinement. UI itself consists of numerous subgroups within the organization such as a technical advisory group and a work group focusing on specific functions designated by the board of directors. The steering committee focuses on the priority of work and oversees work groups. Members in UI include AT&T, Mississippi State University, Sony Corporation, Texas A&M University, Swedish Telecom, Convex Computer Corporation, the U.S. Air Force, and hundreds of others.

Asynchronous Transfer Mode (ATM). This CCITT standard, created in 1991, focuses on the advancement of ATM technology. It now has hundreds of members representing a wide variety of interests around the world. ATM is a cell-based method for data transfer and operates asynchronously at lower layers within a network.

Hewlett-Packard's OpenView. This organization is a technical forum that focuses on network management. According to information supplied to me by Dave Richardson and Tim Jones, Hewlett-Packard's employees whose specialty is in this area, indicated this forum focuses on interacting with users who actually perform heterogeneous network management. This organization also conveys very helpful information to attendees to the annual event. The OpenView product concentrates on managing transmission control protocol/internet protocol (TCP/IP)-, NetWare-, and SNA-based networks. The OpenView forum includes vendors who work with OpenView directly and indirectly, thus making an information exchange between users and vendors direct and helpful.

Other standards organizations exist; this list includes popular and current organizations.

3.2   Network Layers: A Practical Perspective

Networks, regardless of type, can be evaluated by layers. It is possible to identify what happens at each layer within a given network. The International Standards Organization (ISO) created a seven-layer reference model called the *open-systems interconnection* (OSI) model. It is widely used and has been presented in many ways. This section explains the functions and components that operate at each layer.

The idea of a layered network goes farther back in time than the late 1970s when the ISO created the OSI model (see Fig. 3.1). In fact, IBM introduced a layered networking model in 1974; it is called *systems network architecture* (SNA). Layers within a given network may not cor-

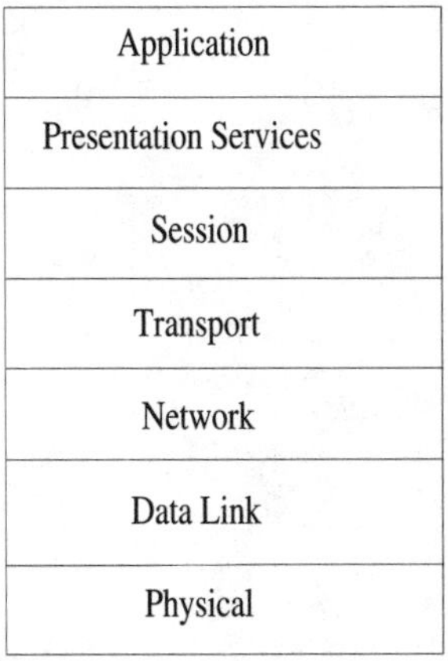

Application
Presentation Services
Session
Transport
Network
Data Link
Physical

Figure 3.1   OSI model.

relate with all seven layers that the OSI model specified, but the OSI model is a good baseline to use for layer evaluation.

These seven OSI layers and their functions and associated components are listed below.

Layer	Function	Components
Physical	Generating electrons or photons to be introduced to the medium or removed from the medium	Hardware interface, software in chips on the board
Data link	Controlling data flow across a link from one node to another	Firmware, hardware interface board, and software
Network	Routing; and, depending on the protocol, flow control	Usually software, sometimes firmware in certain devices
Transport	Moving data from one node to the other; performing retransmissions if required; providing a reliable or unreliable method for data transfer	Software; however, some implementations can implement this in firmware
Session	Providing the logical connection between users and applications or applications if they are peer-to-peer; addressable endpoints reside at this layer	Software; however, in some instances this may be implemented in firmware
Presentation services	Formatting the syntax of data; data representation is performed here	Software, but this can be performed by firmware
Application	Providing applications for users and, depending on the protocols, possibly providing services for applications	Software or firmware

Layers

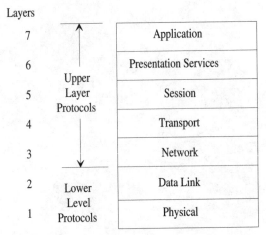

Figure 3.2   Upper- and lower-layer protocols.

These seven layers constitute the OSI model. Viewed differently, this model can be divided into two parts: upper-layer protocols and lower-layer protocols (see Fig. 3.2).

Figure 3.2 depicts the OSI model in two groups: upper- and lower-layer protocols. Parts 2 and 3 of this book are devoted to lower- and upper-layer protocols, respectively.

Application
Presentation Services
Session
Transport
Network
Data Link
Physical
Media

Figure 3.3   Ed's model.

The OSI model shows seven layers when in fact an eighth layer exists. I call this "Ed's layer" because a formal explanation of it is not prevalent, but it must be present. Figure 3.3 depicts this scenario.

This eighth layer is the medium. Types of media were presented earlier. And, in all networks a medium—whether hard or soft—must be present.

3.3 Summary

Protocols are required in computers and networks to maintain order and communication capabilities. Protocols can be traced to how they originate: de facto, de jure, and proprietary. A considerable number of organizations exist working on refining and developing protocols of different sorts.

A network can be divided into layers of functionality. The OSI model is a good reference point. The OSI reference model can be divided into two categories: upper- and lower-layer protocols.

Lower-Layer
Protocols

4

Lower-Layer
Protocols

A Practical Perspective

Network protocols can be divided into upper and lower layers, as explained in Chap. 3. The purpose of this chapter is to orient the reader to the protocols covered in greater detail later in this book.

4.1 Overview

Protocols can be divided into layers based on the OSI reference model. Upper-layer protocols are covered in Part 3; this part covers the lower-layer protocols. Protocols covered in this section are dominant in today's marketplace. They are presented in this chapter and throughout this part of the book in alphabetical order.

4.2 Lower-layer Protocols in General

Before exploring each of the following protocols in depth, as done in the remaining chapters in this part of the book, it is convenient to understand the basics of each protocol. Those basics are provided in this chapter, specifically this section.

Asynchronous transfer mode (ATM)

ATM is a cell-based lower-layer protocol. Another way to say this is that ATM uses small packets; the cell is simply a small packet. To be precise about the concept of a cell versus a packet, a cell is a physical-layer component whereas a packet is a network-layer component from the perspective of the OSI model.

ATM is a broadband protocol and can use switched or nonswitched technology. Its beginnings are in broadband, the Integrated Services Digital Network (ISDN), particularly B-ISDN. ATM is a cell-based technology that has a cell structure of 53 bytes, of which 48 bytes are used for customer data; the remaining 5 bytes constitute the header of an ATM cell.

A particular advantage of ATM is that it can integrate data, voice, and video within the same application without concern for compatibility for the impact on communications in LANs or WANs. ATM is particularly suited to this task because it treats data as small-size packets and relatively simple protocols with low overhead. These two characteristics are the basic requirements for a fast switching technology.

Enterprise system connectivity (ESCON)

ESCON is IBM's fiber-optic channel protocol. It is used with IBM's S/390 architecture. It can be implemented in point-to-point switched or nonswitched configuration. Data is moved serially through fiber cable. Because photons are used to represent data, the length that a cable can run is longer than traditional copper-stranded cable. For example, distances of 23 to 43 km are supported for some devices. According to IBM, ESCON can accommodate up to 17 million bytes per second of data rate transfer.

ESCON is the architectural follow-on to the channels IBM used with prior architectures. However, both are supported, and it appears that IBM will support both for some time to come. ESCON is radically different from parallel channels in two ways: (1) ESCON cables are much lighter in weight than traditional copper-stranded cable and (2) ESCON moves data almost 10 times faster than do parallel channels.

ETHERNET

ETHERNET, a DEC, Intel, and Xerox Corporation specification, is a data-link protocol that uses broadcast technology. It is typically implemented in bus topology, and may use twisted-pair or coaxial cabling. The original specification calls for coaxial cable used with a certain resistance [measured in ohms (Ω)].

ETHERNET is implemented on interface boards which may be used with personal, notebook, midrange, or mainframe computers. ETHERNET technology originated in the mid-1970s. It is an example of a de facto standard. ETHERNET Version 2.0 specifies data transfer speeds at 10 Mbits/s. However, current ETHERNET technology focuses on speeds of 100 Mbits/s; this is also called *fast ETHERNET*.

Fiber distributed data interface (FDDI)

FDDI is a dual-ring-based technology. As its name implies, it uses fiber media; however, copper-stranded media can be used, and if so, it is called CDDI. FDDI data rate transfer is 100 Mbits/s. This standard was created by a committee that is part of the ANSI standards organization. FDDI is implemented by a number of vendors. One vendor implementing this technology is DEC. Other major vendors support FDDI as well.

FDDI is similar to token ring in technology. The architectural structure uses the notion of a token and a station required to capture the token before it can transmit data onto the ring.

FDDI-II is a technology in which FDDI is implemented to operate isochronously and utilizes a 125-µs frame transfer on the ring. The primary difference between FDDI and FDDI-II is that the latter can accommodate voice and/or ISDN traffic.

Frame relay

Frame relay is a lower-layer protocol technology. It is similar to X.25 except that routing decisions are made at the data-link layer within the network. Multiple speeds are supported. For example, in some instances 1.54 Mbits/s can be achieved. Frame relay can also support video and voice transfer in addition to data transfer.

Some of the standards that frame relay supports are ANSI T1.602, T1S1/90-75, CCITT I.122, and Q.933. Frame relay is supported by a number of vendors; IBM is one such vendor who supports frame relay in their product line.

Parallel channels

Parallel channels is the term IBM used to rename their existing channels when ESCON was introduced. Two types of parallel channels exist: byte multiplexer and block multiplexer. Both use channel protocols and bus and tag copper-stranded cable.

The nature of these channels and the cables used in conjunction with them restricts cable length to approximately 200 ft when units are daisy-chained. In some instances these cables can be used for greater distances, but restrictions do apply.

Parallel channels have different data rate transfer speed support depending on how data is transferred. Typically, data rates are from 1 to 3.5 Mbits/s. These speeds may vary depending on the device attached because some vendors offer devices varying in regard to channel speed support.

Synchronous data-link control (SDLC)

SDLC is a bit-oriented protocol widely used in the industry. It can be implemented with switched point-to-point, nonswitched (dedicated) point-to-point, nonswitched multipoint, and loop configurations. SDLC frame structure is such that the control field supports multiple formats such as unnumbered, supervisory, and information.

SDLC is implemented in many IBM installations. However, other vendors, such as Hewlett-Packard, Sun Microsystems, and DEC also support SDLC. Many vendors who offer network devices support SDLC because of its presence in the marketplace.

Token ring

Token ring is an IBM product offering. The IEEE organization has an equivalent which will be discussed later in this chapter. Token ring utilizes ring technology with a star implementation. Token ring operates at two speeds: 4 and 16 Mbits/s. This technology is connection-oriented at a data-link layer.

Token ring was announced in the mid-1980s by IBM and since has grown to command a considerable market share. It is a fault-tolerant technology because of the nature of its operation. Devices can be inserted in and removed from the ring without disturbing ring operations.

X.25

X.25 is a CCITT specification. It commands a considerable presence in the worldwide marketplace. X.25 is a *packet-switching technology,* which means that a packet is created at the network layer and placed in a frame at the data-link layer.

X.25 technology uses permanent virtual circuits. These are similar to leased lines in operation. X.25 also uses a *virtual call concept,* which means that a node must request a connection (place a call) to the destination node for communication to take place. A *fast select call concept* is also used by X.25, which is an extension to the data field used in the normal call issued by the node desiring to make a connection. The extension means that up to 128 bytes of data can be used in the field rather than the 16 bytes used in the normal call.

Packet-switching technology has been in use in the United States since approximately 1970. The X.25 packet-switching protocol is used by many public network-oriented service providers.

A Synopsis of the IEEE 802.X Protocols

The IEEE has a subcommittee devoted to a series of protocols called the 802.X series. A list of these protocols, with their associated group names and functions, is presented in this section.

802.1. This is the *high-level interface* group. This group is concerned primarily with network architecture. Additionally, this group focuses on internetworking heterogeneous networks and network management. This group also works with medium access control (MAC) bridges.

802.2. This is called the *logical link control* (LLC) group. The focus of this group is on the data-link layer in the OSI model. The data-link layer can be divided into two sublayers: the MAC and the LLC layers (see Fig. 4.1).

The MAC sublayer varies according to the type of protocol and network implemented. However, the LLC is common among 802.3, 802.4, 802.5, and 802.6 networks, but it is independent of MAC addressing and the actual medium. If this layer is divided into two sublayers, a degree of independence can be achieved by the LLC.

The LLC sublayer supports four types of service:

- Unacknowledged connectionless
- Connection-mode
- Acknowledged connectionless
- All of these

Figure 4.2 shows the correlation between the LLC protocol data unit and the 802.3, 802.4, 802.5, and 802.6 MAC frames.

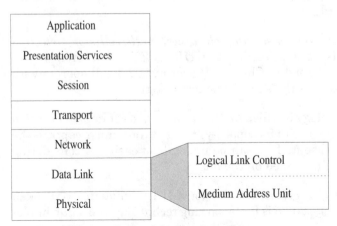

Figure 4.1 Conceptual view of the data-link sublayers.

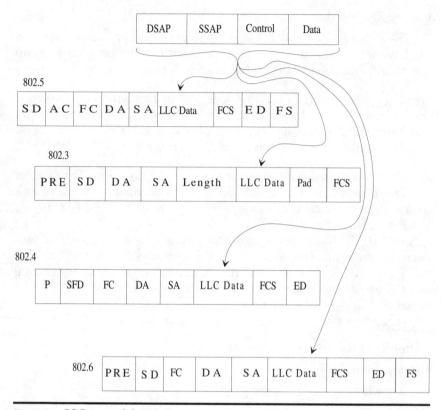

Figure 4.2 LLC protocol data unit.

Each MAC frame shown in Fig. 4.2 contains a LLC protocol data unit. The remaining portion of this section examines these MAC frames individually.

802.3. This is *carrier-sense multiple access with collision detection* (CSMA/CD). This frame is similar to ETHERNET but different in one field, which is explained in Chap. 7. However, 802.3 will operate with ETHERNET and uses the same basic technology.

802.4 This technology is called *token bus*. This type of implementation passes a token but uses bus topology. Each station must capture the token in order to use it. This can be understood easily by considering that a logical ring is formed on the bus.

802.5. This is a *ring-oriented* technology. It is similar to IBM token ring, but not exactly. 802.5 is based on ring technology and a hub implementation via a media access unit. This technology uses token-passing technology, requiring each station to possess the token prior to transmission.

This technology is considered self-healing. Hosts can be put onto and removed from the ring at will without disturbing the ring. Ring speeds are either 4 or 16 Mbytes/s.

802.6. This protocol defines a *metropolitan area network* (MAN). For practical purposes this definition is superimposed with the ANSI X3T9.5 standard. This is a standard defining the FDDI. This specification defines a 100-Mbits/s fiber channel, a 1300-nm signal, distributed clocking, and timed token rotation, for example. The FDDI-II definition claims a fiber channel of up to 620 Mbits/s.

802.7. This group provides support to other working groups, particularly the broadband-oriented groups.

802.8. This group provides support for the optics-oriented technology groups.

802.9. This group is focused on integrated data and voice-oriented networks.

802.10. This group focuses on LAN security. By definition, it is related to all the LAN implementations.

4.3 Summary

Lower-layer protocols are distinctly different from upper-layer protocols. This chapter presented a brief overview of some popular lower-layer protocols; those protocols are discussed in greater detail in subsequent chapters.

Each protocol has its own level of acceptance in the marketplace and is at an individual level of maturity. Some of these protocols are popular and are widely implemented across a variety of vendor products.

The IEEE has a committee known as 802. The 802 is divided into subgroups, each with its own focus. Some of the subgroups are devoted to the advancement and direction of protocols that are already implemented in the marketplace. Other subgroups are more support-oriented to those groups with established protocols in the marketplace. Some of the IEEE definitions parallel protocols by corporations and other organizations. Some also incorporate protocols derived by other entities such as the ANSI fiber standard used for 802.6.

Asynchronous
Transfer Mode (ATM)

ATM has not been on the market as long as some other technologies. However, credit is due to researchers at Bell Laboratories for their work in this area long before it came to the forefront of media attention in the late 1980s to the early 1990s. ATM actually began to take shape in the late 1980s and the driving force behind its development was a need for a fast switching technology that supported data, voice, video, and multimedia in general. This chapter explores various aspects of ATM and serves to orient the reader to it. Just before the summary, a section of references for more in-depth information is provided.

5.1 A Perspective on ATM

Asynchronous transfer mode (ATM) is a cell-switching technology. Specifically, it uses a 53-byte cell and consists of a header with routing and other network-related information and an information field for data, images, voice, and/or video. A major thrust behind ATM is its capability to support multimedia and integrate these services along with data over a signal type of transmission method. Heretofore this was attainable but awkward, and timing problems along with other significant issues hindered it.

ATM can be implemented at the enterprise level (privately), in public networks, and/or both. In reality this is how it is implemented. Realization of ATM throughout a dispersed network environment is twofold. First, it is implemented at the local level, as in a given organization or corporation. Second, it is implemented at the public level (or service provider). ATM usage capabilities can be customized to meet a given environment's needs and it is application-transparent; therefore, integration of multimedia and data does not cause problems.

As of this writing ATM is not new, but neither is it seasoned with regard to implementations. Some trade magazines, forums, and other technical communications predict that ATM will dominate the future of LANs, WANs, and public service providers in the not-too-distant future. ATM growth has been rapid over the past 2 to 3 years. Many vendors are in some phase of support for ATM. Some have ATM products to offer today, while other companies are beginning to create ATM products. The technology itself is being defined, and it appears that refinement of ATM will continue for some time to come.

5.2 ATM Layer Structure

Many network protocols are compared and contrasted against the OSI model. ATM does not fit the structure of the OSI model except for the physical layer.

Figure 5.1 depicts ATM structure in light of a layered approach.

The physical layer of the ATM layered model can be divided into two sublayers: the physical (layer)-media-dependent (PMD) sublayer and the transmission convergence sublayer. The PMD sublayer performs two primary functions: bit timing and line coding. The function of the transmission convergence sublayer is contingent on the interface used beneath it. Some interfaces supported include synchronous optical net-

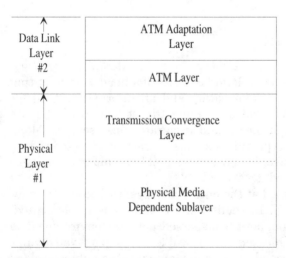

Figure 5.1 Conceptual view of ATM layers.

work (SONET) at 155.52 Mbits/s, DS3 at 44.736 Mbits/s, multimode fiber at 100 Mbits/s, and "pure cells" at 155 Mbits/s.

Functions at the transmission convergence sublayer are contingent on the interface used. Some of the functions at this sublayer include performing *header error correction* (HEC), which covers the entire cell header (this includes generation and verification), multiplexing, frame generation and recovery, and mapping of the ATM cells onto DS3 facilities, if used, by the physical-layer convergence protocol (PLCP). In addition, PLCP framing and delineation is performed if DS3 is used. In addition to these functions, other functions are performed relative to the interface in use.

ATM layer functions include switching, multiplexing, routing, and congestion management. Above this layer is the ATM adaptation layer (AAL). Here different categories of functions are identifiable. Figure 5.2 shows the components at the PMD sublayer and interface supported, the transmission convergence sublayer, the ATM layer functional components, and the AAL.

Figure 5.2 shows functions that occur at the ATM layer, and it also shows categories of functions that occur at the AAL. Details of the AAL functions are explained in Sec. 5.3, but the ATM layer is the focus here.

The ATM layer also defines two virtual connections: a virtual channel connection (VCC) between two ATM VCC endpoints. This may be either

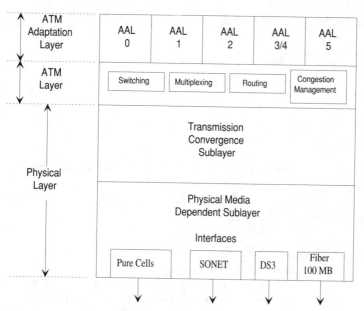

Figure 5.2 ATM adaptation components and ATM layer.

point-to-point or point-to-multipoint configuration. Second, multiple VCCs are carried through virtual path connection (VPC) endpoints. This connection will be either point-to-point or point-to-multipoint.

This layer also translates the virtual path indicators (VPIs) and the virtual channel indicators (VCIs) and is responsible for cell multiplexing and demultiplexing.

The AAL is responsible for mapping data from higher layers within a network into cell fields and conversely. This layer operates on end-to-end functions.

The connection-oriented or connectionless conversion protocol incorporates the broadband data service and includes the B-ISDN, Integrated Switching Data Network (ISDN) standard. This B-ISDN support includes conversational interactive service, such as video, voice, and sounds as well as document messaging for electronic mail (E-mail) multimedia. It also specifies the way text, data, sound, pictures, and video are transmitted and received.

5.3   ATM Adaptation Layer (AAL) Functions

The AAL layer has been explained as that protocol layer itself which maps upper-layer protocols onto ATM. Devices that use this terminate signals. This layer provides functional support for different types of traffic (signals) that come from upper-layer protocols. Consequently, the following categories of functions are explained:

- AAL type 0
- AAL type 1
- AAL type 2
- AAL type 3–4
- AAL type 5

AAL type 0.   AAL type 0 is considered a place holder when customer premises equipment (CPE) perform required functions at this layer. In a sense it is a pass-through capability for cell-oriented service. Figure 5.3 is a detailed view of the AAL.

AAL Type 1.   AAL type 1 functions provide constant bit rate services. This is also known as *unstructured circuit transport between points.* This type function is connection-oriented. The CS component in AAL 1 in Fig. 5.3 is called the *convergence sublayer,* which modulates differences that may occur in the differences of physical interfaces. The SAR

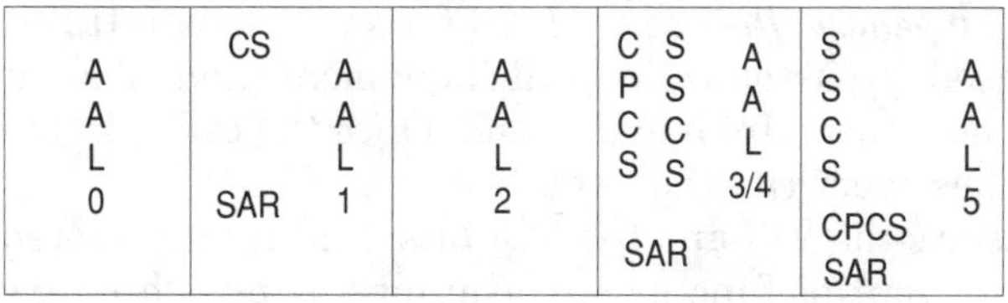

A A L 0	CS SAR	A A L 1	A A L 2	C S P S C C S S SAR	A A L 3/4	S S S C CPCS SAR	A A L 5

Figure 5.3 Conceptual view of functions at the ATM adaptation layer.

function performs *segmentation and reassembly* on data as it moves through AAL 1.

AAL type 2. This service specifies support for isochronous service with varying bit rates. An example of such a user of this function is compressed or packetized video.

AAL type 3–4. AAL type 3–4 function supports LAN traffic. It supports a variable bit rate. Both connection-oriented and connectionless-oriented connections are supported. The SSCS is the *service-specific convergence sublayer*. One of its functions is data translation. It also maps upper-layer services to the ATM layer. CPCS is the *common part convergence sublayer*. This part works in conjunction with switched multimegabit data service. SAR is the component that performs segmentation and reassembly segment.

AAL type 5. AAL type 5 is designed for variable-bit-rate services. It is similar to type 3–4 but is easier to implement. Most ATM LAN devices support this type.

5.4   ATM Cell Structure and Contents

The components of an ATM cell, as it appears at the user network interface (UNI), are shown in Fig. 5.4.

The cell structure shown in Fig. 5.4 contains 5 bytes of header information and 48 bytes of user information. The contents of the cell are

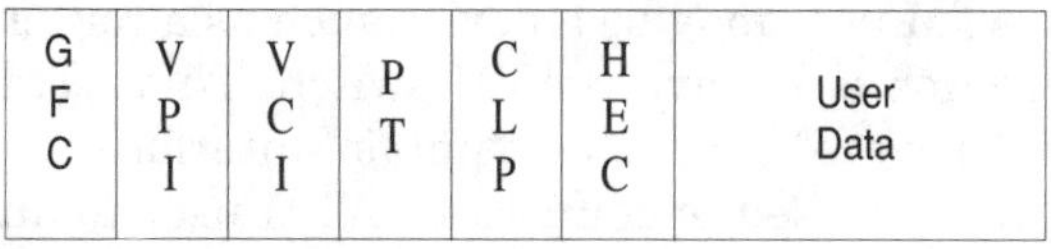

G F C	V P I	V C I	P T	C L P	H E C	User Data

Figure 5.4   ATM cell structure at the UNI.

defined as follows. The *generic flow control* (GFC) controls data traffic locally and can be used to customize a local implementation. The bit value in this field is not moved from end to end. Once the cell is in the network, ATM switches overwrite the fields.

The next two fields are the VPI and the VCI bits. Information stored in these fields performs routing functions. The number of bits here vary because the bits used for VCIs for user-to-user virtual paths are negotiated between the users of that virtual path. These two fields constitute internode communication.

The *payload type* (PT) indicates whether the data being carried is user data or management related information. The field is also used to indicate network congestion.

The *cell loss priority* (CLP) is used to explicitly indicate the cell priority. In short, it indicates whether the cell can be discarded if network congestion occurs.

The HEC field is used by the physical layer. This field is used to detect errors and correct bit errors within the header.

The user data field contains the user data from upper layers within the network. This is a 48-byte field. This field does not have error checking performed on it.

The other cell type, known as the *network node interface* (NNI), is similar to the type shown in Fig. 5.4 except for a difference in the header portion of the cell.

5.5   ATM Interface Types

As indicated in the previous section, two types of nodes are recognized. The user network interface (UNI) can be divided into two groups:

- Private
- Public

Private UNIs can best be described by function. They are typically used to connect an ATM user with an ATM switch to which both are considered local and part of the site. The switch itself may be referred to as "private" and considered to be *customer premises equipment* (CPE), which is that ATM equipment located on a customer site and not in the public arena. An ATM user may be a device such as a router or workstation. The ATM switch is a private ATM switch. This ATM switch is what connects the "private" user to the "public" interface.

The public UNI is the interface used to connect an ATM user to an ATM switch in the domain of the public service provider. Figure 5.5 depicts these concepts.

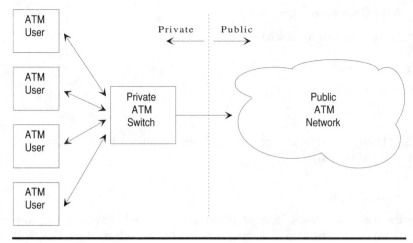

Figure 5.5 Conceptual view of the user network interface.

Figure 5.5 shows private CPE and the public ATM switch. The ATM switches (both private and public) utilize the same functionality at the physical layer; however, different media may be implemented.

The *network node interface* (NNI) refers to the ATM switches in the public service provider network that communicate with one another to achieve routing and thus end-to-end service. Figure 5.6 is an example of this concept.

Figure 5.6 shows the network nodes implemented in a public ATM network, ATM users, an ATM switch, and implicitly the boundaries between the public ATM network and the private ATM switch.

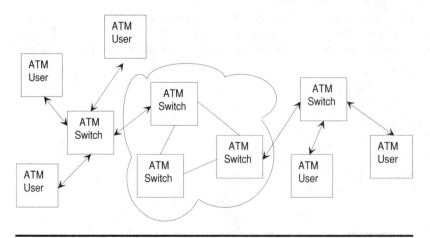

Figure 5.6 Public ATM network.

5.6   ATM Concepts

Three concepts are used with ATM:

- Transmission path
- Virtual path
- Virtual circuit

These three concepts are generally found together because they are part of ATM operation.

Transmission path

The *transmission path* is a physical connection between ATM-supported devices. These paths may have different characteristics, but they nevertheless exist. The physical transmission path between entities is that to which the virtual concepts are mapped.

Virtual path

The idea of a virtual path is derived from having a transmission path on which it can be mapped. The virtual path is mapped to the physical transmission path. Figure 5.7 is an example of this concept.

Virtual circuit

The notion of a virtual circuit exists. The virtual circuit (some refer to this as a *virtual channel*) is mapped to both a virtual path and a transmission path (physical path).

The concepts of transmission path, virtual path, and virtual circuit are implemented at the physical layer in the ATM model. This is merely another method of multiplexing. This structure is part of the

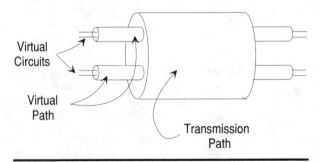

Figure 5.7   Conceptual view of paths and circuits.

intricate nature of the ATM physical layer and is part of the robust nature of ATM.

5.7   ATM Implementation

Where and how is ATM implemented are valid questions. Because its support is versatile and it can accommodate high speeds, multiple possibilities exist as to *how* and *where* ATM is implemented. This section explores four distinct instances of *how* and *where* ATM is implemented:

- Local router and ATM backbone
- ATM-based LANs
- ATM backbone nodes
- ATM LANs and ATM backbone

Local router and ATM backbone

This type of implementation uses routers that support ATM in a local geographic area. Here a LAN exists with other lower-layer protocols and network devices, but a router with an ATM interface is used to connect the LAN into the ATM backbone that serves a much larger geographic area. Figure 5.8 portrays this idea.

Figure 5.8 is a practical example of an ATM implementation. It illustrates a router with an ATM interface in a local implementation along

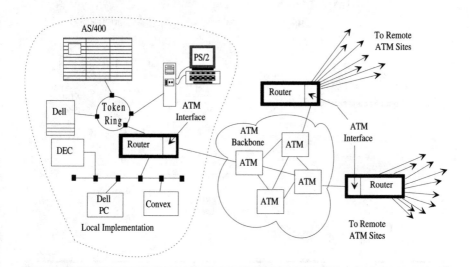

Figure 5.8   Local ATM hub/router and ATM backbone.

with devices attached to it directly or indirectly. It also indicates the local router with an ATM interface connected to an ATM backbone by which other sites connect. The ATM backbone consists of ATM devices. This is one of the simplest examples of those discussed in this section.

ATM backbone LANs

The notion of ATM backbone LANs is as its name implies; the LAN is built around ATM equipment. Figure 5.9 illustrates this idea.

In Fig. 5.9 three ATM-based LANs are shown in addition to the network connected via the router with an ATM interface. Two local implementations are shown; one has two ATM-based LANs. Each local implementation accesses the ATM network backbone which consists of all-ATM equipment.

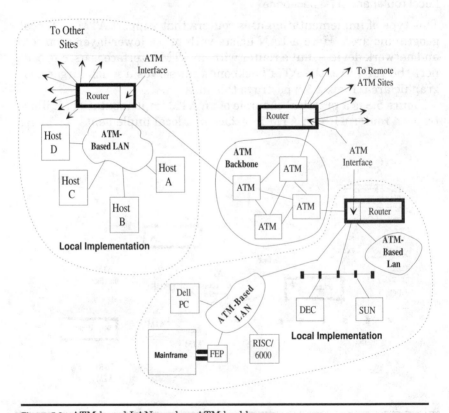

Figure 5.9 ATM-based LANs and an ATM backbone.

ATM backbone nodes

ATM backbone nodes are typically implemented in public environments (in contrast to a private enterprise). Many ATM nodes working
together constitute an ATM backbone. Figure 5.10 shows multiple
ATM nodes, creating a WAN backbone.

Figure 5.10 includes a network attached to the ATM node in Salt
Lake City, Utah. The Salt Lake City network enters the ATM network
via a router. A similar scenario like this could be repeated in other sites
shown in Fig. 5.10.

ATM LANs and backbone

The notion of ATM LANs and an ATM backbone is an example of complete ATM implementation. Figure 5.11 depicts this concept.

Figure 5.11 shows ATM nodes comprising the backbone of the ATM
network and the focal point of the networks in Boise, Idaho, and
Syracuse, New York. This example shows maximum utilization of ATM
locally and in a WAN sense as well.

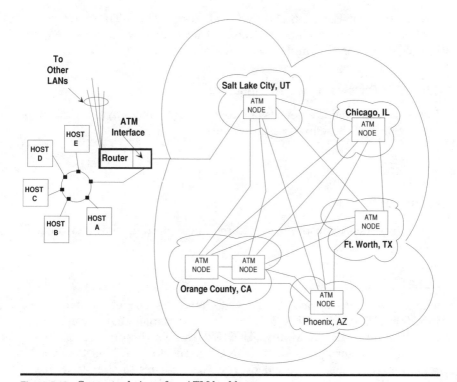

Figure 5.10 Conceptual view of an ATM backbone.

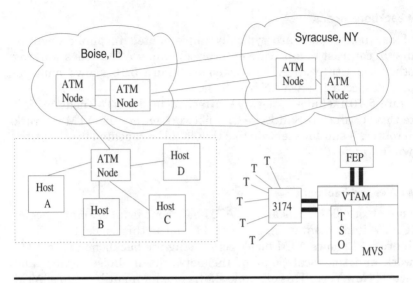

Figure 5.11   Complete ATM implementation.

Variations of Figs. 5.8 through 5.11 are possible. These are representative of likely implementations. Other *hybrid* ways of implementing ATM are possible and depend heavily on the site.

5.8   ATM Physical-Layer Architecture

Figure 5.2 showed the details of supported interfaces by the PMD sublayer. Some of those are presented here:

- SONET
- DS3
- Fiber 100 Mbits

The interfaces and the provided technical characteristics of the previous list are explained here.

SONET

The *synchronous optical network* (SONET) is one of the supported interfaces indicated by Fig. 5.2. Its data rate transfer can accommodate speeds of up to 155.52 Mbits/s. SONET frame structure is such that it can easily accommodate ATM. SONET support for ATM is achieved by mapping ATM cells, aligned by row, correlating to the structure of every SONET byte structure. Figure 5.12 is an example of how SONET and ATM can merge together.

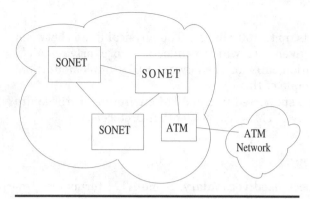

Figure 5.12 Perspective of SONET and ATM.

Figure 5.12 shows an ATM device connecting into a SONET interface. In turn, an ATM network is attached to the ATM device. This
implementation utilizes speeds through the SONET interface.

DS3

DS3 can be used as an interface to carry ATM cells. The data rate of
DS3 is 44.7 Mbits/s. If this is used, a physical-layer convergence protocol (PLCP) must be defined. Once this is complete, then ATM cells are
merely mapped to the DS3 PLCP frame. Figure 5.13 shows an example
of this.

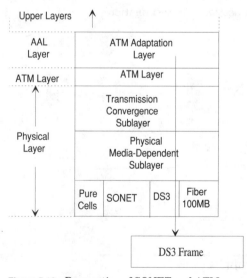

Figure 5.13 Perspective of SONET and ATM.

Fiber 100 Mbits

This interface supports up to 100 Mbits/s. The physical link is between the equipment at a given site which connects it to a private ATM switch. This specification calls for the current FDDI implementation. Figure 5.14 is an example of this.

Other interfaces are supported in the ATM environment; these are good illustrations of how ATM works with other protocols.

5.9   ATM Terminology

ATM has a highly specialized vocabulary. Some ATM terms are presented here to serve as a specific ATM reference source.

AAL connection   The establishment of an association between AAL (ATM adaptation layer) or higher entities.

ATM   A fast packet-switched, cell-based method of moving voice, data, video, and other data and telecommunication-oriented information from one location to another. The ATM cell is 53 bytes in length: 5 bytes of header information and 48 bytes of user information.

ATM layer link   A connection between two ATM layers.

ATM link   A virtual path or a virtual circuit (channel) connection.

cell   An ATM protocol data unit.

connection admission control   The method used to determine whether an ATM link request can be accepted that is based on the origin and destination's attributes.

connection endpoint   A layer connection SAP termination.

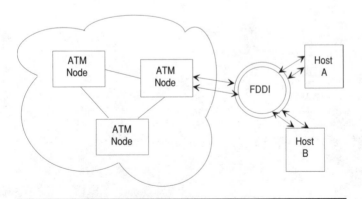

Figure 5.14   Perspective of ATM and FDDI.

connection endpoint identifier   A characteristic of an endpoint used to identify a service access point (SAP).

end system   The place where an ATM connection is terminated.

fairness   The meeting of all specified quality-of-service requirements through control of active connections across an ATM link.

header   That control information which precedes the user information in an ATM cell.

metasignaling   A way of managing virtual circuits and different types of signals.

network node interface   The ATM interface as it relates to the network node.

segment   An ATM link, or group of links, that constitute an ATM connection.

service access point (SAP)   An addressable endpoint at a given layer within a network.

sublayer   The division of a layer logically.

switched connection   A connection established via a signaling method.

symmetric connection   A connection where both directions have the same bandwidth.

virtual channel   Also called a *virtual circuit*. That which an ATM cell can traverse.

virtual circuit   Also called a *virtual channel*. That which an ATM cell can traverse.

virtual path   The logical association of virtual circuits.

virtual path connection   This is a one-way joining of virtual path links.

virtual path link   The connection between points where a virtual path identifier is assigned.

virtual path terminator   The system which processes the virtual circuits (channels) after they are demultiplexed.

5.10   Where to Find Additional Information

Additional ATM information can be obtained from CCITT specifications published in 1992–1993. Some of these include

I.113	I.362
I.121	I.363
I.150	I.413
I.211	I.432
I.311	I.610
I.321	G.707
I.327	G.708
I.361	G.709

5.11 Summary

ATM is a cell-based, fast packet-switching technology. It can be implemented privately or publicly. Many public service providers are opting for ATM installations. ATM technology has received considerable attention since its entry into the technical arena in the late 1980s.

ATM conceptual structure does not correlate one for one with the OSI model. ATM layers begin with a physical layer which has two sublayers. The lowest sublayer is the physical-media-dependent sublayer; above it is the transmission convergence sublayer. Next is the ATM layer which includes four functions as explained previously. The ATM adaptation layer is above the ATM layer. Part of the ATM layer and the ATM adaptation layer make up layer 2. The ATM adaptation layer includes five types of functions.

ATM cells are unique because they are small and can handle voice, data, video, and other multimedia information. The AAL protocols constitute an interface between ATM and upper-layer protocols.

ATM utilizes a transmission path, virtual path, and virtual call as part of its implementation and method of passing information. Virtual paths and virtual calls are mapped to a transmission path.

The CCITT lists ATM specifications, some of which are directly related and some indirectly.

6

Enterprise Systems Connection (ESCON)

ESCON is an IBM product offering. It is the next-generation architecture for IBM channels. As a result, IBM renamed their channels *parallel channels*. According to IBM, the plan for ESCON is considered one of evolution and nondisruptive. This chapter explores ESCON: what it is, how it operates, and how it is implemented.

6.1 Overview

ESCON was announced by IBM on September 5, 1990. To give the reader a perspective of the widespread announcements of ESCON and how far ESCON has come, the following list includes some of the *original* ESCON announcements.

- New input/output (I/O) protocols aiding in increased speed capabilities.
- Switched point-to-point technology.
- ESCON is based on fiber media with data rate transfer at a 10-Mbyte/s rating.
- Increased distances between devices and the I/O connection; specifically up to 9 km.
- Lightweight fiber cables.
- Centralized configuration and management.
- ESCON directors provide dynamic connectivity.
- Updates and reconfiguration during normal data center operations without disruption.

- ESCON converter supporting parallel channels.
- ESCON channels extend the addressing capabilities to support up to 1024 devices.
- ESCON manager software for control of multiple ESCON directors throughout an enterprise complex.
- Support for update and automatic configuration of ESCON directors by maintaining a data file.
- ESCON monitor system provides a centralized method for managing distributed sites.
- ESCON adapters for the 3990 storage subsystem.
- ESCON adapters for the 3490 tape subsystem.
- ESCON support for 3174 models 12L and 22L.
- ESCON support for the 3172 model 1.
- ESCON support through an enhancement to the 3745 communication controller.
- ESCON support for the reduced instruction set computer (RISC)/6000 via microchannel adapter card.
- ESCON channel-to-channel support with up to 9-km distance support.

Since the original announcement, multiple enhancements and new offerings have been made. A few of these are listed below to provide insight to the impact ESCON is having on large IBM systems.

- Sysplex timer facility to provide timing throughout a complex of processors utilizing ESCON fiber links.
- Laser support for driving the fiber.
- Continued LED support for driving fiber links.
- Support for multimode fiber.
- Support for single-mode fiber.
- The Extended Distance Facility (XDF) support.
- ESCON director support for multiple concurrent operations.
- ESCON manager support for SystemView.
- February 9, 1993, announcement for the ESCON 9036 remote channel extender.
- April 6, 1994, announcement for ESCON support on IBM POWER-parallel Systems (SP2).
- 1994 S/390 Parallel Sysplex Offering based on ESCON channels.

6.2   Hypothetical ESCON Environment

Each ESCON environment may require different ESCON products to meet the needs of that environment. In this section most of the components in an ESCON environment are presented and explained.

Processors, directors, and control units

In a typical ESCON environment a processor, ESCON director, and control units will be present. Chapter 16 of this book explores IBM's SNA and explains the characteristics and functions of the devices used in an SNA network; here the explanation will focus on ESCON and how it fits into the picture of a hypothetical environment.

Figure 6.1 represents a possible configuration in a given environment where ESCON is implemented.

Figure 6.1 shows numerous devices in this hypothetical environment. Two processors are shown connected together via an ESCON channel-to-channel connection. This figure also shows a sysplex timer used for timing the multiprocessor environment. Two ESCON directors are shown with attached devices. One shows two tape subsystems

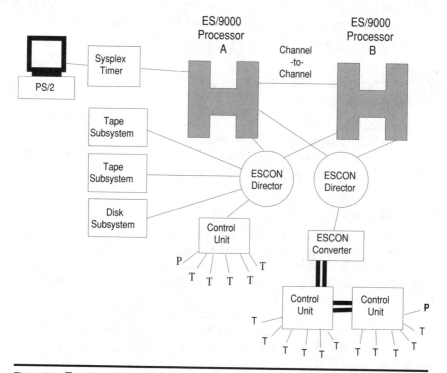

Figure 6.1 Top view of an ESCON complex.

attached and a disk subsystem. The other ESCON director depicts an ESCON converter attached to it with two control units connected via bus and tag cables; these are copper-stranded cables associated with the implementation of parallel channels, which is explained in Chap. 11. Multiple terminals and printers are represented.

This example environment is straightforward, and the devices in the example perform the following functions.

PROCESSORS	These processors can accept direct ESCON attachments. They also support a channel-to-channel connection.
ESCON DIRECTORS	These devices provide a dynamically switched channel environment. They also support distances of up to 43 km. Multiple directors are supported; thus each site may differ.
TAPE SUBSYSTEM	The tape subsystem is the same as with any installation; the difference here is support for ESCON connection.
DISK SUBSYSTEM	The disk subsystem performs the same functions as those prior to ESCON, but now greater throughput can be achieved with ESCON support.
CONTROL UNITS	Control units are standard in SNA environments. They are used to connect terminals and printers.

This environment differs from non-ESCON environments because of increased speed capacity through the data link, management of the ESCON environment via software called the *ESCON manager,* and dynamic switching capabilities of the ESCON directors.

6.3   ESCON Components Found in a Typical Installation

Each site differs with regard to need. Consequently the best way to begin learning about an ESCON environment is to understand the components of any given environment. This section covers the popular ESCON components, explaining their major purpose.

ESCON directors

ESCON directors can be used to dynamically switch logical links between devices such as control units or other devices and a processor or to maintain a dedicated connection between two ports. Figure 6.2 is an example of an ESCON director with multiple ports, any of which can be used dynamically.

Figure 6.2 shows three hosts connected to an ESCON director and three devices attached as well. The ESCON director shows six ports (1, 2, 3, 4, 5, and 6). Any of ports 4, 5, or 6 can request and be connected logically to port 1, 2, or 3.

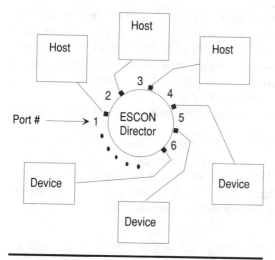

Figure 6.2 ESCON director: dynamic possibilities for connection.

A dedicated connection can be made either by way of a program command or by an operator definition stating it as such.

Figure 6.3 shows a device dedicated to host number 3. This dedicated connection is made between director ports 5 and 3. The figure also shows other ports dynamically assigned on request by either the host or the device.

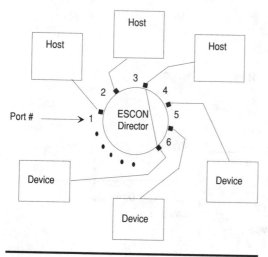

Figure 6.3 Dedicated connection via ESCON director.

Each ESCON director has multiple ports to which devices attach. Selection of a port to pass data to another port en route to its destination is part of the ESCON director function. These port requests are made via the frame that includes information such as the destination point. One function of a director is to establish a connection with the destination if one does not already exist.

Figure 6.4 is an example of what is considered a chained ESCON director environment.

This figure shows directors B and C daisy-chained with a dedicated connection between them. This figure also implies multiple dynamic connectivity. Notice that all the directors have a connection to all hosts. With directors B and C chained, different routes can be realized by devices communicating with hosts and vice versa.

Defining ESCON directors.   Definition of ESCON directors is made to the channel subsystem and director attributes and ports. Each ESCON director must be defined to the channel subsystem via the input/output channel program (IOCP). This process specifies the channel address to which each director is attached and addresses associated with each director and device attached to it. ESCON director attributes can be used to block a port, dedicate a port, and eliminate dynamic connection between ports.

The number of ports supported on the director, console-sharing capability, size, and additional functions depend on the model ESCON director.

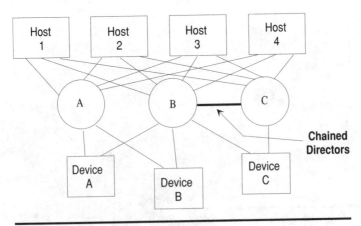

Figure 6.4 Example of chained ESCON directors.

ESCON converters

ESCON converters are used to convert bus and tag cabling and related signals from devices to ESCON signals and fiber cable. Figure 6.5 is an example of the implementation of an ESCON converter installed.

Figure 6.5 shows two ESCON converters. One is connected to an ESCON director, and another is directly connected to a host supporting ESCON channels. Functionality of the converter includes the transmission and reception of fiber signals and the channel program necessary to accommodate parallel channels. Data sent to the host is encoded, and the signal is converted and passed to the ESCON channel. The converse is true with data outbound to a device. In this case, the converter converts the signal, decodes information, and transfers it to the appropriate control unit.

ESCON supervisors

The *ESCON supervisor* is a central point for controlling and maintaining messages and any errors that may occur in an enterprisewide ESCON-based environment. Figure 6.6 is an example of how an ESCON supervisor would appear in a typical ESCON environment.

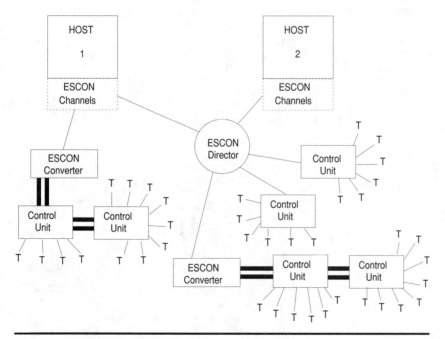

Figure 6.5   Example of an ESCON converter.

This figure shows a token-ring-based network, PS/2 functioning as the ESCON supervisor, and multiple devices in the ESCON-based enterprise network.

In this environment, if the ESCON supervisor is used, the host processor sends data to the ESCON supervisor for maintaining the status of messages that may appear or errors that could occur. This implementation is one way to isolate any possible errors to the ESCON environment itself. All the ESCON links can be supervised from the centralized workstation shown in Fig. 6.6.

Sysplex timer

With the sysplex timer facility, a multiprocessor environment can have the processor's time-of-day clocks synchronized. The benefit of ESCON within a sysplex is this device may be located somewhere other than in a central facility, or it may need to be located in a remote facility that has recently been added to the enterprise. Figure 6.7 shows the sysplex timer attached to host E.

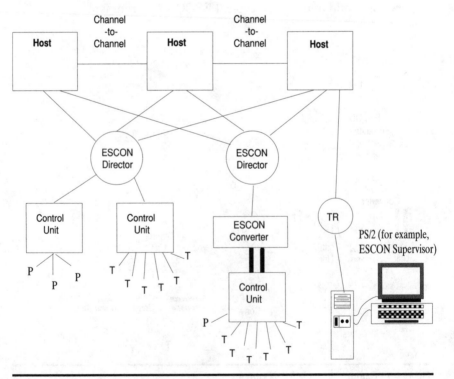

Figure 6.6 Conceptual view of an ESCON supervisor.

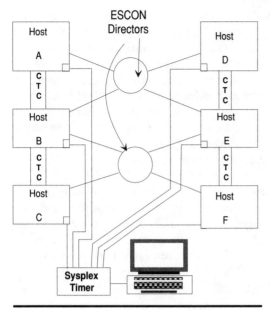

Figure 6.7   Conceptual view of the sysplex timer.

The sysplex timer attaches to a specific port on *each* processor. This port cannot be used to connect an ESCON director. The sysplex timer brings to the enterprise a battery standby, external clock source for the processors. The sysplex unit can synchronize the processors in a complex to within approximately 0.01 s. The console attached to the sysplex unit can also be used to see the status of the port to which it is attached, customization of the screen, and the timing.

Multiple sysplex timers can be used. One would be functioning in real time, and the other on standby ("hot" I/O). If two are used, fiber connections are required. These timers can be set to automatically monitor each processors time and make any changes that long-term operation would require.

6.4   ESCON Manager Program

The *ESCON manager* is a software program that operates under the multiple virtual storage or virtual machine (MVS or VM) operating system. Specifically, the minimum requirements for ESCON manager Release 2 operation are MVS/ESA Version 4.2 or VM/ESA Version 1.1 and ESCON director device support. (Additional requirements may be necessary; to be sure of the requirements for this product, contact your local IBM representative.)

The ESCON manager provides management capability for one or more ESCON directors from a single point. Figure 6.8 is a conceptual view of an installation with four hosts and one ESCON manager program.

In Fig. 6.8 each ESCON director can be controlled and managed via the ESCON manager on host A. In addition to managing these directors, the manager program provides capabilities for audit and control of ESCON manager commands. Configuration information can be obtained, and synchronization of the logical configuration to the physical configuration is also available. The ESCON manager program also permits coordination of directory changes throughout the entire complex.

The ESCON manager program provides other capabilities as well, such as the elimination of some ordinary operator tasks. Automation of changes is possible with the manager program. Remote director configuration capabilities are provided. And, configuration changes to the manager's database can be made to reflect any configuration changes.

The manager program also provides information on the host such as channel path identifiers (CHPIDs), control unit information, results of ESCON manager commands entered, and devices attached.

Interfacing with the manager

The ESCON manager program can be interfaced with via three popular ways. Depending on the installation, one method may be preferred over another.

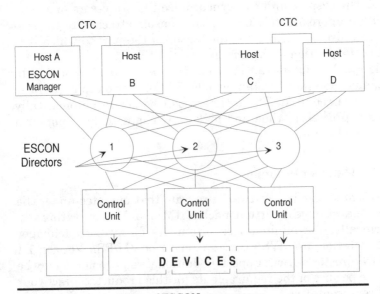

Figure 6.8 Conceptual view of ESCON manager use.

One such way to interface with the program is via the *system console,* which is a popular method for interfacing with the program. At the NetView menu, certain commands can be entered as well.

Another way of interfacing with the program is via the application program interface (API). These executable commands via the REXX (Restructured Extended Executor) language can be interactively worked with via the ISPF (Interactive System Productivity Facility) menu.

The programmable workstation interface is still another way for interaction with the program manager. In this way a windowing environment is used by the workstation with the host to obtain the information requested. This type of interface can provide a graphical representation of the I/O configuration.

6.5 ESCON Orientation

Before investigating ESCON protocols it is best to become familiar with terminology and concepts used in explaining them. First, knowing that the parallel channel and ESCON channel use similar commands, data, and transfer of data is helpful. Distinct differences exist, and those characteristics pertaining to ESCON are covered in this chapter.

The *channel path,* which is a path between a channel and one or more control unit devices, is composed of control units, ESCON directors, the actual channel itself, and physical links between the channel and directors and/or devices.

Control units

Control units may be any number of valid ESCON-supported devices. For example, an interconnect controller might be considered a control unit, or an establishment controller could fit this definition. Other control units are defined in Chap. 16 on SNA.

ESCON directors

ESCON directors are devices that serve as a switch. The director may have dedicated links or may serve the purpose of dynamic switching. Physically, ESCON directors are rectangular devices, but logically they are represented as a circle; however, this does not have to be the case.

Channels

The channel may be part of the processor housing or a separate unit. The channel consists of hardware and firmware; in fact, in its own right it is a "computer." In drawings depicting ESCON environments, most representations include the channel as part of the processor housing.

The physical link

The *physical link* is the fiber cable used in an ESCON environment. It is generally represented by a single line. This, however, does not necessarily imply single or multimode fiber. The link is point-to-point, and optic pulses traverse the cabling. Even though the link appears as point-to-point in a dedicated sense, in reality when a link is established between the channel and a director it is generally considered to be switched point-to-point.

Physical and logical paths

The notion of a *physical and logical path* exists. The *physical path* is as its name implies: the physical path made up of physical components. A *logical path* is a map of a connection between a channel and a control unit (and hence device) via a director in most cases. The logical path is superimposed on top of a valid physical path.

6.6   ESCON Protocols

Link- and device-level protocols are used with ESCON technology. Link-level protocols describe characteristics of a channel path. These characteristics describe the physical aspects of the channel path. A channel path is that path between an actual channel and a control unit (or multiple control units).

Link-level protocols

Link-level protocols establish and maintain both physical and logical paths from a channel to a device. These protocols also provide a mechanism for checking frames for errors that might have occurred during the transmission. Link-level frames are used to transfer this information. Figure 6.9 is an example of a link-level frame.

The link-level frame consists of a header and a trailer. No information field exists. The link header consists of the start-of-field delimiter, the destination address, the source address, and the link control field. The link trailer consists of two fields: the cyclic redundancy check and end-of-field delimiter.

Figure 6.9   Conceptual view of a link–level control frame.

Device-level protocols

Device-level protocols transfer information such as data, commands, control information, and status information between a control unit and a channel. Device-level information is moved via device frames. Figure 6.10 is an example of a device frame.

The device header consists of the information field identifier that specifies device addressing, format identification, and the device header flag field which invokes protocols to control an I/O operation. The device information block part of the frame contains the command flag field, which in turn contains channel command word (CCW) flag bits and a chaining update flag. It also has a command field that contains current CCW information from the current CCW. A count field is also used and serves as a current CCW count.

A variation of the device frame is the frame that contains status frame device information block structure. This type of frame is used to transfer status to a channel.

6.7   ESCON Physical-Layer Specifications

Two physical-layer specifications exist: one for a single-mode physical layer and the other for a multimode physical layer. Differences between these two are significant, especially with respect to the usable fiber-length distances.

Single-mode layer specifications

The single-mode layer specifications can permit cable lengths to run up to at least 20 km. The specification includes a keyed connector for single-mode fiber to prevent cable inversion. Fiber specifications include operating wavelength of 1270 to 1340 nm, cladding diameter of 125 μm, and mode field diameter of 9 to 10 μm. You can obtain further details on complete specifications from your local IBM representative.

Multimode layer specifications

Multimode specifications include definition of physical cable length of up to 3 km without the need for repeating signals. Other specifications

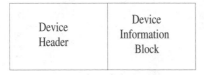

Figure 6.10   Conceptual view of a device frame.

include multimode trunk fiber size of 62.5-μm core and 125-μm cladding, i.e., a fiber-optic rating of 62.5/125. A 50-μm core and 125-μm cladding is also supported. This specification calls for a glass core and cladding with an operating wavelength of 1300 nm. The specification also calls for multimode fiber cable connectors to be keyed to prevent inversion of the connector.

6.8 Summary

ESCON was announced by IBM on September 5, 1990. ESCON is IBM's fiber medium and channel subsystem that has superseded the channels which used copper-stranded cables. The original announcement was robust. The ESCON offering represents a clean break from prior I/O subsystems because it is fiber-based. The lengths to which fiber can be utilized far exceed anything IBM has previously offered.

The ESCON channels brought with them a new name for the existing channels; they were renamed parallel channels. This is accurate because they do transmit data in parallel while ESCON transmits data serially. Both parallel and serial-based channels are supported by IBM; however, it would seem logical to conclude that IBM will eventually replace the parallel channels.

ESCON-based systems include channels, channel paths, directors, control units, devices, an ESCON supervisor, an ESCON manager, fiber cable, and typically a sysplex timer facility. Depending on the site, variations as to what is needed may change.

Multiple ways exist to interface with the ESCON manager. This can be done via a console, via a NetView menu, via an API, or via a workstation. Each interface to the manager provides advantages. Sites will differ as to which is best for implementation.

Physical and logical paths exist. The physical path can be readily understood because it has tangible components. The logical path is a map of data flow through a physical path. It is an abstract view as to how data traverses the system.

Both link- and device-level protocols are used. They perform different functions. The link-level protocols operate to control the link and provide a means for exchanging information on the link. Device-level protocols are particularly related to devices. They are used in device-level addressing and to perform other functions as well.

Chapter

7

ETHERNET

ETHERNET technology is currently one of the most widely used lower-layer networking protocols. ETHERNET has a proven history and presence among numerous supplying vendors and is price-effective. This chapter explores ETHERNET in detail.

7.1 Origins, Evolution, and Versions

ETHERNET can be traced back to Palo Alto, California, in the early 1970s. Specifically, its origins came from the Palo Alto Research Center (PARC). PARC itself is interesting history.

PARC was the outgrowth of a mind-set at the Xerox Corporation. The history behind PARC varies depending on the source, but the most substantiated is the one presented here.

In 1970 Xerox speculated about the future. It was involved in design, manufacturing, and providing copying machines. Xerox was aware of its products and the surrounding technology. Likewise, it was aware of the possibility that future offices could be populated with computers and terminals, and thus that fewer needs for copiers would exist. As a result, PARC was launched.

PARC consisted of a core of individuals whose focus was turning ideas into products. In 1973, Bob Metcalfe worked with a group of individuals whose interest was resolving a fundamental networking problem. That was the amount of time it took to get data from a computer to a printer. The bottleneck was not the computer or printer; it was getting data from the computer to the printer which took more than 10 min—even with high-speed links. This was overcome, and thus the birth of ETHERNET.

The essence of the problem was that if multiple devices were connected to the same medium, collisions occurred when two or more of these devices attempted to communicate. This obstacle of collisions had to be overcome and led to considerations as to possible solutions.

A simple analogy explains the crux of the solution. When a professor teaches students, questions which require the professor's attention inevitably arise. When this is the case some students merely call the professor's name. This is an effective means of communication when only a few students are present, but what happens when the number of students increases? The possibility of more than one student calling the professor's name increases. As a result the potential *collision* rate increases. When such a scenario exists, those students involved in the collision stop, and one typically proceeds with communication with the professor. This example is called "courtesy"; another way to state this is that other students "back off" a random period of time until communication is finished between the student and the professor. Other students continue to listen, and when they sense that communication has stopped, another attempt by one or more students to communicate is made by calling the professor's name. In short, this is called a *back-off algorithm.*

The idea of multiple devices listening to a medium to determine whether communication existed, and then waiting for an arbitrary amount of time before attempting to transmit data again, came to be known as *ETHERNET*. The first implementation of ETHERNET technology delivered a data rate transfer speed of 2.67 Mbits/s. This occurred between 1973 and approximately 1975. This version of ETHERNET is known as *experimental ETHERNET*. Following this a paper, coauthored by Robert Metcalfe, specified a 3-Mbit/s data rate with approximately 100 nodes attached to the medium. Refinement of ETHERNET continued for the coming years.

In 1980 a consortium composed of DEC, Intel Corporation, and Xerox (also known as DIX) came together with the intent of producing a standard for ETHERNET. They did, and it is known as Version 1.0, published in 1981. In 1982, DIX produced an ETHERNET standard stating the following ETHERNET characteristics:

- 10-Mbit/s data rate
- Station separation maximum 2.8 km
- 1024 maximum number of stations
- Coaxial cable using baseband signaling
- Bus topology
- Contention resolution for link procedure
- Variable-size frames

7.2   Theory of Operation

ETHERNET operates using broadcast technology. In short, this is best explained by analogy. Consider a seminar with approximately 150 individuals attending. Assume that someone needs to contact an attendee at the seminar. Assume that the person attempting to contact the attendee goes to the room where the seminar is being held, waits for the speaker to stop speaking, and then asks aloud, "Attendee who works for XYZ corporation, are you in here?" assuming the attendee, normally there, would respond. Granted that everyone else attending the seminar hears the requests, and so does the individual whom someone is attempting to reach.

This analogy, which is similar to ETHERNET operation, uses a broadcast calling the attendee who works for XYZ corporation. Equally, hosts participating in an ETHERNET network do the same. Consider Fig. 7.1.

Figure 7.1 shows a bus topology, with six hosts connected to it via ETHERNET interface boards in each and a transceiver connecting each host to the medium. In this example, suppose that host A wants to communicate with host F. Host A broadcasts onto the medium, "Host F, are you there?" In turn, host F answers positively and communication is established in the ETHERNET interface boards between the two hosts. This is an example of a broadcast because all hosts on the medium *hear* the broadcasts, but only the appropriate host responds.

Transceiver functionality

The transceivers used to connect hosts to the medium perform a specific function. They implement carrier-sense multiple access with collision

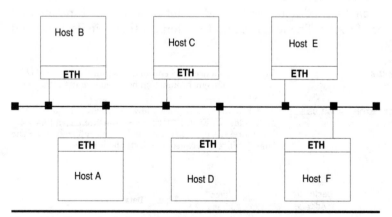

Figure 7.1   Conceptual view of ETHERNET communications.

detection (CSMA/CD) technology. This means that the transceivers themselves are capable of perceiving the carrier on the medium, permit access for multiple hosts (one per transceiver), and can detect collisions.

Heartbeat

The carrier sense that the transceivers listen for on the network medium is actually a voltage. Some receivers require this for operation. The technical term for this is *signal-quality error* (SQE). The importance of this is realized when ETHERNET is mixed with IEEE. 802.3 technology. Delineation between the two is covered later in this chapter. It is important to understand that heartbeat may or may not be required, depending on the vendor and the type of implementation.

Collision detection

Transceivers have the intelligence to determine whether a collision (simultaneous broadcasts) has occurred. If a collision has occurred, a transceiver will invoke a *back-off algorithm* (similar to that described in Sec. 7.1 in the professor-student example), which is a random period of time that the transceiver will wait to attempt another broadcast on the network. This algorithm is implemented in hardware inside the transceiver.

7.3 ETHERNET Frame Components

ETHERNET frames are between 46 and 1500 bytes. Frames less than 60 bytes in the data field are called *runt* frames. Figure 7.2 is an example of an ETHERNET frame.

This frame has six fields. Each field performs a specific function and is of a specific size. The lengths and functions of these fields are listed as follows.

PREAMBLE	This is a sequence of 64 encoded bits that the physical layer uses for clock synchronization between circuits attached to the medium.
DESTINATION ADDRESS	This is a 48-bit hardware address. It is called the ETHERNET address. Each ETHERNET interface board has a 48-bit address; sometimes this address is referred to as the "hard" address because it reflects the hardware.

Preamble	Destination Address	Source Address	Type	Data	FCS

Figure 7.2   ETHERNET frame.

SOURCE ADDRESS	This is a 48-bit hardware address of the sending (broadcasting station). It is the ETHERNET address and is also referred to as a *hardware* address.
TYPE	This is a 2-byte field used to indicate the protocol type if multiple upper-layer protocols occupy the same physical medium.
DATA	This field may contain between 46 and 1500 bytes of data.
FRAME-CHECK SEQUENCE	This is a 32-bit field used to perform cyclic redundancy checking on all fields of the frame except the frame-check sequence (FCS) field itself.

The ETHERNET frame includes the data being passed from one station to another. It is what traverses the medium from source to destination.

Interestingly, the destination and source fields are relative. Any given host may serve as the source in one phase of communication, and then serve as the destination in another phase. The point here is that any given host may be the source or destination depending on which host originates the communications.

7.4   802.3 Frame Components

IEEE 802.3 frames are similar to, but not the same as, ETHERNET frames. Figure 7.3 depicts an IEEE 802.3 frame.

This frame is similar to the ETHERNET field. Each field performs a specific function and is of a specific size. The following list explains the length and function of each field.

PREAMBLE	This is a sequence of 7 byte-encoded bits that the physical layer uses for clock synchronization between circuits attached to the medium.
SFD	The *start frame delimiter* field is binary 10101011. It indicates the start of the frame; thus the receiver can locate the first bit in the frame.
DESTINATION ADDRESS	This is a 48-bit hardware address. It is sometimes referred to as the hard address because it reflects the hardware interface board address.
SOURCE ADDRESS	This is a 48-bit hardware address of the sending (broadcasting station). It is also referred to as a hard address.
LENGTH	This specifies the amount of LLC bytes to follow.

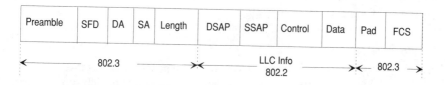

Figure 7.3   IEEE 802.3 frame.

DSAP	This is the *destination service access point* on the destination host.
SSAP	This is the *source service access point* on the originating host.
CONTROL	This specifies connectionless unacknowledged service.
DATA	This field may correlate the data field on the LLC protocol data unit and contains the data.
PAD	This field is variable, but has sufficient room for *padding* to occur so proper collision detection can occur.
FRAME-CHECK SEQUENCE	This is a 32-bit field used to perform cyclic redundancy checking on all fields of the frame except the FCS field itself.

The 802.3 theory of operation is the same as that of ETHERNET except for the function of the different field: specifically, the LENGTH field, which specifies the LLC protocol data unit information to follow. Broadcast technology is used as in ETHERNET. It also calls for 10-Mbit/s baseband signaling on coaxial cable, or unshielded twisted-pair cabling, or broadband signaling at 10 Mbits/s.

7.5 Addressing Schemes

Three different addressing schemes are used with ETHERNET. This section explores them.

Singlecast

A *singlecast address* is the simple implementation where one host communicates with another host (or multiple hosts). In this case a destination and source address is used.

Multicast

A *multicast address* is usually 6 bytes in length. It differs from the singlecast address because it is a special address that particular hosts understand, and they receive it as if the frame were addressed to them with a singlecast address. Hosts that are not concerned with this address do not recognize it; therefore, performance is basically a non-issue. This type of address is used in special cases in which it is necessary to load different hosts on a network and other hosts on the network do not need this information.

Broadcast address

The *broadcast address* can be used to perform isolated broadcasts on networks where multiple subnets are implemented. This type of address is frequently used when the upper-layer protocol is TCP/IP.

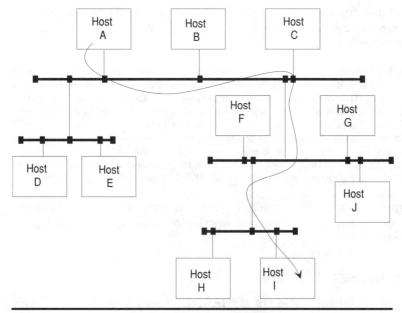

Figure 7.4 Conceptual view of a broadcast address used with subnetworks.

A broadcast address can be used to literally mask off those networks where the host is not located and therefore no broadcast needs to occur. Figure 7.4 is an example of how a broadcast address might be used.

Figure 7.4 depicts a major network backbone where hosts A, B, and C reside. Attached to this backbone are multiple subnets. Assume that host A needs to communicate with host G. The most efficient method of performing this is to *not* broadcast this frame on the major backbone or any other subnets except the one where host I is located. This functions well using what is termed a *broadcast address*.

A broadcast address is also used to broadcast a frame to all networks, and subnetworks and *all* hosts on these networks will read the frame. Consequently the use of this type of address is generally restricted for uses where it is the most efficient method of communicating with all hosts.

7.6   An Implementation Example

ETHERNET is a good lower-layer protocol to use in many environments. ETHERNET technology has been refined, is reliable, is cost-effective, and can be purchased easily, and many individuals understand the technology.

The following example is real. It reflects a corporation whose identity has been concealed for privacy reasons. Here, we will refer to it as the T Bank Corporation.

The T Bank Corporation is located in a major metropolitan city in the United States. The T Bank building has multiple floors with separate departments on each floor. This corporation needed departmental LANs while implementing a corporate solution and the ability to segment some departments as necessary. ETHERNET was selected to meet the needs of a lower-layer protocol because of segmentation capabilities and provisions for a main backbone. Another reason for choosing this lower-layer protocol is that departments could function independently while maintaining the capability to function with other departments in the corporation. Figure 7.5 represents the ETHERNET implementation at T Bank.

Figure 7.5 shows departments located on each floor and each department within a network. It also shows the total solution with all departments together.

7.7 ETHERNET via 10BaseT

A popular implementation of ETHERNET today is with twisted-pair cabling and a hub. When ETHERNET is implemented this way, it is generally referred to as *10BaseT*, meaning 10 Mbits/s using baseband signaling over twisted-pair cabling. Figure 7.6 shows an implementation of a hub, multiple ETHERNET hosts, and twisted-pair cabling.

Figure 7.6 does not show any transceivers, because the functionality of the transceiver is built into the ETHERNET cards installed in each

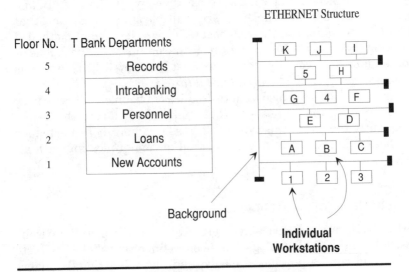

Figure 7.5 Conceptual view of T Bank Corporation solution.

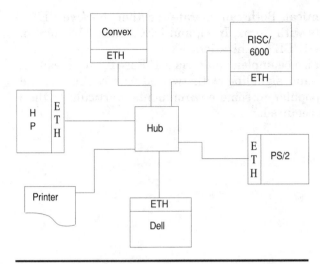

Figure 7.6 ETHERNET implementation via 10BaseT and hub.

host. It is the functionality of CSMA/CD required; how it is implemented is not necessarily important. Additionally, different hosts may participate. For example, the Convex Computer works quite well with a Dell computer and a Hewlett-Packard printer. This implementation also shows a printer as part of the network configuration. As is implied in Fig. 7.6, any host can use the printer. The printer functions on the network because it uses an ETHERNET network interface card to participate on the network.

Figure 7.6 is a good example of how easily an ETHERNET network can be implemented. The Dell Computer Corporation resells some of the equipment required to install a network such as this; however, Dell may not resell all the equipment implied in Fig. 7.6.

In many installations such as that shown in Fig. 7.6, the network is a "plug-n-play" network. Once the network is brought up and is operational, it will probably function well for some time without failure. Figure 7.6 is a good example of how many vendors set up their booths at trade shows where they have a network or network connection from their booth somewhere.

7.8 Summary

ETHERNET technology has been around for approximately two decades. It is a cost-effective solution, is a proven technology, and is understood by many. ETHERNET and the IEEE 802.3 solution are similar but not identical. Both can operate together; however; ETH-ERNET has its roots with Xerox, Intel, and DEC. The 802.3 solution has its origins in the IEEE organization.

ETHERNET can also be implemented via a 10BaseT. This means 10 Mbits/s using baseband signaling with twisted-pair cabling. A hub implementation is popular in some environments, particularly those where portability is required.

8

Fiber Distributed Data Interface (FDDI)

FDDI is a ring-based technology similar to token ring. However, FDDI differs from token ring and can best be understood by examining some of FDDI's characteristics. This chapter explores popular topics of FDDI and presents fundamental FDDI operation.

8.1 Basic FDDI Characteristics

This section lists the highlights of FDDI characteristics and explains their nature of operation with FDDI.

Cable specifications

One characteristic of FDDI is that it is intended primarily for operation with multimode fiber cable 62.5/125 µm (62.5-µm core; 125-µm cladding). However, single-mode fiber can be used. Additionally, FDDI protocols can be implemented over shielded twisted-pair cabling, which some refer to as *shielded distributed data interface* (SDDI). Another implementation is FDDI implemented over unshielded twisted-pair cabling; this is sometimes referred to as *copper-stranded distributed data interface* (CDDI).

Ring speed and ring distance

Another aspect of FDDI is its speed. Its specification is 100 Mbits/s. This is considerably faster than token ring. FDDI specification also

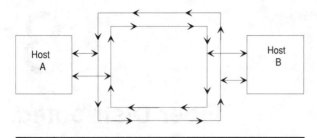

Figure 8.1   FDDI dual rings.

calls for a dual-ring implementation. The dual-ring structure defines data transfer in opposite directions. Figure 8.1 illustrates this concept.

With multimode fiber any given ring segment can be up to 200 km in length. A total of 500 stations can be connected with a maximum separation of 2 km.

FDDI data frame

An FDDI frame can hold up to 4500 bytes of data. This data is carried in a frame or packet. Each frame of data has a header including the origin and destination address of each frame.

Maximum number of stations

Any given ring segment may have a maximum number of 500 stations. However, rings can be segmented, and with a two-segment ring, 1000 stations can be utilized. In this implementation, two physical rings are connected together.

Ring monitor functions

FDDI uses a ring monitor function, but it differs from token ring because the monitoring function is distributed. Stations on an FDDI ring do not require a single monitor.

8.2   FDDI Layer Analysis

The FDDI specification is defined by the ANSI X3T9.5 and can be traced back to its origins in 1982. Since that time the ISO 9314 committee have refined and continue to refine the FDDI standards.

FDDI layer in general

ANSI and the ISO have a layered definition for FDDI operations. These layers include definition of the sublayer function in each layer. Figure

8.2 shows the correlation between layers, functions, and components within those layers.

Figure 8.2 shows the physical and data-link layers divided into two sublayers each. The physical layer also includes the medium interface coupler. Another component is the station management.

FDDI layer components

Four sublayer components are defined by the ANSI specification and are reflected in Fig. 8.2. These sublayers are presented by their individual functions.

Physical (layer)-medium-dependent (PMD) sublayer. The PMD standard specifies the lower portion of the physical layer. It describes the necessary requirements for the fiber cable to connect to the media interface coupler (MIC). This definition includes optical levels, signal requirements, the connector design for the MIC, and permissible bit error

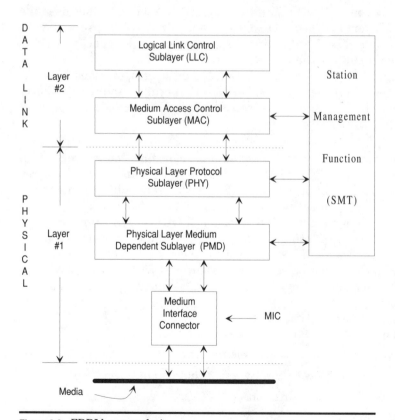

Figure 8.2   FDDI layer analysis.

rates. It provides services required to move encoded bit streams from destination to target station. It also determines when a signal is received by the receiver.

Physical-layer protocol sublayer (PHY). This sublayer is the upper part of the physical layer. It defines data framing, clocking, and data encoding and decoding specifications. This sublayer transmits data received from the MAC sublayer. It also initializes the medium for data transfer.

These two sublayers constitute the *port* on an FDDI interface board. In short, they operate to control and manage the process at the interface level of the connection.

Media access control (MAC). The lower sublayer (MAC) of the data-link layer performs some functions related to data framing. It is also involved in addressing, data checking, token transmission, and data frames on the ring.

Logical link control (LLC). The LLC sublayer of the data link uses the 802.2 sublayer. This is used to enclose data which is encapsulated with a MAC frame. The ANSI and ISO specification do not specify any further functionality of the LLC sublayer.

The data-link layer components focus on peer operations between two entities. The concentration here is on generation and recognition of addresses. This layer is also responsible for proper delivery of valid data to the layer above it.

Station management. The station management function interacts with the FDDI PMD, PHY, and MAC sublayers. The management services receive and collect information from these different sublayers and store them in a management information base (MIB). SMT object information reflects a particular FDDI station or FDDI concentrator.

Each station is identified in the MIB, and therefore collected information can be easily associated with a specific station. As mentioned previously, SMT uses objects to represent an FDDI station or concentrator. Some examples of SMT object management information are as follows.

- A unique identifier for each FDDI station
- SMT version the station is implementing
- The highest and lowest SMT version that can be supported
- Any special manufacturer data
- User-defined data

- Station configuration data
- Number of MACs in the station or the concentrator
- Available paths and their types
- Dual-ring configuration capabilities
- Status report frames queued for transmission
- Current port configuration
- The latest time-stamped event
- Station events that are reported for configuration changes

8.3   FDDI Timers and Frame Formats

The section focuses on the FDDI and token frame. It also presents and explains their contents. First is the FDDI frame.

Timers

Each station on an FDDI network is required to maintain certain timers. These timers perform different operations to coordinate operations of data transfer and maintenance on the FDDI network. This section lists timers and counters and provides a brief description of their functions.

Starting delimiter	This reflects maximum circulation delay for the starting delimiter to traverse the ring. This figure includes the maximum delay encountered in the total cable lengths and the latency encountered with attached stations.
Acquisition time	This reflects the maximum acquisition time for a signal.
Frame transmission time	This is the maximum amount of time required to transmit a maximum-size frame.
MAC frames	This is the maximum number of MAC frames allowed on the network at any given time.
Claim frame	This reflects the time required to transmit a claim frame along with its preamble.
Setup length time	This reflects the amount of time required for setup to transmit after a token is captured.
Insertion time	This is the maximum time required for a station to be physically inserted onto the ring.
Token holding timer	This timer governs how long a station can transmit asynchronous frames onto the ring.
Token rotation timer	This timer is used to control ring scheduling for normal operation on the ring. This timer also contributes to detect errors and aids in the recovery of errors encountered on the ring.
Valid transmission timer	Each station utilizes this timer in the recovery of transient ring errors.

FDDI frame

The FDDI frame appears as shown in Fig. 8.3.

The FDDI frame contents and their meanings include the following.

Preamble

The preamble is the first field in the FDDI frame. This field contains 16 or more symbols (that translates to 64 bits) that cause enough line state changes so that a receiving station's clock can orient its timing and be prepared for the fields that follow.

Start delimiter

This field always contains nondata symbols. The symbols are J and K and are used to indicate the start of the frame.

Frame control (FC)

This field identifies the FDDI frame type. The bit format indicates the different frame types. The structure of this field is

CLFF ZZZZ

In this case the C indicates the class bit. The L indicates the address length bit. The FF indicates the format bits. The ZZZZ are control bits. The C indicates whether the frame is asynchronous or synchronous. The L indicates the number of bits used in the addressing scheme: either 16 or 48 bits. However, the ring may contain both 16 and 48 bit addressing. The FF bits indicate whether an LLC protocol data unit (PDU) is in the information field or whether the frame is a MAC control frame.

The ZZZZ bits provide control information for the frame if it is a MAC frame.

Destination address (DA)

The DA field indicates the station or stations for which the frame is intended. This address may be either a single or broadcast address intended for multiple hosts.

Source address

The source address field contains the address of the sending station.

Information

This field may contain LLC PDU data or indicate that the frame is a MAC control frame.

Frame-check sequence

This field performs a cyclic redundancy check on the FC, DA, single-attachment station (SAS), and information fields.

Ending delimiter

This field indicates the end of the frame. The symbol in this field is a T, and it is 4 bits long.

Frame status

This field contains three indicators reflecting three pieces of information about the frame transmission: an error detected, the address recognized, and the frame-copied indicator; E, A, and C, respectively. Each of these indicators is represented by either an R or an S. The R indicates an "off" state or "false" condition. An S indicates an "on" state or "true" condition. This field uses a T (terminate) to indicate the end of the field.

PA	SD	AC	FC	DA	SA	RI	Info	F C S	ED	FS

Figure 8.3 FDDI frame format.

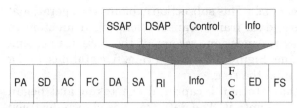

Figure 8.4 Conceptual view of FDDI frame with an LLC PDU.

If the frame contains an LLC PDU, it will appear as shown conceptually in Fig. 8.4.

Figure 8.4 shows a data frame. The data is in the information field which is part of the LLC PDU.

FDDI token

Figure 8.5 shows the format of the token frame used in FDDI networks. The contents of a token include the following.

Preamble	The preamble is the first field in the FDDI frame. This field contains 16 or more symbols (that translates to 64 bits) that cause enough line state changes so that a receiving station's clock can orient its timing and be prepared for the fields that follow.
Start delimiter	This field always contains nondata symbols. The symbols are **J** and **K** and are used to indicate the start of the frame.
Frame control	This field identifies the FDDI frame type. The bit format indicates the different frame types. The structure of this field is

CLFF ZZZZ

In this case the **C** indicates the class bit, the **L** indicates the address length bit, the **FF** indicates the format bits, and the **ZZZZ** are control bits.

The **C** indicates whether the frame is asynchronous or synchronous. The **L** indicates the number of bits used in the addressing scheme: either 16 or 48 bits. However, the ring may contain both 16- and 48-bit addressing. The **FF** bits indicate the frame type: an LLC PDU, MAC control frame, MAC beacon frame, MAC claim frame, VOID frame, restricted or nonrestricted token, or SMT management frame.

The **ZZZZ** bits provide control information for the frame if it is a MAC frame.

Ending delimiter	This field indicates the end of the frame. The symbol in this field is a **T,** and it is 4 bits long.

PA	SD	FC	ED

Figure 8.5 FDDI token format.

Theory of FDDI token operation. This subsection explains the operational nature of token capture, data frame transmission, token function in relation to data frame transmission, receipt of the data frame, and removal of the data frame. Figure 8.6 shows a ring, five stations, and a free token traveling around the ring.

Figure 8.7 shows that station 1 captured the token and began transmitting a data frame onto it. The data frame is shown destined for station 4.

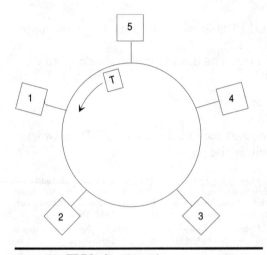

Figure 8.6 FDDI token operation.

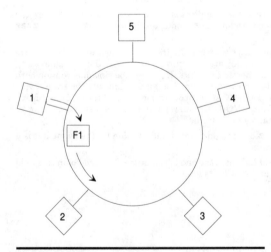

Figure 8.7 FDDI token operation.

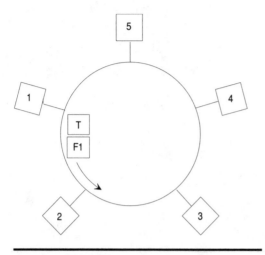

Figure 8.8 FDDI token operation.

Figure 8.8 shows that station 1 finished transmitting the data frame onto the ring and that station 1 appended the token to the end of the transmission frame.

Figure 8.9 shows that the transmission frame and following token frame bypassed stations 2 and 3 because they recognized that the destination address was not theirs. However, station 4 realized that transmission frame 1 had its address so it began receiving the frame into its

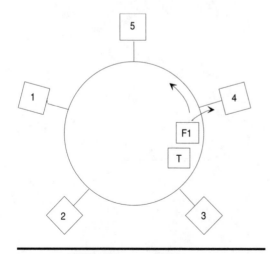

Figure 8.9 FDDI token operation.

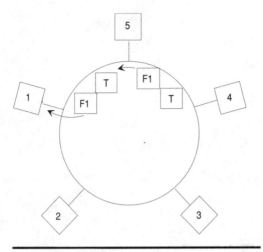

Figure 8.10   FDDI token operation.

buffer as the arrows indicate. The arrows also indicate that frame 1 was continuing its journey around the ring.

Figure 8.10 indicates that station 4 completed copying frame 1 into its buffer and that frame 1 along with the token passed on toward station 5. Figure 8.10 shows the frame and token passing station 5 and station 1 removing the frame from the ring.

Figure 8.11 shows frame 1 removed from the ring and the token free for any station on the ring to capture it and begin transmission.

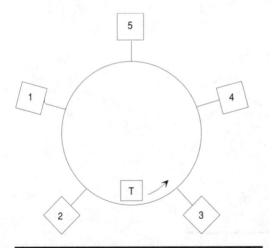

Figure 8.11   FDDI token operation.

Additional FDDI frames

Other frames are identified with FDDI technology. The following is a brief list of those frames and their meanings.

`Nonrestricted token`	This is a frame used in synchronous transmission and for nonrestricted asynchronous transmission.
`Restricted token`	This frame is used to control the transmissions in multi-frame dialogs. It is used in synchronous and restricted asynchronous transmission.
`SMT frame`	This is a management-type frame sent by station management components to control certain operations.
`MAC`	A variety of MAC frames are used for the control of the ring.
`MAC beacon`	This frame is introduced to the ring when a serious error occurs.
`MAC claim`	This frame is used to determine the station which creates a new token and initializes the ring.
`Implementor frame`	These frames are reserved for the implementor.
`Reserved frames`	These frames are reserved for possible future FDDI use.

8.4   Implementing FDDI

Topics for consideration prior to and during FDDI implementation are included in this section. FDDI networks include stations connected together in a manner with transmission media in order to form a physically closed loop. Connecting to the transmission media to add another station is one of the topics covered in this section.

Dual-ring technology

FDDI implements two rings: a *primary* ring that is used to transfer data and a *secondary* ring used as a backup for the primary ring. Each ring consists of two fibers. FDDI accommodates states such as computers, concentrators, bridges, and other network devices. Figure 8.12 is an example of a dual-ring environment.

Notice in Fig. 8.12 that four stations are participating as *dual attachment stations*. Also note that two rings exist with data moving in opposite directions.

In FDDI lingo, a *station* is an addressable device and can generate and receive frames. Each station has two *media interface connectors* (MICs), which are connectors between the station port and the actual media; technically they are called *connector keys*. Technically, the MIC is part of the FDDI interface card. Behind the MIC is the physical layer divided into the PMD sublayer and the physical-layer protocol (PHY) sublayer; these two sublayers constitute a port. Figure 8.13 is a conceptual view of a dual-attachment station having two MICs.

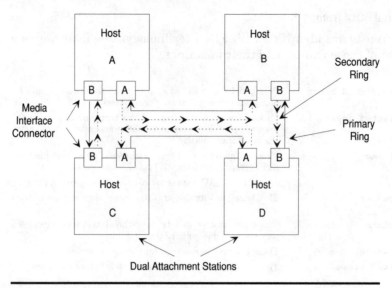

Figure 8.12 Conceptual view of a dual ring.

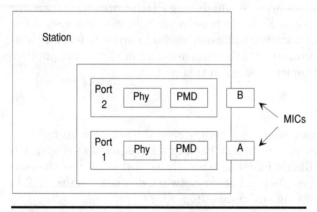

Figure 8.13 Conceptual view of dual-attachment station.

Station types

FDDI specifications identify three types of stations.

- Dual-attachment station
- Concentrator
- Single-attachment station

A *dual-attachment station* can attach to both the primary and secondary rings. This type of station on either side of a failure can "heal" the ring by rerouting traffic in the opposite direction.

A *concentrator* is a device that has ports beyond those required for its own operation. A dual-attachment concentrator is considered fault-tolerant. A single-attachment concentrator connects single-attachment stations in a logical tree topology.

A *single-attachment station* can attach only to the primary ring. If a failure occurs, this type of station cannot recover from the error as can a dual-attachment station. Figure 8.14 is an example of a single-attachment station and a single-attachment concentrator.

Media interface connector (MIC) types

For sake of clarity, four types of MICs are identified.

- Type A
- Type B
- Type M
- Type S

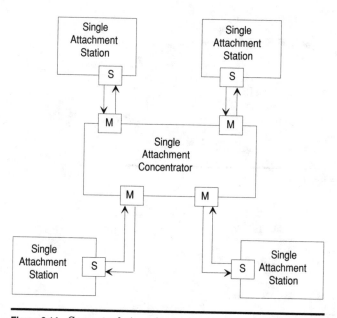

Figure 8.14 Conceptual view of a single-attachment station.

Type A MICs are used for main trunk connections with dual-attachment stations. One serves as the primary input and the other, a secondary output. Type B MICs are main trunk connectors used with dual-attachment stations whereby they function as the secondary input and the primary output. Type M MICs are master connectors used with concentrators for single-attachment stations. Type S MICs are single-attachment station connectors.

Sample topologies

Multiple topologies can be configured because of the flexible nature of FDDI. Some include a dual ring without any trees, a dual ring with trees, a single tree, a wrapped ring without a tree, a wrapped ring with trees, and a subset of dual-ring trees. Figure 8.15 is an example of a dual-ring-based topology.

Figure 8.16 is an example of a tree-based topology with multiple stations.

Figure 8.17 is a hybrid topology reflecting both a ring and tree topology blended together to make one integrated topological network.

The three examples of possible topologies utilize a variety of FDDI station types. Some stations can perform limited functions because of their architectural design. A dual-ring topology does not necessarily have to implement a concentrator; Fig. 8.12 depicted this scenario.

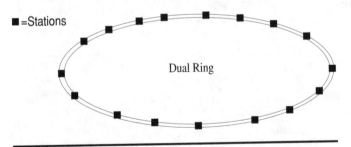

Figure 8.15 Dual-ring topology.

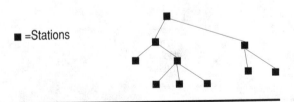

Figure 8.16 Tree topology.

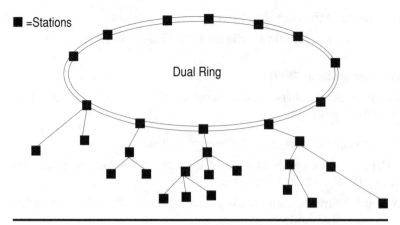

Figure 8.17   Example of ring and tree topology.

However, in certain occasions it is beneficial to implement concentrators and various topologies.

8.5   FDDI Services

Each of the following sublayers performs definable services:

- Physical (layer)-media-dependent (PMD)
- Physical-layer protocol (PHY)
- Medium access control (MAC)

PMD layer

This layer performs two sets of services. One set performs services for PMD PHY sublayers. These services include

- PMD converts bit sequences into optical signals.
- PMD sends a signal to the PHY indicating detected signal status.
- PMD converts optical signals into bit sequences.
- PMD indicates to the PHY that non-return-to-zero-inverted (NRZI) data is transferred.

The other set of services are those between PMD and SMT services.

- This service permits the SMT to manage the PMD operation.
- Requested SMT services have priority over other PMD services.

- SMT generates the signal to leave or join a network.
- PMD generates a signal to indicate to SMT detected signal status.

Physical-layer protocol (PHY)

This layer performs three sets of services. One of these is the PHY-MAC service.

- The PHY request data transfer from the MAC.
- The PHY sends a symbol to the MAC after it has received the bits from the PMD and decoded them.
- PHY sends transmission status information to the MAC and helps synchronize MAC output.
- The PHY informs the MAC when an invalid symbol stream is detected by the PHY.

Another category of services is PHY-PMD services:

- This service permits the transmission of an NRZI bit stream between peer PHY entities.
- The service can also request a transfer of encoded NRZI data be sent to the PMD from the PHY.
- Indication of the transfer of encoded NRZI data from the PMD to the PHY.
- Indicates an optical signal received.

The last category of service is between PHY and SMT and includes

- The service is generated by the PHY but allows SMT to control operation of the PHY.
- Performs a request to send a stream of protocols.
- Indicates line status activity or any change thereof.
- The SMT generates various requests; thus the PHY performs different actions based on these requests.

Medium access control (MAC)

The MAC performs service to three entities. First is the MAC LLC, where the following services are performed:

- A request for a local LLC to transfer service data units to another LLC peer entity.

- The MAC generates a signal to the LLC to inform the LLC that a LLC frame has arrived.
- The MAC indicates results of the previous transfer request to the LLC.
- An LLC entity generates a signal to capture the next available token.

The PHY MAC also provides services; these include

- The transfer of data from the MAC to the local PHY.
- The transfer of data from the PHY to the local MAC.
- Generation of a signal by the PHY to indicate to the MAC that data was received.
- Generation of a signal by the PHY that it cannot present a valid signal stream to the MAC.

The MAC-SMT service performs the following functions:

- Initialization of MAC parameters.
- Indication of errors and/or significant changes.
- Control operations for the MAC.
- Indication of data transfer from the MAC to the local SMT.
- The SMT generates a signal that results in a request to capture a token.

These lists summarize the services of and interactions between the sublayers. Additional details can be found in ANSI and ISO FDDI standards.

8.6 FDDI Management

Station management is defined for the MAC sublayer, PHY sublayer, and PLD sublayer. The function of this management is based on maintaining information gathered in a management information base (MIB). The flexibility of this management structure supports local or remote stations. The remainder of this section presents the core components of SMT system management service.

Station (SMT) object management

The SMT reflects either an FDDI station or FDDI concentrator. Identification includes the operation that the object performs on the FDDI ring. Station configuration information is ascertainable; this will indicate the number of MACs in the concentrator or station, available paths, configuration capabilities for dual rings, and other information. Status information, such as port configuration and the port state, is also ascertainable. Also ascertainable are the current state of the primary and secondary station; time stamps reflecting the latest event; station actions such as disconnect, connect, self-test, and path test; and vendor-specific information. This and other detailed information is manageable via SMT object management.

MAC object management

MAC objects may have multiple instances within a given station. However, the information attainable includes address information of local MAC address attributes; information on operational timers; and configuration information including path, neighbor data, and information reflecting the topology. Frame status information can also be obtained. Counter information for error and nonerror information includes

- Tokens
- Frames not copied
- Error frames
- Transmitted frames
- Received frames
- Expired frames
- Late token

Vendor-specific information can also be obtained. MAC station actions can be determined. Events generated for the MAC can also be gathered. Other detailed information is also available.

Attachment object management

This includes information of attachment resources and ports. The class of attachment can be determined (such as single, dual, or concentrator). Optical bypass information, and vendor-specific information such as actions and notifications defined by a vendor can be obtained.

Port object management

This management information is focused on port attributes such as configuration, status, operational condition, error counts related to the port, or actions and events. The type of PMD associated with the port can be determined along with connection capabilities. A port's operational state is ascertainable. The line state can be determined: that is, halted, idle, quiet, etc. Other detailed information can also be obtained for the port object as well.

Path object management

The path object management can derive information about the following:

- Path configuration of available MACs
- Synchronous bandwidth allocation
- Trace status of the ring
- Ring latency
- Lowest valid point for the transmission timer

Other information can also be obtained, such as the vendor-specific information about the MAC managed object.

The MIB level supported at a given installation dictates to a degree the amount and type of data that can be derived about the objects mentioned above.

8.7   SMT Frame Structure

FDDI stations are managed via SMT frames. This section presents the SMT frame, its header format, and the information field. It also includes a description of each field in the frame.

SMT frame

Figure 8.18 depicts a SMT frame.

The field components and functions of the frame are listed below.

`Preamble`	The *preamble* is the first field in the FDDI frame. This field contains 16 or more symbols to allow a receiving station clock to orient its timing and be prepared for the fields that follow.
`Start delimiter`	This field always contains nondata symbols. The symbols are **J** and **K** and are used to indicate the start of the frame.
`Frame control`	This field identifies the FDDI frame type.
`Destination address`	The DA field indicates the station or stations to which the frame is intended. This address may be either a single or broadcast address intended for multiple hosts.

PA	SD	FC	DA	SA	SMT Header	SMT Info	F C S	ED	FS

Figure 8.18   SMT frame.

`Source address`	The source address field contains the address of the sending station.
`SMT header information`	This field contains a SMT header and information.
`SMT information`	This field contains the SMT information field data.
`Frame-check sequence`	This field performs a cyclic redundancy check.
`Ending delimiter`	This field indicates the end of the frame. The symbol in this field is a `T`, which is 4 bits long.
`Frame status`	This field contains three indicators that reflect three pieces of information about the frame transmission: an error detected, the address recognized, and the frame-copied indicator: `E`, `A`, and `C`, respectively. Each of these indicators is represented by either an `R` or an `S`. The `R` indicates an off state or false condition. An `S` indicates an on state or true condition. This field uses a `T` (terminate) to indicate the end of the field.

SMT header

The SMT header contains information inside each field. Figure 8.19 depicts the fields in this header.

The fields and their functions are as follows.

`Frame class`	This identifies the function of the frame. Some SMT frame classes include ■ Echo frame ■ Resource allocation frame ■ Request-denied frame ■ Status report frame ■ Parameter management frame ■ Station information frame ■ Extended service frame ■ Neighbor information frame
`Frame type`	Three types exist: announcement, response, and request.
`Version ID`	Identifies the structure of the SMT information field.
`Transaction ID`	This field is used to match a response with a previous request.
`Station ID`	A universal station identifier.
`Info length field`	This is the length of the SMT information field.

SMT information field

Figure 8.18 depicted a SMT information field. Figure 8.20 shows the structure of its components.

Frame Header	Frame Type	Version ID	Transaction ID	Station ID	Pad	Info Field Length

Figure 8.19 SMT header.

Parameter Type	Parameter Length	Resource Index	Parameter Value	Parameter Type	- - - -

Figure 8.20 SMT information field.

The contents of the SMT information fields and their meanings are as follows.

Parameter type   The possible components in this field are described by different SMT system management services. Some of the components, such as the station descriptor, the path descriptor, the upstream neighbor address, and the station state, can be mapped directly to the MIB.

Parameter length   This field indicates the length of the parameter.

Resource index   This field identifies the object instance for MAC, port, attachment, and path object classes.

Parameter value   This is the actual parameter value.

This information is repeated in the SMT information field as needed and shown in Fig. 8.20.

Other information is available for management of FDDI. The ANSI X3T9 technical committee document SMT-LBC-177 can provide additional details about this topic.

8.8 FDDI-II

FDDI-II is the next-generation FDDI. Many refer to it as *hybrid FDDI*. The rationale for this is rooted in the architectural change at the data-link layer. Figure 8.21 depicts how FDDI-II appears.

Figure 8.21 shows the architectural differences with FDDI-II. The physical layer remains the same as in FDDI, but a multiplexer is used between the components in the data-link layer and the physical layer.

FDDI-II is capable of two modes of operation: basic and hybrid. Basic operation supports FDDI protocols and operations just as native FDDI. However, the hybrid function of FDDI-II supports isochronous operation, that is, events that have fixed intertransmission time. This is possible because of the hybrid multiplexer function.

The hybrid mode of operation can accommodate voice, video, and other transmissions which are time-sensitive. The basic mode of operation is retained just as in the original FDDI model of operation.

FDDI-II is required for all stations on a network if the network is to operate in FDDI-II hybrid mode. A check is made for this by an FDDI-II monitor station, and if all stations operate in hybrid mode, then the monitor station can start ring initialization.

Layers

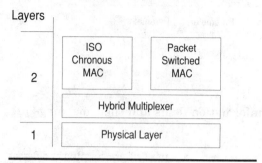

Figure 8.21   Conceptual view of FDDI-II.

Because of the breadth of this topic, further information can be obtained from the ANSI X3.186 specification.

8.9 Summary

FDDI is a ring-based technology with data rate speeds of 100 Mbits/s. The original specification for this technology was a fiber medium; however, testbed environments have implemented FDDI over unshielded twisted-pair cabling and simple copper-stranded cabling. Ring distances are noted to be a maximum of 200 km in length with a maximum of 500 stations per ring segment no more than 2 km apart.

FDDI layers call for a PMD specification along with a PHY protocol. Additionally, the data-link layer is also divided into its typical MAC sublayer and the LLC sublayer. In addition to these, the station management layer is positioned vertically from the physical layer through the data-link layer.

FDDI technology implements timers in each station in order to maintain normal operations. A list of timers, each performing a specialized function, is provided.

FDDI frames are similar to token-ring frames, but differences do exist. Data that can be moved through an FDDI ring is greater than that of a token ring.

The theory of FDDI frame and token operation was provided. A series of steps showing the functions that occur on the ring was presented. One example is that the ring which puts a frame on itself removes that frame after its destination has copied the contents and received the data.

Lists of FDDI frames and their functions were presented. One of these include the beacon frame, which is introduced onto the ring when a serious error occurs such as a break in a cable. The SMT frame is the management frame sent by station management components that control specific operations.

Implementation of FDDI was discussed. Understanding the difference between dual-attachment stations and single-attachment stations, and the role of a concentrator was provided. Explanation of MIC types was provided. MIC types A, B, M, and S were explained.

Sample topologies were explained and their conceptual implementations were shown. FDDI services were also discussed. These services correlate with the PMD sublayer, the PHY protocol, the MC protocol, and the station management component.

FDDI station management and the notion of a station object management were explained. MAC object management, attachment object management, port object management, and path object management were discussed.

SMT frame structure and SMT header contents and structure were presented. The SMT information field was presented and its contents explained.

A brief glance at FDDI-II was presented to orient the reader as to the basic difference between it and FDDI. FDDI-II is capable of operating in either hybrid or basic mode. The former is capable of supporting isochronous traffic.

9

Frame Relay

Frame relay provides real-time communication between end users by serving as an interface into public and/or private networks. Frame relay networks pass frames from origin to destination without intermediate nodes performing packet assembly and disassembly. Frame relay is also considered a protocol. A frame relay control protocol is also defined in this chapter, with explanations as to how users make service requests in the network.

Frame relay is designed to support data in bursts and to provide high speeds. It is *not* a store-and-forward-based technology; rather, it is a bidirectional conversational method of communication. Most frame relay standards are concentrated at layers 1 and 2; however, standards do define the mechanism for upper-layer protocols to "hook" into frame relay.

For example, this means that frame relay operates between users, that is, between origin and destination *networks*. Figure 9.1 best conveys this concept of frame relay operation.

The hypothetical scenario depicted in Fig. 9.1 shows three sites using frame relay as an interface into a dispersed frame relay network. Additionally, it is used to connect multiple geographically different networks. This example shows each location implementing frame relay.

For instance, the Gandy & May Corporation in Florence, Mississippi, is using TCP/IP for an upper-layer protocol. Sauer Enterprises in Atlanta is using NetWare. Information World, Inc., in Dallas, Texas, is a hybrid environment employing a variety of upper-layer protocols. Each physical location uses frame relay for an interface into what is considered a frame relay network.

Frame relay has been defined by at least three noted entities: ANSI, CCITT, and the local management interface (LMI) group. ANSI has a list of specifications. These standards generally have the ISDN name within them. The CCITT has its list of CCITT recommendations. The CCITT has two groups of standards that relate to frame relay: the I and Q groups. Last, the LMI standards were created by Cisco Systems, Digital Equipment Corporation, Northern Telecom, Inc., and StrataCom, Inc. These four vendors collectively created standards that parallel ANSI and the CCITT. Later in this chapter we will

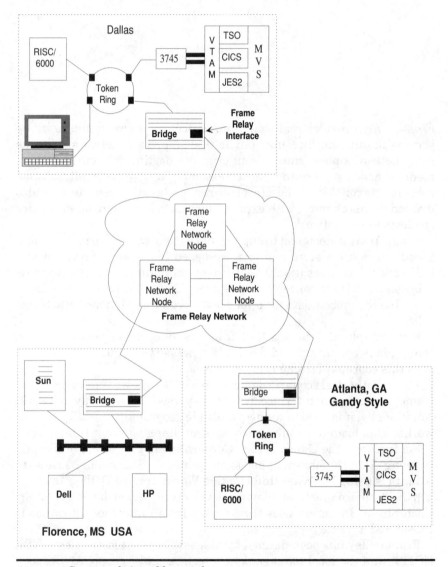

Figure 9.1 Conceptual view of frame relay.

see what information comes from which source and how to obtain additional information.

The remainder of this chapter explains frame relay principles, frame structure, virtual circuits, access devices, consumer tips, and a brief list of reference material.

9.1 Principles of Frame Relay

Frame relay operation is based on a number of principles, including virtual links, permanent virtual connections, and the data-link connection identifier.

The virtual link

A basic frame relay principle is a virtual connection. Anything virtual is not real but has the appearance of reality. A *virtual connection* is a dynamic link and acts like a pipe through which data moves. This connection is permanent in the sense that it remains as long as necessary. Figure 9.2 shows a basic frame relay example.

Figure 9.2 shows three hosts connected to a frame relay node, physical links connecting each host to the node, memory inside the frame relay device, and connection mapping table within the frame relay node. Host number 1's physical link supports two permanent virtual

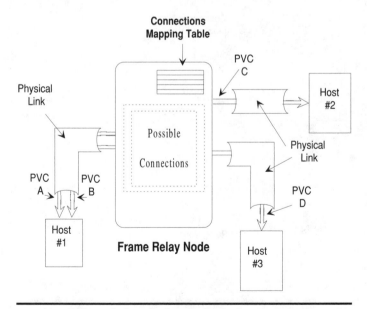

Figure 9.2 Enhanced view of a frame relay node.

connections (PVCs). However, the physical links of hosts 2 and 3 support one PVC each.

The connection mapping table is at the heart of frame relay operations within the frame relay node shown in Fig. 9.2. The connections made are dynamic and based on requests from the incoming data stream. The connection mapping table is responsible for matching the request of source to destination; this constitutes a *route,* or in frame relay lingo, a *virtual connection.*

Figure 9.3 is similar to Fig. 9.2 but shows the frames and a highlighted view of the mapping table.

Figure 9.3 shows PVC A 1 and PVC B 5 originating at host 1 and connecting to PVC C 3 and PVC D 7, respectively. To be more precise, mapping between A 1 and C 3 and B 5 and D 7 is performed inside the frame relay node.

Data-link connection identifier

The inbound frame from host 1 via PVC A 1 contains a *data-link connection identifier* (DCLI), which is the local address in frame relay. The DCLI address is relevant in reference to the particular link. It identifies the frame and its type. This means that the same DCLI value could be used at both ends of the frame relay network where each host connects via a particular link. Differentiating the DCLI is the physical link

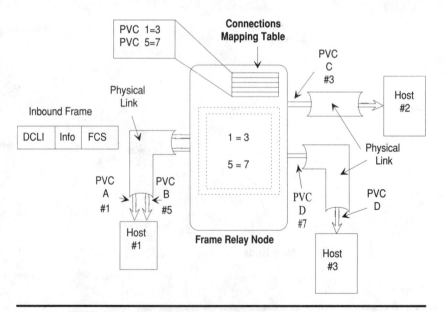

Figure 9.3 Highlighted view of the mapping table.

that the frame traverses from the host. Hence, more than one identifying factor is used for frame identification.

In the connection mapping table each entry includes the following information:

- Node id
- Link id
- DCLI

Frame relay costs

Aside from the hardware and software required to implement a frame relay connection, additional fees are incurred, including the *commitment information rate* (CIR), which is the bandwidth available from one end to another. Another factor is called the *port access rate,* which is access into the frame relay network; and the *network access charge,* which reflects the costs of the line connecting a given site to the access point in the *frame relay network.*

Understanding these aspects of frame relay is important. For example, if data is transmitted in bursts, one must know what the burst data rate is. Another factor to understand that relates to this is the normal or average throughput required on a daily basis.

9.2 Frame Relay Frame Components

The frame components of frame relay are shown in Fig. 9.4. Maximum frame size is 8250 bytes, and the minimum is generally considered 262 bytes.

Figure 9.4 is based on the CCITT I.441 recommendation. Variations of this structure exist, primarily those with different methods of implementing addressing. A significant point to note about frame relay is that it utilizes the same standards as those of ISDN.

The flag field indicates the beginning of the frame.

The address field typically consists of the components shown in Fig. 9.5.

In Fig. 9.5 the DLCI fields identify a logical channel connection in a physical channel or port, thus identifying a predetermined destination. (Simply put, it identifies the connection.) The CR field contains a bit

Flag	Address Field	Control	Information	FCS	Flag

Figure 9.4 CCITT I.441 frame relay frame format.

DLCI	CR	EAB	DLCI	FN	BN	DE	EAB

Figure 9.5 Highlighted view of the address field.

indicating a command response. The EAB field contains a bit set at either 1 or 0. The field indicates extended addressing. The FN field is sometimes referred to as the *forward explicit congestion notification* (FECN). The bit in this field indicates whether congestion was encountered during the transfer from origin to destination. The BN (backward notification) field indicates that congestion was encountered on the return path. The DE field is the discard-eligibility field. This field indicates whether the frame can be disposed of during transfer if congestion is encountered. A bit setting of 1 indicates DE, whereas a bit setting of 0 indicates a higher setting for the frame and should not be discarded.

9.3 Virtual Circuits

Different frame types of virtual circuits have been defined for use with frame relay. In a sense these circuits represent different definable services, such as

- Switched virtual circuit (SVC)
- Permanent virtual circuit (PVC)
- Multicast virtual circuit (MVC)

Switched virtual circuit

The *switched virtual circuit* (SVC) is similar to telephone usage. When the circuit is needed, a request is made. When the circuit is not needed, the circuit is not used. Information is passed from origin to destination to set up the call and to bring it down. Some information provided in the call setup phase includes bandwidth allocation parameters, quality-of-service parameters, and virtual channel identifiers, to name only a few.

Permanent virtual circuit

The *permanent virtual circuit* (PVC) connection is considered a point-to-point configuration. It could be thought of as a leased line in that it is dedicated. This type of circuit is used for long periods of time. Commands are still used to set up the call and to bring it down. The difference between the PVC and a SVC is duration.

Multicast virtual circuit

The *multicast virtual circuit* (MVC) is best described as being a connection between groups of users, through which individual users can use SVC connections as well as PVC connections. Technically, this type of connection is considered permanent and, to date, is generally considered a local management interface (LMI) extension.

9.4   Access Devices

Different devices can be used to connect devices into a frame relay environment. Some of those are examined here.

Switches

Frame relay networks can be accessed via different types of devices. For example, switches similar to those accommodating X.25 provide a way to access frame relay networks. However, these switches are typically implemented in the sense of creating a backbone. Figure 9.6 is an example of this type of device implemented in three environments.

Figure 9.6 shows a network backbone made up of three components: switches in Dallas, Denver, and Bakersfield (Calif.).

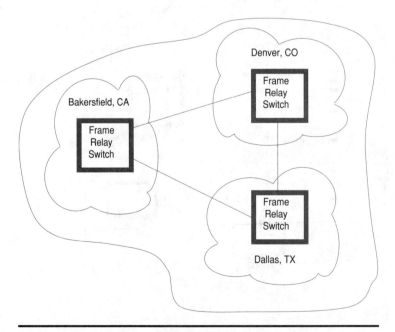

Figure 9.6   Conceptual view of a frame relay network.

Network devices

A more focused view of Bakersfield could be represented by Fig. 9.7.

Figure 9.7 shows a network device (specifically a bridge) connecting a token ring and ETHERNET network into the frame relay environment. It also depicts the lines to Dallas and Denver.

Frame relay access devices

A *frame relay access device* (FRAD) is a particular piece of equipment that typically connotes capabilities including packet assembly/disassembly and speeds of DS0, T1, or fractional T1. Most FRADs can handle multiple protocols and focus network traffic into a centralized managed facility such as that shown in Fig. 9.8.

FRADs can be the best component for concentrating multiple devices into a single unit. Vendors such as Wellfleet and Cisco Systems provide devices such as this.

Other devices may be used to connect a variety of resources into a frame relay network. Many vendors offer frame relay support as an additional function. One such example is IBM's *network control program* (NCP), which operates on a front-end processor (FEP).

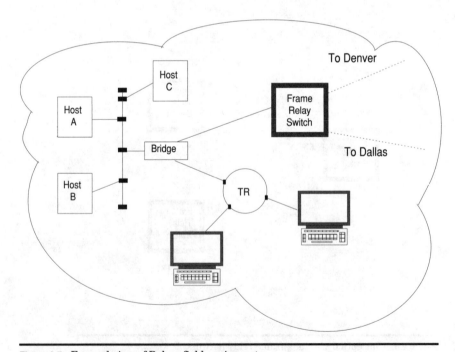

Figure 9.7 Focused view of Bakersfield equipment.

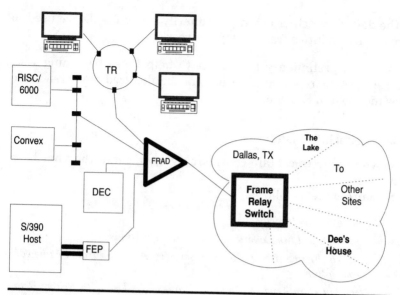

Figure 9.8 Highlighted view of a FRAD.

9.5 Consumer Tips

During my work and research with frame relay I encountered some information that may serve as handy tips for consumers purchasing frame relay equipment. Some of those tips are included in this section.

1. Frame relay standards are still being defined and implemented by vendors differently.

2. When considering a frame relay device, specifically ascertain what it does.

3. Determine whether a frame relay device supports DCLI support, header bits support, FCS, FECN, BECN, and DE bits within the frame. Also determine whether congestion control is performed to standards and if so, which one(s). Determine whether multiple protocols are supported.

4. Find out in what way a given device supports FECN, BECN, and DE bits. How do they function in respect to congestion management? For example, one may say DE is supported, but what does this mean? Does it mean the bit is read, or it can be set? The latter point is important.

5. Is link management supported, and if so, what type?

6. Does the device support the transparent mode?

7. Is the device switch-oriented or primarily an access device used to provide access into a frame relay network?

This is only a preliminary list meant to help those beginning with frame relay. A more exhaustive list can be deduced from reference sources included in Sec. 9.6.

9.6 Additional Information

The following sources contain considerably more detail than has been intended here.
   ANSI sources are as follows:

T1.601	*Basic Access Interface*
T1.602	*ISDN Data Link Layer Signaling Specifications*
T1S1	*/90-75 Frame Relay Bearer Service, Architectural Framework Description*
T1S1	*/90-214 Core Aspects of Frame Protocol for Use with Frame Relay Bearer Service*
T1S1	*/90-213 Signaling Specification for Frame Relay Bearer Service*
T1.606	*ISDN Architectural Framework*
T1.607	*Digital Subscriber Signaling Service*
T1.617	*Standards Concerning Customer Interface*
T1.618	*Standards Concerning Customer Interface*
T1S1	*/90-051R2 Carrier to Customer Interface*

   CCITT standards are

I.122	Q.920	1.320	X.25
I.233	Q.921	1.320	X.31
I.130	Q.922	1.430	X.134
I.441	Q.930	1.431	X.213
I.450	Q.931	1.462	X.300
I.451			

   The following corporations made considerable contributions to the development of the frame relay standards mentioned previously. They should be the point of contact for further information on LMI standards.

   Cisco Systems, Inc.

   Digital Equipment Corporation (DEC)

   Stratacom

   Northern Telecom, Inc.

9.7 Summary

Frame relay principles were explained, including the concept of virtual links, the data-link connection identifier, and the associated costs of frame relay beyond supporting hardware and software. Frame relay frame components were presented and the fields that constitute a frame were explained briefly.

The concept of virtual circuits were explained. The switched virtual circuit (SVC) is similar to a telephone, a permanent virtual circuit (PVC) is similar to a leased-line arrangement, and in a multicast virtual circuit (MVC) a group of users can be reached through one multicast.

Access devices used in frame relay networks were discussed. The basic functions of a switch, network-related device, and frame relay access device were explained. Different implementations may use one or more of these devices. Several access devices were discussed.

A brief list of consumer tips was included for the consumer who is new to frame relay. A list of reference sources was included in Sec. 9.6 for those who desire more in-depth knowledge of this topic.

10

Integrated Services
Digital Network (ISDN)

This chapter answers typical questions that arise when ISDN is mentioned. ISDN is a very comprehensive topic that includes vast amounts
of standards, protocols, and information contributed by a number of
standards-making bodies. The purpose of this chapter is to orient the
reader to ISDN. It begins by clarifying what ISDN is, presents some of
the fundamental standards on which it is based, explains some basic
ISDN terms and concepts, and briefly explains SS7. Interfaces used
with ISDN and their function and some examples of practical services
as a result of ISDN implementation are presented.

10.1 ISDN Theory of Operation

You may have asked yourself—or possibly others if your background is
not related to this technology—"What is ISDN?" The technology is not
new in that it was invented only a few years ago, but rather the implementation and use of ISDN generally keep it one step removed from
most except those who work directly with it.

Working definition

ISDN is the acronym for Integrated Services Digital Network.
Practically speaking, it has more to do with functionality from the perspective of regional telephone companies and service-provider implementations than with an isolated implementation. In fact, ISDN *users*
realize its benefits and are typically not involved in the "implementation" of it. This point is clarified below.

   For example, many telephone companies (in the United States and
other countries) have implemented the ISDN technology in central

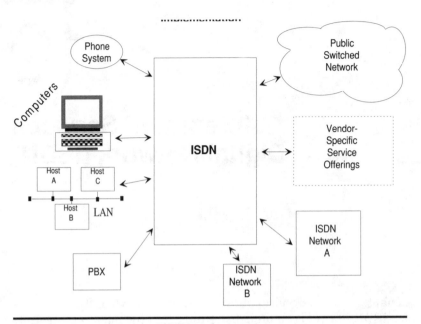

Figure 10.1   Conceptual view of ISDN implementation.

offices, private digital systems, and private branch exchanges (PBXs). This is significant because the technology is based on digital, not analog, signals. Not only are digital signals fundamental to ISDN foundations; ISDN can support voice, data, video, electronic mail (E-mail), and numerous other services which, integrated in unison, make ISDN a powerful technology on which to build. The result is a potential for services offered through the telephone network that were heretofore not possible. Figure 10.1 is a conceptual view of ISDN implementations.

ISDN realization requires more than the telephone system implementation. In fact, ISDN requires the calling party, the network, and the destination party to implement ISDN technology. If these three conditions are not met, some ISDN capabilities are not possible. Further exploration into what this means from a practical standpoint is presented later.

CCITT recommendations

ISDN is a protocol. Its operation, standards, and recommendations have been defined by CCITT. The I series of CCITT recommendations cover a broad range of ISDN technology. The I series is divided into groups of 100. The following list categorically describes the CCITT recommendations.

CCITT number	Function
I.100	Contains general information such as terminology used, structure of I series recommendations, and basic capabilities of ISDN
I.200	Specifies ISDN service capabilities such as circuit, packet, and a variety of digital services
I.300	Includes the principles, protocols, and architecture of ISDN
I.400	Includes specifications of user interfaces and network-layer functions
I.500	Includes interface standards, principles for internetworking ISDN networks, and related topics
I.600	Includes ISDN maintenance recommendations including subscriber access, basic access, and primary rate access

These recommendations, along with many others, detail ISDN according to CCITT. The source of this series of recommendations is from my reference library, which includes all CCITT recommendations through March 1994. Any of these recommendations can be obtained from sources that provide standards, and recommendations from entities such as CCITT.

10.2   ISDN Channels

ISDN has basic concepts that are applicable in the majority of implementations. The issue here is what is actually used in a given ISDN implementation. This section explains the terms and concepts on which ISDN is built.

Channels in general

The term *channel* is used frequently in ISDN to convey the meaning of service provided. Channels are an integral part of ISDN technology, and as this section explains, there are different types of channels.

Channels in general are physical or logical entities through which data, voice, video, or other *information* travels. This is important because different types of channels are defined in ISDN. A definable characteristic about channels is that they can be identified as being either digital or analog. Either way, they carry signals from one entity to another.

The remainder of this section explains the different types of channels available with ISDN.

ISDN channels

The D channel.   In ISDN the D channel is used to convey user signaling messages. This type of channel uses *out-of-band signaling,* meaning that network-related signals are carried on a separate, non-user-data, channel. These signals transmitted over the D channel convey the characteristics of the service on behalf of the user. The term *out of band* originated because the network signal is out of band with the user signal.

The protocol used on the D channel defines logical connection between what is called the *terminal equipment* (TE) and the *local exchange* (LE) via *local loop-termination equipment*. In order to use this arrangement, *customer premise equipment* (CPE), which performs switching functions, is required. Figure 10.2 is a conceptual view of this idea.

The essence of understanding the D channel is knowing that it uses out-of-band signaling but carries user data. It operates at approximately 16 or 64 kbits/s and is used by user equipment to transmit requests and messages within the network. In summary, the D channel provides signaling service between a user and the network and provides packet-mode data transfer.

The B channel. The B channel carries voice, video, and data. This channel, which functions at a constant 64 kbits/s, can be used in packet- and circuit-switching applications. The difference between these two applications is that packet switching utilizes a logical connection through a network and no dedicated facilities exist. This is generally referred to as a *store-and-forward* method of data transfer. Circuit switching differs because its switching technology is based on devices that are connected via some resource for the extent of the call (or communication instance).

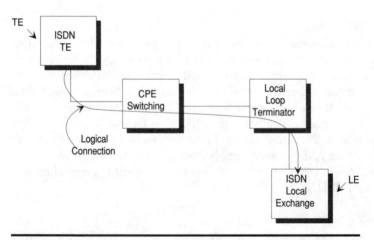

Figure 10.2   Conceptual view of D-channel logical connection between the TE and LE.

B- and D-channel joint operation.   These two channels work in harmony. The D channel is used to transfer requests for services that are delivered on a B channel. Figure 10.3 is an example of B- and D-channel joint operation.

The H channel.   Multiple H channels exist. The basic differences between them are related to the services they offer. As a general rule, H channels have a considerably higher transfer rate than do B channels. These channels effectively meet the needs of real-time videoconferencing, digital-quality audio, and other services requiring a much higher bandwidth.

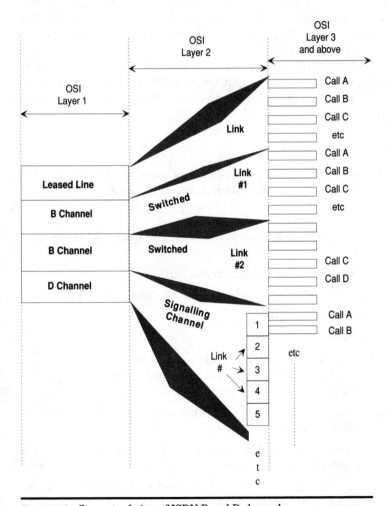

Figure 10.3 Conceptual view of ISDN B and D channels.

The basic H channel, known as H_0 is composed of one channel that can provide rates of 384 kbits/s. H_{11} channels can support throughput rates of approximately 1536 Mbits/s, and the H_{12} channel sustains rates of up to approximately 1920 Mbits/s. This type of channel is most suitable for a trunk where subdivision can be implemented to maximize the effective bandwidth.

10.3 Signaling System 7

Signaling System 7 (SS7) is a standard being implemented by regional and long-distance telephone companies today. Its relationship to ISDN is important. Without SS7 implementation in the telephone systems, ISDN is somewhat inhibited. They are interrelated in many ways. The fundamental reason for this is that SS7 is digital by design and its link capacity outstrips that of SS6.

SS7 is a protocol and method for networks of switching entities to communicate with one another. Although ISDN can be implemented without SS7, with the implementation of SS7 ISDN can be more comprehensive in scope.

Characteristics

SS7 is a complex standard. The CCITT has a series of recommended standards about the signaling system. Some highlights of SS7 characteristics include

- It can accommodate digital communications in networks using digital channels.
- It can also operate with analog communication channels.
- It can be used domestically in the United States or internationally.
- It is a layered architecture.
- Speeds are 56 and 64 kbits/s.
- In regard to information transfer, it is considered reliable because it ensures sequential movement of signals through a network and provides a mechanism to prevent loss or duplicate signals.
- It can operate in point-to-point network implementations and can be used with satellite communications.
- Its method of handling routing and delivery of control messages is also considered reliable.
- It has built-in management capabilities and maintenance and also has a method for call control.

SS7 CCITT recommendations define packet-switched network functions but do not restrict implementation to specific hardware. According to CCITT recommendations, two signaling points are defined: a signaling point and a signaling transfer point. The *signaling point* is a point in the network capable of handling control information. A *signaling transfer point* is an entity in which routing of messages can be achieved.

Protocol components

SS7 protocol can be categorized into three groups:

- Signaling connection control part (SCCP)

- Application

- Message transfer part (MTP)

The SCCP protocol specifies five classes of network service:

0. Unsequenced connectionless

1. Sequenced connectionless

2. Connection-oriented

3. Flow control connection-oriented

4. Flow control connection-oriented with error recovery

The SCCP protocol supports OSI addressing capabilities and in a sense functions to deliver messages intended for a specific user once the message reaches the destination signaling point. The SCCP protocol works with the MTP protocol to achieve OSI network-layer support.

The MTP protocol has multiple layers of signaling. The first level corresponds to the physical layer of the OSI model. It functions as the signaling data link. The second level of the protocol operates at the data-link layer of the OSI model. It is by design a bit-oriented protocol; consequently, it is robust in its capabilities. The third level corresponds roughly to the network layer of the OSI model. It is concerned with routing and link management.

The SS7 application protocol correlates to upper layers in the OSI model. The structure at these layers is divided into a telephone user part which focuses on signaling processes required for voice communications. The data user part is related to requirements for circuit-oriented networks and has been usurped by the ISDN user part.

Additional information

The CCITT and ANSI standards for SS7 are comprehensive in scope. The following list is not exhaustive but provides an excellent starting point for those who need additional information. A more complete list can be obtained by standards source suppliers who maintain lists of standards-making bodies such as the ANSI and the CCITT.

ANSI T1.110	*SS7 Overview*
ANSI T1.111	*Overview of the Message Transfer Part*
ANSI T1.113	*ISDN User Part*
CCITT Q.700	*SS7 Overview*
CCITT Q.701	*Message Transfer Part*
CCITT Q.702	*Signaling Data Link*
CCITT Q.711–Q.716	*Signaling Connection Control Part*
CCITT Q.730	*ISDN Supplementary Services*
CCITT Q.761–Q.766	*ISDN User Part*

10.4   ISDN Interfaces and How They Are Used

Two ISDN user interfaces are explained in this section: the basic rate interface and the primary rate interface.

Basic rate interface

This is a means of accessing ISDN. It consists of two B channels and one D channel according to the CCITT recommendation number I.430. Hence, it is sometimes referred to as the *2B+D interface.*

No specific protocol specification restrictions are applicable here. This interface utilizes circuit-switched transparent "pipes" to make a connection between two end users by way of the ISDN network. Examples of where this interface is implemented would be PBXs, individual terminals, videoconference units, personal computers, and workstations.

Primary rate interface

This is also a means of accessing ISDN. This interface calls for one of the following implementations:

- 23 B channels and 1 D channel
- 24 B channels and 1 D channel
- 30 B channels and 1 D channel

These originate from the CCITT I.431 recommendation. In some ways this is basically the bandwidth equivalent to T1s used in the United States. This interface calls for point-to-point, serial, or synchronous communications. A fundamental difference between the primary rate interface and the basic rate interface is that the former can support H channels as well as B channels. As a result, bandwidth is greatly enhanced.

How interfaces are used

These two interfaces are used by a business, personal user, or other party to connect them directly to the local telephone company's central office. In large scenarios the basic rate interface is used to connect individual users to the organization's PBX, and in turn the PBX is connected to the local telephone central office. However, this could be achieved by connecting this latter scenario to an interchange carrier via a broadband interface. The connection is made via an ISDN interface board, ISDN controller, or an external terminal adapter, and the connection is physically established.

10.5 Practical Uses of ISDN

Like many technologies used in networks that may not reside inside a user's workplace environment, ISDN is somewhat abstract. ISDN uses can be understood more easily than the internal operations for those who are not technically adept. This section hones in on practical services that are based directly or indirectly on ISDN technology.

Automatic number identification

Automatic number identification (ANI) is a service that provides the individual being called the telephone number of the caller prior to the party answering the call. This service is beneficial for a company who prides itself on customer service.

Simply the telephone number and a well-designed and well-implemented database can be powerful tools for a company to better serve its customers. The examples below are based on the ANI service and an updated database. Consider the implications.

- Identification of the customer: name, address, phone number(s), and other information pertinent to serving the customer.

- Identification of the customer's language preference. Our diverse culture today is a blend of individuals who speak different languages. This is important information for companies who have international customers and those who do not speak English. With a good database and user network interface (UNI), a caller speaking a language other than English can be routed immediately to a service representative who can speak the caller's native language. This in itself communicates to the customer that the company cares enough about its customers to have individuals who can communicate in a variety of languages.

- Ascertaining the telephone numbers of callers who terminate a call before a representative can respond because they have been on hold for an inordinate amount of time. With ANI, a representative can return the customer's call.

- Identifying callers via their phone numbers and database(s) in a particular category. For example, assume that a customer purchases large quantities of *widgets* and has had a longstanding relationship with a particular sales representative because the representative understands the needs of the caller. In this case a caller and the caller's files can be directed to the appropriate representative.

- A customer's preferred method of payment can be determined on the basis of prior payment history, and therefore repeating the same numbers and information again can be avoided except to verify appropriate information.

These ANI-based services are available today for large system computing to personal computer systems. Regardless of the implementation, the results of such a service generally make a positive impression on a customer.

Electronic library interconnections and manual access

With the help of ISDN and supporting technology today it is possible for libraries in geographically remote regions of the United States to exchange more information electronically than would have been practical a decade ago.

Many companies such as Phillips, Apple Computer, and Microsoft have or are beginning to put manuals in electronic media and distributing them to databases through ISDN networks. These manuals span topics such as hardware maintenance and software changes.

Image retrieval

A major advantage of ISDN and supporting systems is the ability to transmit images, full-motion video, text, graphics, data, and prerecorded video information. This is particularly significant in the realm of medicine. Now CAT (computer-assisted tomography) scans can be moved in entirety from participating ISDN customers in the medical community.

Other practical everyday uses of ISDN are not generally presented in technical form and are explained to or are easily understood by the recipients who utilize the capabilities it offers. Suffice it to say that

ISDN is growing in vendor products and telephone system implementations, and is even coming close to the houses we live in.

Companies such as AT&T, Sprint, and MCI are implementing ISDN and variations thereof. The end result will be services via networks to which a widespread customer base will have access.

10.6 Summary

Actual ISDN technology is not accessible to most users. The services of ISDN are the result of the implementation. ISDN services provide capabilities for moving voice, data, live video, text, imaging, and variations of multimedia technology from one location to another in real time.

The CCITT and ANSI have numerous recommendations and proposed standards related to ISDN and supporting peripheral equipment. These standards are lengthy; my references on the topic consume many linear feet of bookshelf space. A list of CCITT and ANSI specifications was provided for the reader who needs additional information on the topic.

ISDN channels were explained. The B, D, and H channels were explained in light of their features and differences. SS7 characteristics and protocol components were presented, and references for additional information were provided.

ISDN basic rate and primary rate interfaces were explained. A brief explanation of how interfaces are used was also presented.

A section on the practical uses of ISDN was presented to help the reader understand some of the services that are based either directly or indirectly on ISDN technology.

11

Parallel Channels

Parallel channels is the term IBM used to rename what was called *channels* prior to the introduction of Enterprise Systems Connection (ESCON). Since IBM introduced ESCON two types of channels are identified: serial (ESCON) and parallel (those prior to the introduction of ESCON) channels. Parallel channels are still supported by IBM at the time of this writing, and over the past few years the number of IBM products that support parallel channels has increased. In fact, some companies other than IBM have supported these channels in time past. But the focus of this chapter is on parallel channels.

Parallel channel characteristics, implementations, and basic operations are explained. (*Note:* In this chapter parallel channels will be referred to simply as *channels*.

11.1 Orientation to Parallel Channels

This section explains terms and concepts used in discussion of parallel channels. Understanding them is necessary before exploring more in-depth topics about parallel channels.

Channel subsystem concepts

A *channel subsystem* is identified in certain IBM processor complexes. Parallel channel implementations are directly tied to the hardware architecture used. As processor architecture is discussed in detail in Chap. 16, it will not be covered exhaustively here.

A channel subsystem is used to control the direction of data in the processor complex between storage and I/O devices. The purpose of this design is to remove as many I/O operations from the processor as possible. The channel subsystem consists of components such as channel paths, subchannels, and control system facilities. Figure 11.1 is a conceptual view of a channel subsystem in relation to other components in a typical environment.

Figure 11.1 shows a processor complex with the channel subsystem part of the physical housing of the complex. It also shows actual channel paths. Channel paths are generally referred to as *channels*. The lines shown connecting devices from a channel to a device are called bus and tag cables.

Channel paths and subchannels

A *channel path* is simply a channel. The number of channels in a given system is contingent on the architecture. Another way of viewing a channel is that it is a link between a device and the processor complex.

Subchannels are device-related. For example, one channel may support 25 devices. In this example the 25 devices share a common channel, but they each have different subchannels.

The addressing scheme for this scenario is a channel address and a subchannel address. Figure 11.2 illustrates these concepts.

Figure 11.2 is an example of a channel subsystem with eight channels. Channel 1 has three control units attached to it in a daisy-chained

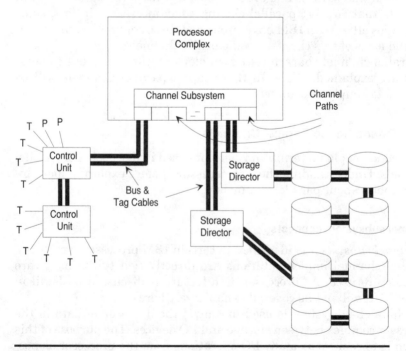

Figure 11.1   Conceptual view of a channel subsystem.

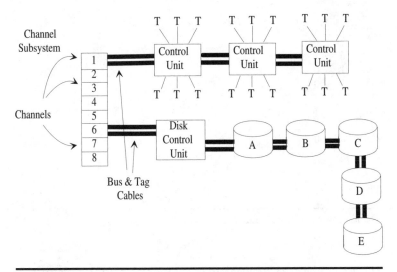

Figure 11.2 Conceptual view of channels and subchannels.

manner. Channel 6 has an attached disk controller to which five disk drives are, in turn, attached.

Parallel data flow

As shown in Fig. 11.2, all these devices are connected via bus and tag cables. These are large copper-stranded cables. with 132-pin connectors. The bus cable is used to transport data. The tag cable is used to control data traffic on the bus. These cables are called *parallel* because they move data in parallel. For example, there are eight inbound signal lines and eight outbound signal lines: hence, 1 byte in and 1 byte out. Figure 11.3 is an example of parallel data transfer.

Figure 11.3 shows a channel moving data in parallel toward a host. Notice that 8 bits are aligned and three different eight-bit rows are being transmitted. In short, 1 byte is moved at a time. This is the concept behind parallel data flow.

Basic view of channel and subchannel addressing

Addressing of channels and subchannels is simple. It is like a street with houses located on it. For instance, consider a street named XYZ and five houses residing on that street. To address any one of those houses, you would always identify the street first and then identify the house. The house addresses could be A, B, C, D, and E. This analogy applies to the channel and subchannel addressing.

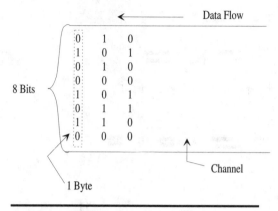

Figure 11.3 Example of parallel data flow.

In Fig. 11.2 the channel address was 6 and the disk drives on the channel were A, B, C, D, and E. The same holds true for other devices attached to a channel such as the control units where terminals and printers are connected as shown in Fig. 11.2.

11.2 Types of Parallel Channels

Two modes of operation exist with parallel channels in the marketplace today; these include byte and block multiplexers. This section explores the functionality of and differences between byte and block multiplexers.

Byte multiplexer mode

In *byte multiplexer mode,* a device on the channel is logically attached for a short period of time because other devices will most likely be sharing the channel also; henceforth sharing the channel for data transmission is a necessity. This is simply a form of timesharing the channel. Typically, slower devices operate in byte multiplex mode. Figure 11.4 provides a conceptual view of this scenario.

Figure 11.4 shows an example of seven channels, an enhanced view of channel 4 (the one to which devices are attached), bytes of data multiplexed, and five devices: X, Y, Z, T, and V. These devices share the same channel for data transfer. This mode of operation is called *multiplexed.*

Block multiplexer mode

Block multiplexer mode is another mode of operation that channels support; actually the mode of operation with this channel characteristic is called *burst* mode. Generally, fast devices use this mode of data transfer.

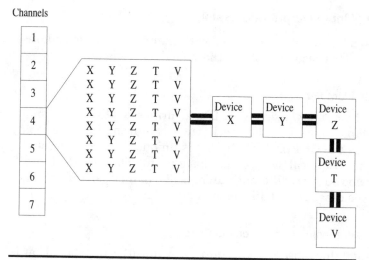

Figure 11.4   Conceptual view of byte multiplexer mode.

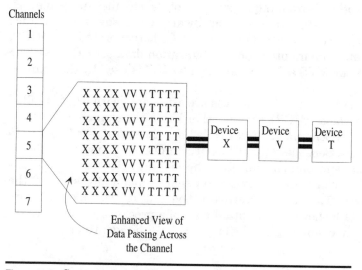

Figure 11.5   Conceptual view of burst mode operation.

For example, some third-party devices that attach to a channel operate in burst mode. Figure 11.5 is a conceptual view of this operation.

Figure 11.5 shows channels, an enhanced view of channel 5 with data passing across it, data moving across the channel in burst rates, and three devices, X, V, and T. Notice in Fig. 11.5 that the data moving across the channel is in the form of *blocks,* as the name of this type channel support implies.

11.3   Formal Input/Output Addressing

Four formal addresses are associated with channel subsystem address-ing schemes. This section explores those four components.

Channel path identifier

The *channel path identifier* (CHPID) identifies a physical channel. Among technical people who work with channels it is commonly referred to as the CHPID. The CHPID is unique to each channel of a given system. A typical system may have up to 256 CHPIDs, but different systems support different numbers of channels, so this may not always be the case. The CHPID is an 8-bit value.

Subchannel addressing and device numbers

To understand this concept it is easiest to begin with a question. How is a channel-attached device perceived under the MVS operating system? The answer is twofold: One is from the perspective of the operating system and the other is from the perspective of the channel subsystem.

After physical installation of the hardware device, say, a control unit, an input/output configuration program (IOCP) is generated. The IOCP, in turn, creates the input/output configuration data set (IOCDS). The IOCP is used by MVS software, and the IOCDS is used by the channel subsystem.

From an MVS perspective, devices are associated with what is called a *unit control block* (UCB). On the other hand, the channel subsystem uses a 16-bit subchannel address called a *unit control word* (UCW) to control devices. Consequently, the device number as known to the IOCP is mapped to a subchannel address in the channel subsystem. The channel subsystem can manage devices via subchannel addresses, and the operating system and user perceive the device as having a "device" number. Figure 11.6 shows a conceptual view of a processor complex, channel subsystem, channels, and devices with a 16-bit identifier.

11.4   I/O Operations at a Glance

This section provides reference information about I/O execution and channel command words (CCWs). As a reference, the IBM *ESA/370 Principles of Operation,* document no. SA22-7200 has been the source for this section. This section does not attempt to replace any IBM documentation. For additional details on topics covered in this section, contact your local IBM representative for information that may be pertinent to your installation.

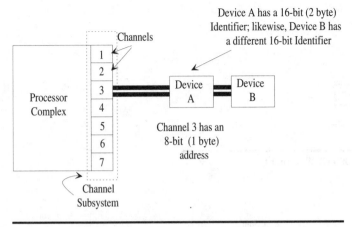

Device A has a 16-bit (2 byte) Identifier; likewise, Device B has a different 16-bit Identifier

Channels

Processor Complex

Channel 3 has an 8-bit (1 byte) address

Channel Subsystem

Figure 11.6   Conceptual view of channel and subchannel addressing.

I/O operations are executed and controlled by the following:

- Start subchannel
- Channel command words (CCWs)
- Orders

Start subchannel

The *start subchannel* instruction execution parameters are passed to the appropriate subchannel to advise it to perform a start function with the I/O device associated with that subchannel. The next function is decoding and execution of CCWs by the subchannel and I/O device. A group of CCWs form a channel program and are executed as such.

ESA architecture channel command words

Channel command words (CCWs) are commands that perform a specific function and are executed to perform specific tasks. CCWs can be linked together, and a channel program can be created and then executed.

According to the IBM manual *ESA/390 Principles of Operation,* document no. SA22-7201, two CCW formats exist: format 0 and format 1. Format 0 CCWs can be located anywhere in the first 16 Mbytes of main storage. Format 1 CCWs can be located anywhere in main storage.

Figure 11.7 is an example of a CCW format 0.

Figure 11.8 is an example of a CCW format 1.

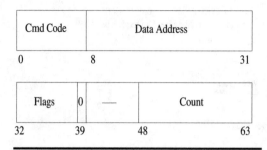

Figure 11.7   ESA/390 CCW format 0.

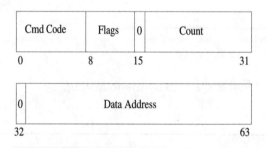

Figure 11.8   ESA/390 CCW format 1.

Seven CCWs are defined; their formats are presented here along with the meanings of the contents in each field. The CCWs covered in this section include

- Control
- Read
- Read backward
- Sense
- Sense ID
- Transfer in channel
- Write

Control CCW.   Control CCWs originate at the I/O device. The subchannel is consequently set up to move data *from* memory to the device originating the control CCW. This command can also originate at the device to initiate an I/O operation such as rewinding a tape or positioning a disk-access mechanism. Generally, it is the transfer com-

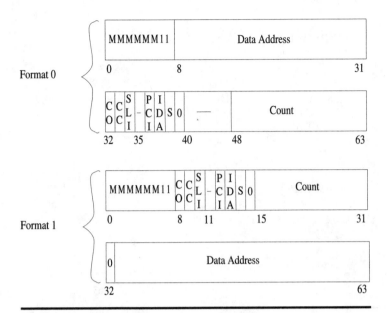

Figure 11.9 Control CCW format.

mand to the I/O device from main storage. Figure 11.9 is an example of format 0 and format 1, respectively, for this CCW.

Read CCW. The read CCW is used by the I/O device to move data from the device to main storage. The structure of format 0 and format 1 CCWs for this command are presented in figure 11.10.

Read-backward CCW. This CCW is initiated at an I/O device and the subchannel prepares to move data *from* the device to main storage. The data is stored in reverse order in storage with the execution of this CCW. Figure 11.11 shows both formats for this CCW.

Sense CCW. This command is initiated at the I/O device, and the sub-channel is set up to transfer sense data *from* the device to main storage. The command transfers status information from the device to main storage. Figure 11.12 depicts both formats for this CCW.

Sense ID CCW. This command is used to request device-type identification and other information. This command differs from the read command because it acquires its information from sense indicators rather than a record resource. Figure 11.13 shows both formats of this command.

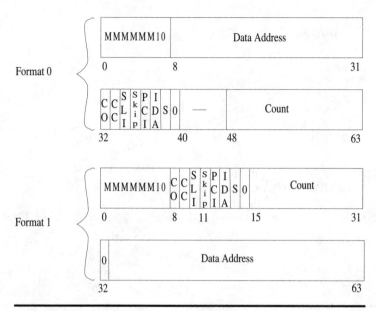

Figure 11.10   Read CCW format.

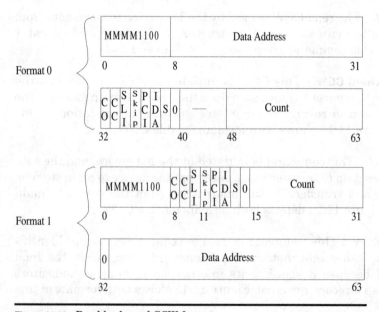

Figure 11.11   Read-backward CCW format.

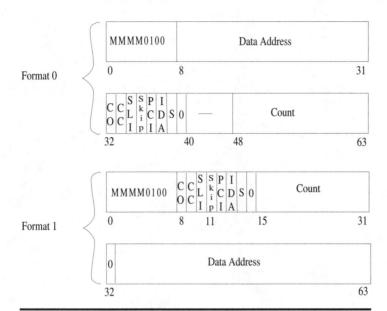

Figure 11.12   Sense CCW format.

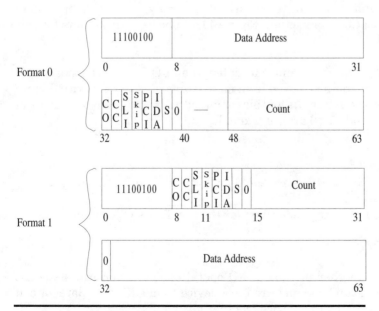

Figure 11.13   Sense ID CCW format.

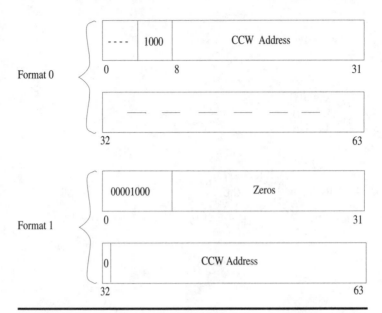

Figure 11.14   Transfer-in-channel CCW format.

Transfer-in-channel CCW. This command is used to provide CCW chaining not located in adjacent locations. The command can be used for data and command chaining. Put simply, this is a branch to another location instruction. Figure 11.14 shows both formats for this command.

Write CCW. This command is initiated by an I/O device, a subchannel is prepared, and data is moved from main storage to the I/O device. Figure 11.15 shows both formats for this command. (Abbreviations appearing in Figs. 11.15 and 11.16 are defined in legend to Fig. 11.16.)

Reference table

Figure 11.16 is a reference table for formats 0 and 1 in the previous examples.

Figure 11.16 is the code chart for the previous CCW commands.

Orders

Orders are another method of I/O operation. They perform a specific function and can be transferred to a device as modifiers or data, or can be made available to a device by other means.

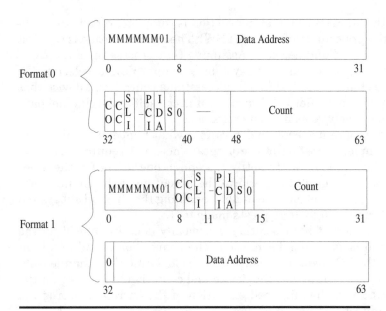

Figure 11.15 Write CCW format.

CCW Name	Command Code	Flags						
Write	MMMM MM10	CD	CC	SLI		PCI	IDA	S
Read	MMMM MM01	CD	CC	SLI	SK	PCI	IDA	S
Read Backward	MMMM 1100	CD	CC	SLI	SK	PCI	IDA	S
Control	MMMM MM11	CD	CC	SLI		PCI	IDA	S
Sense	MMMM0100	CD	CC	SLI	SK	PCI	IDA	S
Sense ID	1110 0100	CD	CC	SLI	SK	PCI	IDA	S
Transfer In Channel	XXXX 1000							

Figure 11.16 CCW code chart: CC—chain command; CD—chain data; IDA—indirect data addressing; M—modifier bit; PCI—program-controlled interruption; S—suspend; SLI—suppress-length indication; SK—skip.

11.5 I/O Execution Operation Synopsis

As previously explained, I/O operations begin and are controlled by the start subchannel, channel command words (CCWs), and orders. Specifically, the start subchannel operation is executed by a processor and supervises the I/O flow of requests among other functions.

When the start subchannel instruction is complete, results are indicated in the program status word (PSW). The next operation is the first CCW is retrieved and decoded. Assuming that processing of the CCW is successful, the channel subsystem attempts device selection by selecting a channel path. If this is successful, a control unit device that recognizes the identifier will logically connect itself to the channel path and will positively respond to its selection.

Channel program execution determines what happens at this point. If, for example, execution of the program does not require any data, then the device may simply notify the end of operation. On the other hand, if data is required to perform the execution of the channel program, a subchannel is made available, enabling the channel subsystem to respond accordingly to requests made to it.

The conclusion of I/O execution is typically done by indicating the channel-end or device-end condition. The channel-end condition indicates that the I/O device is finished and no longer requires channel subsystem facilities. However, the device-end condition indicates that the I/O control device has finished execution of the current I/O program and is ready for another operation.

11.6 Summary

The *parallel channel* is IBM's new name for what was previously considered *channels*. This was done when ESCON was announced. Channels, historically in IBM, have been parallel, so the name is fitting; conversely, ESCON channels operate serially.

Channel subsystems have channel paths which for practical purposes are the channels. Parallel channels use bus and tag cables to connect devices such as tape and disk drives, control units of different sorts, and even connect other processors via a channel-to-channel connection.

Parallel data flow moves bytes of data in parallel. Two types of channels (modes) were explained: byte multiplexer mode and block multiplexer mode. Burst mode operation and its purpose were explained. The addressing scheme for channels, subchannels, and the correlation of devices were also explained.

A number of CCW formats were presented and explained briefly. These CCWs are a key operational characteristic of the channel subsystem. These CCWs have standards that can be traced back many years; hence upward compatibility with new equipment has been possible. A CCW code chart is provided to aid in discerning the meaning of fields in the CCW frames.

Synchronous Data-Link Control (SDLC)

Synchronous data-link control (SDLC) is a data-link control protocol. As of this writing SDLC is prevalent in the marketplace and is supported by vendors other than IBM. SDLC protocol is used predominantly by vendors whose equipment provides access into system network architecture (SNA) and advanced peer-to-peer networking (APPN) environments.

This chapter covers some basic SDLC concepts and provides reference material to that end.

12.1 SDLC Operation

Some explanations of SDLC lend themselves to a broad perspective; however, this section presents a close-up view. Consider Fig. 12.1 showing where SDLC operates in respect to other components in a network environment.

In Fig. 12.1, notice a processor, two front-end processors (FEPs), 3174 control units, terminals, and a printer. This figure portrays the data link itself occurring between two data terminal equipment (DTE) components with two data-circuit-terminating equipment (DCE) components operating with the two DTEs. More is involved componentwise than this, but the concept to understand here is that SDLC operates between the two DTEs which are inside these FEPs.

SDLC can operate among equipment other than FEPs; Fig. 12.1 is an example of where SDLC operates with FEPs. Also Fig. 12.1 does not indicate a geographic distance because that is relatively unimportant. These FEPs could be located in the same city or in a distant location.

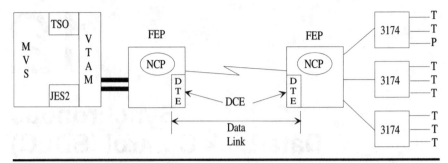

Figure 12.1   SDLC operation.

12.2   Components Used in Data-Link Operation

SDLC data-link operation can be described by components used and their functionalities. First, a DCE component is needed. A DCE can be characterized as being that component which converts signals and codes between the DTE and the medium used for transmission. Simply put, the DCE converts digital signals to analog and vice versa; that is, it performs modem operations.

The DCE can also be part of the component where the DTE is located, or it may be a separate piece of equipment. The DTE is a component within an entity that serves as a data repository, or resource, for the DCE. Another way to explain a DTE is by what it does; it converts control commands and data into binary digits: ones and zeros (1s and 0s). Consider Fig. 12.2.

In Fig. 12.2 entities A and B can be any device that supports DCE and DTE functions. In this example, it could be the FEPs as shown in Fig. 12.1.

Figure 12.3 introduces the link station and the link. The *link station* (LS) is part of an entity which permits attachment and control over a

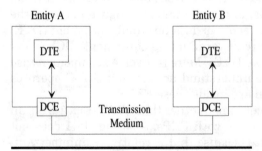

Figure 12.2   Conceptual view of the DCE and DTE.

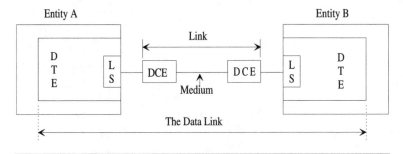

Figure 12.3 Components used in a SDLC implementation.

medium. A *link* is the connection between the DCEs. This figure also shows the DCE as separate from the entity. This is different from the scenario depicted in Fig. 12.2 but represents another possible implementation.

The LS is physically part of the DTE. In a practical way, one could consider the DTE as an interface board. Likewise, the LS could be thought of as hardware, firmware, and software because it is through the LS that a link (the connection between the DCEs) is managed. Notice in Fig. 12.3 that the actual data-link connection is made between the DTEs using the LS, link between the DCEs, and the medium.

12.3   Link Station Types and Implementations

Two types of link stations are recognized:

- Primary
- Secondary

A *primary link station* is responsible for controlling a link. It does this through the use of commands issued to a secondary link station. The *secondary link station* responds to a primary link station by executing instructions issued to it.

A good way to approach the concept of primary and secondary link stations is by examining link connection implementations and the characteristics thereof. Consider Fig. 12.4.

Figure 12.4 shows a dedicated link connection. Some refer to this type of connection as a *leased line*. Whatever it is called, the meaning is that the *link* is not broken between DCEs. The link may not be used but nevertheless is maintained. In such a scenario, entity A would be the primary link station and entity B, the secondary link station.

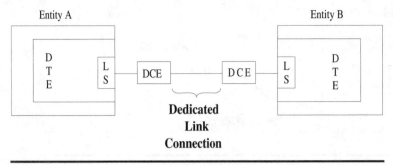

Figure 12.4 Example of a dedicated link connection.

Figure 12.5 is an example of a switched link connection. Most refer to this type connection as a *point-to-point link.* A connection such as this is best explained by analogy.

The telephone operates on a switched basis. This means that when someone wants to use the telephone it is placed off-hook (switched), a number is entered, and connection is made with the destination. But the telephone is unlike the connection shown in Fig. 12.5 because any given telephone can be used to contact other telephones, not necessarily only one other particular telephone. Thus, the analogy is that a connection is made on an as-needed basis.

The concept of a switched multipoint connection exists. This differs from the previous example because a primary link station exists and multiple secondary link stations to which the primary link station may be connected also exist. Figure 12.6 depicts such a scenario.

Figure 12.6 shows four entities capable of making a connection with one another. In this example entity C is shown communicating with entity B. This does not necessarily have to be the case. Entity C could communicate with entity A, B, or D because they share a common connectivity point indicated via the cloud.

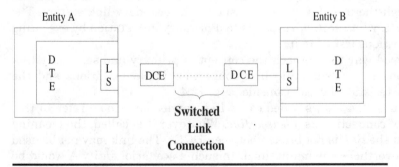

Figure 12.5 Example of a switched link connection.

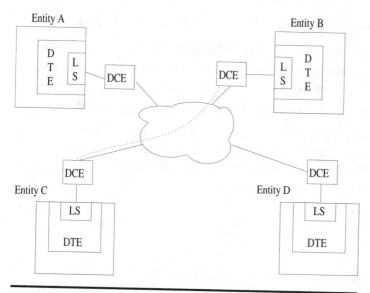

Figure 12.6   Example of a multipoint connection.

These are popular implementations, but variations of these may exist. These three are fundamental examples of implementations.

12.4 SDLC Frame, Format, and Contents

When SDLC is implemented, data and control information is moved via a frame. This frame structure appears as indicated in Fig. 12.7.

The first and ending part of the frame is called a *flag*. The beginning flag consists of a reference point for the address and control fields farther back into the frame. The flag of SDLC frames is represented in binary code as 01111110 or in hexadecimal code as 7E.

The second field is the address field. Two possibilities for addresses exist with SDLC. If a secondary link station is transmitting, it includes its own address (which is the origin address). If a primary link station is transmitting, it includes the target destination (which is the secondary link station's address). In other words, the address included in this field is the secondary link station's address.

Figure 12.7   SDLC frame structure.

The secondary link station supports three types of addresses:

- *Station address:* This is the individual address of the link station.

- *Broadcast address:* This is a broadcast address to which any stations on the link will respond.

- *Group address:* This address is common to multiple stations.

The purpose for having such addresses is to support secondary link station applications that may need such addressing capabilities.

The control field in the frame defines how the frame functions. Three SDLC frame formats are

- Information

- Supervisory

- Unnumbered

An *information format* (also called an *I format*) is used to communicate the send and receive count of information frames. This frame includes a poll bit and a final bit as well. The poll bit is sent to the secondary station so that it will initiate transmission. A final bit is sent to the primary station from the secondary station to indicate that this is the last frame in the transmission. However, this bit is not the same as the last part of the SDLC frame known as the *flag*. This bit indicates that this is the last transmission from the secondary link station to the primary link station but that it is not necessarily the last frame to be sent.

The *supervisory format* (also called the *S format*) frame structure is used to acknowledge frames that were received. This type of frame format also communicates a busy state or ready condition. This frame format can also be used to indicate that frame order was out of sequence when it was received. This frame format has a poll bit and a final bit as well. Its function is like that of the information format frame.

An *unnumbered format* (also called a *U format*) frame is used to establish and terminate a data link. It is also used to transfer data and report some errors. This format uses the poll bit and final bit also.

The information field part of the SDLC frame follows the control field. The information field contains data if data is passed. This field is formed in multiples of eight (8). This field does not have to contain data; nor does it have to be an information field. When a supervisory frame format is used, no information exists.

Next in the frame is the *frame-check sequence* (FCS) field, which performs error checking on the frame to ensure that no errors have been incurred during transfer over the link. The method of checking is *cyclic redundancy checking* (CRC). This checking is performed at the point of transmission. Similarly, a check is made on the receiving end of the link by the receiving station.

The last part of the SDLC frame is the flag. It is similar to the beginning flag; however, this delimits the frame itself. Some refer to this flag as the *ending* or *closing* flag.

12.5 Transmission States

The concept of a state exists with respect to transmission of SDLC frames and link stations. Three states are identified:

- Active

- Idle

- Transient

According to IBM documentation, in an *active state* a link station is transmitting or receiving control signals or information over a link. A link is considered active when flags are transmitted as well. An *idle state* is defined as an operational link with no SDLC information or control data being passed over the link. A link station determines when a link is idle by continuing to receive binary ones after 15 consecutive ones have been detected. A *transient state* is the condition of the link connection when it is being prepared for initial transmission or after a transmit reversal in the transmission. In a sense it is measured in lapsed time, even though this amount is fractional.

12.6 Frame Format Command and Responses

The frame formats mentioned previously have related commands and responses. A brief listing and explanation of each is provided here.

Information format

The information frame format numbers its frames by NS or NR (indicating the number of sends and the number of receives, respectively). This numbering mechanism presents sequential numbers indicating the position in the transmission sequence.

Supervisory format

The supervisory frame format includes the following commands and responses along with their meanings:

REJ   The *reject* command or response is sent to request the transmission or retransmission of numbered information frames.

RR   The *receive-ready* command can be sent by either primary or secondary stations. This command confirms frames received and is used to indicate that the originating station is ready to receive more information frames.

RNR   The *receive-not-ready* command can be sent by a primary or secondary link station. This state may be due to buffers being temporarily full or other internal component restrictions at the current time.

Unnumbered format

Some unnumbered format commands are supported by different vendors. Those listed below are considered part of the set for SDLC; a particular vendor implementation may not use all of them.

BCN   The *backward congestion notification* command is used in a loop configuration. It is used to indicate a loss of input. When this occurs the secondary link station puts out a BCN command to the primary link station.

CFGR The *configuration* command is used by secondary link stations in response to primary link station configuration commands.

DISC The *disconnect* command is used by secondary link stations (those that are receiving data and commands). It functions by eliminating any other modes that may be present. When this occurs, the expected response to it is an unnumbered acknowledgment.

DM   This command is used by a secondary link station to indicate that it is in *disconnected mode*.

FRMR When a secondary link station receives an invalid frame, it sends the *frame-reject* command to the primary link station.

RD   Secondary link stations use this command to send to a primary link station to *request a disconnect* command.

RIM   Secondary link stations send this command to a primary link station to *request* a set *initialization mode* command.

SIM   The *set initialization mode* command, used by primary link stations, initiates link-level initialization.

SNRM The *set normal response mode* command is used by a primary link station to put a secondary link station in a normal response mode. This means that information received by it will be in modulo 8, not modulo 128.

SNRME The *set normal response mode extended* command is used by the primary link station to put the secondary link station in response mode so that it can receive in modulo 128, which is the maximum number of information frames sent before the secondary station responds.

TEST This command is sent by a primary link station to the secondary link station to request a *test* response from that link station.

UA   This command is an unnumbered acknowledgment. It is used to indicate a positive response to DISC, SNRM, SNRME, or SIM commands.

UI   The *unnumbered information* command is used to transmit information without sequence numbers.

UP   This is the *unnumbered poll* command. When this command has the poll bit set to zero (0) the primary link station sends optional poll responses in a loop configuration. When the poll bit is set to one (1), a response is sent to all secondary link stations with the expectation of a response.

XID   This is the *exchange identification* command. This is used by the primary link station to request a secondary link station's identification. The information field in the response from the secondary link station includes the identification of the responding secondary link station.

12.7   SDLC Concepts

Certain SDLC concepts are commonly used during conversations about SDLC; these concepts are considered here.

Bit stuffing

The notion of bit stuffing is technically referred to as *zero-bit insertion*. Before explaining what bit stuffing is, explanation of normal operation of SDLC is in order.

An SDLC frame begins and ends with a flag on both ends. This beginning and ending flag is always binary 01111110 or hexadecimal (hex) 7E. No flags with this pattern are permitted between the beginning and ending flag; hence the result is a bit-oriented protocol. The nature of this bit-oriented protocol means a more powerful method for moving data.

Timeouts

Timeouts are used by a primary link station. This is the case because the primary link station is responsible for continuous operation of a data link in an orderly fashion. As such, it must check for responses for commands it issues to determine whether execution has been achieved. Two timeouts are used by the primary link station:

- Nonproductive receive
- Idle detect

A *nonproductive receive* is defined as reception of bits in such a fashion that a frame is the received result. Consequently, the primary link station is informed by the secondary link station of such a state, and the primary link station then waits for a determined number of seconds. If the condition cannot be resolved at the data link, other layers in the network are required to intervene.

An *idle detect* condition occurs when a primary link station transmits a frame to a secondary link station and a response is not received by the primary link station before a timeout period. Considerations in correcting a potential problem such as this include awareness of propagation delay time to and from the primary link station and the secondary link station, adequate time for clear-to-send on the secondary station's DCE, and processing procedures at the secondary link station.

Frame numbering

The concept of *frame numbering* is a method for the receiving link station to maintain a count of the sequence. This concept is implemented by the transmitting link station numbering information frames and putting the number of the frame inside the count field of the frame. This numbering method works because the receiving station counts these numbers and knows whether sequential reception is in order. Thus a receiving station anticipates the next frame.

If a sequence is misplaced, the receiving station does not accept a frame out of sequence. However, the receiving station does accept a receive count for verification purposes. Retransmissions are performed if sequence errors are discovered.

SDLC loop configuration

SDLC can operate in a loop configuration where one primary link station and multiple secondary link stations exist. In such a configuration one-way transmission is performed (some refer to this as *simplex operation*). Figure 12.8 shows such a configuration.

Figure 12.8 shows multiple secondary link stations and one primary link station. When SDLC is implemented in this fashion, it transmits data in one direction, and the responses move in that direction as well. Additionally, transmissions occur one at a time.

In this type of configuration the primary link station can transmit a frame to one or all secondary link stations in the SDLC loop configuration. This is achieved because the frame incorporates an indicator for the type of addressing used, which is a designated station or a broad-

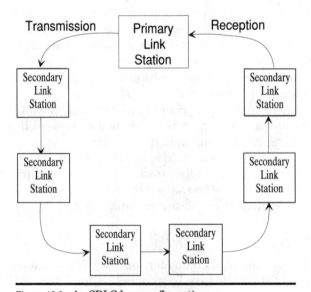

Figure 12.8   An SDLC loop configuration.

cast for all stations. The consequence of this operation is that all participating secondary link stations must examine the address field in each frame to determine frame destination.

Secondary link stations can transmit in this configuration after they have received a frame addressed to them. In reality each secondary link station acts as a repeater, generating the frame to put back on the medium.

12.8 Summary

SDLC is a bit-oriented protocol that operates at a data-link layer. It uses primary and secondary link stations in a variety of configurations.

SDLC employs the concept of a link state; they are active, idle, and transient. SDLC also uses three formats in the control field of the frame: information, supervisory, and unnumbered formats.

SDLC frames begin and end with a flag which has a hex value of 7E. This value is not used within the beginning or ending flag because the concept of bit stuffing is employed to eliminate this.

Chapter

13

Token Ring

Token ring is a protocol implemented at the lower two layers of a network. It uses ring technology with a hub or star implementation. Token ring operates at 4 or 16 Mbits/s. This chapter explores the major characteristics of token ring.

13.1 Physical Components

Physical components that constitute a token-ring network include

- Media access unit (MAU)
- Token-ring network interface cards (NICs)
- Cables (lobes)
- Participating hosts
- Configuration software

Media access unit

Token-ring technology is implemented via a media access unit (MAU). This device appears as shown in Fig. 13.1.

Figure 13.1 Front view of media access unit.

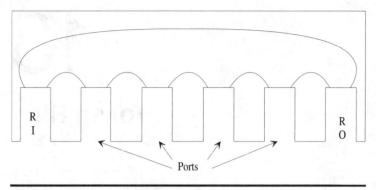

Figure 13.2   Top view inside a MAU.

The MAU is the device where the *ring* is located. Inside the MAU each port is connected to this ring, including the ring-in (RI) and ring-out (RO) ports. Figure 13.2 shows the inside of a MAU.

The RI and RO ports are used to daisy-chain MAUs together, thus increasing the possible number of nodes on the ring. Figure 13.3 shows multiple rings connected together.

Token-ring network interface card

Network interface cards (NICs) are required for devices that participate in token-ring networks. Regardless of the host, token-ring interface cards are required for participation in a token-ring network. Each token-ring interface card performs three functions: receiving a packet, transmitting a packet, and performing normal repeat mode functions,

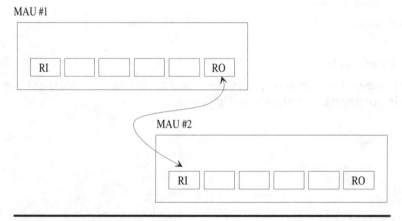

Figure 13.3   Daisy-chained MAUs.

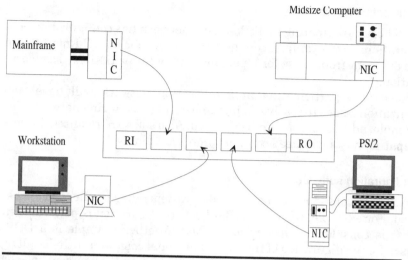

Figure 13.4 Host NICs in token-ring processing.

which involve checking the data in tokens and frames that it receives and then setting the appropriate bit address recognized, frame copied, or error detected. Figure 13.4 shows multiple-size hosts participating in a token-ring network.

Configuration differs on these hosts because of their internal operation, but a token-ring interface board is required for participation in the network.

Cables (lobes)

The cable used to connect a device to a MAU is called a *lobe,* which is the cable between the interface board and the MAU itself. Consider Fig. 13.5.

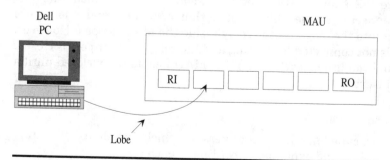

Figure 13.5 Conceptual view of a lobe.

Participating hosts

Each host participating in a token-ring network has configuration considerations. For example, configuration issues for an IBM mainframe are different from those for a personal computer or UNIX-based workstation.

Each host participating in a token-ring network will have customization issues to resolve on installation of a token-ring network. For example, addressing *should* be planned. Also software required for participating hosts will need configuration accordingly.

Configuration software

Token-ring configuration software differs depending on the type of host used. For example, with large IBM hosts configuration is performed in what is considered a *generation* (GEN). Another example is a DOS-based PC; configuration of this type of personal computer includes alterations to the `autoexec.bat,` `config.sys,` and token-ring-specific software files. Other configuration changes may be necessary, and these are contingent on the device participating in the token-ring network.

13.2 Media Access Unit Theory of Operation

Token ring is considered *self-healing technology,* meaning that devices can be inserted and removed from the MAU without disrupting the network. This ability is possible because of token-ring component operation. Before exploring the details of the official steps for devices inserted onto the ring, some basic information is important.

The MAU relay

Each port in the MAU has a relay that functions during insertion and removal. This insertion and removal is technically called *lobe* insertion and removal. Consider Fig. 13.6.

Figure 13.6 shows a MAU and four devices, three of which are physically connected to the MAU. Notice that node 1, 2, and 3 lobes are inserted into the ring and the relay inside the MAU is open. However, node 4 is not connected to the ring and the relay for that port is closed. In normal conditions each port of the MAU has its relay closed until a lobe is inserted.

Station ring insertion

This is also called *the five-step process,* in which five identifiable steps can be isolated and explained according to each step.

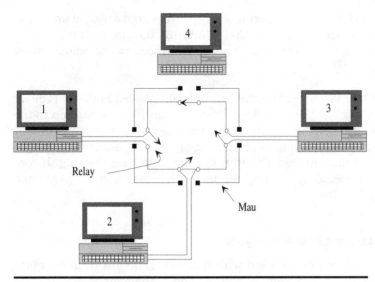

Figure 13.6 Port relays on a MAU.

Step 1. First, a lobe test is performed. This test does not affect the internal ring or the operations on it. The lobe test checks the cable between the interface card and the MAU. It also performs an internal diagnostic check. It transmits media access control (MAC) frames from the station to the MAU. Assuming that the lobe test passes, the relay is opened, the station is inserted onto the ring, and a request for a token is generated. If the lobe test fails, an error message is generated.

Step 2. The next function assumes that the station is attached to the ring, and this step begins with a timed period to determine whether certain frames are present on the ring, such as active monitor present and standby monitor present. If the station receives any of these or other frames from the ring, it proceeds to the next step. On the other hand, if the station does not receive any required frames prior to timer expiration, the station concludes that no active monitor is present—a problem incurred by the station inserting itself onto the ring—or that it is the first station on the ring.

Step 3. In this step the station's interface card determines whether a duplicate physical address is already in use on the ring. If one is, the station informs the software serving as the network manager, and the station will remove itself from the ring. If no duplicate address is found, the station remains on the ring.

Step 4. The station next determines its nearest active upstream neighbor (NAUN) at the same time that it informs the closest downstream station of its identity. This function is related to network management on the ring.

Step 5. In this last step, the station's interface board makes a request for initialization on the ring. This function identifies the station to the ring manager. The ring manager, in turn, checks for any conflicts that might arise by that station participating on the ring. Assuming that no problems are encountered, the station is permitted on the ring. If conflicts are discovered by the ring manager, the station is not permitted on the ring.

13.3   Token-Ring Frame Analysis

Three types of frames are used with token ring: a data frame, a management frame, and a control frame. The management and control frames are called *media access control* (MAC) frames. The difference between a data frame and the management or control frame is what appears in the information field of the frame. In a MAC frame data appears via a logical link control protocol data unit (LLC-PDU) which resides in the information field. In a MAC management or control frame a MAC-PDU is located in the information field.

The two other frames used in token ring are a token frame and abort frame. These frames and their components and functions are presented in this section.

Token-ring LLC frame

The *token-ring LLC frame* is the frame by which data is moved through the ring. Some refer to this frame as a MAC frame. This is valid assuming that the MAC frame has an LLC-PDU within the information field. An example of a basic MAC frame, without indicating whether it is an LLC-PDU or MAC-PDU, appears as in Fig. 13.7. In a sense this can be considered a generic MAC frame.

The field components in the MAC frame include a *start delimiter* (SD), which begins the MAC frame and contains a pattern represent-

SD	AC	FC	DA	SA	Info	F C S	ED	FS

Figure 13.7   Generic MAC frame.

ing nondata so that the field can be recognized by the token-ring controller. The significance of this field is to indicate the next field. However, the pattern of this field is

JKOJKOOO

This pattern is meaningful because it violates a signal encoding scheme. By doing this, the interface card knows that the next field is the access control field.

The *access control* (AC) frame has four components: (1) the priority indicator, (2) the token indicator, (3) the monitor indicator, and (4) the reservation bits. This is represented by

PPPTMRRR

The priority indicator has eight possible bit settings that indicate the nature of the priority. The following table lists the bit settings and their meanings.

Bit setting	Meaning
PPP	
000	Low priority
001	Low priority
010	Low priority
011	High priority
100	IBM bridge
101	Reserved
110	Reserved
111	Network management
T	
0	Token
1	Data frame
M	
0	Set by the sending station
1	Set by the active monitor
RRR	
000	Reservation bits

The next field in the frame is *frame control* (FC) field, which indicates whether the frame is data or a MAC frame. If it is a MAC frame, the bits within the frame indicate the type. Bit settings for two components of the frame and six possible MAC control frames and their functions are as follows.

Bit setting	Meaning	MAC control frame function
	FF	
00	MAC control frame	Duplicate address test
01	LLC frame	Beacon
10	Reserved	Claim token
11	Reserved	Purge
		Active monitor present
		Standby monitor present
	ZZZZZZ	
000000		
000010		
000011		
000100		
000101		
000110		

The next field is the *destination address* (DA), which specifies the station or stations to which the frame is intended. If the frame is intended for one station, this address will be unique on the ring. Or the frame could be intended for all stations on the ring, in which case it is considered a multicast address.

The next field is the *source address* (SA). This address identifies the sending station.

The next field is the information (I) field. This field may contain either an LLC-PDU or a MAC-PDU. If the MAC frame has an LLC-PDU inside, it appears as shown in Fig. 13.8.

Figure 13.8 not only shows the MAC frame with MAC addresses but also shows the LLC-PDU that contains the source service access point (SSAP) as well as the destination service access point (DSAP). Assume

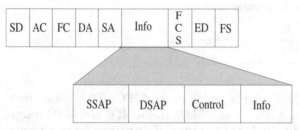

Figure 13.8   MAC frame with an LLC-PDU.

that the information field has LLC-PDU data. If so, the LLC-PDU contains the following:

SSAP	The *source service access point* indicates the originating source.
DSAP	The *destination service access point* indicates the target for the data.
Control	This field indicates whether it is a command or response PDU. It also indicates whether the PDU is information transfer, a supervisory frame, or unnumbered PDU.
Information	This field carries user data or management data within.

MAC frames that do not contain user data perform ring management and control functions. This type of frame has a MAC-PDU inside the information field and appears as shown in Fig. 13.9.

Figure 13.9 shows the MAC frame with a MAC-PDU inside the information field.

A closer look at the MAC-PDU reveals five components (see Fig. 13.10).

These fields in the MAC-PDU are defined as

MVL	Major vector length
MVI	Major vector identifier
SVL	Subvector length
SVI	Subvector identifier
SVP	Subvector parameters

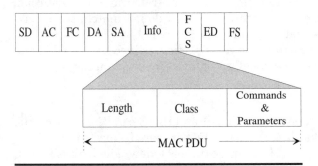

Figure 13.9   MAC frame with a MAU-PDU.

MVL	MVI	SVL	SVI	SVP

Figure 13.10   MAC-PDU format.

These fields inside the MAC-PDU are referred to as *vectors* and *subvectors*. *Each of these is a parameter and relates to the third field in the frame*; that is, the frame control (FC) field.

Examples of some MAC-PDU control frames and their functions include

Active monitor present: This frame is sent by the active monitor approximately every 3 to 4 seconds to indicate its presence to all hosts connected to the ring.

Standby monitor present: This frame is sent by standby monitors. Any host on the ring may serve as a standby monitor. A standby monitor frame is sent approximately every 7 seconds to indicate that it is ready to assume the active monitor role if necessary.

Claim token: This frame is issued onto the ring by a station attempting to become an active monitor.

Ring purge: This frame is issued by the active monitor to release a new token on the ring after error recovery has been performed.

Beacon: This frame is put onto the ring by any station that detects a lost signal.

Following the information field is the frame-check sequence (FCS) field. The sending station performs a cyclic redundancy check (CRC) and stores the result in the frame. The receiving station performs another CRC and compares it to the one sent from the sending station. If they match, no error has occurred; if they do not match, the receiving station assumes that an error has occurred.

The next field is the *end delimiter* (ED) field, which uses special bits that only the interface board will recognize. The sequence is

```
JK1JK1IE
```

where J and K are nondata bits. The I and the E can be changed. The I bit indicates whether it is the last frame or more frames will follow. If I = 0, it is the last frame. If I =1 , more frames follow. The E bit is known as the *error bit*. If the I bit = 1, then an error was detected by a station on the ring, and any station can change this bit to 1.

The last field is frame status (FS). The bits in this field appear as

```
ACrrACrr
```

The meaning of the A is that it is an address recognized bit set by the station which receives the frame, realizing that the frame is for that station. The C indicates that a station copied the bit. The rr bits are

SD	AC	ED

Figure 13.11   Token-ring format.

reserved for future use. Two ACrrs appear because this field is outside
the frame-check bits. By sending two sequences, a redundancy is
achieved.

MAC frames can be evaluated by either a header and its components
or a trailer, as shown in Fig. 13.11.

The frame can also be evaluated with respect to components that are
validated in the FCS.

Token frame structure

The *token frame* is one of the components that makes token ring
unique. No station can transmit data unless it first possesses the token
frame. The structure of this frame appears as shown in Fig. 13.11.

The start delimiter (SD) of this frame contains the following:

    JKOJKOOO

This pattern indicates the start of the frame. The purpose of this pattern is that it violates signal phase patterns and consequently the NIC
realizes that the next field is the access control field. This information
is static and indicates nondata so that it will be recognized by the
token-ring controller.

Token frame operation

Token ring is named so because of the nature of its operation; a token
travels around the ring.

To transmit data, a station on the ring must capture the token frame.
Once the token frame is captured, the station desiring to transmit on
the ring changes the token bit in the access control (AC) field. This
changes the token in the start-of-frame sequence. The network interface card constructs the necessary fields around the remaining part of
the frame and then sends the frame back onto the ring.

Since no *token* is on the ring, other stations desiring to pass data
through the ring must wait for a token frame to appear. Each station
is monitoring the ring and thus is aware of the absence of a token. After
the frame or frames of data are received by the destination, the destination host puts the frame back onto the ring. Once the transmitting

Figure 13.12   Abort frame format.

station receives the frame(s) originally sent and transmission is complete, the transmitting station generates a new token for the ring.

Abort frame

The *abort frame,* or as some call it, the *abort delimiter,* is shown in Fig. 13.12.

This token format can be issued by the station that transmits a frame. For example, if a transmitting station needs to abort the sequence, it merely needs to issue an abort token.

MAC control frames

Token ring uses numerous MAC frames to control the ring. Some of those frames and their functions are listed below.

Frame	Function
Claim token	This token is used by a station on the ring to become the active monitor on the ring when no active monitor is present
Duplicate address test	This frame is issued on initial insertion of a station to the ring to ensure that no duplicate addresses exist
Active monitor present	This frame is sent around the ring to indicate to all stations that an active monitor is present
Standby monitor present	This is a frame sent around the ring by a station to indicate that a standby monitor is present
Beacon	This frame is issued when a serious fault occurs, such as a break in a cable
Purge	This is a frame that is issued by the active monitor following a claim token or to reinitialize the ring

13.4   Token-Ring Concepts and Functions

Token ring implements concepts of stations attached to the ring performing certain functions. For example, the following is a brief list of some concepts and functions implemented with token ring.

- Active monitor
- Standby monitor

- Nearest active upstream neighbor
- Beacon

Active monitor

Each ring has an *active monitor,* which is the station on the ring with the highest address. Therefore, if a different station is desired to perform this function, the address of that station will have to become the numerically high station on the ring. The purpose of an active monitor on the ring is to perform management-related functions on the ring. For example, the active monitor checks the AC field in a frame to determine whether a busy frame is continuously on the ring. This scenario could be caused by a station failing to take a frame it transmitted off the ring. The active monitor removes a frame from the ring if it determines that the frame indicates it is busy and the monitor count is on. Once this action is performed, the active monitor issues a free token.

Maintaining appropriate ring delay is another key function of the active monitor. The token frame is 24 bits in length; consequently, the minimum delay for a token to traverse the ring is 24 bits. If the value is less than 24 bits, the token cannot traverse the ring. The purpose of an active monitor is to ensure this by inserting a 24-bit buffer and therefore guaranteeing its required delay.

The active monitor performs other functions such as

- Controlling transmission priority
- Compensating for jitter (a delay distortion caused by velocity of signal propagation through a medium; termed "jitter" because it can cause pulse position deviation and hence clocking problems)
- Issuing an "I'm alive" message in regular intervals
- Controlling tokens
- Performing an autoremoval function if multiple active monitors happen to occur

Standby monitor

Each token-ring network has potential standby monitor stations. The *standby monitor* function is necessary in case something should happen to the active monitor. The procedure in determining the standby monitor is as follows. The active monitor issues an ACTIVE_ MONITOR_PRESENT frame at regular intervals. However, if it fails to do so, a station can issue a CLAIM_TOKEN and take over the role of active monitor.

Nearest active upstream neighbor

The notion exists in token ring that each station *knows* who is its nearest active upstream neighbor (NAUN). This is determined as follows.

The active monitor issues a frame with a broadcast address inside the destination field. When the frame reaches the first station, it recognizes the broadcast address and determines that the address-recognized indicator bit and frame-copied bits are set to zero. This station realizes that it is the first station to receive the frame and deduces who is its NAUN. This station remembers this, sets the ARI and FCI (address-recognized indicator and frame-copied indicator) bits to one, and then puts the frame back onto the ring.

When the station that received the frame from the active monitor receives a free token, it transmits a standby monitor present frame with a broadcast address. Consequently, the next station on the ring receives the frame and determines who is its NAUN. Once this procedure is complete, the active monitor receives the frame with the standby-monitor-present address and the process is complete.

Beacon

The notion of a beacon condition exists with token ring. A *beacon* occurs when a station attempts to claim a token and eventually performs a timeout if no token is received. Consequently, the station issues a beacon indicating a default. Normally, when this occurs it is the result of a hard error such as a break in a cable or ring itself. When the beacon is issued the beacon frame contains the beaconing station address along with its NAUN address. Figure 13.13 illustrates this concept.

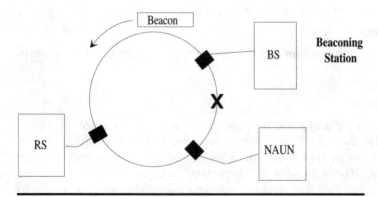

Figure 13.13 Conceptual view of the beacon function.

13.5   IBM's Token-Ring and IEEE 802.5 Frames

IBM token-ring frame structure differs from the IEEE 802.5 structure. So far this chapter has focused on the IEEE 802.5 frame structure. Concise information is presented here to delineate the two.

IBM token-ring frame structure

IBM token-ring frame structure (Fig. 13.14) and IEEE 802.5 frame structure differ with respect to one field.

The IBM token-ring frame has a field for routing information; it is indicated by the RI field. The information contained in this field includes routing control information and one or more segment numbers. Figure 13.15 shows those components within the RI field of the token-ring frame.

This RI field permits broadcast by the source station and includes routing information on every intraring frame.

IEEE 802.5 frame structure

The IEEE organization is responsible for the 802.5 standard. It is widely implemented in the marketplace. Figure 13.16 reiterates the IEEE 802.5 frame structure explained above.

| S D | A C | F C | D A | S A | RI | Info | F C S | E D | F S |

Figure 13.14   IBM token-ring frame structure.

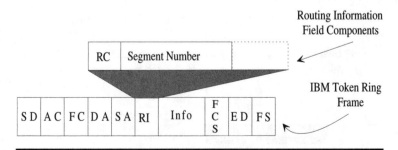

Routing Information Field Components

IBM Token Ring Frame

Figure 13.15   Components of ring-out field of IBM token-ring frame.

S D	A C	F C	D A	S A	R I	Info	F C S	E D	F S

Figure 13.16   IEEE 802.5 frame structure.

13.6   Token-Ring Addressing

Token-ring addressing is straightforward. This section covers those aspects that relate to it. Two basic types of addresses compose the token-ring addressing scheme: the interface card address and the interface card service access point addresses.

Interface card address

Each token-ring interface card has a unique 12-digit hexadecimal address that reflects the interface card itself. This is sometimes referred to as a "hard" address. Any device participating in a token-ring network must have an interface card, or the equivalent of one, to generate this 12-digit hexadecimal address. Figure 13.17 depicts an interface card.

Interface card SAP addresses

Each token-ring interface board (also called a "card") has what is considered *service access points* (SAPs). These SAPs are also addressable. Token-ring interface cards have approximately 128 addressable SAPs per card. SAPs do not conflict with SAPs on other interface cards because *each* interface card has a unique 12-digit hex address. Figure 13.18 is a conceptual view of token-ring SAPs.

SAP addresses are configured in increments of 4. The beginning SAP address on each card begins with 04. Each SAP is locally administered

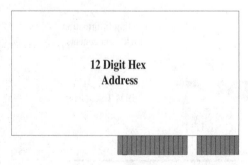

12 Digit Hex Address

Figure 13.17   Token-ring interface card.

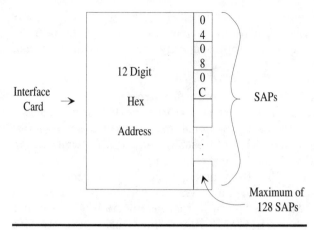

Figure 13.18   Token-ring interface card SAP addressing.

by a network administrator. The addresses then increment by 4 up to approximately 128 SAPs per card as Fig. 13.18 shows.

Link connections. Connections are made between SAPS by what is called a *link*. Multiple link connections can be made against one SAP. For example, Fig. 13.19 shows two link connections made against token-ring interface card A, SAP 04 from token-ring card B, and SAP 04 and 08.

The link connections shown in Fig. 13.19 is between two different interface cards; hence the existence of two different hosts is implied. Focus on the two link connections made from card B to card A SAP 04. This is possible because link connections at the SAP level are dynami-

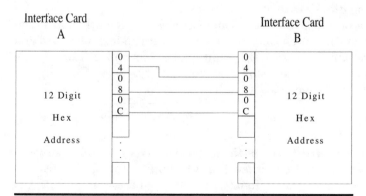

Figure 13.19   Conceptual view of link connections between SAPs.

cally allocated by the interface card itself. Another possible reason is that it is architecturally built into the operational design structure of the token-ring interface cards. Approximately 255 link connections can be allocated per SAP.

A general word about SAPs. The term *service access point* can be used two ways. The term *destination service access point* (DSAP) indicates a target SAP address. However, the term *source service access point* (SSAP) indicates the origin SAP address. The use of the terms DSAP and SSAP is relative and can be very confusing.

For example, examine Fig. 13.19 again. Is SAP 04 on interface card A a DSAP or an SSAP? It can be both. The accurate answer to a question such as this is *which* host is initiating a connection with another host. In the example shown in Fig. 13.19 interface card A and interface card B SAPs can be both DSAPs and SSAPs depending on the particular transmission. The point here is the mere occurrence of the term DSAP or SSAP in a conversation does not mean much unless those participating in the conversation understand the originator of the conversations between participating token-ring hosts.

Interface card configuration within a network

The network implementation will dictate what is required for customization and configuration to make the token-ring interface cards operational. For example, customization in IBM's SNA mainframe environment requires knowledge of the 12-digit hexadecimal address and two SAP addresses. The 12-digit hex address must be supplied to the person performing the configuration on the mainframe. Likewise, the SAP addresses must be supplied as well if the device being connected to the network is not part of the native SNA environment. Configuration differs with other networks.

Consider a network with Dell Computer Corporation systems implementing token ring. Each Dell host would require an interface card and software customization as well. Additionally, basic modifications are required to the `autoexec.bat` and `config.sys` files. Other minor modifications may be needed as well.

13.7 Summary

Token ring is a protocol used at the lower two layers within a network. Token-ring speeds can be either 4 or 16 Mbits/s. Token ring utilizes ring technology in a star configuration. The media access unit (MAU) is the device to which participating hosts connect. Cables that connect the interface card to the MAU are called lobes.

Token ring is considered self-healing technology. Because of the relay inside the MAU, a station can be inserted and removed without disrupting operation on the ring. The five-step insertion process was explained in terms of the functions performed at each step.

Token ring utilizes multiple frames, some of which were presented here: the MAC frame with LLC-PDU data, the MAC frame with MAC-PDU data, a token frame, and an abort frame. The fields in each frame have specific meanings for the operation of a token-ring network. The meanings of all fields were supplied for the generic MAC frame. The MAC-PDU components were also presented. A brief list of MAC control frames and their functions was presented.

Some token-ring concepts and functions were provided such as the active monitor, the standby monitor, the nearest active upstream neighbor, and the beaconing function. Each was explained according to its function.

IBM's token-ring frame was contrasted to the IEEE 802.5 frame. The IBM frame differed because it contains a field for routing information.

Token-ring addressing was explained. The meanings of a 12-digit hex address, SAP addresses, and link connection addresses were explained. Information was also provided to explain why SAP addresses can have the same number in a network yet not violate addressing structure. A concluding word about SAPs was provided to enlighten the reader about a very confusing point.

14

X.25

X.25 is a CCITT recommendation. It is a packet-switching technology that dates back to the mid- to late 1960s and early 1970s. Around that time many public network service providers were in somewhat of a dilemma. On one hand, a networking technology was beginning, but no interface for connecting computers or other equipment dominated. Hence, X.25 originated from that need to have an interface into the data network.

The CCITT drafted its first specification for X.25 in 1974. Since that time, the CCITT specifications have been updated every 4 years. They are published, and now available in whole on CD-ROM (compact disk–read-only memory) and are available for anyone to purchase. Demystifying X.25 is easy. It is merely a definition that explains how data is to be exchanged between data terminal equipment (DTE) and data circuit-terminating equipment (DCE).

The remainder of this chapter explores a number of aspects about X.25 technology, and one section includes a list of references for those individuals needing further information.

14.1 A Perspective on Switching Technology

Different switching technologies are available. Three basic types are presented here.

Packet switch

Fundamental to *packet-switched* networks is that data traversing the network is divided into packets and moved from origin to destination.

Each packet contains user data, control, and identification information. A premise of packet-switching technology is that since packets are relatively small and each contains control and identification information, a computer port or line is occupied only when packets are moving through it. This means that port(s) and line(s) are available for others to use. Another premise of packet-switching technology is that a packet-switched network has multiple nodes in the network and hence multiple routes to reach any given destination from any given origin. Figure 14.1 is a conceptual view of a packet-switched network.

In Fig. 14.1 numerous packet-switching devices are shown. They are also linked together in multiple ways, providing multiple paths from any given point to any other given point. Three examples of moving packets through the network are shown. First, host 1 sends data to host 2, host 3 sends data to host 4, and host 2 sends data to host 5.

The packet-switching node D is traversed by multiple traffic paths. This is possible because in a practical sense it is merely routing relatively small packets from one point to another.

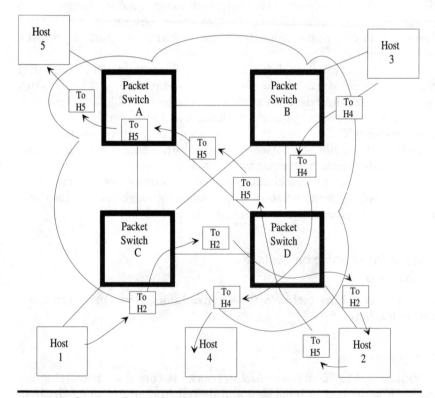

Figure 14.1 Conceptual view of packet switching.

Circuit switch

A *circuit-switch-based* network is best described by analogy. For example, the telephone network depicts this scenario well. For instance, if
two users so desire, they can occupy a point-to-point line as long as they
wish (reasonably, of course). Circuit switching has its own characteristics. If a host utilizes such a service to communicate with another host,
numerous issues are the responsibility of the transmitting and receiving host.

Issues such as error checking, protocols in use, and flow control are
the responsibility of the sending and receiving host. Figure 14.2 is an
example of two hosts using a switched circuit.

Figure 14.2 shows the Gray Corporation and Northern Enterprises,
respectively. This example shows these companies' systems communicating via circuit switching. Here, each customer information communication subsystem (CICS) passes data to its respective transaction
program(s): This scenario shows one modem in each site; however, in
reality multiple modems may exist, thus providing redundancy.
Moreover, multiple sites like Northern Enterprises and Gray may exist
in the circuit-switched network.

Message switch

Message switching is best described as a store-and-forward network.
This type of network environment is used primarily for data. Figure
14.3 is an example of a message-switch system with a corporation having a chain of stores covering a large geographic area.

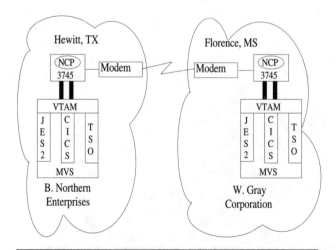

Figure 14.2 Conceptual view of a circuit-switching network.

Figure 14.3 shows four physical locations, each with a dedicated host used in this message-switched network. Notice that each host in Jackson (Miss.), Atlanta (Ga.), Detroit (Mich.), and Orlando (Fla.) has multiple disks attached to the hosts involved in the entire message-switched network.

Each location has satellite drugstores that connect to the host that is part of the backbone network. These typical examples include a bus

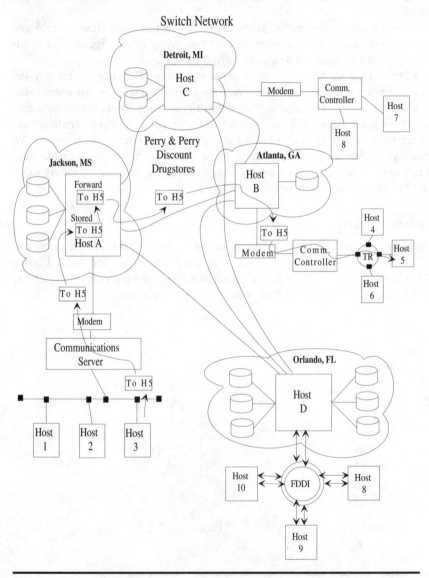

Figure 14.3 Conceptual view of a message-switch network.

network at the drugstore location located in the town of Hoover Lake just outside Jackson, the token-ring network in the Marietta hotel just outside Atlanta, the FDDI network in the backbone network for the Orlando metropolitan area, and two hosts serving the Detroit metropolitan area on the east side. Each network or host connects to a central system in the cities shown. These hosts located in the four major cities are dedicated and constitute the backbone of the message-switch network. However, this does not have to be the case. Examine Fig. 14.3 again; the backbone host could easily be located at the satellite offices.

The network works as shown in Fig. 14.3. Host 2 has a message for data destined for host 5 just outside Atlanta. Notice that when the data leaves host 2 it is stored in host A and is then forwarded to host B and routed to host 5. Notice each host has multiple disk drives connected to them because they serve as the central repository for customer archive files.

The notion of store and forward as in this network means that the message examines the packet for the destination address and then routes the packet onto the appropriate link for the destination. Put simply, in Fig. 14.3 computers (hosts) A, B, C, and D perform the *switching* function. The amount of time elapsed in the store-and-forward-type network for moving a data packet varies relative to a number of variables. The amount of load, size of hosts, and other aspects factor into the lapsed-time equation. In some cases, data packets may be stored on disk and forwarded at a later time.

Each type of network is complex and has its own nuances of how it operates, but efforts here are focused on packet-switching technology.

14.2 X.25 Layer Analysis

X.25 layers do not correlate to the OSI model exactly. Interestingly, the structure X.25 layers predate the OSI model by about 4 to 6 years. These layers and their functions are explained below.

Highlights of X.25 layers

X.25 layers make up the lower three layers in a protocol stack. Different methods of technology are supported at each layer. Figure 14.4 is an example of X.25.

Notice in Fig. 14.4 that multiple interfaces are supported at the physical layer. At layer 2 there are two operating definitions, and at layer 3 the packet layer protocol is supported.

Physical layer

In this layer a host interfaces with media of some sort, either hard or soft. This is where the electrical transmission occurs. Data at this level

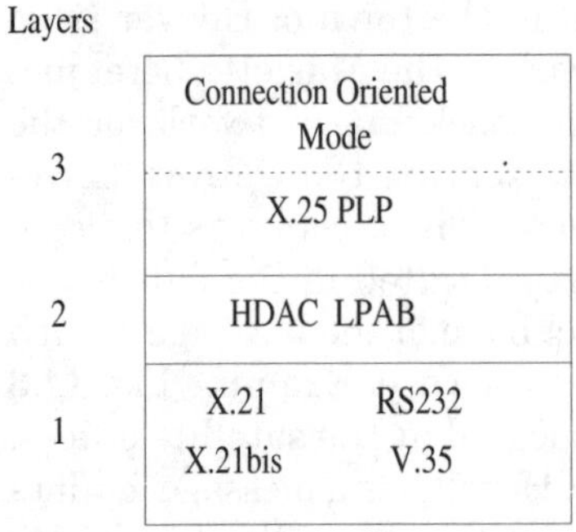

Layers

```
        ┌─────────────────────────────┐
        │    Connection Oriented      │
        │          Mode               │
  3     │ ·········:···········       │
        │        X.25 PLP             │
        ├─────────────────────────────┤
  2     │      HDAC  LPAB             │
        ├─────────────────────────────┤
        │   X.21        RS232         │
  1     │   X.21bis     V.35          │
        └─────────────────────────────┘
```

Figure 14.4   X.25 layers.

is represented by voltages which image binary 1s and 0s. According to Fig. 14.4 four voltages are defined: X.21, X.21bis, RS-232, and V.35.

X.21.   X.21 is a CCITT recommendation used by X.25. X.21 defines the interface for synchronous operation between a DCE and DTE used in public data networks. This recommendation specifies more than just physical-layer interface standards; for example, it defines the numbers of pins on the interface and the action(s) that each pin performs. X.25 can use this physical-layer recommendation.

According to the CCITT 1992 recommendations which I use as a reference, X.21 is defined for a 15-pin connector and supports data transfer rates of up to 9.6 kbits/s, operating in unbalanced mode. In balanced mode the recommendation calls for data rates of up to 64 kbits/s. The major point of this specification is synchronous operations.

X.21bis.   X.21bis is similar to the V.24 recommendation. X.21 defines leased-line operation between a DTE and a packet-switched network. It particularly states the interface parameters for performing synchronous operations with the V series modems.

The V.24 recommendation is a widely implemented standard for modems. The recommendation itself defines data transfer, flow control, and timing with other interface circuits. The importance of the standard is that it defines DTE and DCE interaction and functions.

CCITT intended to phase out X.21bis and replace it with X.21. However, as I have said earlier, we live in a de facto world. X.21bis utilizes the standards set forth for V.24. These standards include data transfer, control, and timing through its circuits.

RS-232 and V.35.   X.25 can also use the RS-232 interface. The EIA RS-232 standard does not specify the physical connector; rather, it specifies the electrical, procedural, and functional characteristics of the interface.

The V.35 specification calls for higher throughput than in RS-232. It can be used for duplex operation over leased lines. The CCITT recommendations detail the meaning of interfaces.

Data-link layer

Basic to the data-link layer is its purpose of transferring frames from a DTE to a DCE error-free across a link. Another function of the data link is flow control and monitoring for errors and/or link failures. X.25 employs two techniques at the data-link layer: high-level data-link control (HDLC) and link-access procedure (LAP). The latter is a subset of HDLC. Both are presented here to show their similarities and differences.

HDLC. *High-level data-link control* (HDLC) is a data-link-layer protocol used by X.25. HDLC supports half- and full-duplex transmission. It also supports switched, nonswitched, point-to-point, and multipoint communication links. A node using HDLC can be categorized into one of the following groups:

- *Primary station:* This category defines nodes in control of the data link.

- *Secondary station:* Nodes in this category require a primary station for operation. In a sense they are dependent on the primary station.

- *Peer station:* This type of station is a *peer* station because it can function as a primary- or secondary-type node.

HDLC is a bit-oriented protocol and is robust because this means that it has a level of transparency; this means that either ASCII or EBCDIC data can be in the data field and it is insignificant for transmission purposes. Figure 14.5 is an example of an HDLC frame.

The beginning flag field in the HDLC frame is set to binary 01111110. The same is true for the ending flag of the HDLC frame. Nodes attached to the data link monitor the data link, examining the flags as they pass. The reason for this is that continuously transmitted frames are needed to keep the link active.

The next field in the frame is the address field. Here are the primary and secondary node addresses. Each node has a unique address and is located here during frame transmission.

Flag	Address	Control	Information	FCS	Flag

Figure 14.5 HDLC frame structure.

The control field is next in the HDLC frame structure. This field contains commands, responses, and sequence numbers to maintain frame data flow between primary and secondary nodes. Three different formats of control frames exist: information, supervisory, and unnumbered. The information field holds user data. The supervisory control frame contains control information such as request for retransmission of frames, acknowledgment of frame receipts, and other control information such as requesting temporary suspension of frame transmission. The unnumbered frame performs functions such as initialization of the link, disconnection of the link, and other link-related functions. The information field exists if the information format is used in the control field for user data. If so, it contains user data. The frame-check sequence performs error checking to determine whether an error has occurred between the sending and receiving nodes. The ending flag is the same as the beginning flag field.

LAPB. *Link-access protocol balanced* (LAPB) is considered a subset of HDLC and is prevalent in X.25 implementations. Other subsets of HDLC exist, but the focus here is on LAPB because of its widespread use with X.25. The basic difference between LAPB and HDLC is that the former considers DTEs and DCEs as peers, meaning that either can initiate communications with the other. LAPB can operate in this manner because it is balanced and hence considers each node capable of initiating the link.

Figure 14.6 shows the frame structure of LAPB. Notice that it is the same as that for HDLC.

The difference between HDLC and LAPB frames is the structure and contents of the control field. In either case the X.25 packet and data from higher layers are carried in the information field of both HDLC and LAPB frames.

Network layer

At the network layer, data from the upper layers is inserted into the X.25 packet. This is achieved because the network layer is actually split into two sublayers: (1) the connection-oriented sublayer, which is closest to the transport layer; and (2) the X.25 packet-layer protocol (PLP) sublayer. At this layer the connection-oriented data is mapped to

Flag	Address	Control	Information	FCS	Flag

Figure 14.6 LAPB frame structure.

Figure 14.7 X.25 packet structure.

the PLP sublayer and then moved via the X.25 structure. A simple view of an X.25 format can be divided into the header and user data portions as shown in Fig. 14.7.

Figure 14.7 is a simple view of a packet structure containing data. Multiple types of packets are used in X.25. For example, different packet types are used for call setup, flow control, restart, data and interrupt, diagnostic, and registration purposes.

Figure 14.8 is an example of a packet used for call setup.

The general format identifier indicates information such as whether modulo 8 or 128 is used, whether user data or control data is contained, and whether the end-to-end acknowledgment bit is set to binary 1 or 0. This is where the Q bit (distinguishing between different types of data) is set or the L bit (distinguishing the addressing scheme).

The logical channel group number and the logical channel number indicate the user's session identification.

The packet type identifier indicates whether the packet is a call setup, reset, data transfer, or other type of packet. If the packet does contain data and more is to follow, then an M bit is set to indicate that more data will follow.

The calling DTE address length, the called DTE address length, and the called DTE address that is being called and is calling indicate the DTE addresses of the respective entities placing and receiving these calls. Also, the first two fields mentioned here indicate the length of these addresses, respectively.

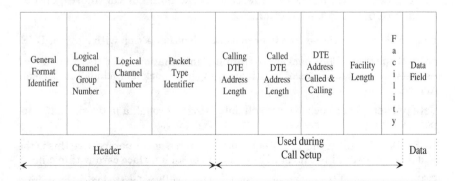

Figure 14.8 Example of an X.25 call setup packet.

The facility length field is used to specify the length of the facility field because multiple facilities can be requested. Certain DTE-specified facilities may be put in this field or facilities offered by a public data network, for example. A few specific examples make this clear. A facility exists for a user to obtain information about the charges related to the call. Another example is call redirection; in this instance a call can be forwarded to another DTE. This is similar to the call-forwarding feature available through some telephone service providers. Regardless of the facility, it is placed in the facility field of the packet.

The data field is user data from upper layers.

After a call is set up the calling DTE address length, the called DTE address length, DTE address called and calling, along with the facility length field and the facility field are no longer needed. The header and the data portion are used.

Generally, approximately eight groups of packets are identifiable:

- Call setup
- Call clearing
- Data
- Diagnostic
- Interrupt
- Flow control
- Reset
- Restart

14.3   Popular X.25 Terminology

As do the other technologies discussed previously and later in this book, X.25 technology has fairly specialized terms that need to be understood by its users for effective work. Some of the more popular terms are listed in this section.

call   A request initiated by a user to establish a connection with a target DTE.

control packet   The packet containing information that is transferred between DTEs. Examples of information in these packets include reset, clear, and call.

data packet   That packet by which data is sent through a network. X.25 supports varying sizes of packets.

DCE   Data circuit-terminating equipment. From a user's perspective this is the end of the network link. It could be a modem or an interface card within a node.

DSE   Data-switching exchange. This term generally refers to logical switching that occurs within a node, most notably a computer.

DTE   Data terminal equipment. Generically, this could be anything that uses an X.25 interface. Typically, it is a terminal or a host.

interrupt packet   These packets are small and have a low priority for transmission through the network. They carry information required for an interrupt.

LCN   Logical channel number. This refers to a specific DTE connection within an X.25 network.

link level   Reference to line control or the link protocol that transfers frames from one network to another.

logical channel   Reference to a virtual circuit that is mapped to a physical link.

packet   That unit which carries data through the network between source to destination DTEs.

packet level   Multiplexing of the physical channel occurs here thus creating multiple logical channels.

PAD   Packet assembler/disassembler. This is hardware and/or software that performs the function of formatting incoming data into X.25 packets. In other words, it converts X.25 packets into other communication protocols.

permanent virtual circuit   A relationship between a logical channel on one port and another on a different port in another part of a network.

physical level   The lowest level referred to within a network. It is the level that provides the electrical interface into a network. (However, an interface can support optical fiber, thus producing an optical signal.) This is the layer where interface standards such as RS-232, V.35, and others operate.

qualified data packet   This type of packet has the Q bit set to the on position. This lets the DTEs know that a different type of data is in the packet and can be used any way the DTEs require.

signaling terminal exchange   This is an interface that supports DTEs between different networks using the X.25 CCITT recommendation.

switched virtual circuit   A circuit set up by a calling user that sends a special type of packet called a *call packet.*

14.4   X.25 Concepts

X.25 networking utilizes numerous concepts. Some of these are explained in this section because of their prevalence and common use in the networking arena.

Channels and circuits

These two terms are often confusing, but a simple example demystifies this semantic difference. Consider Fig. 14.9.

Figure 14.9 shows two hosts connected to the X.25 network. Notice that the logical channels are identified with each host and the modem

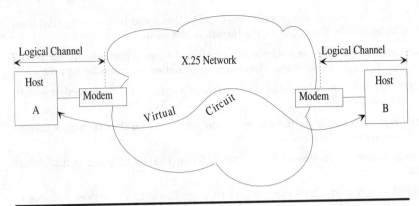

Figure 14.9   Conceptual view of channels and circuits.

in use. This means that logical channels refer to this local connection. Notice the line indicating a connection from one host to the other. This refers to the virtual channel. The channel is virtual because it may traverse a number of different hosts within the network before reaching its destination; therefore, these connections within the X.25 "cloud" are considered virtual.

PAD types and functions

A *packet assembler / disassembler* (PAD) can be defined as either hardware or software that performs the function of putting non-X.25 data into X.25 packets. Three different PADs are presented here: X.3, X.28, and X.29.

X.3.   The CCITT X.3 recommendation calls for characters to be put into packets to pass through the X.25 network. It also calls for the virtual call setup, the clear function, and the reset functions, as well as other required actions. In short, this recommendation is responsible for the appropriate parameters in call setup between a non-X.25 host and a target DTE.

X.28.   This CCITT recommendation specifies the data flow between the non-X.25 device and the X.3 PAD. The X.28 parameters specify use of X or V series specifications with modems. The basic purpose of this PAD is to transmit the correct call request packet.

X.29.   The CCITT recommendation for X.29 specifies control messages used during the communication between the calling DTE and the called DTE. It specifically explains support for information exchange during any phase of the connection.

Figure 14.10 best represents a conceptual view of these three PADs.

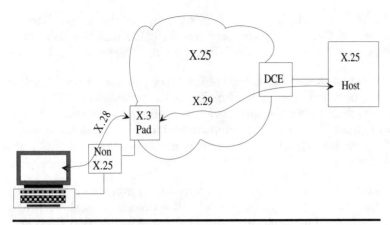

Figure 14.10   Conceptual view of X.3, X.28, and X.29 pads.

X.25 and a System Network Architecture environment

IBM's System Network Architecture (SNA) environment supports connectivity with X.25 networks. This functionality is achieved by a program called *network packet-switched interface* (NPSI). It runs on what is termed a *front-end processor* (FEP) or a *communication controller*. This program permits bidirectional connections between SNA and an X.25 environment. Figure 14.11 illustrates this.

Figure 14.11 shows multiple X.25 networks communicating and a connection into the SNA network via NPSI on the FEP.

X.25 interfaces

Communication within X.25 networks occurs via *logical channels,* in X.25 lingo. As defined previously, this is a reference to a virtual circuit

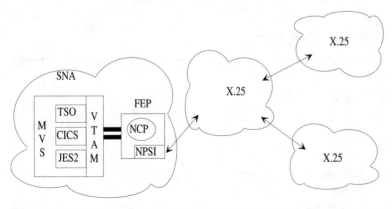

Figure 14.11   Conceptual view of NPSI.

mapped to a physical link. From a user's viewpoint, this is a dedicated physical circuit. Communication between the caller and the called is a function performed through one of the following *interfaces*.

Permanent virtual circuit.   This type of interface provides a service similar to what is called a *dedicated line* in the telephone arena. This scenario means that a user or users have been defined to the network and the logical channel number is predetermined and is part of the header of the X.25 packet. This type of circuit is always considered "up" and available for use.

Virtual circuit.   A *virtual circuit* is the opposite of a permanent virtual circuit. The virtual circuit requires that call setup and bringdown functions are performed for each call. Logical channel numbers (LCNs) are thereby distributed dynamically. Using the telephone analogy, this is similar to placing a telephone call; that is, the telephone does not go into the off-hook position until a user is ready to enter the number of the target party. Likewise, when the conversation is over, the telephone receiver is placed in the on-hook position.

Fast select.   *Fast select* permits a call request packet to contain user data up to 128 bytes. In turn, the called DTE can respond to the calling DTE and can also carry user data in the packet. If the call is accepted and transmitted, then continuation of the call takes on the characteristics of a switched virtual call.

Fast select with an immediate-clear function.   This type of call is similar to the fast select, in that it contains user data. In short, a packet is sent to a target point, and once the return packet is received, the virtual call is brought down. This is in contrast to the fast select, which is similar to the virtual call.

14.5   Additional Information

The following list represents CCITT recommendations up through early 1994:

X.21	X.25-6
X.21bis	X.25-A
X.25-1	X.25-D1
X.25-2	X.25.D2
X.25-3	X.25-D3
X.25-4	X.25-I
X.25-5	

14.6 Summary

X.25 is a switching technology that has been proved over a period of at least two decades. Three types of switching technologies were described, and X.25 was described as a packet switching technology. The CCITT has been a driving force behind X.25 recommendations, and it specifies all aspects of its operation.

X.25 can be examined by layers; however, it does not fit exactly to the OSI model. For example, at the physical layer four recommendations are specified at the time of this writing: the X.21, X.21bis, RS-232, and V.35. The data-link layer specifies HDLC and LAPB. The network layer is divided into sublayers with the bottom sublayer—packet layer protocol (PLP)—and immediately above it is a connection-oriented sublayer.

A section of X.25-specific terms for the user who may not be familiar with some of the terminology is followed by a section on X.25 concepts that includes an explanation of channels and circuits, and PAD types and functions along with examples, and an example of how X.25 fits into a popular network such as systems network architecture. X.25 interfaces were also explained; they included a permanent virtual circuit, virtual circuit, fast select, and a fast select with an immediate clear function.

Upper-Layer Protocols

Upper-Layer
Protocols

15

Upper-Layer Protocols

A Practical Perspective

This chapter presents a brief technical overview of the networking environments covered later in this part of the book. The order in which these protocols are presented was chosen by random selection. No significance is placed on the order of presentation.

15.1 Overview

Networks use protocols at all layers. For example, a protocol is used to govern the data transfer over a transmission medium regardless of whether that medium is a coaxial cable or microwave. Another example of protocol usage in networks is the protocols that applications use to fit into the system on which they operate, or in some cases the *remote* systems in which they operate. So, regardless of the level of operation, protocols govern network operations. Without them networks would not be possible, maybe—but would certainly be less orderly.

Networks are generally the offerings of corporations. Some of these corporations are larger than others; some networks have their roots in government agencies, and others in forums or organizations or the like. Some protocols are *de facto*; this means in essence that something was created, done, proven to work, and then a "standard" of this protocol was documented, explaining how the protocol works and what must be done to work with it. The implication is that, to operate with such a protocol whatever is done must follow the specifications recorded about the protocol.

In another type of protocol, known as *de jure*—which is the opposite of de facto—*the plan* is created, and then the protocol (something) is brought forth. Typically, the individuals who work with both de jure and de facto protocols have different philosophical approaches to the technology. These terms, *de facto* and *de jure,* are best understood when used in reference to developers of both software and hardware. On the basis of my own observation and experience, these mind-sets clearly exist in development circles.

15.2   Upper-Layer Protocols at a Glance

Most networks can be described in terms of their lower- and upper-layer protocols. Lower-layer protocols have been discussed; here the focus is on upper-layer protocols. Basic insights are presented here to provide a foundation for deeper discussion about these protocols, which are discussed in detail in Chaps. 16 to 22. Even though network protocols can be categorically divided into layers, sometimes a clear delineation of function versus layer can be skewed.

Many are familiar with the OSI reference model. It can be used as a yardstick, so to speak, to discuss other network protocols. When a contrast or comparison of different protocols is evaluated in light of the OSI reference model, it is sometimes clear that not all network protocols fit all layers defined by the OSI model. A good example of this is the transmission control protocol/internet protocol (TCP/IP) suite of protocols.

The remainder of Sec. 15.2 will concentrate on individual protocols.

Systems Network Architecture

Systems Network Architecture (SNA, as it is typically called) is IBM's networking solution for medium to large networks. Its origins go back to 1974, when it was introduced, and it has been through many iterations since that time. It began as a layered network architecture. And until 1992 this layered architecture was IBM's primary networking solution.

In spring of 1992, IBM announced what they call their *networking blueprint.* The meaning of this announcement was a radical break from the past with regard to IBM's approach to networking solutions. It not only embraced SNA, but it incorporated other protocols as well and will be examined in detail in Chap. 16.

SNA is based on terms, concepts, hardware, and software architecture. Many terms used in Chap. 16 were coined and defined by IBM. Until 1993, IBM produced its own dictionary of terms defining them and explaining how they fit into the SNA environment. SNA is built on concepts as well. These concepts are, for the most part, abstract. However, some SNA concepts are used in actual software and hardware definitions to make the network functional. IBM's hardware and software are the tangible components on which SNA is built.

To learn SNA is a challenge. To focus on learning SNA concepts without understanding the terms is difficult at best. On the other hand, to memorize terms and not attempt to understand the abstract concepts is little more than memorization. The best way to learn SNA is from a topical perspective. In time as one learns topics (consisting of terms and concepts), a cumulative effect takes place and the whole picture begins to come into focus.

Advanced peer-to-peer networking architecture

Advanced peer-to-peer networking (APPN) is another network solution offered by IBM. It originated around 1983. Since that time APPN has undergone approximately two versions. As of this writing APPN is at Version 2, according to the APPN architecture manual.

APPN is peer-oriented, whereas traditional SNA is hierarchical in nature. Simply put, APPN is designed for interprogram communication without an intervening entity to aid in this communication—as would be the case in traditional SNA.

APPN is a proprietary networking technology. IBM has not made it so that other vendors can implement it in its fullness according to their architectural manuals.

Another characteristic of APPN is that some of its functions have been incorporated into SNA. Some of SNA's core components, namely, the virtual telecommunications access method (VTAM), began official support of some APPN functions with VTAM Version 4 Release 1. The indication of VTAM Version 4 Release 2 currently, according to general knowledge, seems to point more to an integration between SNA functionality and APPN.

This author tries not to predict future events that may or may not occur with respect to any companies, so it would be incorrect to conclude from this text that IBM is planning to merge SNA and APPN. They are based on philosophical and functional differences, and they have been positioned to meet different needs. Simply because IBM has incorporated some APPN functions into SNA logically means that. Other conclusions are speculations.

Open-systems interconnection

Open-systems interconnection (OSI) is the culmination of the International Standards Organization (ISO). Around 1977, the ISO chartered a committee to create a standard for networking. The committee did, and the standard became known as the *OSI reference model*.

This OSI model stated that at a minimum the seven layers identified should exist in a network. The committee defined what occurs at each layer, but did not define the layer that is the medium. Part of the rationale for this was that different network protocols are capable of using different media types.

The ISO has come to embrace CCITT and other specifications that it works within its networking model as explained in Chap. 18. In a real sense the OSI model is an example of a de jure standard. Some may disagree with this, but generally this is true.

OSI products are implemented by vendors, because the ISO is an organization, not a company in the sense of production (at least at the time of this writing). Consequently, some companies are more involved than others in developing OSI-compliant offerings.

Some parts of adopted OSI protocols are used more than others. For example, the electronic-mail (E-mail) and directory service protocols are more widespread than some other ISO OSI specifications.

Transmission control protocol/internet protocol

Transmission control protocol/internet protocol (TCP/IP, as it is better known in the technical community) originated in 1975. It was then, and a little before, that TCP/IP's birth began. It was the follow-on to the ARPANET, which was government-related [supported by U.S. Defense Advanced Research Projects Agency (DARPA)]. In 1978 a public demonstration of TCP/IP was given, and in 1983 the U.S. government made a statement to the effect that if connection was to be made to what had become known as the *Internet*, it must be done with the TCP/IP protocol suite.

TCP/IP has been refined and been through many changes since 1975. It is unlike IBM's SNA in the sense that no particular corporation or entity charts its course and maintains proprietary aspects of it. TCP/IP is a prime example of a de facto network protocol.

Even though TCP/IP has been funded by the U.S. government at different times, its basic direction has come from citizens. TCP/IP is a protocol suite, a collection of protocols. The protocols have been the contributions and enhancements of individuals and corporations alike. Granted that an overseeing body does provide guidance and maintain order, the protocols themselves that make TCP/IP what it is are not *owned* by any one body. This is commonly considered public domain.

Similar to OSI, TCP/IP is implemented by corporations who chose to make them available to prospective customers. If no other fuel was added to the growth of TCP/IP, the mandate made by the U.S. government in 1983 was sufficient motivation to get many vendors interested to the point of making that protocol available to government agencies and other ancillary agencies.

TCP/IP is considered a *client / server protocol*. This term comes from the fact that two of the most popular applications in the TCP/IP protocol suite are client/server-based. Clients initiate something; servers serve the request of clients. In a sense, this concept is similar to the peer concept in APPN.

TCP/IP is unique in one sense; it has a windowing system as a part of the protocol specification. Another interesting characteristic of TCP/IP, and the windowing system, is their hardware and software independence. This means that TCP/IP can operate on practically any vendor equipment; although exceptions exist, for the most part this is the case. Another aspect of TCP/IP is that it does not have to be implemented in a full protocol suite; pieces of it can be implemented.

NetWare

NetWare protocol is the product that originated with the Novell Corporation. It began offering print and file services and has evolved into a full network protocol that is now at Version 4. NetWare has proliferated throughout the market for multiple reasons. First, it operates on PCs, and the Novell Corporation has kept upgrading the product to be as robust as possible.

Second, Novell has ported NetWare to some UNIX operating platforms. This is significant because of the prevalence of UNIX throughout the marketplace today. A third reason for its popularity is its user-friendliness. It operates as a peer environment which makes some aspects of the installation easier to implement.

Another aspect of NetWare that keeps it in the circle of popularity is the protocol support it provides. For example, it supports FDDI, token ring, and ETHERNET as data-link protocols. It also supports connectivity into IBM's AS/400 series. Additionally, Novell supports NetWare TCP/IP on a NetWare file server with services such as IP routing, tunneling, and simple network management protocol (SNMP).

Beyond this, NetWare supports IBM's RISC/6000 AIX (Advanced Interactive eXecutive) operating system as well as OS/2. IBM included the NetWare functionality in the blueprint introduced in spring 1992. Even in NetWare's early years it was broad in scope. NetWare Version 2.01 had support for DEC's Virtual Address eXtended (VAX) operating system.

NetWare is different from the other protocols mentioned because its original design intent was different. However, today it has evolved into a competitive upper-layer network protocol. It has grown in breadth to support large networks and a diverse mix of operating systems.

Digital Network Architecture

DEC is completing the transition from the Digital Network Architecture (DNA) Phase IV method of networking to what their documentation calls, "DECnet/OSI for OpenVMS." This is DEC's implementation of OpenVMS, which includes

- Support for OSI communication specifications
- Digital Network Architecture (DNA) Phase V that is backward-compatible with Phase IV
- Standards provided by the CCITT
- Standards provided by the IEEE

According to DEC documentation, this phase of their network offering supports more systems than do previous versions of DNA, provides distributed management, and maintains a multivendor network support environment.

DEC's DNA network architecture is not new. According to DEC documentation (DECnet/OSI for OpenVMS *Introduction and Planning*, part no. AA-PNHTB-TE), the first phase of DNA was introduced in 1976. Since then DEC has revised it and made enhancements to bring it to Phase V. Some call DNA Phase V, DECnet/OSI. But whatever name may be attached, it is clear that the structure of DEC's network offering supports multiple protocols and is open to standards not available previously.

From a practical standpoint, DEC has managed to provide support for file transfer and access management (FTAM) commitment, concurrency, and recovery (CCR), virtual terminal support, and OSI application support through the VAX OSI application kernel (OSAK). Multiple transport-layer protocols are supported, along with OSI network-layer addressing support. At the data-link layer, DEC's DECnet/OSI for OpenVMS supports ETHERNET, HDLC, FDDI, and X.25 support, to name only a few. At the physical layer, the appropriate drivers are available to support a variety of interfaces. From a network management standpoint, DECnet/OSI for OpenVMS supports what DEC calls *enterprise management architecture* (EMA).

The EMA defines a way to manage heterogeneous networking environments with distributed computing. This approach offers wide support for large enterprise environments.

Another feature of DECnet/OSI for OpenVMS is its scalar capabilities. It can meet the needs of small-base operations, or it can accommodate large mission-critical data centers that require large amounts of processing power and versatility.

AppleTalk

AppleTalk is a protocol used to connect Macintosh computers together to make a network. AppleTalk is a proprietary protocol. The Apple Corporation itself had a line of Apple computers before the Macintosh line was introduced. In approximately 1984, the Macintosh was introduced. In many ways the rest is history—if you are familiar with Macintosh computers.

AppleTalk has had two phases, or versions, Versions 1 and 2. Prior to the Macintosh, Apple did not have a networking solution for their computers. However, this is not a negative statement; it is merely factual. In the late 1970s basically three "personal computers" existed; they included: the Apple, Tandy Corporation's TRS-80, and Commodore's Computer. As history revealed, in 1981 that changed when IBM introduced their first personal computer. So, the point is at that time what are now considered personal computers were not prevalent in the world of networking, as Ted Taylor explained.

Apple introduced Phase I of AppleTalk after the introduction of the Macintosh. Apple's solution (or idea) to networking was simply to meet the needs of a local area where numerous Macintoshes might be located, as in a documentation department or such. This phase of AppleTalk supported defining a node by using 8 bits to do so. This alone meant that 254 nodes could be used; and remember that at that time, in a personal computing environment that was a lot. Only 254 rather than 256 nodes were supported because AppleTalk used 2 bits in a reserved manner.

Phase II of AppleTalk is the latest at the time of this writing. The primary difference between it and Phase I is that Phase II supports more than 254 nodes because of the use of a network number that employs 16 bits; hence, a given node is identified by the network node number *and* the 8-bit node assignment. Another feature of Phase II is the multiple zone support. A zone according to Apple documentation is the logical grouping of a subset of nodes on an internet (note that this refers to a local internet, not the "big I" Internet).

One of the greatest strengths behind AppleTalk is its simplicity from a user's perspective. Another strength behind the protocol is the underlying architecture and how the software and hardware have been designed with AppleTalk. It is a peer-oriented protocol, and it does support multiple connections. It can be integrated into other networks with some effort but usually less than the effort needed for other network protocols.

Apple computers are interesting in terms of their original design intent. Because AppleTalk and Apple computers, including Macintoshes, are proprietary, a certain tendency seemed to have developed. On the basis of observation, it seems that Apple computers have evolved over the past decade to fit into a stratum of organizations. And, as they evolve it seems that they are becoming more integrated in a variety of organizations.

A simple point sums up the thought concerning Macintoshes (which use AppleTalk) and other vendor personal computers that use network protocols; in the early years of Tandy's, Apple's, Commodore's, and even IBM's personal computers, the concept of networking them in the sense that they are being networked today was not prevalent. Times have changed, and so have personal computers from all vendors.

15.3 Summary

The intent of this chapter was to present a brief overview of the seven upper-layer protocols. Each of these network protocols has strengths and weaknesses. Evaluating them should be based on their original design intent. Any of the protocols discussed may be the "best" solution for any given situation. Networking requirements in a given situation should determine the selection of a network protocol. If the requirements are understood and the network protocols evaluated are understood, then a decision can be reached on the basis of facts.

Chapters 16 to 22 explore these seven protocols in detail. The material presented is based on technical facts from the sources of these protocols and/or credible sources on the protocol being discussed.

16

Systems Network Architecture (SNA)

Systems Network Architecture (SNA) is a proprietary IBM protocol, and SNA networks are built from hardware and software components. SNA networks vary in size. They can be small, but the nature of SNA is robust enough to accommodate the world's networking needs.

This chapter focuses on IBM hardware, software, SNA terms, concepts, and protocols. Each section can be used individually as a reference, or the chapter can be read from beginning to end, and the reader will have a thorough understanding of SNA.

16.1 Hardware Architecture

An evolutionary perspective

Much development occurred at IBM during the 1940s and 1950s. History has shown that these two decades led to the creation of what became known as the biggest gamble in the history of the IBM corporation, up to that time: the System/360 (S/360) architecture. Before exploring the S/360 architecture, a brief view of hardware offerings preceding the S/360 is in order.

During the 1940s, 1950s, and even into the 1960s IBM had approximately six popular mainframe computers, each having its own strength. Other systems existed, but six were fundamentally solid-state devices. A fundamental problem with these systems was that little was interchangeable among them.

This meant that their programming, support, sales, and technical sales advisors concentrated only on their areas of expertise. This usually meant no overlap from one type of system to another. This was costly and increasingly a problem for IBM. Not only was this scenario a problem for IBM; their customers had to contend with this if they had more than one type of machine to meet the needs throughout a given corporation. Some examples of these machines and the forte of each include

- *The 604:* This was an electronic calculating punched-card machine. It was first available in approximately 1948. The strength of this machine and its major selling points were speed, a pluggable circuitry, and its concentrated components in such a small physical location. Certainly from the users' point of view at the time the major forte of the 604 was its speed.

- *The 650:* This machine became available in approximately 1954; however, it was announced in 1953. This machine's strength was twofold: (1) it was a magnetic drum storage machine and (2) its strength was focused on meeting the needs of general computing. This machine was extremely successful after its introduction into the marketplace.

- *The 701:* This machine was announced in 1952. Its strengths were multiple. Its input/output was fast. It, too, focused on increased processing speed over some of its predecessors. The 701 system had its power in scientific and related areas of computation. It was not a general-purpose machine such as the 650, for example.

- *The 702:* This system, announced in 1953, made its debut in approximately 1955. Interestingly, this system originated in the late 1940s. This large system focused on the ease of character handling.

- *The 1401:* This system was announced in 1959. It soon grasped a large market share after shipping began in 1960. It had increased speed, a fast printer, and other peripherals such as tape and card processing capabilities that made it a popular system. Its major strengths were versatility and cost-effectiveness; it had multicomponent selective capability with respect to the pricing scale at that time.

These systems, and others IBM offered during the 1940s through the early 1960s, proved IBM's ability to meet a diversity of needs. However, such diversity in systems led to complexity in terms of one corporation attempting to maintain this posture in a technical environment. IBM was not unaware of these matters, both the positive and negative contributions the diverse systems offered.

After years of planning, designing, and reengineering 1964 proved to be a pivotal point in the history of the IBM corporation. It was 1964 when the System/360 was introduced. This system was unique for multiple reasons, but at the heart of the S/360 was now one architecture capable of accommodating what previously was achieved by different systems. In addition, the S/360 architecture had different models so that customers could begin with a purchase to meet immediate needs while having an architecture that provided a migration path to accommodate consumers' growing needs.

Another appeal behind the S/360 architecture was IBM's commitment to users for upgradability to later architectures as they were introduced. At this time, IBM is four architectures removed from the S/360, and on the basis of a personal account, a program originally written for the S/360 has been executed successfully on the S/390 architecture.

Technical highlights of each of IBM's hardware architectures are presented below.

S/360

The S/360 (System/360) was successful, to say the least. Some of the characteristics and functions of this hardware architecture are listed below. The S/360 hardware architecture is presented by component with a synopsis of the characteristics and functions of each component.

S/360 components included

1. Central processing unit (CPU)
2. Channels
3. Control unit(s)
4. Devices such as terminals, printers, tape and disk drives, as well as card punch and readers
5. Main storage

CPU characteristics and functions

- Single-processor architecture, but models introduced later supported multiple processors.
- Five classes of interrupts.
- Interrupt priority.
- 16 general-purpose 32-bit registers.
- Four optional 64-bit floating-point registers.
- Dynamic address translation on selected models.
- Supervisor facilities including a timer, direct control capabilities, storage protection, and support for multisystem operation.
- ASCII and EBCDIC character set support.
- 24-bit addressing.
- Channel-to-channel adapters used for interconnection of multiple processors.

Channel characteristics and functions

- Provides a data path to and from control units and devices.

- Uses a protocol for data transfer.

- Selector channels used with tape and disk devices for high-speed data transfer. Uses one subchannel.

- Byte multiplexer channels interleave I/O operations. Slower operating devices are used with this type of channel. It can logically support up to 256 subchannels.

Control unit(s).   These serve as an *interface between* devices such as a terminal, card reader, card punch, and printer.

Devices.   These are terminals, printers, card punches, and card readers, which serve as I/O devices. Terminals are used interactively, whereas punched cards are used in batch processing.

Main storage.   Main storage in the early models emphasized speed and size. Virtual storage was available in later models.

General notes on S/360.   Figure 16.1 is a logical view of the S/360. Figure 16.2 is a physical perspective of the S/360.

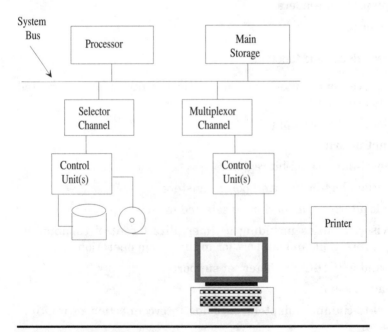

Figure 16.1   S/360 logical perspective.

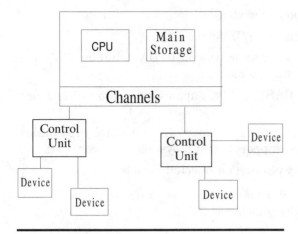

Figure 16.2   S/360 physical view.

In later years of the S/360 some features and functionality built into some S/360 models were carried over to the next hardware architectural generation, known as the S/370.

S/370

The System/370 architecture, announced in 1970, was the successor to the S/360. During the next six years IBM refined more than 14 models based on this architecture. Examination here includes features and functions based on the S/370 architecture as a whole, not from inception with all incremental enhancements. Components and their characteristics and functions are listed below.

S/370 components included

1. Central processing units (CPUs)
2. Channels
3. Control unit(s)
4. Devices
5. Storage: main and virtual

CPU characteristics and functions

- User program upgrade support from S/360
- Multiple processor support
- Six classes of interrupts
- Interrupt priority
- 16 general-purpose 32-bit registers

- Four 64-bit floating-point registers
- Dynamic address translation (DAT) facility
- Extended real addressing, an extension to DAT, making 64-Mbyte addressability of real storage possible
- Dual-address space (DAS) facility, supporting semiprivileged programs
- Supervisor facilities including a timer, direct control capabilities, storage protection, and support for multisystem operation
- Optional vector facility offering on selected models
- EBCDIC character set support
- Removal of ASCII support as in the S/360
- 24-bit addressing
- Two page sizes: 2 and 4 kbytes
- Two segment sizes: 64 kbytes and 1 Mbyte
- Multiprocessing with one operating system
- Approximately 50 new instructions over the S/360
- Translation look-aside buffer minimizing DAT use

Channel characteristics and functions

- Three types of channels supported: selector, byte multiplexer, and block multiplexer
- Support for 2-byte channel-bus-width extension
- An associated set of subchannels for each channel
- Uses channel protocol for data transfer
- Data transfer rates of 1.5 and 3 Mbytes achievable, depending on the channel
- Suspend and resume facility for programmed control of channel program execution
- Removal of 16-byte channel prefetching from S/360 channels

Control unit(s). These serve as an *interface between* devices such as a terminal, card reader, card punch, and printer.

Devices. These are terminals, printers, card punches, and card readers, all of which serve as I/O devices.

Terminals are used interactively, whereas punched cards are used in batch processing.

Storage

- Main: addressability of up to 64 Mbytes
- Virtual: addressability of up to 16 Mbytes beyond that of main storage

Figure 16.3 is a representation of the S/370 architecture.

370/XA

The 370/eXtended Architecture was announced in 1981. It followed the S/370 and what IBM called "S/370-compatible realized in the 4300 series of systems." From 1981 until 1988 370/XA was IBM's hardware architecture implemented in their mainframes. The focus here follows the same scheme as the prior two architectures, that is, features and functions of 370/XA as a whole.

370/XA components included

1. Central processing units (CPUs)

2. Channel subsystem

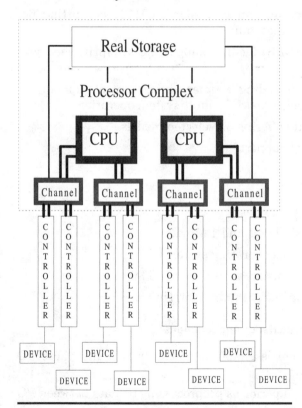

Figure 16.3 A conceptual view of S/370.

3. Control unit(s)

4. Devices

5. Storage (addressing types): absolute, real, and virtual

CPU characteristics and functions

- Two addressing modes of operation: 24- and 31-bit
- 2-Gbyte addressability with 31-bit addressing mode
- 13 new instructions
- Multiple processor support
- Six classes of interrupts
- Interrupt priority
- 16 general-purpose 32-bit registers
- Four 64-bit floating-point registers
- Dynamic address translation (DAT) facility
- Extended real addressing, an extension to DAT, permitting 64-Mbyte real storage addressability
- Dual-address space (DAS) facility, supporting semiprivileged programs
- Supervisor facilities including a timer, direct control capabilities, storage protection, and support for multisystem operation
- Optional vector facility offering on selected models
- EBCDIC character set support
- Dynamic I/O reconnect
- 24-bit addressing
- Two page sizes: 2 and 4 kbytes
- Two segment sizes: 64 kbytes and 1 Mbyte
- Multiprocessing with one operating system
- Approximately 50 new instructions over the S/360
- Translation look-aside buffer minimizing DAT use

Channel subsystem characteristics and functions

- The channel subsystem is simply a processor that interfaces I/O devices and processors.
- The channel subsystem performs preprocessing on data between I/O devices and processors.

- A *channel path* in 370/XA refers to the physical path between the channel subsystem and a device.
- Subchannel numbers have a one-to-one relationship with an I/O device.
- Path-independent addressing for I/O devices.
- The implementation of paths enables dynamic data routing from I/O device to processor.
- A channel path identifier (CHPID) is associated with devices such as control units.
- Path management is performed by the channel subsystem.
- Increased channel command word (CCW) support for direct use of 31-bit addressing in channel programs.
- 13 I/O instructions added.
- Two types of channels supported: byte multiplexer and block multiplexer.
- Subchannels are not owned by the channel as in S/370.
- Uses channel protocol for data transfer.
- Data transfer rates of 1.5 and 3 Mbytes can be achieved, depending on the channel.
- Suspend and resume facility for programmed control of channel program execution.
- Removal of 16-byte channel prefetching.

Figure 16.4 is a view of a channel subsystem.

Control unit(s). These serve as an *interface between* devices such as a terminal, card reader, card punch, and printer.

Devices. These are terminals, printers, card punches, and card readers, all of which serve as I/O devices. Terminals are used interactively, whereas punched cards are used in batch processing.

Storage (by addressing scheme)

- *Absolute:* This is an address to main storage. In essence, the term is synonymous with *main storage*. This type of address is storage where no transformations are performed on the contents. Two gigabytes of absolute storage is possible.
- *Real:* This also refers to an address in main storage. This term is used when multiple processors are accessing the same main storage.

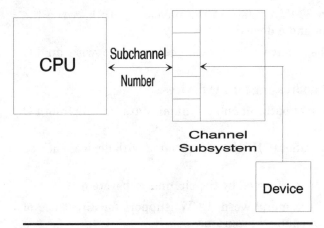

Figure 16.4 370/XA channel subsystem example.

A CPU prefix number distinguishes processors, so storage order is maintained.

- *Virtual:* This address reflects an abstract location. Virtual storage does not exist in reality; it is a concept, achieved by main storage, secondary storage, and processor speed as the fundamental components that make virtual storage possible.

Figure 16.5 is a representation of 370/XA architecture.

Other functionalities were added to 370/XA until the announcement of Enterprise System Architecture (ESA); however, the highlights listed above constitute the bulk of additions to the XA architecture.

ESA/370

370/Enterprise System Architecture was announced in 1988 and succeeded 370/XA. This architecture built upon advances made in 370/XA and S/370. Consequently, different hardware-related functions were implemented. However, from a system standpoint considerable advances were made, including an operating system, as will be shown in a later section. Highlights of ESA/370 include

1. Addressing
2. Storage
3. Machine-dependent support

Addressing. Key enhancements in ESA/370 include 16 new access registers. These registers provide the hardware capability for a program

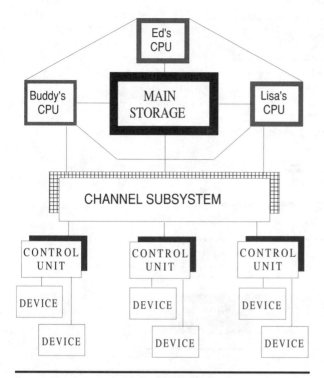

Figure 16.5 A conceptual view of 370/XA architecture.

to address up to 16 spaces. This is the *home address space,* which is the name of a translation mode that permits the control program to gain control quickly where principal control blocks are maintained. A private space is supported to prevent use of the translation look-aside buffer for common segments. This private space support enhances security functions.

Storage. In ESA/370 a major improvement was in the system's storage-handling capacity. The storage management subsystem is designed to stage data. An I/O boundary exists, and multiple storage locations for data are possible. With a storage subsystem, the framework is in place for a particular processor complex to capitalize on this feature. Figure 16.6 depicts the storage hierarchy.

In addition, working with the storage hierarchy is the *system control element* (SCE), which routes data through the CPU's main and expanded storage and the channels. The SCE keeps track of changes made to data and is the key component for moving data throughout the hierarchy. Figure 16.7 is a logical view of the SCE.

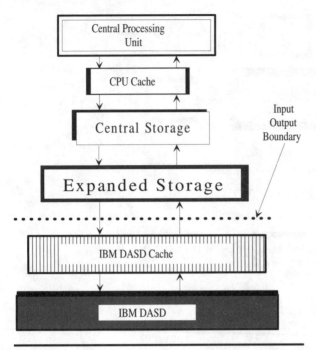

Figure 16.6 IBM's storage hierarchy.

Machine-dependent support. A feature known as the *logical partitioned mode* (commonly called LPAR) is a machine-dependent function, but is tied to hardware and a software component known as the *processor resource/system manager* (PR/SM), pronounced *prism*. To function in LPAR mode, a machine must be supported by that system and selected on power-up time.

LPAR permits a system to run multiple logical partitions, each running an operating system simultaneously and all independent of one another. For example, Fig. 16.8 depicts a logical view of a LPAR-configured system.

The logical partitioning of a processor includes the processor's resources such as storage, channels, and the processor itself. Isolation of the LPARs is enforced via hardware. PR/SM itself is an IBM-supplied option that some machines can take advantage of to offer LPAR. PR/SM is implemented in microcode. *Microcode* is an IBM term that is synonymous with firmware.

ESA/370 architecture differs from 370/XA because of the storage, addressing, and machine-dependent enhancements offered. Physically little differs; however, Fig. 16.9 depicts ESA/370 with two processors.

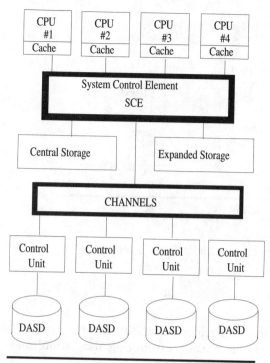

Figure 16.7   A logical view of the SCE.

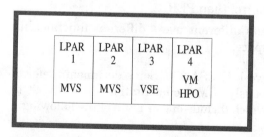

Figure 16.8   A logical view of LPAR.

S/390

S/390, also known as *Enterprise System/390,* was announced in September 1990. This announcement was broad in scope, encompassing a new processor line, channel subsystem, many software announcements, and wide-sweeping networking-related support. A brief view of the hardware highlights includes

1. S/390 enhancements

2. ES/9000 processors

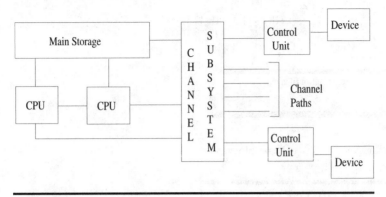

Figure 16.9   A conceptual view of ESA/370.

3. Enterprise System Connection (ESCON) architecture

4. Networking support

S/390 enhancements

- *Storage override protection:*   Provides reliability of executing programs by keeping a different application from executing simultaneously within the same address space.

- *Program event recording (PER) facility Version 2:*   Provides a more focused method of event control than PER 1.

- *Access list control:*   Permits different users different functionality within the same address space.

S/390 builds on the framework of ESA/370. Some documents refer to S/390 and ESA/390 simultaneously. IBM documentation uses ESA/390 to refer to those environments which include one or more of the following:

- ESCON

- Common cryptographic architecture

- An environment providing data spaces for VM

ES/9000 processors.   Initially 18 processors were announced. These processors included both air- and water-cooled models. A view of an ES/9000 processor physically divided into two logical independent complexes is shown in Fig. 16.10.

ESCON.   IBM's greatest contribution to the S/390 announcement was the new fiber channel subsystem called *Enterprise System Connection*

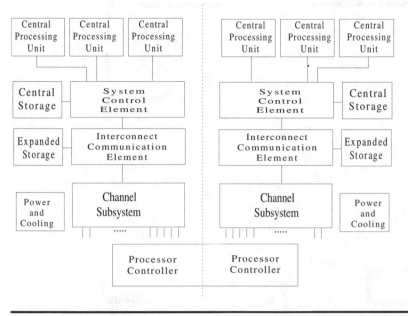

Figure 16.10   A conceptual view of an ES/9000 partitioned.

(ESCON). Figure 16.11 depicts an example of a possible ESCON configuration.

In Fig. 16.11 two ESCON directors are shown. An ESCON director is the focal point for dynamic connectivity used in IBM's fiber channel environment. The director shown in Fig. 16.11 is attached to two processors and control units with attached devices connected back to the director.

Some fundamental highlights of ESCON include

- Switched point-to-point connections

- Point-to-point connections

- ESCON cable distance support 23 km

- ESCON cable distance support with daisy-chained directors 43 km

- Data transfer rates of up to 17 Mbytes/s

- 1024 devices assignable to any addressable path

- Lighter cable weights

- Serial data transfer

- Photon transfer versus electron transfer as with parallel channels

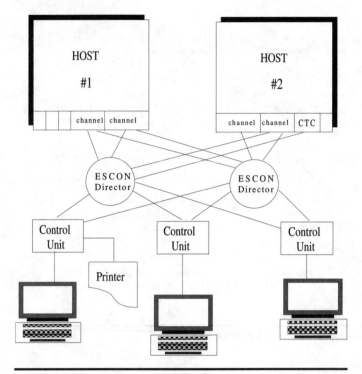

Figure 16.11   Possible ESCON configuration.

Networking support. The S/390 announcement includes devices and support for networking technologies for other than native IBM networking solutions. Following is a brief, but not exhaustive, list of these offerings and some of their features.

- VTAM Version 3 Release 4:

  T2.1 node multitail support

  Dynamic NETID (network identification)

  FDDI support

  ESCON channel support

  T2.1 boundary function support

  Dynamic I/O

  Support for program selectivity of data encryption

- 3745 enhancements:

  Buffer chaining channel adapter

  32-line capacity

  Memory expansion

- 3174 enhancements:

  X.21/X.25 switched autocall and autodisconnect
  Systems management support with NetView
  Multihost token-ring gateway support

- 3172 enhancements:

  Model 002
  Up to two ESCON connections
  Up to two parallel channel connections
  FDDI support
  Remote CTC (channel-to-channel) controller function
  LAN gateway functionality

- NCP Version 5 Releases 3.1 and 4:

  Buffer chaining channel adapter support
  Increased network management with token ring
  Dynamic NETID
  Reduced storage requirements for explicit routes
  Additional options for T1 support

- 8209 token-ring attachment module

  Connects two token-ring networks into one logical ring
  Source and destination filtering
  Custom filter support

Other software and hardware support was introduced with the S/390 announcement; however, there were too many products to list here. For a detailed list of these product offerings as part of this announcement, contact your local IBM representative.

16.2   Hardware Components

Processors

IBM processors are grouped into what IBM calls *series*. IBM has different series based on different architectures, such as the aforementioned. For example, the ES/9000 series is based on ESA/390 or S/390 architecture. Some previous and present IBM processor series are

ES/9000	303X
3090	308X
4300	9370

Processor series have models. For example, the announcement of the ES/9000 included 18 models. Some of these models included

120	260	500
130	320	580
150	330	620
170	440	720
190	480	820
210	340	900

These models have different support levels in terms of processing capability, and some of them are water-cooled while others are air-cooled.

Some of the processor series IBM has offered provide different functionalities. For example, some models in the 9370 series support direct ETHERNET network attachment.

Some 308X series have model numbers indicating a uniprocessor and others indicating dual-processor capabilities. In general, the model number of a particular series indicates a significant amount about the processor.

Understanding this numbering scheme helps break the number barrier for those new to IBM equipment and environments.

Channels

The saying in IBM circles is that all data inbound to a processor must go through a channel to get to the processor. So far, this author has found that to be the case. Channels, a channel subsystem, and channel paths are deeply rooted in the hardware architecture dictating how data is manipulated at the lowest layers in a system. The channels used prior to ESCON used copper-stranded cables called bus and tag cables. These cables are weighty. The operating distance for parallel channels (those prior to ESCON) was approximately 200 ft if devices were daisy-chained. A straight run of 400 ft might be obtained under ideal conditions. The bus cable contains signal lines used to transport data. Tag cables control data traffic on the bus.

Prior to the ESCON offering, IBM had parallel channels to move data in parallel from a source to a destination point. Three types of channels—selector channel and byte and block multiplexer channels—represented this offering; however, the selector channel support was discontinued in the 1970s.

Selector channel. The *selector channel* has only one subchannel. However, multiple devices, such as tape and disk devices, can be connected to this subchannel. Characteristics of the selector channel include the following:

- It is intended for operation with high-speed devices only.
- This channel can accommodate only one data transfer at a time.
- Once a logical connection is established between a device and the channel, no interruptions occur for the duration of the data transfer.

Figure 16.12 shows a selector channel and attached devices.

Byte multiplexer channels. This type of channel differs from selector channels because

- It is intended for operation with low-speed devices.
- One channel can address up to 256 subchannels.
- The subchannels operate in burst mode, meaning that once a logical connection is established between a device and the channel, the data is pushed to the channel. After release, the next data transfer can occur. This permits interleaving of data at a byte level. Since these channels are designed to operate with slow-speed devices, when compared to selector channels this operation works quite well, exploiting bandwidth and utilization of the available resources.

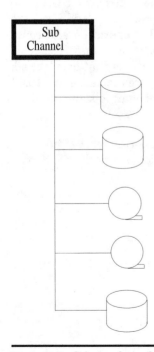

Figure 16.12   Selector channel.

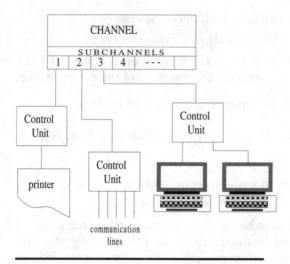

Figure 16.13 View of a byte-multiplexer channel, subchannels, and devices.

The byte multiplexer channel is shown in Fig. 16.13.

Block multiplexer channel. This type of channel, introduced with the S/370 hardware architecture, had the ability to record the address, byte count, status, and control information for an I/O operation and could perform a disconnect from a device if no data is being transferred. This meant that high-speed devices could have operations that overlapped. In this sense a greater utilization of resources is realized. Figure 16.14 depicts the block multiplexer channel and its subchannels and also how data is interleaved as it is passed to the channel.

ESCON channel. Another type of channel is supported by IBM now: the ESCON channel. Simply put, it is a fiber-optic data path. ESCON is

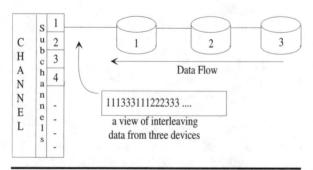

Figure 16.14 View of block multiplexer channel and data flow.

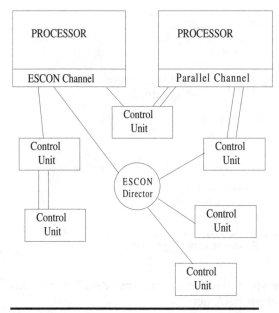

Figure 16.15 Example of serial and parallel channels.

referred to as a *serial channel,* and its channel protocols differ from those of the selector and byte and block multiplexer channels, which have been renamed *parallel channels.*

ESCON channels support greater physical cable length because photons, not electrons, are moved; hence no voltage drop is realized. ESCON channels also support dynamic connectivity with the use of ESCON directors. Additionally, ESCON extenders can be used to increase the operating distance of these cables to approximately 43 km. Figure 16.15 is an example of a processor with ESCON channels and a processor in the same complex with parallel channels.

Communication controller

The *communication controller,* also known as a *front-end processor* (FEP), is where the network control program (NCP) is located; the NCP is discussed shortly. Communication controllers are available in different sizes and models; this dictates the abilities and limitations of the controllers. Communication controllers perform multiple tasks, including

- Routing
- Flow control
- The point where communication lines connect

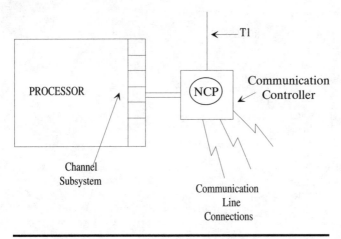

Figure 16.16 Processors, channels, and communication controller.

- The location where some specialized programs operate allowing non-SNA equipment to access an SNA network

A processor, channels, and a communication controller are shown in Fig. 16.16.

Cluster controller

The *cluster controller* is the forerunner of the establishment controller, explained next. In SNA environments, cluster controllers are used to attach terminals and printers. Different models of cluster controllers exist. A cluster controller model dictates how many devices can be attached to the controller. In SNA drawings where a cluster controller is used it is typically represented as in Fig. 16.17.

The cluster controller is known in SNA environments as a 3274 control unit. The 3274 control unit family has different models; some of them are

- 1A
- 21A
- 31
- 41

3274

Figure 16.17 Cluster controller.

Depending on the model, two possible modes of operation are possible: local and remote. Beyond this, there are three methods of identifying the operational properties of a particular control unit.

- An A indicator means that the control unit is functioning as a channel-attached local device.

- The B and D designators mean that the control units are channel-attached using the processor channel program.

- The C designator indicates that the control unit is operating as a remote unit with SDLC or BSC data-link-layer protocols.

The 3274 is still in use today even though it has been replaced by the 3174. Both work well together, and depending on the need, justification for a 3274 may be equal to that for a 3174.

Establishment controller

The *establishment controller,* which succeeded the cluster controller, provides services offered by the cluster controller and more, such as those used in networking. An establishment controller is shown in Fig. 16.18.

The establishment controller has numerous models in its family. Certain models are capable of performing functions that others cannot. A brief list of these includes

1L	12R	51R
1R	13R	52R
2R	21L	61R
3R	21R	62R
11L	22L	63R
11R	22R	
12L	23R	

Some characteristics and functions offered by the establishment controller offers are

- Token-ring support

- Token-ring gateway support

- Multihost support via concurrent communication adapter and the single-link multihost support

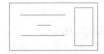

Figure 16.18   Establishment controller.

- ISDN support
- PU2.0 support
- PU2.1 support
- APPN support
- Control unit terminal (CUT) support
- Distributed function terminal (DFT) support
- SDLC support
- X.21 and X.25 support
- Parallel channel support
- ESCON support
- BSC support
- 3270 data stream support
- Printer support
- Response-time monitor support
- Common management information protocol (CMIP) support
- Generic alert support
- T2.1 channel command support

The APPN support offered by the establishment controller is a considerable enhancement over the 3274 cluster controllers, which did not have this support. This device covers a broader support than its predecessor, and is positioned to fit into either SNA or APPN networks or both.

Interconnect controller

The *interconnect controller,* which succeeded the cluster controller, provides services of the cluster controller and more, such as those used in networking. Consider Fig. 16.19.

The interconnect controller is known numerically as the 3172. Three models have thus far been introduced to the market:

- 001
- 002
- 003

The latter models' forte is its versatility. The model 003 supports the TCP/IP off-load function offered by IBM. This means that a customer can purchase a TCP/IP to run as a VTAM application but select the

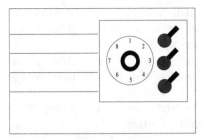

Figure 16.19 Interconnect controller.

applications or functions desired and offload TCP, UDP (user datagram protocol), and IP to the 3172 model 003. In turn, only the desired portion of TCP/IP resides as an active application under VTAM. The result is conservation of resources from a processor standpoint.

The off-load function of TCP/IP to the model 003 means that protocol conversion of TCP and/or UDP is performed on the 3172, not the processor. Additionally, this means that IP performs routing functions on the 3172. Again the benefit is no work on the processor for this function.

The 3172 model 003 communicates effectively with a processor via what IBM calls the *common link access to workstation* (CLAW) protocol. The 3172 model 003 can achieve this because it implements a CLAW driver to do so. The benefit of this is how the subchannel is utilized with regard to data rate transfer.

Other benefits of the 3172 model 003 include its support for datalink-layer protocols. This model supports

- FDDI
- ETHERNET Version 2
- 802.3
- Token ring
- Channel protocol (to the processor)

Another powerful support aspect of the 3172 is its ability to support not only NetView but also the simple network management protocol (SNMP). It also provides a system log facility via the interconnect control program (ICP), which can be used for debugging if necessary.

Direct-access storage device

The *direct-access storage device* (commonly known as DASD) is IBM's term for a disk drive. Significant here are the five delineable DASD offerings. Many IBM DASD units are still working in companies and corporations around the world. The most important ones and their fundamental significance are mentioned here.

The first DASD devices can be characterized by removable media, known as *platters,* which are that segment of the drive where data is stored.

The second DASD devices to appear had more intelligence than their predecessors. These had removable platters also, but a great improvement was the increased storage capacity of some models in these offerings.

Third came the DASD offering which brought significant improvement of storage in terms of density and also better diagnostic capabilities. Significant performance was also achieved.

The fourth category of DASD offerings from IBM dominated the 1980s. A resounding word recurring about this category of DASD is *reliability.* Performance, speed, and flexibility in regard to implementation also characterized this category.

The fifth category of IBM DASD appeared in the 1990s. Many in the technical community, including IBM, consider it the DASD architectural foundation for this decade and into the twenty-first century. Strength of this category includes a robust number of models from which to select. Software support used in processors supporting the IBM storage hierarchy has also caught up significantly. Consequently, leveraging advanced hardware along with supporting software brings synergy to the storage subsystem offered in this current category.

Figure 16.20 depicts how reference to DASD is made in SNA generally. Most references are not to a specific DASD model, unless an in-depth discussion is required about the topic at hand.

Tape devices

IBM has two basic groups of tape devices. The fundamental difference between the two is that one is reel-based and the other uses cartridges. In many IBM shops tape is a method of backup for data that may be archived. Figure 16.21 shows the general symbols used in reference to a disk drive.

Figure 16.20   Direct-access storage device.

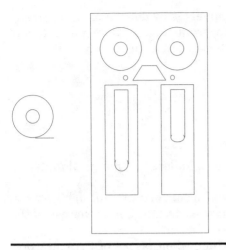

Figure 16.21 Representation of a tape drive.

Printers

IBM has different types of printers. They can be categorized by speed and type of technology. Figure 16.22 shows a generic example of a printer. In most cases this symbol suffices to convey the information being discussed.

Terminals

Different types of terminals are used in SNA, but most have one commonality: the use of a 3270 data stream. This feature will be discussed in greater detail later, but the important point here is to understand the general nature of IBM terminals used in SNA.

There are two general categories of terminals:

- Those which do not support graphics.
- Those which do support graphics.

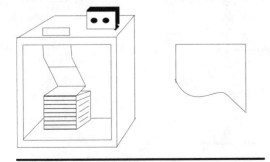

Figure 16.22 Representation of a printer.

Another way of categorizing terminals is by determining how many columns and rows they support. For example, a 3278 terminal family exists. Four types of 3278 terminals constitute this family:

- Model 2
- Model 3
- Model 4
- Model 5

The basic difference between these terminals is how many columns and rows are supported.

Another terminal is the 3179G, which supports graphic applications. A terminal type known as a 3279 also exists. It is a later version of the 3278 terminal.

Terminals will be discussed further when issues of SNA are presented, such as the type data stream they support. References in this book to the 3270 terminal appear as shown in Fig. 16.23.

16.3 IBM Operating Systems

Three IBM operating systems are discussed here. The author is aware of IBM's other operating systems that function in the hardware architectural environments explained previously. Some of these operating systems have been excluded purposely because the ones presented here are dominant in regard to market share and their presence in the SNA networking environment. The three for our concern are

- Multiple virtual storage
- Virtual machine
- Virtual storage extended

Each of these is at a different version and release and varies according to the underlying hardware architecture on which they operate. They

Figure 16.23 Terminal.

have different architectural characteristics. Each operating system and its basic characteristics, functions, and other pertinent information that seem to be agreed on in the industry are presented here. After the operating systems have been explained, software subsystems that operate under their control will be presented. These software subsystems have considerable market share, but others exist; however, the intent is not to present them all but rather to focus on those dominant ones. This section, as does Sec. 16.2, provides the basics to understand some of the SNA-centered terminology and concepts.

Multiple virtual storage

Multiple virtual storage (MVS) is rooted in the first versions of Operating System/360 (OS/360) designed in the early 1960s. It is generally considered a production-oriented operating system. From a historical viewpoint, the operating system has evolved through many versions and releases, but for our purposes, suffice it to say that MVS's immediate predecessor was *m*ultiprogramming with a *v*ariable number of *t*asks (OS/MVT).

Many books have been written about MVS, both from IBM and authors not affiliated with the IBM corporation. The intent of this section is to focus on the highlights of MVS. Not all its features and functions are presented, just those which this author believes have made significant impact through the versions and releases MVS has realized thus far. Conceptually, representing it would appear as Fig. 16.24.

MVS operated with S/370 hardware architecture and followed two earlier versions of operating systems known as OS/VS1 and OS/VS2. MVS could address up to 16 Mbytes of virtual storage. It utilized program areas which were names. These names referred to an *address space,* which is the identifiable amount of memory addresses a program can use. Tied to this was the concept of an *address space identifier,* which identifies a particular address space. Data areas also existed. Here, data areas, precisely known as *common data areas,* provide capabilities for the user program messaging and communicate with the

Figure 16.24   Conceptual view of MVS processor.

operating supervisor, and the operating system interacts with the program. I/O buffers were also utilized to pass data from a telecommunication subsystem to an address space where that data would be moved for operations to be performed.

A major change for MVS came in 1981 with respect to addressability. The significance of MVS/XA was that it broke the 16-Mbyte boundary and that virtual storage addressability was supported up to 2 Gbytes; numerically this is 2,174,484,684 bytes!

MVS/XA also provided each user with the perception of having a unique address space and was capable of distinguishing between programs and user data in each address space. Cross-memory services brought the ability for a user to access other address spaces as necessary.

The concept of task management in MVS/XA meant that MVS/XA divided jobs into smaller pieces and processed each piece "task" as efficiently as possible. Control of this process lies within the supervisor.

MVS/XA was unique. It supported both 24- and 31-bit addressing. Along with MVS/XA came changes in the I/O facilities. The greatest change with MVS/XA was the I/O subsystem. It handles I/O operations independently of the processor. With this came other changes.

First, the channel, and not the operating system, handled channel path selection. Additionally, dynamic reconnection was added to the I/O portion to support dynamic path selection. Another enhancement to I/O facilities that fit into MVS/XA was that the number of supported devices increased to 4096, but a contingent factor was dependence on the I/O configuration program. Support for up to 256 channel paths with 8 paths per device was also supported under MVS/XA.

MVS/XA brought with it three types of tracing:

- Address space
- Branch
- Explicit software

The generalized trace facility (GTF) was changed to support 31-bit addressing. Parallel to this was the improved dump facility.

MVS/XA controlled work or managed a resource by identifying a control block. Three types of control blocks are associated with MVS/XA:

- *Resource:* Represented a DASD or MVS processor, for example
- *System:* Contained systemwide information
- *Task:* Represented one unit of work

The *system resource manager* (SRM) is used in MVS/XA to make decisions concerning where an address space should remain: in real storage or in DASD. The SRM is the primary method for managing "resources" in a MVS/XA environment.

In 1988, MVS/ESA was introduced. Its virtual storage addressability is 16 Tbytes. New features and functions made this software architecture the most vigorous to date.

With MVS/ESA a data space is available. It contains data only, no programs. Since data spaces are available, they must be managed. A *data space manager* keeps track of those in use in the system and manages them accordingly. A data space may be assigned to one program only or shared. This new type of space available further extends the abilities of virtual storage.

Along with the data space is the concept of a *hiperspace*. This is data in an address space that has a special classification. Although it is processed in an address space, it typically uses expanded storage or migrates to auxiliary storage.

An *access list* is also part of MVS/ESA. It is responsible for determining which data spaces a program is authorized to use. This facility is also used by hardware for operations with the segment table descriptor.

The *linkage stack facility* is simply a facility used to retain information about the state a program is in during execution when call is given to another program.

The *virtual look-aside facility* (VLF) can be characterized to curb time for partitioned data set (PDS) searches and reads. It can also perform multiple reads against the same object. VLF operates in its own address space and assigns names to objects it manages. The naming convention uses three levels: class, major, and minor. The purpose of this object naming is to increase data retrieval by VLF.

Advanced program-to-program communication (APPC) support under MVS enables peer connectivity outside the MVS environment; for example

- AIX for the RISC/6000
- S/38
- OS/400
- VM/ESA
- VSE
- OS/2

Figure 16.25 shows conceptually how APPC under MVS by APPC/MVS applications, Time Sharing Option/Extension (TSO/E) users, the IBM Information Management System, and even APPC batch jobs would appear.

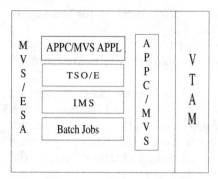

Figure 16.25   Conceptual view of APPC/MVS.

Other functions in MVS/ESA for examination include

■ Sysplex

■ Cross-system coupling facility

■ Automatic reconfiguration facility

■ Hardware configuration definition

Sysplex is the concept of bringing together one or more MVS systems via hardware and software services. In this scenario the systems are treated as a single complex and are initialized as such. To achieve this sysplex concept, an external clock—called the *sysplex timer*—must be used to synchronize the systems brought together as well as the time across the entire central processing complex.

The *cross-system coupling facility* (also known as XCF) is software providing control of members and groups, intercommunication among members, and monitoring of members. Put another way, it provides the services necessary for programs in a multisystem environment to communicate successfully.

The *automatic reconfiguration facility* (ARF) can be used in a single or multiprocessor configuration environment. In either case it serves to redistribute a workload in the event of failure without operator intervention.

The *hardware configuration definition* (HCD) is a facility that permits communication between the channel subsystem and I/O definitions at the same time. Prior to MVS Version 4 this process was a two-step function: (1) defining the I/O subsystem via the I/O control program (IOCP) and (2) defining software by the MVS control program (MVSCP). Now, with HCD the two are combined.

Other functions of MVS exist. There are too many to list here. For further information about MVS, contact your local IBM representative.

Virtual machine

VM/ESA combines the VM/System Product (VM/SP), VM/SP High Performance Option (VM/SP HPO), and VM/SP XA. Some VM/ESA facilities include

- *APPC / VM VTAM support* (*AVS*): This is a facility of VM/ESA that converts APPC/VM into APPC/VTAM protocol for communication throughout a SNA network.

- *Group control system* (*GCS*): This facility manages subsystems that permit VM/ESA to interact within a SNA environment.

- *Interactive problem control system* (*IPCS*): This component of VM/ESA provides an operator with online capability for problem diagnosis and management.

- *Transparent services access facility* (*TSAF*): This is a component of VM/ESA that permits communication between programs by name specification instead of user or node id.

The forte of VM/ESA is its ability to run multiple operating systems on the same processor. This concept is known as *multiple preferred guests* (MPG). Figure 16.26 shows this concept.

The idea behind the concept of multiple preferred guests is that totally different operating systems can be executing simultaneously and that one MPG fails but does not cause any other MPG to fail.

This type of environment can be advantageous if a production operating system such as MVS/ESA is needed and a development-oriented operating system like VM/ESA is needed. Both of these can execute under the control of a VM/ESA operating system.

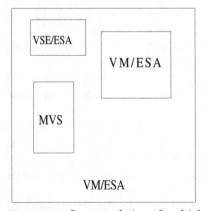

Figure 16.26 Conceptual view of multiple preferred guests.

VM, as did MVS, originated during the 1970s. Technically, VM dates back to 1965 with what is known as *Control Program 40* (CP-40). By the late 1960s evolution of the Control Program was shaping up to what would become known as VM/370. VM/370 was made available in 1972 according to the *IBM Systems Journal,* vol. 18, issue no. 1, 1979.

A closer view of VM/ESA reveals these components, which are the focus of the remainder of this section:

- Control program (CP)
- Conversational monitoring system (CMS)
- Group control system (GCS)
- Interuser communication vehicle (IUCV)

The CP is that component used to manage the actual hardware. It communicates with resources such as the real processor, I/O subsystem, and storage. CP is responsible for management of these resources. It is the CP that utilizes the physical resources to create the logical (virtual) machines.

CP exploits the hardware architecture mentioned previously. In the case of S/390 architecture, many of the resources available with the MVS operating system are also available through the platform where CP is in control.

CP has a set of commands that permit the manipulation of resources, both physical and logical. Some of the commands can be used immediately after log-on to perform system management functions.

Some aspects of CP should be mentioned here. First, the CP is software. It is transferred to real storage when the system is booted. One way CP works is by defining internal objects by what is called a *control block.* An example of an important control block is the VM definition block (VMDBK). This control block represents a logged-on virtual machine; it is created by the CP when one logs onto VM.

A trace table exists in VM, particularly to maintain any events related to problems, system crashes, or the like.

The CP knows the real I/O configuration and correlates real I/O device capabilities to virtual devices. CP with VM/ESA architecture also supports ESCON I/O architecture.

The *conversational monitoring system* (CMS) is another part of VM/ESA. This component of VM/ESA operates with the CP but is used as a two-way communication processor between users. IBM has a large manual of valid CMS commands that can be used to perform a number of functions. CMS is an operating system for users of VM.

CMS communication can occur between users and CMS. Users can enter commands to CMS, and CMS can issue messages to users. CMS permits a number of user functions, including

- Communication with other users
- Creation, testing, and debugging of programs by users
- Control of work flow in a VM/ESA system
- File sharing

The concept of file sharing under CMS is achieved via the *shared file system* (SFS) facility, which is an extension of CMS and provides file management capabilities. According to IBM's *CMS Shared File System Primer,* document no. GG24-3709, the features of SFS include

- A hierarchical file system.
- Files can be stored in pools.
- User space can be assigned to a spool file.
- A file can be located in more than one directory.
- Files and directories can be shared with users on other systems.
- File and directory locks ensure data integrity when multiple users are involved with the same file and/or directory.
- Users can have concurrent access to files and directories.
- Files in a file space are stored in a directory.

CMS also has a facility known as *XEDIT,* which is a full-screen editor that permits users the ability to create, edit, and manipulate files. It runs under the control of CMS. Some functions possible under XEDIT include

- CMS file creation
- File editing
- Joining existing files
- Performing searches in files for specific data
- Creating XEDIT macros
- Performing a sort function on data in a file
- Providing HELP for users

The *group control system* (GCS) is a VM component shipped by IBM with each VM/ESA system. It is the component that manages subsystems that permit interoperation with IBM's SNA. Actually, GCS is a supervisor. IBM's *virtual telecommunication access method* (VTAM) runs under a GCS group which, in turn, has a supervisor. Programs in a VM/ESA machine or complex and devices outside it in a SNA network use VTAM to communicate.

The *interuser communication vehicle* (IUCV) facilitates communication between programs running in different virtual machines. It is also aided by CMS. The IUCV serves the function of facilitating the communication between a virtual machine and a CP service. Some examples of the latter include

- Logging errors
- Communication with the system console
- Performing message system service functions
- Other services such as a spool system service

Figure 16.27 is a conceptual view of these two IUCV communication functions.

Much more could be said about the VM environment, but for our purposes here this information suffices. IBM has exhaustive documentation about VM and its components, and this author recommends contacting IBM directly for further information if the needs dictate such.

Virtual storage extended

Virtual storage extended (VSE), i.e., VSE/ESA, is another of IBM's operating systems that operate on some S/370 architectures and the ESA/390 architecture. Its strength is in batch processing capabilities. Also, according to IBM documentation, it can support high transaction

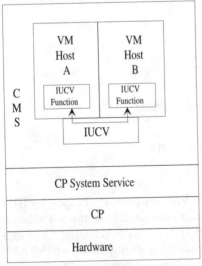

Figure 16.27 Conceptual view of IUCV communications.

volumes. VSE/ESA is *pregenerated,* meaning that a generation is not required; IBM ships object code. This is in contrast to MVS/ESA and VM/ESA, which must be generated at installation time for site-specific customization. Because VSE/ESA is supplied this way, much planning can be eliminated.

Before examining the functionality of VSE/ESA, consider its lineage according to IBM's *VSE/ESA Version 1.3: An Introduction Presentation Foil Master,* document no. GG24-4008.

Date	Offering
1965	DOS
1972	DOS/Virtual Storage (DOS/VS)
1979	DOS/Virtual Storage Extended (DOS/VSE)
1985	VSE/System Product (VSE/SP) Version 2
1989	VSE/SP Version 4
1990	VSE/ESA Version 1.1
1991	VSE/ESA Version 1.2
1993	VSE/ESA Version 1.3

VSE/ESA can operate in multiple environments. Examples of some of those environments are

- As the sole operating system on a processor. As a result, both local and remote operations can be realized.
- It can run under a LPAR.
- It can run as a standalone system. This means that it can be run without human intervention, in some cases functioning as a node within a network.
- VSE/ESA can also operate under VM as a preferred guest.

Major enhancements and features of VSE/ESA include some of the following. This list includes inherent features and functions with VSE/ESA, but some optional facilities as well. The purpose here is to understand the breadth of change that VSE has undergone with the advent of its support in ESA. The list is not exhaustive, but the core aspects of this release are presented.

- 31-bit virtual addressing
- Support for data spaces
- Support for more I/O devices
- ESCON support
- ESCON director support

- VTAM Version 3 Release 4 features for VSE/ESA
- National language support, including

  Spanish

  German

  Japanese

- Addressability up to 2 Gbytes of real storage
- Addressability up to 2 Gbytes of virtual storage
- VSE/ESA POWER operates in a private address space
- Support for the 3172 via VTAM 3.4
- ES/9000 processor support
- ESCON CTC (channel-to-channel) adapter support
- Support for virtual disks in storage
- Support for extended functions of the 3990 DASD
- NetView for VES/ESA
- Support for SQL/DS making a VSE/ESA capable of functioning with *Distributed Relational Database Architecture* (DRDA) as a server
- 3174 ESCON connectivity

So many enhancements were made that the VSE/ESA operating system has become more popular in the past few years (according to reports in credible trade magazines). Beyond the aforementioned enhancements to VSE, some of its system components need to be presented and are briefly discussed for those new to the VSE/ESA environment. The following VSE system component functions will be discussed succinctly and sequentially.

- Librarian
- ICCF
- POWER
- VSAM

The Librarian is a utility program that is used to manipulate libraries. It aids in the creation, maintenance, and use of libraries. With VSE/ESA the following were included with the Librarian:

- Copy and Compare commands
- Move command
- LISTD command, which is a list including the date

- Search command

- Lock and Unlock command to provide data integrity to single library members that may be in the process of being updated

- Backup command, which now supports the backup of an individual member

Libraries have one or more sublibraries which, in turn, consists of members. These members are where data, programs, source code, and other "data" exist in a VSE/ESA system.

Two types of libraries exist in VSE/ESA: System and ICCF libraries. System library's sublibraries can include the following members:

- *Source:*   This is source code to be processed.

- *Object:*   These modules are generated by the output of a language translator and are used for input by a linkage editor.

- *Dump:*   If an abnormal termination results, the contents will be sent to this type of member in a sublibrary.

- *Procedure:*   This is a set of procedures, such as a set of job control statements.

- *Phase:*   This type of member has a program or section(s) of a program stored in it which is ready to run.

Other system-oriented libraries exist, and if further information is needed, acquire the appropriate IBM VSE/ESA manual for an explanation. A list of VSE/ESA documentation can be obtained through your local IBM representative.

VSE/Interactive Computing and Control Facility (VSE/ICCF) uses libraries. A considerable number of these libraries exist, but the ones most frequently used and of concern here include

- *Public:*   This type of library consists of data that may need to be accessed by a number of users systemwide.

- *Common:*   This type of library contains data in which all users are interested.

- *Main:*   This is the library attached to an individual once it is logged on.

- *Private:*   This type of library contains data limited to one or a few users.

The *interactive computing and control facility* (ICCF) in a sense is the interface between a user and a VSE/ESA system. Technically, it is the subsystem for program development and system administration. Through ICCF libraries, sublibraries and members can be created.

Priority output writers and execution processors and input readers (POWER) is the subsystem that provides networking support, batch job processing, and spooling functions.

The *virtual storage access method* (VSAM) is the access method (or way) by which data, programs, and the like are stored. Put another way, it is a method of data management. Its supported data organization includes

- Entry sequenced files (similar to a sequential file)

- Key sequenced data sets (similar to an indexed file)

- Relative record data sets (similar to direct-access files)

- Variable-length relative record data sets

VSE/ESA has been put on equal ground with MVS/ESA and VM/ESA. A full discussion of all aspects of VSE/ESA is not intended here; these operating systems are mentioned briefly only to convey some of their strengths and because of their presence in the marketplace. The enhancements necessary to place VSE/ESA on the same level as MVS and VM took time. But now IBM has three refined and tested operating systems to offer a wide variety of customer needs.

16.4 IBM Software Offerings

IBM has many software subsystems that operate with the previously presented operating systems. The focus in this section includes the following:

- Virtual telecommunication access method (VTAM)

- Job Entry Subsystem 2 (JES2)

- Network control program (NCP)

- NetView

- Time Sharing Option (TSO)

- Customer information control system (CICS)

- DATABASE 2 (DB2)

- Remote spooling communication system (RSCS)

- Local Area Network (LAN) Resource Extension Services (LANRES)

Virtual Telecommunication Access Method (VTAM)

VTAM is a software subsystem that operates under the previously described operating systems. It is a critical component in traditional SNA and functions in new ways with IBM's APPN. Practically every application that runs in a MVS host runs as a VTAM application. Applications must be defined to VTAM in such a way that they can be used. The same holds true for hardware and components outside the processor. Figure 16.28 provides a conceptual view of VTAM with some peripherals attached.

One of VTAM's major roles in traditional SNA is aiding in session establishment. When a terminal user requests to use a software subsystem, the request is first interpreted by VTAM and then passed to the appropriate subsystem.

VTAM is also the centralized point for network component activation and deactivation. VTAM's component, known as the *system services control point* (SSCP), plays a vital role in the area of network management. This role should not be confused with the software that enables network management to be realized; however, they work together. The SSCP network management role involves communicating with hardware and other components throughout the network.

VTAM must "know" (have defined) software and hardware that operate within a SNA network. Some VTAM components that must be defined for software and hardware to operate include

- Application(s)

- Device(s)

- Session operating parameters

- Log-on menu (if utilized)

- Communication controller and attached devices

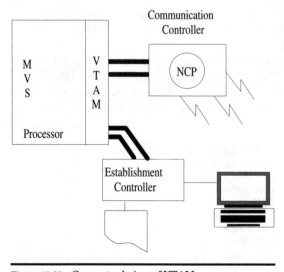

Figure 16.28 Conceptual view of VTAM.

Applications must be defined to VTAM for the application to work. For example, if application Z is loaded into a system, then certain parameters of the application must be defined. Different applications have requirements for definition. In order to determine whether this applies, you should consult the IBM *VTAM Resource Definition* manuals. Figure 16.29 shows a basic example of an application definition.

Figure 16.29 shows the following in left-to-right order:

The name of the application	MYAPP
The VBUILD statement	TYPE=APPL
The name of the application	MYAPP
Definition statement	APPL
Operands	ACBNAME
	AUTH
	DLGMODE
	EAS
	MODETAB
	SSCPFM
	USSTAB

Figure 16.29 is an example of how each application that operates under VTAM would need to be defined. Not all the parameters will be the same. The application and VTAM requirements will dictate how the application should be defined. Moreover, the site requirements will dictate this.

Devices must be defined to VTAM as well. A number of factors dictate how a device is defined. Some examples of this, among sundry others, concern how the device is physically attached and the data-link protocol used and its role in the SNA environment (according to SNA definitions). Figure 16.30 is an example of a 3174 establishment controller defined to VTAM.

In Fig. 16.30 the definition of the device can be divided into three parts. VTAM requires a VBUILD statement, *physical unit* (PU) statement, and *logical unit* (LU) statement for those LUs to be used. More information will be provided about the PUs and LUs forthwith.

```
MYAPP      VBUILD    TYPE=APPL
MYAPP      APPL      ACBNAME=MYAPP
                     AUTH=
                     DLOGMODE=
                     EAS=
                     MODETAB=
                     SSCPFM=
                     USSTAB=
```

Figure 16.29 Example of an application defined.

```
MY3174        VBUILD      TYPE=LOCAL

MY3174        PU          CUADDR=
                          DLOGMODE=
                          DISCNT=
                          ISTATUS=
                          MAXBFUR=
                          PUTYPE=
                          USSTAB=

YOUR327401    LU          LOCADDR=
  .             .
  .             .
  .             .
  .             .
  .             .
YOUR317406    LU          LOCADDR=
```

Figure 16.30 Some establishment controller definitions.

The statements and parameters in this example are typical of how a device is defined to VTAM. Three factors determine how a device should be defined to VTAM: the architectural capability of the device, how VTAM dictates that it be defined, and how the device is used in a given site.

Session operating parameters must be customized. If a terminal user desires to work with an application, the terminal parameters must be programmed. The location for these session parameters is known as the *log-on mode table* (*LOGMODE table* for short).

The LOGMODE table consists of numerous entries, each of which define the session parameters for a particular type of session. Sessions will be described in greater detail later in this chapter, but Fig. 16.31 is an example of how a LOGMODE table appears.

The log-on menu, if utilized, is what users see on viewing a terminal under the control of VTAM. This log-on menu is formally called the *unformatted system services* (USS) table. Figure 16.32 is an example of a USS table.

The USS table shown in Fig. 16.32 can be divided into three parts. The arrangement of this table is based on IBM's *VTAM Resource Definition* manual. Some flexibility exists, but generally these three parts are present.

First, the name of the application a user wants to access is listed and beside it, the parameters required to pass the request to that application. Second, VTAM has 15 messages which constitute the second part. These are messages that may be generated if certain conditions exist. Third, the contents of what is displayed on the menu must be coded. As

```
****************************************************************
*           THIS IS THE LOGMODE TABLE FOR MY3174           *
****************************************************************
*         LOGMODE TABLE ENTRY FOR 3278-M2 EMULATION          *
****************************************************************
EDMODE     MODETAB

EDMODE2    MODEENT    L   O   G   M   O   D   E   =   E   D   M   2   ,
FMPROF=X'03',
TSPROF=X'03',
PRIPROT=X'B1',
SECPROT=X'90',
COMPROT=X'3080',
RUSIZES=X'F8F8',
P S E R V I C = X ' 0 2 8 0 0 0 0 0 0 0 0 0 0 0 0 0 0 0 0 0 0 0 0 2 0 0 '
****************************************************************
*         LOGMODE TABLE ENTRY FOR 3278-M3 EMULATION          *
****************************************************************
EDMODE3    MODEENT    L   O   G   M   O   D   E   =   E   D   M   3   ,
FMPROF=X'03',
TSPROF=X'03',
PRIPROT=X'B1',
SECPROT=X'90',
COMPROT=X'3080',
RUSIZES=X'F8F8',
P S E R V I C = X ' 0 2 8 0 0 0 0 0 0 0 0 0 0 1 8 5 0 2 0 5 0 7 F 0 0 '
****************************************************************
          LOGMODE TABLE ENTRY FOR 3278-M4 EMULATION          *
****************************************************************
EDMODE4    MODEENT    L   O   G   M   O   D   E   =   E   D   M   4   ,
FMPROF=X'03',
TSPROF=X'03',
PRIPROT=X'B1',
SECPROT=X'90',
COMPROT=X'3080',
RUSIZES=X'F8F8',
P S E R V I C = X ' 0 2 8 0 0 0 0 0 0 0 0 0 0 1 8 5 0 2 B 5 0 7 F 0 0 '
****************************************************************
          LOGMODE TABLE ENTRY FOR 3278-M5 EMULATION          *
****************************************************************
EDMODE5    MODEENT    L   O   G   M   O   D   E   =   E   D   M   5   ,
FMPROF=X'03',
TSPROF=X'03',
PRIPROT=X'B1',
SECPROT=X'90',
COMPROT=X'3080',
RUSIZES=X'F8F8',
P S E R V I C = X ' 0 2 8 0 0 0 0 0 0 0 0 0 0 1 8 5 0 1 B 8 4 7 F 0 0 '
****************************************************************
EDMODE     MODEEND                                           *
****************************************************************
```

Figure 16.31 LOGMODE table.

```
*********************** TOP OF DATA *********************************
*                                                                   *
************* USSTAB TITLE 'ETUSS TABLE' ***********************

ETUSS      USSTAB

APPLICATIONS

LOGON      USSCMD   CMD=LOGON,FORMAT=PL1
           USSPARM  PARM=APPLID
           USSPARM  PARM=LOGMODE
           USSPARM  PARM=DATA

TSO        USSCMD   CMD=TSO,REP=LOGON,FORMAT=PL1
           USSPARM  PARM=APPLID,DEFAULT=A01TSO
           USSPARM  PARM=LOGMODE
           USSPARM  PARM=DATA

CICS       USSCMD   CMD=CICS,REP=LOGON,FORMAT=PL1
           USSPARM  PARM=APPLID,DEFAULT=DETTCCICS
           USSPARM  PARM=LOGMODE
           USSPARM  PARM=DATA

JES2       USSCMD   CMD=JES2,REP=LOGON,FORMAT=PL1
           USSPARM  PARM=APPLID,DEFAULT=JES2
           USSPARM  PARM=LOGMODE
           USSPARM  PARM=DATA


VTAM MESSAGES

USSMSGS    USSMSG MSG=0,TEXT='USSMSG0:   @@LUNAME LOGON/LOGOFF IN PROGRESS'
           USSMSG MSG=1,TEXT='USSMSG1:   @@LUNAME INVALID COMMAND SYNTAX'
           USSMSG MSG=2,TEXT='USSMSG2:   @@LUNAME % COMMAND UNRECOGNIZED'
           USSMSG MSG=3,TEXT='USSMSG3:   @@LUNAME % PARAMETER UNRECOGNIZED'
           USSMSG MSG=4,TEXT='USSMSG4:   @@LUNAME % PARAMETER INVALID'
           USSMSG MSG=5,TEXT='USSMSG5:   @@LUNAME UNSUPPORTED FUNCTION'
           USSMSG MSG=6,TEXT='USSMSG6:   @@LUNAME SEQUENCE ERROR'
           USSMSG MSG=7,TEXT='USSMSG7:   @@LUNAME SESSION NOT BOUND'
           USSMSG MSG=8,TEXT='USSMSG8:   @@LUNAME INSUFFICIENT STORAGE'
           USSMSG MSG=9,TEXT='USSMSG9:   @@LUNAME MAGNETIC CARD DATA ERROR'
           USSMSG MSG=10,BUFFER=MSG10
           USSMSG MSG=11,TEXT='USSMSG11:  @@LUNAME SESSION ENDED'
           USSMSG MSG=12,TEXT='USSMSG12:  @@LUNAME REQ PARAMETER OMITTED'
           USSMSG MSG=13,TEXT='USSMSG13:  @@LUNAME IBMECHO %'
           USSMSG MSG=14,TEXT='USSMSG14:  @@LUNAME USS MESSAGE % NOT DEFINED'
```

Figure 16.32 USS table.

long as the table is created and does not violate any VTAM regulations, much flexibility exists.

A communication controller [also called a *front-end processor* (FEP)] has special definitions to VTAM because it has a software program known as the *network control program* (NCP) operating within it. Because this is so, knowing what is attached directly and indirectly to the FEP is required.

Creating a NCP is site-dependent, as is creating a VTAM definition. However, restrictions do apply regarding how the NCP can be gener-

```
CUSTOMIZED BANNER SCREEN MESSAGE

MSGBUFF

MSG10     DC     (MSG10E-MSG10-2)

          DC     C'                                      ',X'15'
          DC     C'                                      '
          DC     C'          USING THE CORRECT           ',X'15'
          DC     C'                                      '
          DC     C'          VTAM SYNTAX                  ',X'15'
          DC     C'                                      '
          DC     C'          THE MENU OF                 ',X'15'
          DC     C'                                      '
          DC     C'          CHOICE CAN                  ',X'15'
          DC     C'                                      '
          DC     C'          BE DISPLAYED                ',X'15'
          DC     C'                                      '
          DC     C'                                      '
          DC     C' --------------------------------- ''X'15'
          DC     C' --------------------------------- ',X'15'
          DC     C'     YOU CAN CREATE YOUR OWN          ',X'15'
          DC     C'                                      '
          DC     C'             MENU                     ',X'15'
          DC     C'                                      '
          DC     C'          SO USERS LOGON              ',X'15'
          DC     C'                                      '
          DC     C'        BY APPLICATION NAME           ',X'15'
          DC     C'                                      '
          DC     C'             SUCH AS                  ',X'15'
          DC     C' --------------------------------- '
          DC     C'             TSO                      ',X'15'
          DC     C'                                      '
          DC     C'             CICS                     ',X'15'
          DC     C'                                      '
          DC     C'             JES2                     ',X'15'
          DC     C'                                      '
          DC     C'                                      '
          DC     C' ---------------------------------',X'15'
          DC     C' ---------------------------------',X'15'
          DC     C'                                      '
          DC     C'                                      ',X'15'

END       USSEND
**************************** BOTTOM OF DATA ****************************
```

Figure 16.32 *(Continued)*

---

ated. The term *generate* is often used in a deprecated form known as *GEN*. Figure 16.33 shows the location of a NCP with reference to other components in the network.

VTAM is a critical component in a SNA network. Understanding it is no trivial task. I have said, and believe it is true that, "VTAM is the heart and soul of SNA."

Job Entry Subsystems 2 (JES2)

JES2 is a spooling subsystem. According to IBM, it is a primary subsystem. An interesting concept about software that operates under the

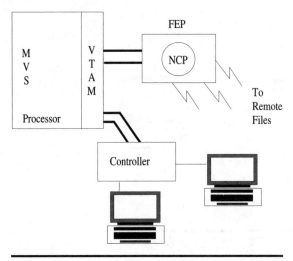

Figure 16.33   Conceptual view of a NCP location.

control of VTAM (with the exception of third-party software) is that this software is referred to as *subsystems*. One reason for this is the sheer size of some of them. As we shall see, some subsystems are practically operating systems themselves with sub-subsystems!

JES2 is a spooling subsystem that receives jobs, schedules them, and controls their output. In essence, it serves as an interface to the operating system for job processing. An example of how JES2 is used could be a job that is to be printed. Once the command is issued to print, the job goes to JES2 and then is subsequently printed. Figure 16.34 is a conceptual view of JES2.

JES2 functions so that data to be processed (whatever that data may be) may be literally spooled to a DASD for buffer storage. In this way delays can be minimized because processing continues and jobs are queued to be processed.

The term *remote job entry* (RJE) refers to a site not in the same physical facility as the processor complex that has terminal capabilities and a link to the processor so that jobs may be submitted to JES2. *Network job entry* (NJE) is another term used to refer to a complex of processors where they are dispersed but connected together in a network. NJE functions to provide the ability for multiple JES2 subsystems to communicate as peers in a network environment. This is in contrast to RJE, which permits JES2 interaction with a remote workstation.

According to James L. Brown, former IBM instructor for large systems hardware and software, "JES2 is a descendant of the Houston Automatic Spooling Priority (HASP)." He said, "HASP was originated by two IBM employees in Houston to expedite job processing in a typical university

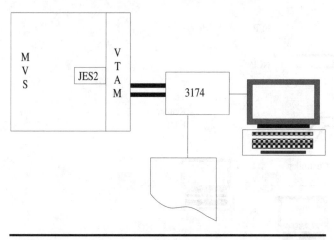

Figure 16.34   Conceptual view of JES2.

environment—many jobs, short execution times. The early OS/360 could spend more time scheduling a job than for its execution, particularly in this environment. HASP provided a means of achieving a great overlap in scheduling and execution through spooling."

Another job management subsystem is JES3. Originating from IBM's *attached support processor* (ASP) system, JES3 broadens the scope of job management. It manages jobs in the job queue, during the execution phase, and after execution for proper outputting; this management is accomplished by specially tailored algorithms, furnished by the installation. A fundamental difference between JES2 and JES3 that JES3 operates in an environment in which one copy of JES3 is the "master" (called the *GLOBAL*) and other copies (called *LOCALs*) operate throughout a processor complex in communication with this master copy.

Network control program (NCP)

The *NCP* is a program that operates on IBM's communication controllers. It, as does VTAM, plays a critical role in IBM's SNA. The NCP serves two primary functions: routing and flow control. Granted it serves other functions as well, but these seem to be the primary ones. A NCP must have devices, paths, communication lines, and connections defined to it.

The NCP performs two major functions: it controls data flow (1) through a network with other NCPs and VTAMs and (2) between itself and VTAM.

The NCP also routes data throughout a network. Within a NCP generation a list of routes (explicit and virtual) is defined. Also, a class-of-service can be used to aid in routing data. Depending on how the COS has been implemented, particular data from one source may have higher priority than data from another source.

The NCP also has programs that operate with it on a communications controller. For example, the *network packet-switched interface* (NPSI) option is a program used to permit connectivity with an X.25 network. Another example is the *network terminal option* (NTO), which works in conjunction with a NCP to permit non-SNA devices to connect to the SNA network. NTO's fundamental purpose is to convert this non-SNA protocol into SNA protocol before the data gets routed to the processor.

NetView

NetView is IBM's tool for managing SNA. It originated back in the 1970s. NetView was announced in 1986. Prior to that time components were used to selectively manage a network. For example, the *network communications control facility* (NCCF), the command-line interface for NetView, was announced in 1978. At the same time the *network problem determination application* (NPDA) program was announced. Both were released in 1979.

Today NetView has grown to consist of

- NCCF
- NPDA
- Network logical data manager (NLDM)
- Browse facility
- Status monitor
- Graphic monitor
- Resource object data manager (RODM)

Figure 16.35 is a conceptual view of the VTAM and SNA network.

All NetView, VTAM, and MVS system commands can be entered at the command line from NCCF, which is the base of NetView; thus it is possible to control a SNA network from a remote location.

Typically, SNA management is accomplished via a console (generally inside a data center). However, with the capability that NCCF provides, remote operation is possible. As long as a connection can be made to the system running NetView and an individual has the authority to execute system-oriented commands, then NetView, VTAM, and MVS commands can be issued against the NCCF command prompt.

NPDA is that part of NetView used to manage hardware. It can collect and maintain data about devices throughout the network. NPDA has the capability to request data about a particular piece of hardware or accept data sent to it from a given hardware device.

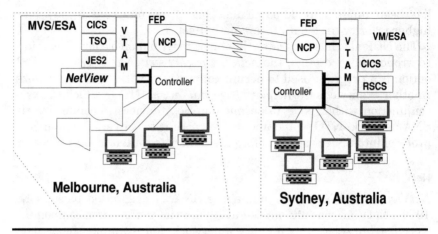

Figure 16.35 Conceptual view of NetView.

NLDM is the session monitor. A *session* is a logical connection between two endpoints. With NLDM the following information can be obtained about a session:

- Availability
- Configuration
- Error
- Event
- Explicit route
- Response time
- Session partner
- Trace
- Virtual route

The *browse* facility enables an operator to view NetView log data, VTAM definitions of devices within the network, command lists, and systemwide definitions.

The *status monitor* collects information about parts of a SNA network. It displays that data in columnar form, or it can be routed to the graphic monitor where the data can be displayed graphically.

The *graphic monitor* is a menu-driven method for monitoring network operations. It has pull-down menus and displays selected parts of a network in color. Certain colors are also used to highlight problem areas.

The *RODM* operates in its own address space in memory. It serves as a central repository for storing and retrieving information about resources throughout the network. It can obtain execution information, configuration information, and/or status information. Because of its object-oriented structure, it can make this data available to those applications which need it.

NetView can operate under MVS, VM, and VSE. Two commonly used functions of NetView are

- Alerts

- Response-time monitor

Alerts are messages about the status of a given device. These alerts use IBM's *network management vector transport* (NMVT) protocol. Information such as the day, date, and time of a failure can be included in an alert. Other status information is also included and can be site-specific.

The *response-time monitor* (RTM) is a tool used to measure the time it takes for data to leave a terminal after an *attention identifier* (AID) key is pressed, reach the host application, and return.

NetView has been expanded by IBM to support other platforms now such as the RISC/6000. Other capabilities are possible with NetView, even participating in network management with TCP/IP-based networks.

Time Sharing Option (TSO)

TSO is IBM's interactive facility that operates under MVS. TSO has three modes of operation:

- Interactive System Productivity Facility/Program Development Facility (ISPF/PDF)

- Information Center Facility

- Line Mode

The ISPF/PDF main menu is typically what users see once they log onto TSO. Through the ISPF/PDF main menu many choices exist which lead, in turn, to submenus. Some of the main menu choices and their basic function include

- *ISPF PARMS:* Specify parameters for use with ISPF.

- *Browse:* Permits viewing only of authorized data sets.

- *Edit:* Invokes the editor. With it a user can create a memo, program, or basically anything for which one would use an editor.

- *Utilities:* This leads to another menu which has a number of selections that permit disk, data set, and other types of maintenance and utility functions.

- *Foreground:* Causes the language processor to move to the foreground.

- *Batch:* Enables a user to submit jobs for batch processing.

- *Command:* If chosen, this takes the user to TSO line mode of operation. Valid TSO commands can be entered here.

- *Dialog test:* This function provides a user the ability to perform dialog testing.

- *LM utility:* This provides an individual with capabilities to perform maintenance utility functions.

- *Exit:* If selected, this causes ISPF/PDF to terminate.

Other functions are available from the main menu. Some of the ones listed above have submenus that provide additional capabilities.

The *information center facility* (ICF) provides users with a main menu similar in appearance to the ISPF/PDF main menu. Some of the functions available to users via ICF include

- *News:* Obtain news from the system.

- *Names:* This provides a list of names and phone numbers.

- *Chart:* This permits a user to create a chart or graph.

- *PDF:* If selected, this enables a user to use program development services.

- *EXIT:* This causes ICF to terminate.

TSO line mode can be used for multiple reasons. First, valid TSO commands such as LISTCAT can be entered. For example, this command is used to list data sets that are cataloged by and accessible to a user.

Another function of TSO line mode is that custom programs can be executed here. If a customized program needs a certain protocol for operation, such as terminal interaction, then a program can be created to execute under line mode.

Figure 16.36 depicts a conceptual view of TSO and other subsystems mentioned above.

Customer Information Control System (CICS)

CICS is an online transaction processing system. It is supported under MVS, VM, and VSE operating systems. Its focus is generally toward

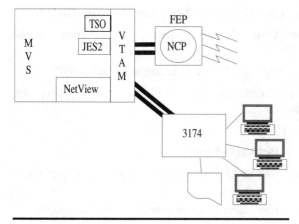

Figure 16.36   Conceptual view of TSO.

business-oriented implementations rather than scientific or engineering computations.

An example of where CICS would typically be implemented is in a banking environment. For example, if a customer has a bank account and wants to access this account via an automatic teller machine (ATM), this could be achieved with transaction programming using CICS. In this case a program in the ATM communicates with a program running under CICS control in the bank's computer. The program at the ATM is communicating with the CICS program in real time. The program running under CICS, in turn, accesses a database that maintains the requesting ATM account balance. After the program under CICS verifies the requesting party's account, it communicates with the ATM, sending it the appropriate response. Assuming that funds are available, the ATM dispenses the cash.

Customized programming is possible under CICS, thus making it attractive to users who need to create online processing programs. CICS also supports communication between transaction programs within one CICS subsystem.

DATABASE 2 (DB2)

DB2 is IBM's relational database application that provides users with flexibility and power via the functions it supports. Some of those functions include

- Utilization of a single VTAM conversation to manipulate multiple requests and responses with other DB2 applications throughout a complex.

- Support for distributed relational database architecture (DRDA).
- Support for Structured Query Language (SQL) request from remote locations.
- Site independence but capacity for interaction with other DB2 sites.
- Multiuser support for concurrent access including making updates, deletions, and insertions.
- DB2 fits into SAA via CPI-SQL.
- An audit trail can be selectively chosen.
- DB2 can be used in an XRF [cross-system relational (database) facility] environment.
- Support for 10,000 open concurrent data sets per address space.
- Maximum number of columns in a DB2 table is 750.
- Multiple simultaneous index recovery can be performed on the same table space.

Remote spooling communication subsystem (RSCS)

RSCS, as it is commonly known, is an application that operates under VM to provide data transfer capabilities. According to IBM, RSCS has the following support:

- File transfer, message, commands, and mail between
  VM
  MVS
  VSE
  NJE
  OS/400
- ASCII support to printers and plotters
- Supports IPDS and SCS data streams
- Provides a gateway programming interface for protocols such as TCP/IP
- Supports 3270-type printers with form control buffering
- Provides the capability to share printers
- Plus other functions

RSCS provides VM users with the ability to send mail, specific messages, and jobs to other users within a SNA network. VM users use

RSCS for printing purposes. The basic function of RSCS is that the origin node starts communication with a destination node, and multiple devices may be along the path between the two. Because of the wide protocol support, RSCS can function over multiple types of links.

Local Area Network Resource Extension and Services (LANRES)

LANRES is an application that operates in MVS and VM environments. According to IBM sources, it brings the power behind the S/390 architecture to a NetWare environment. LANRES achieves this by making DASD available to NetWare servers and S/390-based printers available to NetWare clients.

LANRES also permits authorized MVS users to move data to and from a NetWare server. Additionally, NetWare server files and directories can be listed, created, and/or deleted.

LANRES also makes LAN printers available to MVS users. In effect, it brings together NetWare environments with S/390 seamlessly to take advantage of right-positioning of workloads. Another function LANRES offers is centralization of LAN management to the MVS host, if desired. It also permits MVS users to send a PostScript file to a PostScript printer on a LAN.

Conceptually, a LANRES environment would appear as depicted in Fig. 16.37.

LANRES is a versatile tool because of the connectivity solutions it supports. The following connectivity solutions are supported by LANRES:

- ESCON
- Parallel channels
- APPC connection
- Host TCP/IP connection
- VM programmable workstation services (VM PWSCS)

The method of connectivity dictates how LANRES is configured on the host. Because of the breadth in support for connectivity solutions, the requirements, installation, and definitions are site-dependent and are directly related to how the product is used. For example, if the product is used with TCP/IP under MVS, LANRES uses sockets and TCP for connectivity. However, if APPC is used, then LANRES connects to APPC MVS via CPI-C, conforming to SAA standards.

IBM has many other software products, too many to list here. If your needs have not been covered in this section, contact your local IBM representative for additional information.

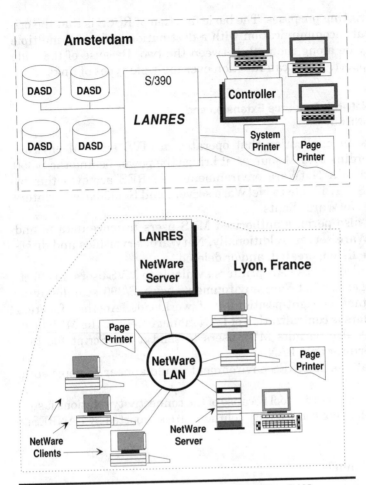

Figure 16.37 Conceptual view of LANRES and NetWare LANs.

16.5 SNA by Traditional Layers

In 1974 IBM announced SNA. It began as a layered architecture, and traditional SNA remains as such. *Traditional SNA* is defined as being more hierarchical than *peer-oriented SNA*. Succinctly, it is pre-1992 with the "Networking Blueprint" announcement. This does not mean that the Networking Blueprint replaces the functionality of traditional SNA; rather, it provides a different approach to networking in general. More information on the Networking Blueprint will be presented later in this chapter.

By layers, SNA appears conceptually as shown in Fig. 16.38. The basic functions at these layers include

7	Transaction Services
6	Presentation Services
5	Data Flow Control
4	Transmission Control
3	Path Control
2	Data Link Control
1	Physical Control

Figure 16.38 SNA traditional layers.

1. That point where a link is made between two or more nodes.

2. Responsibility for moving data across a link.

3. Routing and flow control.

4. Throttling data movement and performing security functions if required.

5. Synchronizing, correlating, and grouping data.

6. Formatting data to protocol.

7. Providing application-required services.

In many installations this list describes the functionality of data flow within the network. This model of networking is considered traditional SNA.

16.6   IBM's Blueprint for Networking

On September 15, 1992 IBM announced their Networking Blueprint. This blueprint is a new approach to IBM networking. It provides a framework for selecting features or parameters to fit the needs of a particular networking situation. Figure 16.39 is a representation of that framework.

As Fig. 16.39 shows, the structure of the blueprint is entirely different from the seven-layer traditional SNA. According to IBM, the structure of the blueprint consists of four major layers, three switching boundaries, and what IBM calls the *systems management plane*. The four layers are addressed first, then switching boundaries and the systems management plane.

Subnetwork layer

This layer constitutes the lower layer. IBM divides this layer into four categories:

- Local area network (LAN)
- Wide area network (WAN)

- Channel

- Emerging

These categories can be further divided into protocols. For example, protocols that would generally fall into the LAN category include

- ETHERNET

- Token ring

- FDDI

Protocols applicable to the WAN category could include

- FDDI

- Frame relay

- SDLC

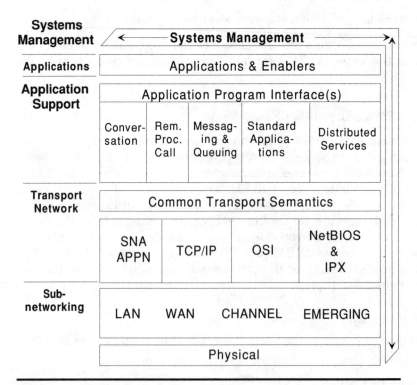

Figure 16.39 IBM's Networking Blueprint.

Protocols applicable to the channel category could include

- Byte multiplexer channel
- Block multiplexer channel
- ESCON

Protocols applicable to the emerging technology category include

- Asynchronous transfer mode (ATM)
- Fast ETHERNET

This list is not exhaustive, but it conveys the idea behind the supported protocols at this layer. And to a considerable degree these can be selected for what best fits the site requirements.

Transport/network layer

This layer is represented by six networking capabilities, which apply differently according to the following supported protocols:

- SNA
- APPN
- TCP/IP
- OSI
- NetBIOS
- NetWare

How this layer functions is contingent on the protocol selected. This means that TCP/IP works in a particular fashion whereas APPN works according to its structural definitions at this layer. Users have the option to select the protocol of choice.

Application support layer

This layer provides service support for applications. According to IBM, the prevalent interfaces and services that work at this layer are

- *Conversational:* Known as CPI-C; this deals with streams of related interactions.
- *Remote procedure call:* Known as RPC; this is capable of passing parameters to a subroutine.
- *Message queue interface:* Known as MQI; this manages the queues that relate messages.

The following services may also apply at this layer:

- Distributed system services.

- Different vendor applications such as

  TELNET for remote log-in

  FTP for file transfer

  SNMP for network management

- Other non-transport-layer-dependent applications

Application layer

Applications inherent to the protocols available are at this layer. These applications include print, mail, file transfer, remote log-on, and other services.

Switching boundaries

The applicable switching boundaries include

- Application program interface (API)

- Common transport semantics (CTS)

- Subnetwork-access boundary (SAB)

The *API* switching boundary serves a primary purpose; that is, to make the underlying architecture transparent. The *CTS* switching boundary enables any protocol above it to access any protocol below it. The *SAB* switching boundary resides between the transport/network part and physical part of the blueprint. It serves to make link services available to the protocols driving the network.

IBM's stance with this blueprint indicates a more flexible networking approach than does traditional SNA. IBM has two documents that are good references on the Networking Blueprint:

#GC31-7057   *Networking Blueprint Executive Overview*

#GC31-7074   *Multiprotocol Transport Networking (MPTN) Architecture: Formats*

To obtain these and other IBM documents, contact your local IBM representative.

16.7 Traditional Concepts

This section explains terms and concepts that make up the core of SNA. SNA as a network protocol that is implemented via hardware and software.

Nodes

The term *node* is used in IBM documentation, and depending on the context, it can take on different meanings. In traditional SNA, which is sometimes referred to as *subarea SNA,* different types of nodes exist. Those popular nodes in subarea SNA include

- *Host node:* Also known as a *subarea node*; it provides end-user services. It is a type 5 node. (These "types" will be explained later.)
- *Communication controller node:* This refers to a communication controller [also known as a front-end processor (FEP)]. It is a type 4 node.
- *Peripheral node:* This is a cluster controller or establishment controller. Depending on the device, it may be a type 2.0 or 2.1 node.

Figure 16.40 shows all three node types.

Subareas

In traditional SNA, subareas exist. No areas are defined. A *subarea* is defined as one of the following:

- A subarea node and peripheral node(s)
- A subarea node
- A subarea node and communication controller node

Figure 16.41 depicts these three subarea types.

Network-accessible units

IBM defines a network-accessible unit (NAU) categorically as such

- System services control point (SSCP)
- Physical unit (PU)
- Logical unit (LU)

These NAUs are explained by characteristic, function, and location.

System services control point (SSCP). The *SSCP* is a controlling point in SNA. It is located in the virtual telecommunications access method (VTAM). Some of its characteristics and functions include

- Network control
- Session management
- Resource activation
- Resource deactivation

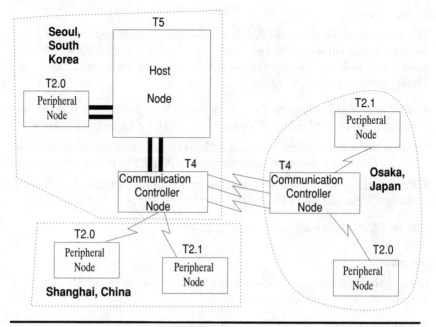

Figure 16.40   Conceptual view of nodes.

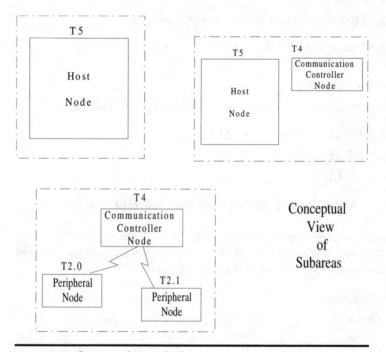

Conceptual
View
of
Subareas

Figure 16.41   Conceptual view of subareas.

- Focal point for receipt of PU data
- Passing data to and from NetView
- Executing commands

Conceptually, the SSCP and VTAM would appear as shown in Fig. 16.42.

Physical unit. A *physical unit* (PU) pertains more to the functions and capabilities of a node than to those of a particular hardware device. The PU type is architecturally related to and part of microcode and software. Some basic characteristics and functions of PUs include

- A PU is defined in software or microcode.
- A PU receives messages from an SSCP.
- PUs provide internal network functions, not user-related functions.
- Participating entities in a SNA network are known by their node type.
- A PU manages links and link stations.
- A PU sets up virtual and explicit routes in certain nodes.
- A PU communicates with one or more control points

Four types of PUs are presented here and are best described by their functional characteristics.

- PU type 5
- PU type 4
- PU type 2.0
- PU type 2.1

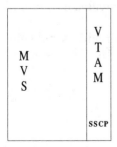

Figure 16.42   Conceptual view of a SSCP.

A PU type 5 (T5 or PU5) node is a host subarea node. It is a processor, practically speaking, or can provide T5 functions, including

- Managing subarea resources
- Facilitating session establishment
- Monitoring resources

A PU type 4 (T4 or PU4) node is a communication controller node. It is a FEP or has the capability to emulate PU4 functions, which are

- Communicating with the physical unit control point (PUCP)
- Managing its peripheral nodes
- Communicating with a SSCP

A PU type 2.0 (T2.0 or PU2.0) node is a peripheral node. This type of node is totally dependent on a T5 node for session establishment. Type 2.0 node functions include

- Communicating with a T4 PUCP or a SSCP
- Monitoring its local resources
- Sending status-related data to a SSCP

For practical purposes it is correct to associate a physical device with a PU, but this can be elusive because some devices can act as different PU types depending on how they are GENed (generated). Also, the location where terminals and printers connect on controllers can be considered a LU. Granted that IBM documentation does not say this (at least I have not seen any), but it is accurate. The same holds true for PU2.1 devices.

A PU2.1 node is also a peripheral node. Its control point is called just that: a *control point*. This node differs from a T2.0 node because it supports peer communications to some degree. It can perform the functions of a T2.0 node, but it can also perform functions native to T2.1 architecture. Additional details on T2.1 nodes are provided in Chap. 17.

Logical units. IBM defines a *logical unit* (LU) as an addressable endpoint. This applies to hardware and software. Six LUs of concern here and their functions (protocol support) are as follows:

LU0   Create your own program.

LU1   SNA character string (SCS) printing.

LU2   A 3270 data stream.

LU3   A 3270 data stream for printers.

LU6.2   Advanced program-to-program communication (APPC) protocol.

LU7   5250 data stream for AS/400 systems.

Two categories of LUs exist: dependent and independent. The fundamental difference between the two—after initial download of tables—is that the former require VTAM for session establishment and the latter do not.

Figure 16.43 is an example of hardware, software, and concepts presented to this point. It is a good example to help the reader get a holistic perspective of SNA.

Figure 16.43 shows two processors, with software on each; communication controllers; an establishment controller; cluster controller;

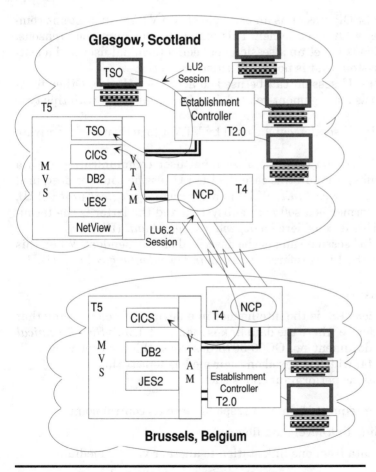

Figure 16.43 Conceptual view of a typical SNA network.

printers; and terminals. It also shows two CICS subsystems communicating with one another via a LU6.2 session. Additionally, it shows a terminal communicating with TSO via LU2 protocol.

Sessions

As mentioned previously, sessions are defined as a logical connection between two endpoints. Four types of sessions are considered here:

- SSCP-SSCP
- SSCP-PU
- SSCP-LU
- LU-LU

The SSCP-SSCP session is an example of two VTAM subsystems communicating with one another. This may occur for numerous reasons, one of which is to set up a session between a terminal user and a software subsystem that is not in the same processor.

An SSCP-PU session can be used to activate a device. Other functions, such as management-related data flows, can occur with this type of session.

An SSCP-LU session can be used by VTAM to activate or deactivate a LU.

A LU-LU session could be a terminal user communicating with a software subsystem. In this case the LU-LU session could be described as a *primary logical unit* (PLU) and a *secondary logical unit* (SLU), where the former is a software subsystem and the latter is the terminal user. This is considered a *dependent logical unit* (DLU).

The LU-LU session can also be described as independent. When this is the case, the LU is referred to as an *independent logical unit* (ILU).

Link stations

A *link station* (LS) is the intelligence in a device defined as being that point in a device where the data link is managed. IBM's *SNA Technical Overview*, document no. GC30-3073 describes this concept very well. Figure 16.44, which is based on their model, depicts this.

Link stations perform functions such as

- Receiving requests from and responding to its control point
- Controlling link-level data flow
- Moving data from one link station to another via the medium
- Managing error recovery at the link level on the node

Link

In SNA a *link* refers to the data link. There are several types of links, including

- Parallel channel
- ESCON
- Frame relay
- Token ring
- SDLC
- ETHERNET

Technically, IBM defines a link as that connection between two link stations. Hence, this includes the medium, DCEs, and link connection. The *link connection* is defined as that part consisting of the DCEs and transmission medium.

Domains

Another concept in SNA is the *domain,* which is defined simply as that area whose components have a single point for control. In a T5 node the control point is the SSCP. In a T4 node this is the NCP.

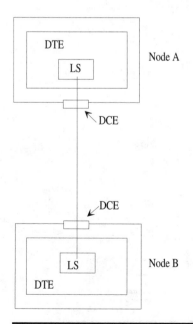

Figure 16.44 Conceptual view of a link station.

Parallel to the concept of domains is that of *ownership.* In SNA all resources are owned; the question is by which device. This concept of ownership is related to the domain. Resources in a given domain are normally owned by a control point in that domain.

Now that the concept of a domain is understood, it is reasonable to understand the concept of the *cross-domain resource,* which is defined as a resource in a domain other than where the requesting party is located.

Figure 16.45 is an example of two different domains.

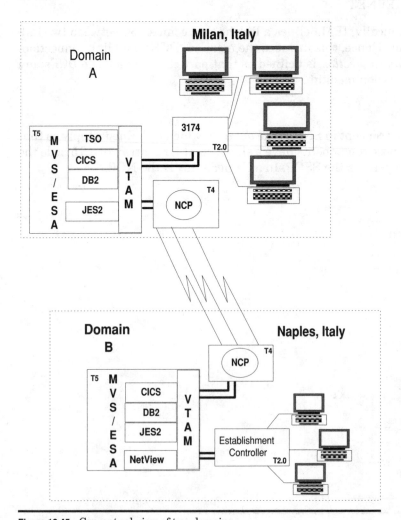

Figure 16.45 Conceptual view of two domains.

In Fig. 16.45 two processors have application subsystems. Each processor owns the applications on that host. If defined appropriately to VTAM, users in domain A can access application subsystems in domain B and vice versa.

16.8 SNA Protocol Structure

SNA protocol structure can be explained by layers. Figure 16.46 is a view of what SNA considers as a message unit.

Three distinct components of the message unit are

- Basic link unit
- Path information unit (PIU)
- Basic information unit (BIU)

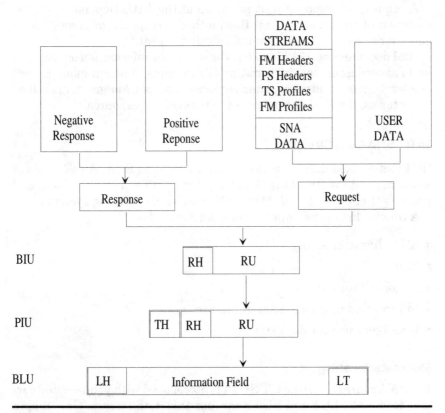

Figure 16.46 Structured view of the message unit.

Basic link unit

The *basic link unit* (BLU) is assembled at the data-link layer of the network. It includes data and protocols that have been passed from layers above it. In front of the BLU is a *link header* (LH). The information field is next with all other fields above it. Next, the link trailer is the end of the message unit.

Path information unit

The *path information unit* (PIU) consists of the *transmission header* (TH), *request* (or *response*) *header* (RH), and the *request* (or *response*) *unit* (RU). The PIU operates at that layer in the network responsible for routing data (message units) through SNA.

Basic information unit

The *basic information unit* (BIU) consists of a request or response header and a request or response unit.

A request or response unit is on top of the BIU. Depending on the direction of the message unit flow, either a response or some type of end-user or SNA data stream information is sent.

IBM document no. GA27-3136 is the source for information on message units as well as on data streams, profile concepts, function management header concepts, and request and response concepts. For greater detail on these topics, the IBM document is the best reference source.

16.9   SNA Data Streams

IBM has defined data streams which are used in SNA. An example of a dominant data stream is the 3270, which is used by terminals and printers that operate with MVS, VM, and VSE operating systems.

A concise list of the data streams used includes

- SNA character string
- 3270
- General data stream
- Information interchange architecture
- Intelligent printer data stream

SNA character string

The *SNA character string* (SCS) is a protocol used with printers and certain terminals. LU1 and LU6.2 can use this data stream. One unique aspect of this data stream is its lack of data flow control functions.

3270

The 3270 data stream is characterized by containing user-defined data. It also includes commands that aid in LU-LU control. LU2, and LU3 use this data stream for terminals and printers, respectively. It can also be used by LU6.2 as an optional data stream.

General data stream

The *general data stream* (GDS) is used by transaction programs to interpret data records as they were sent and received. This data stream is used by LU6.2.

Information interchange architecture

The *information interchange architecture* (IIA) data stream is used to define a collection of data streams. It is used by applications exchanging programs. This means that *open-document architecture* (ODA) can be used. *Document-content architecture* (DCA) can also be used.

Intelligent printer data stream

The *intelligent printer data stream* (IPDS) is used between a host and a printer. It is a data stream that is used with an all-points addressable printer. IPDS can intermix text and graphics—both vector- and raster-based.

16.10   Profile Concepts

Two profiles for examination here include the transmission service and function management profile.

Transmission service profiles

These profiles are used at the transmission layer in the network. They represent protocols that may be selected at session activation.

TS1	Used with SSCP-PU and SSCP-LU sessions
TS2	Used with LU-LU sessions
TS3	Also used on LU-LU sessions
TS4	Used on LU-LU sessions
TS5	Used on SSCP-PU sessions
TS7	Used on LU-LU sessions
TS17	Used on SSCP-SSCP sessions

These profiles provide a variety of services. They are used at a developmental and debugging level.

Function management profiles

FM0   Used on SSCP-PU and SSCP-LU sessions

FM2   Used on LU-LU sessions

FM3   Used on LU-LU sessions

FM4   Used on LU-LU sessions

FM5   Used on SSCP-PU types 5 and 4 sessions

FM6   Used on SSCP-LU session

FM7   Used on LU-LU sessions

FM17   Used on SSCP-SSCP sessions

FM18   Used on LU-LU sessions

FM19   Used on LU-LU sessions

Like the transmission service profiles, the function management profiles provide a variety of service based on the session type and need. For in-depth information about these profiles, refer to IBM manual no. GA27-3136.

16.11   Function Management Header Concepts

The concept underlying a function management header (FMH) is if a session supports these headers a request header can contain a option indicating that a FMH is present. If present, they indicate specific functionality. Consider the following:

FMH1   Used to select a destination logical unit.

FMH2   Used to handle data management for a task.

FMH3   Used for the same purposes as FMH2, but does not have a stack reference.

FMH4   Carries logical block commands that are used to define different parameters.

FMH5   A LU6.2 ATTATCH header; used to carry a request for a conversation. With non-LU6.2 ATTATCH this originates from the sending half-session program to the destination manager.

FMH6   Used during an active transaction program conversation.

FMH7   Provides error information for LU6.2; used for non-LU6.2 in a similar fashion.

FMH8   Used with IMS applications with LU6.1 protocols.

FMH10   Prepares a session for sync-point processing.

FMH12   Used with LU6.2 for security.

16.12   Request/Response Header (RH) Concept

These headers are used to perform bit-level operations inside message units. IBM identifies an exhaustive list of RHs to accomplish a variety of tasks. Some of the functions they perform include

- Providing a format indicator
- Indicating sense data
- Indicating the beginning of a chain
- Indicating the end of a chain
- Indicating the type of response
- Requesting a larger window
- Indicating the beginning and end of a bracket

The formats of the RH are dependent on the type of session used. Details provided by these formats are used in the formatting of SNA data.

16.13   SNA Commands

SNA commands vary according to the type LU and session used. Some commonalities exist in theory, but specific commands can differ. Consider the following details on command flow and SNA commands.

Theory of command flow

Assume a terminal user wants to sign on to TSO. What are the theoretical operation and the commands that flow between the two? Figure 16.47 depicts this scenario.

A brief summary of some commands that flow between the terminal user and TSO includes

1. A user enters TSO and it is received by VTAM. VTAM sees this as a character-coded log-on.

2. A log-on exit is scheduled for the primary logical unit.

3. After the PLU receives control of the log-on exit, the PLU passes an open-session request to the SSCP.

4. As a result of the open-session request, the BIND command is sent to the secondary logical unit (SLU).

5. Assuming that the terminal sends back a positive response, the session is bound.

6. If a negative response is returned, a BIND failure command is generated.

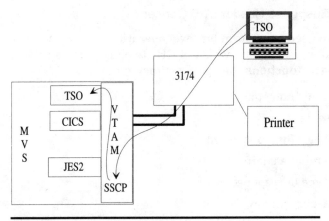

Figure 16.47   Conceptual view of a TSO user.

SNA commands

IBM has an entire manual describing SNA commands: *SNA Formats,* document no. GA27-3136, but SNA commands fit into the request or response structure explained previously. An example of some SNA commands and their functions includes

ACTLU	Activate logical unit
ACTPU	Activate physical unit
DACTLU	Deactivate logical unit
DACTPU	Deactivate physical unit
BIND	Activate a session between LUs
CDINIT	Cross-domain initiate sent between two SSCPs
CINIT	Control initiate; request the PLU to send a BIND
LUSTAT	Used to send status information
NOTIFY	Used to synchronize awareness of an SSCP and PLU
SDT	Start data traffic
SESSEND	LU notification to the SSCP that a session has ended
UNBIND	Send two LUs to UNBIND

16.14   Flow Control

SNA has a method for data control. SNA controls data partially by the following three methods.

Explicit route

In SNA, an *explicit route* (ER) is a defined set of nodes and transmission groups (TGs) of a path. For example, an explicit route could be subarea node X, TG2; subarea node T, TG2; and subarea node D. An explicit route is the definition of a path in subarea SNA. It is a physical connection.

Virtual route

A *virtual route* (VR) is a logical connection between two endpoints. A virtual route is mapped to explicit routes. Consequently, a virtual route reflects the characteristics of an explicit route. For example, in most scenarios where multiple FEPs are installed, multiple links connect them. These links are physical and are defined as explicit. The logical route is then mapped to the route that best fits the need of the session.

Class-of-Service

Class-of-service (COS) includes characteristics such as transmission priority, bandwidth, and security. With a class-of-service the following can be defined:

- Providing response times reflecting high priority
- A class reflecting best availability
- A class with higher levels of security
- A class for batch processing

Transmission priority is also used with flow control. The combination of these abilities makes flow control possible in the network.

16.15   Advanced Program-to-Program Communication

Advanced program-to-program communication (APPC) is IBM's premier peer-oriented protocol. It is based on LU6.2. The flexibility of the protocol enables it to be implemented across a variety of platforms.

Origins and evolution

APPC originated in the early 1980s. It evolved from limited support to its current support by MVS/ESA and many application susbsystems operating under VTAM. Some benefits of using APPC include

- One protocol can be used across a variety of architectures.
- It provides security.
- It offers a distributed approach to transaction processing.
- It offers multiple ways to create transaction programs.

APPC now is widespread among MVS, VM, and VSE operating systems. It is also fundamental to APPN. APPC is widespread in the marketplace with third-party vendors.

Conceptual overview

The idea behind APPC is peer communication between programs. This means that customized programs can be written to utilize the power behind APPC. Consider an example of two banks, one in Dallas and the other in Jackson (Miss.). Now assume that daily information needs to be exchanged between bank D in Dallas and bank T in Jackson. Figure 16.48 shows an example of two programs exchanging information between the two banking institutions.

Conversations

Sessions have been explained as being a logical connection between two endpoints. A *conversation* is communication between two or more transaction programs using a LU6.2 session through a defined independent logical unit (ILU). Figure 16.49 explains this concept.

Figure 16.49 shows the following components:

- Node A
- Multiple transaction programs (TPs)
- A LU6.2 session
- A point where an ILU is defined
- A conversation between two TPs

Figure 16.49 shows the idea behind a conversation. In this figure any of the transaction programs on node A can communicate with any of the TPs on node B.

Two types of conversation exist: basic and mapped. A *basic conversation* provides a low-level interface for those transaction programs needing support for privileged functions. A *mapped* conversation is the protocol boundary like the basic conversation, except that it enables arbitrary transmission of message format. System- or user-defined mappers can be used.

Transaction programs

IBM defines a *transaction program* as an application that is executed within the LU6.2 protocol. It is a type of application, typically user-written, designed to meet the needs of a specific installation.

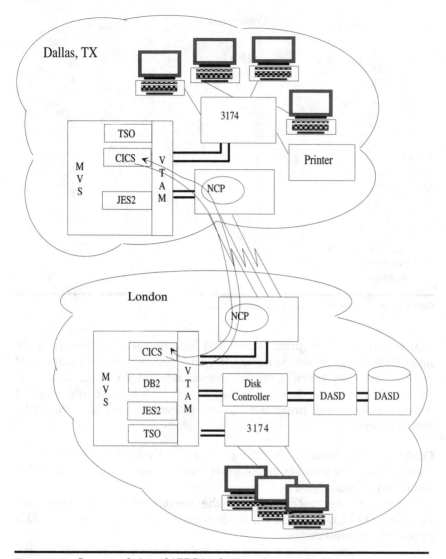

Figure 16.48 Conceptual view of APPC implementation.

Types of verbs

APPC is a high-level language and uses "verbs" to achieve communication. Two categories of verbs exist: they are conversation and control operators.

Conversation verbs. The verbs in this category include mapped, basic, and type-independent. The first, mapped verbs, are used by applica-

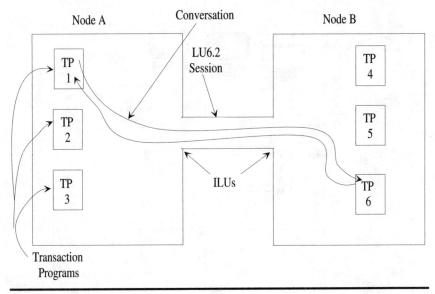

Figure 16.49   Conceptual view of a conversation.

tion programs. They provide services for programs written in high-level languages. Basic verbs are used by LU service programs that provide end-user services or protocol boundaries for application programs. Type-independent verbs can be used with basic and mapped conversations. They provide a variety of generic services needed by both conversations.

Control operator verbs.   This category of verbs includes subcategories, including change number of session (CNOS), session control, LU definition, and miscellaneous.

CNOS verbs are used to change the session limit that controls the number of LU-LU sessions per mode name available between two LUs allocated for conversations.

Session control verbs are used for session control. This includes activation and deactivation sessions and deactivate-conversation groups.

LU definition verbs define or modify local LU operating parameters.

Miscellaneous verbs are those verbs needed but not defined to another category

LU6.2 session considerations

The following concepts and functions are a part of APPC:

- Parallel sessions
- Single sessions

- Session pools
- Session selection
- Session limits
- The concept of contention
- Winners and losers

Parallel sessions are based on the concept of multiple pairs of sessions communicating with the same pair of LUs. Typically, one pair of TPs use a session at a given instance. LU6.2 supports multiple concurrent sessions. Applications must be capable of multiple session support, including the processors and workstations. The concept of multiple session support is called *parallel sessions.*

Single sessions are defined as LUs that cannot support more than one session against a given LU in a given instance.

Session pools are a collection of named LUs that contains active sessions which can be allocated to different conversations if required.

Session selection is the way a transaction program controls selection of a session. TPs cannot control session selection directly, but via the mode name parameter they can map this to a set of characteristics.

A *session limit* is simply the maximum number of sessions that can be active at a given LU at one time.

The *concept of contention polarity* is the theory that two LUs attempt to initiate a session simultaneously. Contention polarity is a method of preventing this by defining multiple LUs for operation to function as a contention winner or contention loser. Typically, multiple winners are defined and losers are defined in each node to prevent a state of contention.

LU6.2 sync-point processing

In LU6.2 sync-point processing lets transaction programs synchronize their resources at specified time periods called *sync points.* This is important because multiple transaction programs are exchanging data; thus TPs must be in "sync."

Many additional concepts exist with LU6.2. Many books have been written on this topic. A helpful one is IBM's *SNA Transaction Programmer's Reference Manual for LU Type 6.2,* document no. GC30-3084.

16.16   Summary

This chapter covers a lot of information about the components that make up SNA and SNA terms and concepts. SNA can be very large in implementation. Volumes of manuals are available from IBM about any of the topics presented in this chapter.

However, the purpose of this chapter is to familiarize you with the architecture, hardware and software, SNA terms, concepts, and popular topics that arise when SNA is discussed. The chapter's intent is not to replace the original manufacturers' manuals, but rather provide helpful reference information to those working with SNA.

17

Advanced
Peer-to-Peer
Networking (APPN)

In 1986 IBM announced T2.1 node support with the System 36 (S/36) and also introduced its *SNA Type 2.1 Node Reference* manual, document no. SC30-3422. It defined the beginnings of APPN implementations. Nodes can be understood best by knowing the services they provide or characteristics that represent them. Both architectural design and characteristics supported define a node type.

It is best to view APPN from this abstract viewpoint because it does not matter (to a certain degree) which host or model number a particular device is or is not. The core of APPN is that it uses LU6.2, not that a certain device exists. The question to ask is: "Which functions or features are supported by a certain device with respect to APPN?"

17.1 Origins and Evolution

Version 1

Version 1 can be characterized by low-entry networking nodes, end nodes, and APPN end nodes. According to the IBM *Type 2.1 Architecture Manual* no. SC30-3422, published in 1991, these two references to an end node are synonymous.

IBM documentation also states, "a T2.1 node is that node which uses protocols that require less system requirements." A T2.1 node provides peer connectivity and session-level connectivity using LU6.2 protocols.

APPN Version 1 can best be described as an evolution. According to the *Type 2.1 Architecture* manual, three types of nodes were identifiable with T2.1 architecture:

- APPN network node (NN)
- APPN end node (APPN EN)
- Low-entry networking end node (LEN EN)

APPN NNs functions are

- Performing intermediate session routing
- Performing directory searches and route selection
- Providing LU-LU service for its local LUs

Consider Fig. 17.1.
   APPN ENs can

- Perform limited directory services
- Register their LUs with a NN
- Be attached to multiple NNs

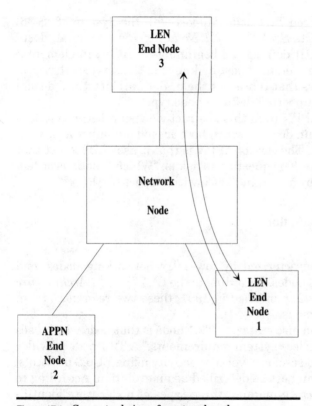

Figure 17.1   Conceptual view of a network node.

Figure 17.2   Conceptual view of two APPN end nodes.

Consider Fig. 17.2.
   LEN ENs

- Cannot register its LUs with a NN
- Must have predefined remote LUs via the system
- Use T2.1 protocols *without* APPN enhancements

Consider Fig. 17.3.
   Traces of Version 1 can be seen through IBM product offerings from 1986 to 1992. An example of one such offering supporting APPN is the AS/400, announced in 1988. Other products were announced during that period, but a significant break with Version 1 came in 1992.

Version 2

APPN Version 2 can be identified by IBM's *APPN Architecture Reference* no. SC30-3422, published in March 1993. In these 1000-plus pages, IBM clearly defines extensions to earlier APPN and LEN networking.

   It seems that IBM started to bring APPN together in 1992 with the announcement of VTAM Version 4 Release 1. In short, VTAM 4.1 (shipped in May 1993) permits VTAM to participate with other nodes in an APPN network and appear as another peer node. Technically this capability was available in VTAM Version 3 Release 2; then support for causal connections followed in VTAM 3.3, according to VTAM manuals.

Figure 17.3   Conceptual view of two LEN end nodes.

Correlations with T2.1 architecture

APPN has roots in T2.1 architecture. It has extended beyond that, but nevertheless its beginnings can be traced back to the mid-1980s with T2.1 node architecture. APPN has evolved into its present offering, and it seems that IBM will continue to enhance it, thus perpetuating its evolution.

APPN simplifies network definition. It also permits dynamic route selection. APPN also provides a distributed directory service. This function determines remote LUs that may be known locally only by name. This means that manual definitions for routes or location of remote LUs are not required.

APPN implementations can select routes based on user-defined criteria. A component in each NN called a *control point* (CP) is used to determine the best route from the initiating LU to the destination LU. APPN also supports *intermediate session routing,* which is routing data through the NN for sessions that do not originate or terminate with that NN. Transmission priority is established according to the user's class-of-service specification.

17.2 Node Types

APPN's approach to networking parallels that of the client/server. This architectural nature lends APPN to a "peer"-oriented network. The pivotal issues are what node types exist, what functions they perform, and what additional APPN option sets exist. These issues are explained in the remainder of this section.

APPN network node (NN)

A major role of the NN is performing the function of *server.* In this context, other nodes participate as *clients.* This concept of client/server is similar, but not identical, to that of TCP/IP client/servers. A NN functions as a server to the end nodes attached to it. The NN and its attached end nodes are considered a domain. Some NN services include

- Directory services

- Route services

- Intermediate LU-LU routing

- End node management services

- LU-LU session services

- Support for any APPN or LEN node attachment with the same network identifier

- Functioning as a *server* for its clients
- Support for SNA subarea boundary nodes

APPN end node (EN)

In light of the client/server parallel that APPN purports, an end node functions as a client. End nodes support LU6.2. Without a NN, ENs can communicate via LU-LU sessions only with the partner LU located in an adjacent node. However, with a NN an EN can communicate with remote LUs. Such nodes have the ability to inform a NN of their local LUs. ENs can have active links to multiple nodes at any given time, but an EN can have CP-CP sessions with only one NN at a given instance. ENs can have attachments to multiple NNs in case one NN fails. Examine Fig. 17.4.

This type of node can make a connection to any LEN or APPN node. An EN cannot have CP-CP sessions with another EN.

LEN end node

This node implements basic T2.1 protocols; no APPN enhancements are included. This type of node is not capable of having a CP-CP session. Connections with destinations must be predefined. A LEN EN communicates with remote LUs by system definition. At system defin-

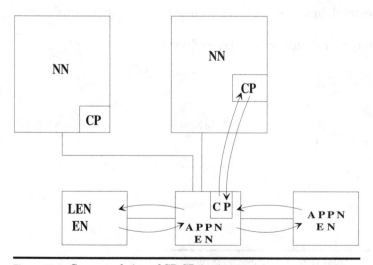

Figure 17.4   Conceptual view of CP-CP session.

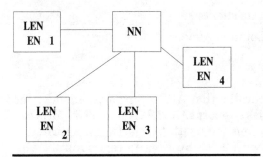

Figure 17.5   LEN ENs defined to the NN.

ition time the CP name of its adjacent node is defined and determines how the local LUs can be accessed. The LEN EN accesses some remote LUs by the services provided via the NN server functions. Consider Fig. 17.5.

Peripheral border node

These nodes do not support intermediate network routing, but they do support the following:

- Directory services
- Session establishment
- Route selection
- Session establishment of LUs between adjacent subnetworks

Figure 17.6 is an example of this type of node.

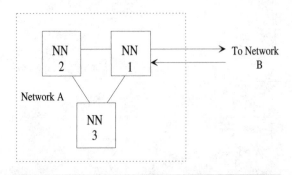

Figure 17.6   Conceptual view of a peripheral border node.

Extended border node

These nodes support intermediate network routing, but the subnetworks must be predefined. Additionally, they provide the following:

- Directory services
- Session establishment
- Route selection
- Partitioning of a subnetwork into two or more clusters

Figure 17.7 depicts four networks with a LU-LU session between peripheral networks via intermediate networks with extended boundary node function support.

APPN-subarea interchange node (IN)

This is a T5 node in SNA feature and function. It permits connectivity between APPN networks and SNA networks. An IN achieves this by mapping routing, directory, session setup, and route selection for both network types. Figure 17.8 is an example of this.

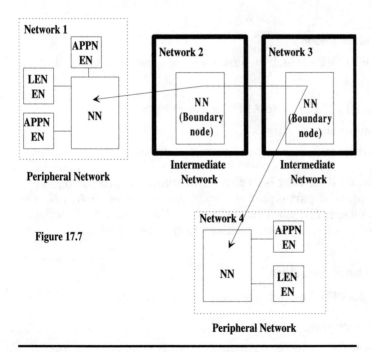

Figure 17.7

Figure 17.7 Conceptual view of an extended border node.

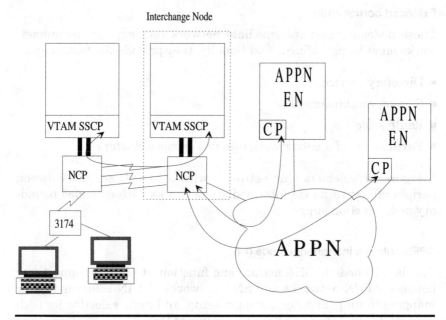

Figure 17.8 APPN and subarea SNA via an interchange node.

Migration data host node

Beginning with VTAM 4.1, a host node can perform the following functions:

- Emulate a APPN EN in an APPN network
- Support subarea connections
- Perform cross-domain resource manager functions

The migration data host is dedicated primarily to processing applications; it does not participate in broadcast searches in APPN networks. However, it functions not only as an EN in an APPN network but also to support subarea connections in traditional SNA.

17.3 APPN Node Structure

At the core of APPN lie the three nodes mentioned previously:

- APPN network node
- APPN end node
- LEN end node

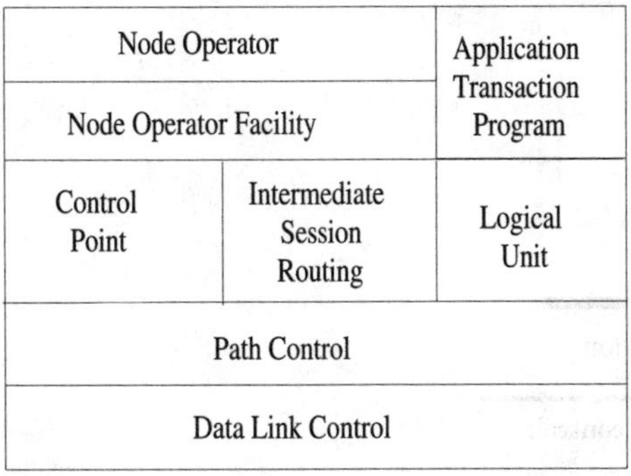

Node Operator		Application Transaction Program
Node Operator Facility		
Control Point	Intermediate Session Routing	Logical Unit
Path Control		
Data Link Control		

Figure 17.9 Conceptual view of APPN node structure.

These nodes are built around the structure displayed in Fig. 17.9. Components of the node are

- Data-link control
- Path control
- Logical unit
- Intermediate session routing
- Control point
- Node operator facility

Each of these is explained below. These components have subcomponents and multiple functions. They, too, are explained by the aforementioned category.

Data-link control

The *data-link control* (DLC) is the interface with the link connection. It provides data-link protocols and establishes communication between nodes, maintains the synchronization between the two, gives acknowledgments, performs error recovery when required, and sequences data flow. Figure 17.10 shows how the DLC appears conceptually.

This part of the node is responsible for communication with the physical link. It consists of two components. The DLC element is responsible for the following functions:

- Moving data to the physical medium
- Performing retransmissions
- Moving data to and from other DLC elements

```
┌─────────────────────────────────────────┐
│                 DLC                       │
│  ┌──────────────┐   ┌──────────────┐     │
│  │ DLC Manager  │   │ DLC Element  │     │
│  └──────────────┘   └──────────────┘     │
│                            ▲              │
└────────────────────────────┼─────────────┘
                             ↕
                    ━━━━━━━━━━━━━━━━━
                Physical Link Connection
```

Figure 17.10   Conceptual view of data-link control.

- Managing the DLC and path control (PC) boundary
- Receiving traffic from the session

Figure 17.11 depicts the position of the DLC element within the node. The DLC manager performs the following functions:

- Activation of the DLC element
- Deactivation of the DLC element
- Activation of links
- Deactivation of links
- Passing parameters to the CP when a station becomes operative or otherwise controls the boundary between the CP and the DLC

Figure 17.12 depicts the relationship between the DLC manager and other node components.
    Figure 17.13 shows how the DLC communicates with the medium, internally, with the session, and with the CP.

Path control

The node's PC component, like the DLC component, has two subcomponents: an element and a manager. The main functions of these PC

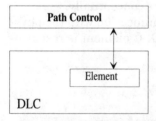

Figure 17.11   The role of the DLC element.

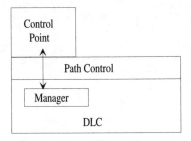

Figure 17.12 Functionality of the DLC manager.

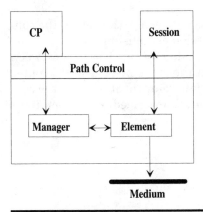

Figure 17.13 Functionality of DLC as a whole.

components are listed below. Figure 17.14 depicts the PC components and their relationship with the DLC component.

- Element functions:

  Error checking in messages received from the DLC element
  Generation of segments for outbound messages
  Message conversion from messages received from the DLC
  Prioritization of message transmission to the DLC component
  Route messages between the PC manager, half session (HS), session connector (SC), and the DLC component

- Manager functions:

  Session connection
  Session disconnection
  Stopping outbound data traffic on notification
  Interaction with the CP

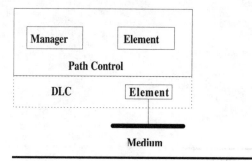

Figure 17.14 Conceptual view of the path control.

Figure 17.15 shows the relationships between path control, the control point, and the parts that aid in session establishment.

General characteristics of the PC include routing messages between destination nodes and LUs residing in the same node. The PC routes messages from the DLC to the appropriate component, such as the control point (CP), the logical unit (LU), or the intermediate session routing (ISR) component.

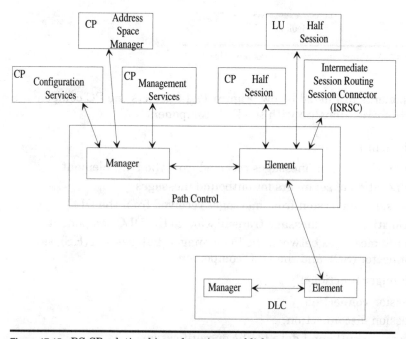

Figure 17.15 PC-CP relationship and session establishment.

Logical unit

The LU has many parts that enable it to communicate with the PC, LUs in another node, and other components within the same node. Figure 17.16 is a conceptual view of an LU in relation to the APPN node structure as a whole.

The function of the LU is to serve as a port (addressable point) for application transaction programs. The components that make up a LU include

- LU-LU half session
- Session manager
- Presentation service component
- Resource manager
- Service transaction programs

LU-LU half session. This part of the LU comprises two components: data flow control and transmission control. The half-session component controls local and remote LU communications. The data flow control part of the half session performs the following functions:

- Creates request or response headers (RHs)
- Ensures that proper RH parameters are in place
- Ensures proper function management (FM) profile for the session

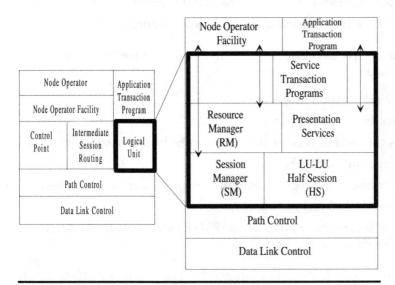

Figure 17.16 The LU with respect to overall node structure.

- Manages bracket protocol
- Flushes rejected brackets
- Is responsible for generating chaining

The transmission control part of the LU performs these functions:

- Performs session-level pacing
- Examines received sequence numbers for possible BIU errors
- Reassembles request or response units (RUs)
- Enforces the exchange of cryptography verification when it is used
- Enciphers session cryptography when used
- Deciphers session cryptography when required
- Provides reassembly for RUs that have been segmented

Session manager.   This component performs these functions:

- Sends the BIND
- Is the recipient of the BIND
- Creates half-session instances
- Connects half sessions to the path control
- Supplies session parameters during the BIND exchange
- Negotiates parameters during the BIND exchange
- Informs the resource manager when session outage occurs

Presentation services component.   This component performs the following functions:

- Calls a transaction program
- Loads a transaction program
- Keeps the send or receive state alive with the transaction program
- Puts data into logical records
- Maps transaction program data into mapped conversation records
- Confirms logical record length
- Generates function management (FM) headers for an ATTACH and provides error information

Although other LU presentation service component functions exist, these are the major ones.

Resource manager (RM). This component works in conjunction with presentation services and conversations flowing between transaction programs. Some basic functions of the RM are that it

- Creates presentation service instances
- Destroys presentation service instances
- Creates conversation resources
- Connects conversation resources to the half session and to presentation services
- Destroys conversation resources
- Maintains data structures
- Enforces session-level security
- Generates the FM Header 12 (security header)

These represent the primary RM functions.

Service transaction programs (STPs). These programs make up the transaction services layer. They can be used to *change the number of sessions* (CNOS) and also interact with the node operator facility.

Intermediate session routing

Figure 17.17 shows the structure of the intermediate session routing (ISR) facility.

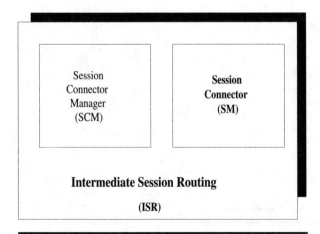

Figure 17.17 Conceptual view of ISR component.

The ISR consists of two components: the session connector (SC) and the session connector manager (SCM). A SC is allocated for each session. It performs the following functions:

- Routing
- Pacing
- Reassembling basic information units (BIUs)
- Monitoring the session for errors
- Performing intermediate reassembly

This component is responsible for routing session traffic through intermediate nodes. The companion component of the ISR is the *session connection manager* (SCM), which performs the following functions:

- *Intermediate* BIND and UNBIND processing
- Creating, initializing, and eliminating session connectors
- Connecting SCs to the PC
- Buffer reservation

Figure 17.18 shows the correlation between the ISR component and other components in the node.

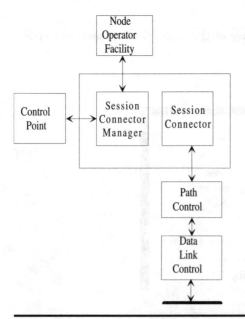

Figure 17.18   ISR communication with other components.

Control point

The *control point* (CP) manages resources within a node. The CP uses CP-to-CP (CP-CP) sessions to exchange management information. According to IBM documentation, the CP may be merged with the LU; this is an implementation issue. If this is done, certain implications do apply. For our purposes here, the CP is treated as a separate entity.

Another way of defining the CP is based on the IBM *APPN Architecture* manual, document no. SC30-3422. The definition in the glossary states:

> (1) A component of an APPN or LEN node that manages the resources of that node. In an APPN node, the CP is capable of engaging in CP-CP sessions with other APPN nodes. In an APPN network node, the CP provides services to adjacent end nodes in the APPN network. (2) A component of a node that manages resources of that node and optionally provides services to other nodes in the network. Examples are a system services control point (SSCP) in a type 5 node, a physical unit control point (PUCP) in a type 4 subarea node, a network node control point (NNCP) in an APPN network node, and an end node control point (ENCP) in an APPN or LEN end node. An SSCP and an NNCP can provide services to other nodes.

The focus here is on the components within the CP and their functions. Figure 17.19 is a conceptual example of how the CP appears.

Functions of the CP components shown in Fig. 17.19 are described in the following paragraphs. (In Fig. 17.19, TRS represents topology and routing services, described later, in Sec. 17.5.)

Configuration services (CS). This component manages physical-link connections of the node itself. CS characteristics and functions include the following:

- *Link activation:* The CS exchanges information with the DLC. Exchange identification (XID) parameters are passed to ensure that the abilities of each node are understood by the other nodes. Figure 17.19 shows the CS relative to the DLC and the PC.

- *Link deactivation:* When a link is deactivated, the CS performs cleanup functions and then notifies the appropriate components within the node.

- Link queries are performed.

- *Exchange of XIDs:* This can occur if link or node characteristics change while the link is active.

- *Link definition verification:* The CS saves this definition when activating and deactivating the link is performed.

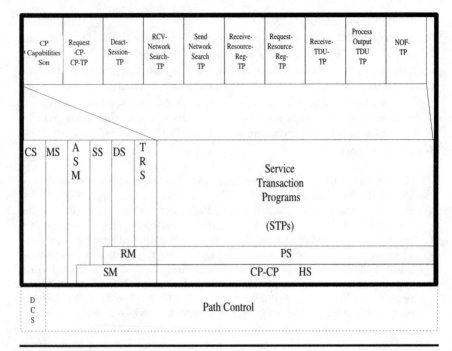

Figure 17.19 Control point components.

Management services (MS). This component handles the alert level at a local level. Other components in the node log alert information in the local node. This component communicates with a number of other components within the node such as the ISR, NOF, and the PC.

Address space manager (ASM). The ASM handles the address spaces related to each local transmission group (TG) within a node. In APPN, a TG is synonymous with a *link*. In each address space a local form session identifier (LFSID) is defined. Additionally, one address space correlates with a TG (link).

The ASM performs these functions:

- Designates session addresses
- Activates an address space
- Deactivates an address space
- Frees address spaces on request
- Routes nonrelated session data
- Paces BINDs via adaptive pacing
- If BINDs are received segmented, the ASM assembles them

The ASM communicates with the PC, NOF, and other components within the node.

Session services (SS). These services focus primarily on initialization and termination of sessions: both CP-CP and LU-LU sessions. Specific SS functions include

- Procedure correlation identifiers [also known as *fully qualified procedure correlation identifiers* (FQPCIDs)]
- Initiation of sessions
- CP-CP session activation
- CP-CP session deactivation
- Termination of sessions
- Monitoring of the active number of sessions

Directory services (DS). DS functions depend on the node type. NNs and ENs function differently. Basically, the DS maintains the node directory within that node, but this is not the case in LEN end nodes. It does, however, provide the ability to search and update directories in other nodes throughout the network. Other functions it performs include

- *EN searches:* Three types of searches can be identified in ENs: searches initiated locally, searches initiated from a remote node, and the sending of a search request to a NN. In this latter case the NN actually performs the search; in a sense it functions as a proxy.
- *NN searches:* Examines its own directory for resources owned by that node or if a resource is owned by its associated end nodes.
- *EN updates:* Local directory updates are performed via system definition.
- *NN updates:* There are three ways to achieve this: (1) as a result of systems definition, (2) dynamic updates occurring as a result of communication with the NN server's client nodes, and (3) after network search completion updates may be performed by data cached during the search.
- *EN registration and deletion:* This function occurs only if the end node is authorized for this function.
- *NN directory maintenance:* A NN performs updates and deletions based on information received from nodes wherein their registration applies.

Topology and routing services. This component of the CP performs three primary functions:

- *Class-of-service manager (COSM)*: The COSM keeps the database updated. It notifies route selection services (RSS) when a service class changes.

- *Route selection service (RSS)*: This component functions differently in NNs and ENs. In a NN the RSS determines preferred routes within the network and determines what transmission priority to use on selected routes. It also updates the topology database to reflect the most current topology and specifies routes computed by the COS requested. The TG from origin to destination is selected by the RSS. In an EN, RSS selects a TG and transmission priority.

- *Topology database manager (TDM)*: TDM functions vary per the node type. In an EN it maintains the topology database. In a NN it broadcasts to the network once changes are made locally. It performs what is called a *periodic broadcast,* which is a broadcast throughout the network in intervals of approximately five 24-h days. It also *deletes* resources from the database if no data has been received about a resource in 15 days. It also responds to remote queries.

Service transaction programs

Service transaction programs (STPs) exist in a APPN node. They exchange information over CP-CP sessions. They do not exist in LEN ENs. Ten transaction programs (TPs) are explained here:

- *CP capabilities sent outside the node (SON)*: This is the only SS TP used to attach a remote node. Its purpose is twofold: it sends the PC capabilities to another participating node and performs processing for session outage.

- `Request_CP_CP_TP`: When the SS wishes to start a CP-CP session with another node, this TP is used to invoke the request CP capabilities TP.

- `Deact_Session_TP`: This deactivates a CP-CP session to a specific node.

- `Receive_Network_Search_TP`: Receives a locate search from an adjacent APPN node.

- `Send_Network_Search_TP`: This is an APPN node that sends a locate search request or reply to a remote directory service.

- `Receive_Resource_Registration_TP`: This is where an APPN NN sends registration and deletion variables to an APPN EN pertaining to specified resources.

- `Request_Resource_Registration_TP`: An instance where an APPN EN sends registration and delete information to the target APPN NN.

- `Receive_TDU_TP`: Where a NN TRS communicates with a `TDP_TP` receiving information from an adjacent node.

- `Process_Output_TDU_TP`: TDM sends a signal to the `Process_ Output_TDU_TP` when a local node wants to broadcast topology information.

- `NOF_TP`: This is involved in the startup of a node operator facility.

Node operator facility

This component initializes the CP and ISR on starting the node. It is the user interface for the CP, ISR, and LUs. The following functions can be achieved through the node operator facility (NOF):

- Activating links
- Deactivating links
- Creating LUs
- Deleting LUs
- Ascertaining status information
- Retrieving database information
- Defining:

  Directory information
  Local and remote LUs
  Node characteristics
  Session limits
  Transaction programs
  Links
  Start TPs
  Other CP names

More details are provided about the NOF later in this chapter.

17.4 Directory Services

The directory services (DS) component is responsible for resource searches for the local node as well as those throughout an APPN network and is also responsible for the registration of resources NNs where they function as servers for directory services.

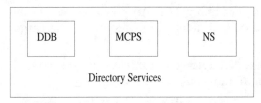

Figure 17.20   Conceptual view of DS components.

The DS component of the CP itself has three major subcomponents:

1. Directory database function

2. Maintaining CP status function

3. Network search function

Figure 17.20 shows these three components of the directory services.

The directory database (DDB) function performs lookup and maintenance for directory services.

The maintain-CP-status function keeps a log of other CPs that wish to communicate with the CP in that node. In NNs it keeps track of end nodes (clients) and also maintains information as to other NNs with which it can establish CP-CP sessions.

The network search function component sends and receives resource search requests to and from other nodes in the network.

Understanding the terminology is important for examining the DS component.

DS terminology

- *Authorized node:* When this term is used in conjunction with DS, it means that information sent about itself is accepted. An *unauthorized node* cannot use certain protocols and furthermore is refused the ability to register its LUs in the distributed directory.

- *Border node:* This is a NN connecting APPN networks that maintain different databases reflecting their topology. Peripheral border nodes support directory services, session setup, and route selection between networks of different identifiers. This type of node does not support intermediate routing. An extended border node provides session setup, directory services, and routing through a boundary of different networks with different topologies.

- *Central directory server (CDS):* This function resides in a NN. It differs from DS function because the CDS maintains all resource information within a network. More than one CDS can exist in a network.

- *CP send-and-receive session:* The DS uses CP-CP sessions between APPN EN and the NN server. This session carries the `Locate_Search` function.

- *Distributed directory database (DDDB):* Directories of resources exist throughout the APPN network. The concept of DDDB is the collective whole of the databases throughout the network.

- *Local directory database (LDDB):* This refers specifically to the local directory database in a given node.

- *Locate search:* This is how a DS finds resources not in that node. A locate search can be one of the following:

  Broadcast search: This is sent throughout the entire network.

  Directed search: This is sent to a known location for verification.

  Domain search: This is a NN communication with its client ENs to verify resources in a given location.

- *DS user:* This is a component in a node which uses the DS.

- *Subnetwork:* A collection of nodes which have common characteristics such as a common network address or database.

DS functions

The function of the DS component is present in each T2.1 node; the degree to which it is exploited is contingent on the node itself. The DS functions in a NN are as follows:

1. A database is maintained of local resources and resources that have been cached because of a locate search function.

2. The DS provides the ability to determine the location of a specified resource.

3. It registers resources in a domain via the NOF or a end node (client).

4. It deactivates CP-CP sessions if the node is in a state of deadlock with respect to sending a locate search.

5. It provides support for intermediate nodes in locate searches.

Directory service in an APPN EN provides these functions:

1. Once notification of a CP-CP session failure has occurred, it cleans up any outstanding searches.

2. Sends and receives locate resource searches with the NN.

3. Registers resources with the NN.

4. Supports CP-CP sessions with the NN.

5. Maintains a DDB of local resources and adjacent node resources.

Directory database function

The *directory database* (DDB) is a distributed database containing lists of resources throughout the APPN network. For example, ENs keep information about their resources in their DDB. A NN maintains a node operator facility (NOF). The NOF defines directory entries of resources in that node and in the nodes it serves. A major function is keeping the database within the storage requirements for that node. Different entries exist in the DDB.

Types of DDB entries. Fundamentally, the DDB maintains information about its own resources. Additionally, the following types of entries may be found:

Domain entries include resources that are in that domain but are located in one of the client nodes (end nodes).

Other domain entries maintain information about resources in other domains, as the term implies.

Other network entries keeps information about resources that can be reached by a different net ID.

Origin of DDB entries. As mentioned previously, a *distributed directory* is the individual, local databases viewed as a whole. Information gets into these directories by one of the following means:

1. NOF definition

2. APPN EN registration with a NN server via a CP-CP session

3. Those entered by the caching function as a result of the locate-search function

Network search function (NSF)

The *network search function* (NSF) maintains the protocols used while searching the DDB. This function also maintains control of the transport directory services and enforces logic with regard to the sending of DS messages; however, its primary purpose is to locate network resources and control flow throughout the network of request and replies.

Within the NSF it may become necessary to send a request to another node asking for information about the directory in that node; in such cases the message that flows is called a *locate search*. Three types of locate searches can be identified:

1. One-hop search

2. Directed search

3. Broadcast search

The one-hop search is a locate-search request exchanged between an APPN EN and a NN. A *directed search* traverses a predefined path from one NN to another NN. In this case the originating NN calculates a CP-CP path to the target node and adds routing information to the search. This works because each NN on the path uses this information to select the next hop. By functioning in this manner it ensures that the most direct route to the destination node is obtained. The *broadcast search* is used by NN to send a locate-search request to multiple CPs. Two dominant types of broadcast searches are

- Domain

- Network

The *domain* broadcast search sends a locate-search request for the resource to adjacent APPN ENs. Because more than one reply may return, the directory service used the first positive reply. The *network* broadcast search is sent to all NN nodes. It is used to ascertain a resource location when it cannot be found otherwise. It is used as the last resort because this type of search permeates the network with request for the location of the resource.

Locate searches can carry non-DS information. If this is the case, the locate search is used by other CP components normally to transport control data. When this non-DS information is used, the user is presumably an application. For example, session services could use this to transport information variables. Examples of some information capable of being transported as non-DS information are

- Fully qualified procedure correlation identifier (FQPCID)

- Destination LU

- Mode name

- Class-of-service (COS) information

- Originating LU

- Endpoint vectors

Central directory server

The *central directory server* (CDS) resides in a APPN network. More than one CDS can exist. The CDS accepts registration of resources from other network nodes. Once information is received from the registrations of other nodes, a central directory is maintained.

Directory entry contents

The contents of a directory entry depend on node type. The following is a brief list of the contents found in directory entries.

- Resource name

- Resource type: NN control point, EN control point, LU information

- EN control point using a pointer to an adjacent CP status control block

- Information about the hierarchy: LU entries for NN servers, adjacent entries for an LU entry, the "child" LU entries for adjacent EN control points

- Classification: home, cached, or registered

- Information regarding whether the resource can be registered with the CDS

- Status of registration (whether it has been registered)

- Status of registrability (whether it can be registered)

17.5   APPN Topology and Routing Services

Topology and routing services (TRS) are present in each NN and in a lesser form (with respect to functionality) in APPN ENs and LEN ENs. In NNs TRS creates and maintains the class-of-service database and is responsible for maintaining a copy of the network topology database.

In ENs the TRS is created and maintains the local topology database. It is also responsible for the class-of-service table.

The TRS consists of three components:

1. Class-of-service manager

2. Route selection service

3. Topology database manager

Figure 17.21 depicts these three components.

The class-of-services manager enables translation of a mode name to a COS name. This is a base function for NNs; however, it is optional for ENs.

Route selection service computes routes. It selects the path from origin to destination. Technically it computes the most efficient route between nodes in an APPN network.

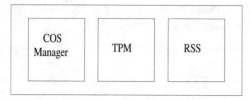

Figure 17.21 Conceptual view of topology and routing services components.

The topology database manager maintains the COS and topology databases. In NNs the TDM maintains the network topology database, and on one or the other end it maintains local topology information.

Class-of-service database

The COS database exists in all NNs and in those ENs which support them. The COS database contains

- Mode names which include a pointer to a COS name

- COS names which have COS definitions representing characteristics of the node, transmission priority, and weight assigned

- Weight index structure for computing the actual transmission group weight

  Each COS entry in the COS database contains some basic information, such as

- COS name
- Transmission priority
- Transmission group characteristics
- Security level
- Cost per byte
- Propagation delay and other characteristics

Route selection service

A *route* in APPN is a path between two endpoints. This includes the intermediate components that may exist. For example, this includes links, NNs, domains, and transmission groups, to name only a few.

  The minimum criteria for determining the best route in an APPN network are

- Route characteristics must be known.
- All possible routes must be calculated.

- If a resource is not acceptable, it must be excluded; hence, determining this factor about resources must be performed.
- All resources that will be used during the route must be accounted for and calculated accordingly.

Topology database

The topology database contains information on the logical structure of the APPN network as well as on all nodes in the network, transmission groups, intermediate transmission groups, and other pertinent information.

There are two types of topology databases:

1. Network topology database
2. Local topology database

Network topology database. A *network topology database* is maintained in all NNs. This database includes information on NN connections to other NNs and connections to virtual routing nodes. Each NN participating in the APPN network is aware of this database because the database is on each NN.

The structure of the network topology database includes two categories:

- Node table, including node information such as

  CP name

  Network ID

  Characteristics

  Resource sequence number
- Transmission group table, including TG information such as

  CP-CP session support

  Status

  A pointer to the TG vector

  A pointer to the weight (amount of resource requirement)

Local topology database. *Local topology databases* are located in end nodes. This database contains information about each endpoint attached to that node. It is created and maintained by the topology database manager. The local topology database is used in the following situations:

- When no CP-CP session exists to a NN
- When establishing sessions to predefined LUs
- When sending information to the NN for the route selection process

17.6 APPN Configuration Services

This component of the control point (CP), known as *configuration services* (CS), is responsible for managing local node resources such as links to other nodes. A conceptual view of configuration services is shown in Fig. 17.22.

CS performs a number of functions, including

- Node configuration definition

  Data-link type

  Ports

  Adjacent nodes

  Adjacent links

- Link activation

- Link deactivation

- The nonactivation of an XID exchange

On node initialization, CSs receive the node name, network identification, link station support information, and information concerning TGs.

Through CS the NOF defines the basic node configuration, the first to be defined is the data-link type. This CP component communicates with the data-link control manager for definition purposes. Ports are also defined by the NOF via CSs. They are considered hardware.

Port type is defined; that is, whether switched, nonswitched, or defined as a shared-access facility. Information includes buffer size, limits, timeout values, TG characteristics, and any associated DLC process.

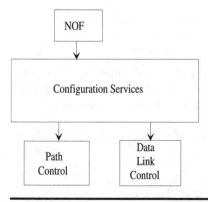

Figure 17.22 Conceptual view of configuration services.

Link stations may be defined at activation time, or their parameters can be negotiated; either way they must match. Nodes require system definition for their local link station. These definitions include

- Role: primary or secondary
- Address, or link station defined as negotiable
- Inactivity timer
- Retry limit on mode setting
- Modem delay limits

17.7 APPN Management Services

The concept of management services is implemented in each T2.1 node and is known as the *control point management services* (CPMS). The CPMS is a component of the CP in T2.1 nodes. The functionality of CPMS is straightforward.

The NOF sends messages to the CPMS. These messages are converted into local management services where they are carried out by the receiving component in the T2.1 node.

With respect to the CP, management services interacts with each of the following:

1. Session services
2. Configuration services
3. Session manager
4. Resource manager
5. Address space manager
6. Topology and routing services
7. Directory services
8. Presentation services

Some information CPMS receives from CP components; other components are listed below in the order in which they are usually received. Some information must be solicited; some does not have to be.

1. Information about currently active sessions, such as problems detected in a node by the CP
2. Domain information
3. Information about currently active LU6.2 sessions
4. Information about session conversations

5. Unsolicited information about problems related to this component

6. Information about routing

7. Locating network resources for the CPMS

8. LU6.2 protocol boundary information

The LU session manager provides information about currently active sessions. The LU resource manager provides information about conversations across a LU6.2 session. The LU management service component provides unsolicited information about the LU. The session connection manager provides information about data passing through the node.

The path control manager reports unsolicited information about any problems detected at this component.

The data-link control manager provides a vehicle for testing resources such as links and modems.

Other management information is ascertainable, but the point here is that management services actually interact with all T2.1 node components, not only with the CP.

17.8 Address Space Manager

Overview

The *address space manager* (ASM) resides in NNs and APPN ENs. Fundamental functions of the ASM include

- Managing session addresses [also known as *local form session identifiers* (LFSIDs)], which are used for routing data traffic and local path control

- Managing flow control of session activation messages (the BIND)

- Informing the appropriate session manager component when a link fails

- Routing session activation and deactivation messages

The ASM is created on initialization of the node by the NOF. Once this is performed, the ASM is notified by the NOF of the CP name, network ID, and nature of BIND assembly supported.

ASM functions

For communication to occur between LUs or CPs and other CPs, a LFSID must be allocated by the ASM. By performing this function the ASM achieves address control in the node. The ASM bases the LFSID on the PC instance identifier for that session. The ASM maintains an address space list and a list of assigned LFSIDs in use.

Address spaces are defined by the ASM in relation to the transmission group attached to the node. The ASM assigns an address space consisting of a number large enough to allocate enough LFSIDs for that TG.

Whenever a TG is activated or deactivated, the ASM is informed. At that time the ASM creates or destroys the tables used to control that TG's address space. ASM handles the assigned address space by dividing it into groups. These groups consist of 256 LFSIDs. However, the ASM allocates the LFSIDs only as necessary.

LFSIDs

Local form session identifiers (LFSIDs) are 17-bit session identifiers used by the path control to route session traffic. It is composed of two components:

1. *A 1-bit assignor indicator:* Every ASM in the nodes connected by a TG selects a LFSID from the TG address space with a different value to prevent duplication.

2. *A 16-bit session identifier:* This is further broken into an 8-bit identifier considered high and an 8-bit identifier considered low.

LFSIDs assigned to a session maintain its active state as long as the session exists. The ASM is the component that terminates the association of the LFSID and the session once the ASM receives notification that an UNBIND or a response to UNBIND is delivered from the PC component.

This LFSID is used because, on each session hop between two endpoints, each node uses distinct session identifiers to identify a session; consequently, the term *local form session identifier* (LFSID) is used.

Considerably more detail accompanies this topic, but this is not a section on programming! For further information on the topic, refer to IBM's *APPN Reference* manual no. SC30-3422.

17.9   APPN Session Services

In general, the session services (SS) functional part of the control point (CP) aids in

- Generating unique session identifiers
- LU-LU session initiation
- LU-LU session deactivation
- CP-CP session activation
- CP-CP session deactivation

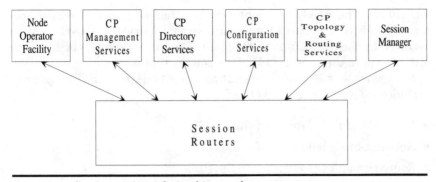

Figure 17.23   Session services relationships to other components.

Figure 17.23 identifies the relationship between SSs and other components.

Fully qualified path control identifier

The session identifiers generated by this part of the CP should not be confused with the LFSIDs generated by the ASM mentioned previously. The session identifiers mentioned here are called *fully qualified procedure correlation identifiers* (FQPCIDs).

Session services assign a *network*-unique session identifier, known more commonly as the *FQPCID*. This FQPCID performs the following functions:

- Correlates request and replies between APPN nodes
- Identifies a session for problem determination
- Identifies a session for auditing, accounting, performance, and other purposes
- Identifies a session for cleanup or on which to perform recovery actions

FQPCID is assigned at the originating node. It is of fixed length. It contains an 8-byte session identifier field that includes the network-qualified name which generated it.

LU-LU session initiation and deactivation

With APPN NNs, APPN ENs, and LEN ENs LUs can initiate sessions and respond to the session initiation request from another LU, or from a CP, for that matter. The session activation request (also called a *BIND*) is sent by a particular LU. That LU is considered the primary

LU (PLU). The BIND recipient, on the other hand, is called the *secondary LU* (SLU). The LUs go into session once the BIND is sent, received by the target, and the target LU sends a response (RSP) to the BIND back to the sending LU. This is an active session.

An example of some information specified in a BIND request includes information such as the following:

- Network-qualified name of the PLU
- Network-qualified name of the SLU
- Route through the network to the SLU
- The FQPCID
- Maximum request or response unit (RU) size

On the other hand, a session is stopped when an UNBIND is sent to the target and the target responds with a RSP back to the originator of the UNBIND. This is session deactivation, or an inactive session.

Sometimes the terms *PLU* and *SLU* are substituted with the terms *origin LU* (OLU) and *destination LU* (DLU).

CP-CP session activation and deactivation

CP sessions are always LU6.2. This means that given a CP in two nodes, the possibility for contention exists. *Contention* occurs when both CPs attempt to establish a session with the other at the same time. Since this possibility exists, the question of how to overcome this scenario is in order.

Contention can be overcome by what are called contention "winners" and "losers." Each CP has contention winner LUs defined and generally the same number of contention losers defined. Because CP-CP sessions are established in parallel, each CP has winner and loser LUs defined. With this configuration, contention can be overcome.

Establishment of CP-CP sessions begins when the session services notifies directory services that a session is pending active. Then the directory service queues network operations that may involve the CP session LU. Session services also notifies the resource manager to attempt activation of a winner LU with the destination LU in the target node. Session services is once again invoked to assign a FQPCID having a mode name of CPSVCMG (control point service message).

The following information is part of information that flows across CP-CP sessions:

- Topology database updates
- Session activity

- Request for data management support
- Reply to a request for data management support
- A resource search capability

CP-CP session deactivation may occur for one of two reasons: (1) normal deactivation may occur—if so, it usually means that the node, or its partner, no longer requires the session; or (2) an abnormal CP-CP session termination can occur—if so, it could be the result of either protocol violation during the session or a link failure or, in remote cases, both.

17.10   Node Operator Facility

Perspective

The *node operator facility* (NOF) is the interface between an operator and the T2.1 node. Its purpose is to permit operators to control node operation. A node operator can be one of the following:

- Human operator
- Command list for execution
- A transaction program

Either of these node operators can perform node operator functions. A human operator can execute a specific dialog between the NOF and the individual and make changes possibly not anticipated or executable by a program.

A command list is simply a file with a list of node operator commands to be executed. The NOF interpreter logs the commands and responses from the NOF and maintains this for future reference.

Transaction program control is used in remote operations. This works by a transaction program actually issuing commands against the NOF which is in a remote location.

A graphic illustration of the three ways to communicate with the NOF is presented in Fig. 17.24.

NOF functions

A brief list of NOF functions includes

- Creating other components in the node
- Issuing commands to initialize the node
- Converting commands to signals capable of being understood by components within the node

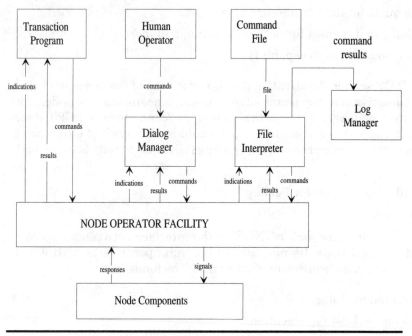

Figure 17.24   A perspective of the node operator facility.

- Starting a log of commands issued and the results of issuing these commands
- Receiving results from node components
- Routing signals to the appropriate node components
- Managing unsolicited messages from any node components

On initialization the NOF creates the following components in the order indicated:

1. Address space manager
2. Session services
3. Directory services
4. Configuration services
5. Management services
6. Topology and routing services
7. Session connector manager
8. Session manager of the control point (CP)

Commands listing and function

When a command is entered and received, the NOF parses the command into a form understood by the NOF. At this time the syntax is verified.

The following are node operator commands and their functions, defined architecturally:

CHANGE_SESSION_LIMIT	Changes the session limit
DEFINE_ADJACENT_NODE	Defines an adjacent node to a local node
DEFINE_CLASS_OF_SERVICE	Changes or updates the COS
DEFINE_CONNECTION_NETWORK	Defines a connection network to the local APPN node
DEFINE_DIRECTORY_ENTRY	Defines or updates directory entries
DEFINE_DLC	Defines a data-link control
DEFINE_ISR_TUNING	Adds or updates the session connector manager
DEFINE_LINK_STATION	Defines a connection to a link station
DEFINE_LOCAL_LU	Defines a new LU
DEFINE_MODE	Creates a new mode definition for a local LU
DEFINE_NODE_CHARS	Defines or updates current characteristics in the local node
DEFINE_PARTNET_LU	Defines or changes a local LU
DEFINE_PORT	Defines a port to a local node
DEFINE_TP	Defines or changes a local LU's operation with a local transaction program
DELETE_ADJACENT_NODE	Removes a definition of an adjacent node
DELETE_CLASS_OF_SERVICE	Removes a COS definition
DELETE_CONNECTION_NETWORK	Removes connection network from a local node
DELETE_DIRECTORY_ENTRY	Removes an entry from the DS directory
DELETE_DLC	Removes a data-link control instance
DELETE_ISR_TUNING	Removes one or more session connector managers
DELETE_LINK_STATION	Removes adjacent link station definition
DELETE_LOCAL_LU	Removes a local LU from a node
DELETE_MODE	Removes a mode definition from a local LU
DELETE_PARTNER_LU	Removes a definition a local LU uses with a remote LU
DELETE_PORT	Removes a port definition in the local node
DELETE_TP	Removes a local transaction program definition
INITIALIZE_SESSION_LIMIT	Initializes the number of sessions allowed
QUERY_CLASS_OF_SERVICE	Obtains the values defined for a COS
QUERY_CONNECTION_NETWORK	Obtains the status of a connection network
QUERY_DLC	Obtains the status of a specific DLC instance within a node
QUERY_ISR_TUNING	Ascertains information about the session connector manager
QUERY_LINK_STATION	Obtains the status within the node of an adjacent link station
QUERY_PORT	Obtains status of a port within the node
QUERY_STATISTICS	Obtains detailed information about a link station
RESET_SESSION_LIMIT	Resets the number of sessions allowed

START_DLC	Starts a specified data link
START_LINK_STATION	Establishes communication between a local link station and an adjacent link station
START_NODE	Brings up the SNA node
START_PORT	Starts a specified port and local link stations
START_TP	Requests a local LU to start a TP in a node
STOP_DLC	Stops the named data-link control
STOP_LINK_STATION	Stops communication with a specified adjacent link station
STOP_PORT	Stops a specified port and associated local link stations

17.11   APPN Concepts and Traditional SNA

APPN and SNA are philosophically different. APPN is peer-oriented, using LU6.2 protocols, and is implemented across a variety of equipment. SNA has been hierarchical in nature. This meant that VTAM was involved in practically all session establishment. This began changing with VTAM Version 3 Release 2 and is more prevalent in VTAM Version 4 Release 1, and VTAM Version 4 Release Z.

APPN and traditional SNA are becoming less clearcut. They are evolving into a cooperative way of networking when both are present in one environment. This section explores some differences between the two and also examples of areas where they are coming together in a sense.

APPN structure

APPN builds on different types of nodes that provide services such as routing, database maintenance, directory services, and end user services. The growth in different types of nodes that constitute APPN has, and is, changing. No longer is an APPN network considered implemented with midrange and PS/2 systems. Now, APPN can be implemented with SNA via VTAM support.

APPN uses LU6.2, an independent LU, capable of initiating a BIND (request for a session with another LU).

SNA structure

SNA has been built around hardware and software architectures, and VTAM is becoming the centerpiece of software for the network. SNA has been hierarchical (some also call it *subarea SNA*) in nature, but it supports peer operations. Now, those peer operations are expanding to embrace APPN via VTAM and the NCP.

SNA uses LU6.2, but primarily it utilizes other LU types such as LU 0, 1, 2, 3 and 7. These LUs are dependent on VTAM for session establishment. As a result, the question becomes: "How can LU2s be implemented into an APPN network and access a VTAM host?"

APPN-SNA mixture

Subarea SNA supports dependent logical units. SNA's roots are in this functionality. This means that a LU requesting a session with a VTAM application must have the services of VTAM (SSCP) or aid via the NCP boundary function. For LUs residing on adjacent nodes to VTAM or the NCP, they traverse the VTAM or NCP boundary function.

If VTAM is configured as an end node, VTAM cannot perform intermediate session routing. However, nodes can attach to VTAM using the boundary function of SNA. Consequently, dependent LUs must access VTAM via this boundary function.

Two terms need clarification and explanation: dependent logical unit requestor (DLUR) and dependent logical unit server (DLUS). An implementation where DLUR and DLUS are implemented provides the following scenario.

When APPN nodes are mixed in networks with nodes such as a PU2.0 device, the need exists for this PU2.0 to have SSCP-PU sessions and SSCP-LU sessions. Once support for these two sessions is achieved, a dependent LU-LU session can be achieved from the PU2.0 device and a subsystem application.

To realize this in APPN and mixed subarea SNA, this data must be encapsulated within a LU6.2 session and passed to the SSCP and PU, respectively. When this is realized, the need for a T2.0 node to be directly attached, or data link to be switched, to the SSCP—thus providing SSCP access—is removed. Hence, integration of PU2.0 and PU2.1 APPN and subarea-dependent LU can be achieved.

Figure 17.25 depicts an APPN network and a subarea network with VTAM functioning as a composite network node.

As a result of the implementations shown in Fig. 17.25 session establishment can occur between any LU in the subarea network and any LU in the APPN network. In this case APPN VTAM must be implemented to convert subarea to APPN protocols and vice versa.

17.12 Summary

APPN has its own architecture. It is peer-oriented, whereas SNA is hierarchical in nature. VTAM Version 4 Release 1 was the first formal "merge" of SNA and APPN.

The T2.1 architecture consists of many components. T2.1 node components can be viewed two ways: the components that make up the node itself and those that constitute the CP.

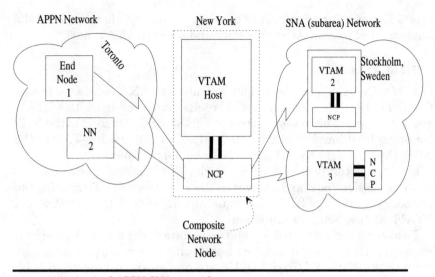

Figure 17.25 A mixed APPN-SNA network.

APPN and subarea SNA are different. Neither is better; they are different. Much more information on APPN is available; your local IBM representative can tell you how and where to obtain more information on this topic.

Chapter

18

Open-Systems Interconnection (OSI)

The International Standards Organization (ISO) has been in existence for approximately four and a half decades now. Its model proposed in 1977–1978, called the *open-system interconnection* (OSI) model, has probably affected the use of the phrase "open systems" more than any other phrase since. A brief look at the use of this phrase provides insight about possible meanings to be conveyed. Consider the decade of the 1980s; three characteristics are evident.

First (but not necessarily in chronological order) a phenomenon began to occur in the business community; it was called *downsizing*. Corporate consolidation and utilization of new technology began. While this occurred, technology of all sorts continually sprouted.

The year 1981 brought the first personal computer (PC) from IBM. Shortly many companies began competing with IBM in the PC market. The PC alone had caused radical changes in the way many businesses performed work by the mid-1980s. The names of some high-tech(nology) companies came to be well recognized by the end of the 1980s, and some of these were barely known at the turn of 1980.

By any measure, the 1980s witnessed an avalanche of technological innovations. Technology seemed to be a catalyst for itself. With technology of all sorts entering the market monthly, the concept of local area networks (LANs), then network integration, began to sweep the business community.

The idea of open systems permeated published periodicals on computing: weeklies, magazines, and journals. It seems that the phrase was defined in as many ways as there were vendors. Nevertheless, the

ISO kept moving the OSI specifications forward with the intent of realizing true open systems.

To provide an example of the power behind the ISO OSI model and what it meant, the U.S. government made a statement of direction in late 1988 or early 1989 about this matter, officially referring to this direction as the *government OSI profile* (GOSIP). Its roots were in the OSI protocols. Not only was the United States involved in OSI; there were also British, Swedish, and other GOSIPs. OSI was making significant impact.

18.1   Origins and Evolution

International Standards Organization (ISO)

The ISO is an international organization focusing on standards in various areas. In 1977–1978 the ISO announced a direction for creating a framework for what would become the OSI network reference model.

The OSI model did not appear *overnight*. It took time, and there were contributions from numerous sources. In fact, it is accurate to say that OSI architecture and protocols are a collection from sources such as CCITT, IEEE, ISO committees, and other contributing parties. The ISO committee meets twice yearly, and in between it exchanges information created between meetings. The OSI model itself began as a statement of direction and evolved therefrom. It is still evolving.

OSI architecture and protocols are listed by letters and numbers indicating its purpose. For example, ISO 8571-1 is *FTAM-Part 1: General Description of File Transfer Access and Management*. Many such documents are ISO specifications; however, not all must be adhered to to realize an ISO-oriented network. The flexibility within the standards make it possible for selective choices.

OSI impact on the U.S. government

The U.S. government imposed a standard in the late 1980s stating that OSI compliance must be met for the purchase of products for government purposes. This statement meant phasing in OSI-compliant technology to government agencies. However, an interesting observation is that in 1983, 5 years before the U.S. government announcement of GOSIP, it made a similar proclamation about TCP/IP, stipulating that anyone in the United States who connects to the Internet must use TCP/IP. This spilled over into government agencies as well.

The irony is that moving to TCP/IP is taking years, and now another statement of direction. When the GOSIP announcement appeared, many in the technical community wondered to which specification suppliers for the government would capitulate. It seems as of this writing that both TCP/IP and OSI products are being produced and implemented.

In a sense, the GOSIP announcement caught some vendors by surprise. Many vendors were focusing efforts to meet the direction for TCP/IP set forth in 1983.

At the time of this writing, a considerable number of vendors supply TCP/IP products. At the same time many are doing the same with OSI products. However, the lag time to get the number of OSI products on the market is there simply because of logistics. For example, the Digital Equipment Corporation (DEC) has put considerable resources in moving to a more open networking environment by embracing the OSI products they offer. The IBM corporation and other well-known corporations have also done so.

OSI products are present in the marketplace today. How much? It is difficult to quantify. Three observations are sensical: (1) the OSI product growth rate is increasing, (2) it appears that complete OSI-based networks are not as dominant as other network protocols, and (3) parts of OSI are being implemented to varying degrees.

OSI is evolving and gaining market share. The extent of worldwide penetration it will make and how fast this will occur remains to be seen.

18.2 Open-Systems Interconnection Model

Better known as the OSI model, this consists of seven layers as shown in Fig. 18.1.

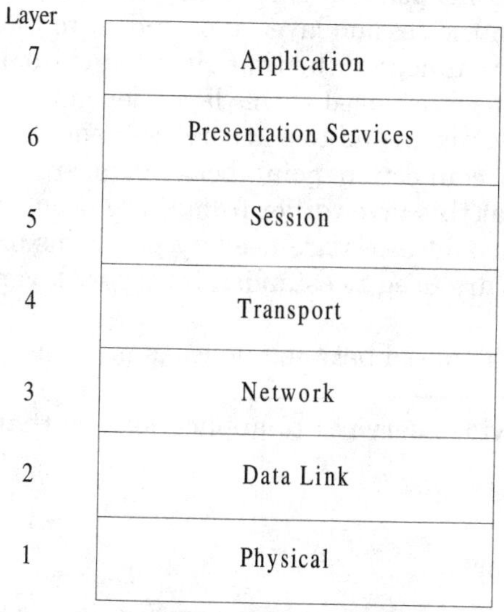

Layer	
7	Application
6	Presentation Services
5	Session
4	Transport
3	Network
2	Data Link
1	Physical

Figure 18.1 Conceptual view of the OSI model.

The following is an overview of basic functions of the seven OSI layers.

1. *Physical:* This layer is concerned with bit-level representation of data. Its primary purpose is to provide the connection between the host and the medium. This layer may concern itself with voltages or photons, depending on which medium is used. This layer sends and receives bits (voltages or photons).

2. *Data link:* This layer, implemented primarily in interface cards, provides a mechanism for reliable data transfer across a physical link. Data-link protocols that govern this process are implemented here. Data is put into frames at this level and is sent sequentially from the sending node to the receiving node. This layer is also responsible for responding back to the sender once frames have been received. This layer is divided into sublayers: the medium access control (MAC) and the logical link control (LLC). The MAC sublayer frames the data to be sent and on the receiving end, disassembles the data before passing it up to the LLC. The LLC is responsible for governing the connection between the two endpoints. It provides the protocol used at this layer (token ring, ETHERNET, etc.).

3. *Network:* Components of this perform specific functions, but the primary function is routing. The internetworking sublayer is responsible for global message routing. The intranetwork sublayer routes messages within one type of network. The harmonizing sublayer is an interface for adjacent sublayers. The access sublayer is an interface to the data-link layer.

4. *Transport:* This is a network-independent service provided to the session layer. It receives data from the session layer and sends it to the destination. It performs the converse function when data is inbound from a target host. The type or extent of service used at this layer depends on the reliability of the layer. This layer is an end-to-end service provider.

5. *Session:* This is where the connection point between users is established. The addresses used at this layer differ from those used at the transport layer. Session-layer addresses are used by programs or users. Transport-layer addresses are used to establish transport-layer connections.

6. *Presentation:* Syntax on data passed between nodes is performed at this layer.

7. *Application:* This layer provides services to applications so that they can access OSI capabilities.

18.3   OSI Layers: A Detailed View

Many have seen Fig. 18.1 or likenesses thereof; however, Fig. 18.2 is a more detailed view of the OSI model.

After much research, this author believes that Fig. 18.2 is an adequate representation of the OSI model. The fundamental philosophy of OSI is openness. OSI has specifications that software and hardware developers implement in products. A question to be considered is: "How is OSI implemented?"

Figure 18.2 shows numerous components at the application layer. Above the application layer are *applications*. These applications use association control service elements (ACSEs) and other components as well. The remainder of this chapter is devoted to providing a perspective on this detailed view of the OSI model.

OSI is abstract. The theories are abstract, and few of them yield themselves to clear, understandable drawings to convey the idea. Core

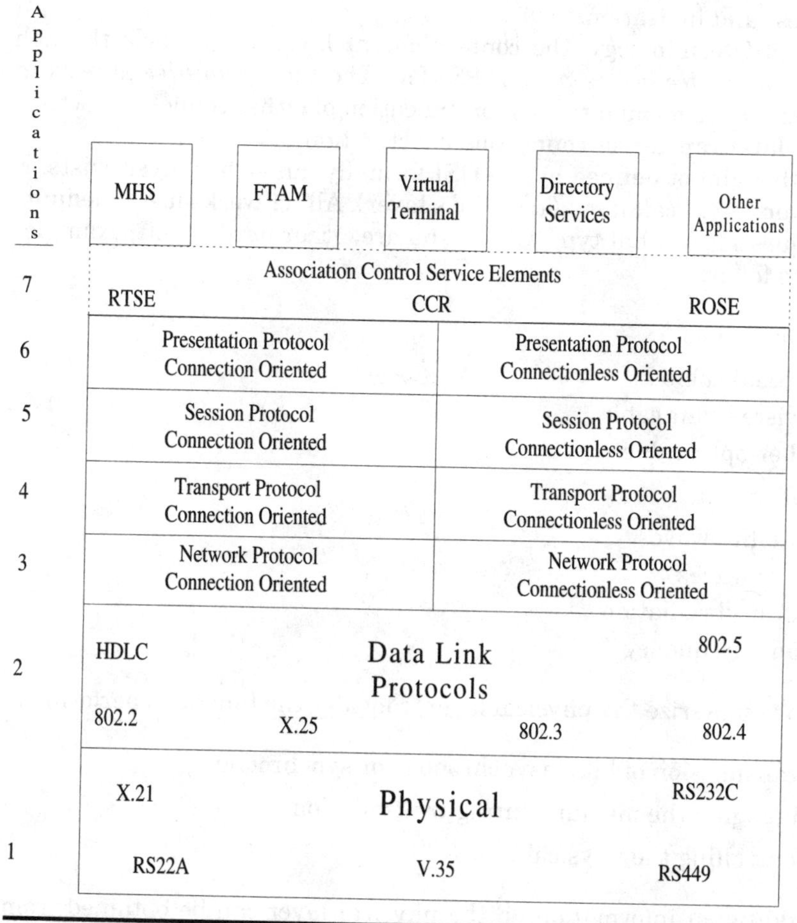

Applications				
MHS	FTAM	Virtual Terminal	Directory Services	Other Applications

7	Association Control Service Elements		
	RTSE	CCR	ROSE

6	Presentation Protocol Connection Oriented	Presentation Protocol Connectionless Oriented
5	Session Protocol Connection Oriented	Session Protocol Connectionless Oriented
4	Transport Protocol Connection Oriented	Transport Protocol Connectionless Oriented
3	Network Protocol Connection Oriented	Network Protocol Connectionless Oriented

2	HDLC	Data Link Protocols	802.5
	802.2	X.25	802.3	802.4

1	X.21	Physical	RS232C
	RS22A	V.35	RS449

Figure 18.2 OSI model components.

components of each layer are presented, but additional information can be obtained from the sources listed at the end of each section.

Physical layer

The physical layer represents an interface. Specifically, it is the component between the host and the medium. Certain interfaces are recommended by the ISO. Some of those include X.21, RS-232-C, and IEEE 802.3 (also X.25, which involves more than just the physical layer).

The physical layer represents details such as the pin-outs on the interface and the operations on the interface itself. Additionally, the physical layer specifies services such as connection activation and deactivation and data transfer. Each of these services has two modes: request and indication.

In OSI terminology, the connections at layer 1 are made through *physical service access points* (PSAPs). The term *primitive* is used in OSI circles to mean a request or indication of either connection activation, data transfer, or connection deactivation.

Although not defined by the OSI formally, an eighth layer exists. It is layer zero. I call it *media* (or Ed's layer). All networks use a medium; the question is what type. OSI media are either hard or soft; examples are as follows:

- Hard media

  Coaxial cable

  Twisted-pair cable

  Fiber-optic cable

- Soft media

  Satellite waves

  Microwaves

  Infrared radiation

  Radio frequency

To summarize the physical layer, consider the functions performed:

- Transmission of bits (asynchronous or synchronous)
- Managing the medium during transmission
- Controlling the physical link

Additional information on the physical layer can be obtained from the following specifications.

ISO no.	Document title
8802-1	*LANs Part 1: Introduction*
8802-2	*Logical Link Control*
8803	
8802-5	*Part 5: Token Ring*
10022	*Physical Service Definition*
8480	*DTE-DCE Interface Backup Control Operation Using the 25-Pin Connector*
8481	*DTE-DTE Physical Connection*
9543	*Synchronous Transmission Signal Quality at DTE-DCE Interfaces*

Data-link layer

The data-link layer consists of protocols that perform multiple functions. More than one data-link protocol is included in ISO recommendations. Two examples will suffice here: IEEE 802.3 and IEEE 802.5.

Consider the following functions identified at the data-link layer:

- Establishing a data-link connection between network entities

- Framing bits prior to transmission

- Passing control information between entities

- Providing retransmission service of protocol data units (PDUs)

- Providing a method of addressing

- Maintaining an ordered delivery of bits

- Providing a mechanism for network-layer control

- Managing the link between the two entities

- Providing error detection

- Providing flow control

- Releasing the data-link connection between entities

Data-link protocols can be divided into two sublayers. Figure 18.3 depicts these sublayers.

Implementation of the MAC sublayer depends on the protocol employed at this layer. For example, two IEEE standards supported by the ISO model define the MAC sublayer and physical-layer characteristics:

Characteristics	ISO 802.3	ISO 802.5
Encoding scheme	Manchester	Differential Manchester
Transmission	Baseband	Baseband
Topology	BUS	Ring
Medium characteristics	Coaxial cable 50 Ω	Twisted-pair cable
Throughput	10 Mbits/s	4 or 16 Mbits/s
Access method	Broadcast	Token passing

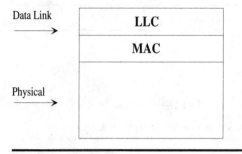

Figure 18.3   View of data-link sublayers.

The logical link control (LLC) sublayer provides unacknowledged connectionless-oriented, connection-oriented, and acknowledged connectionless service. Both data-link protocols just mentioned use LLC protocols, as do all the IEEE 802.X protocols.

The 802.3 and 802.5 differ with respect to data-link frame technology. The 802.3 is connectionless-oriented, and the 802.5 is connection-oriented.

Additional data-link-layer specifications are as follows:

ISO no.	Document title
8886	*Data-Link Service Definition*
3309	*High-Level Data-Link Control*
11575	*Protocol Mappings for the OSI Data-Link Service*

Network layer

The network layer embraces two modes of connection: connection-oriented and connectionless-oriented. The function of this layer is to form data packets and route them from one host to another. Examples of network layer protocols are

- X.25
- ISDN

X.25 is an example of how OSI can be implemented at the network, data-link, and physical layers. Figure 18.4 represents how X.25 appears at these levels.

Figure 18.4 shows the physical layer implementing X.21 on the interface. Layer 2 uses link-access protocol balanced (also known as LAPB). At the third layer X.25 is implemented and is known as *packet-layer*

Layers

4	Transport Layer
3	Connection-oriented Network Service
	X.25 Packet Layer Protocol (PLP)
2	Link Access Protocol Balanced (LAPB)
1	X.21

Figure 18.4 X.25 in an OSI network.

protocol (PLP). Immediately above PLP in layer 3 is the OSI connection-oriented network service. According to ISO standard 8208, X.25 has been used for a packet-switched network interface.

X.25 uses a collection of network service elements to perform the following functions:

1. Establish a connection

2. Release a connection

3. Transfer data

4. Acknowledge receipt of data

5. Reinitialize a connection

When X.25 is implemented, the data-link layer functions as a point-to-point connection. At the data-link layer synchronization between end systems occurs. At the network layer X.25 creates a virtual circuit (also called a *path*) through the network and maintains this circuit until data transfer is complete. Hence, it is connection-oriented.

The virtual circuit X.25 establishes includes three distinct parts. First is the setup part (where the call is made), in which a connection is established between end systems. Functionality of this part rests in the DTEs and the DCEs. Second is the part of the connection that moves data from one point to another. Data is transferred in full-duplex mode through the network. The third part is where the call is cleared. This last part is complete once the clear confirmation is acknowledged.

The OSI network layer also supports connectionless-oriented connections. When this is the case, X.25 is not used. This connectionless network protocol (CLNP) is similarly implemented between end systems.

Connectionless-oriented network service, unlike its counterpart, connection-oriented network service, does not require a connection to be established prior to data transfer. The CLNP characteristics can be summarized as follows: having a time-control mechanism for the packet and the ability to break up the packet into smaller pieces and to identify each piece of a packet for reassembly and identify the host in the routing system that sends the packet, which determines where the packet travels next.

Additional information on the network layer can be obtained from the following specifications:

ISO no.	Document title
8348	*Network Service Definition*
8348-1	*Connectionless Mode Transmission*
8348-2	*Network-Layer Addressing*
8348-3	*Additional Features of Network Service*
8648	*Internal Organization of the Network Layer*
8880-1	*Protocol Combinations to Provide and Support the OSI Network Service Part 1: General Principles*
8880-2	*Part 2: Provision and Support of the Connection Mode Network Service*

Transport layer

OSI supports connection-oriented and connectionless transport service. They are referred to as COTS and CLTS, respectively.

Connection-oriented transport service (COTS). COTS uses four service elements (an "element" is an OSI abstraction of a service). For practical purposes "elements" here means function. The four COTS services are

T_CONNECT	The function that establishes a full-duplex, transport-layer connection with the target system.
T_DATA	The data transfer function between origin and destination. It does not use confirmations, but does notify the user if an error has occurred.
T_EXPEDITED_DATA	Data is transferred with a higher priority.
T_DISCONNECT	Issued to release the connection.

Connectionless transport service (CLTS). CLTS uses one service element (function): T_UNIT_DATA. It has four parameters associated with it: destination address, user data, source address, and quality of service. CLTS does not guarantee reliable data transfer. Mechanisms guaranteeing reliable data transfer are left to the application above.

Types of network connections. OSI network connections (NCs) can be categorized into three groups based on the following characteristics:

Group	Characteristics
A	Low residual errors and an acceptable signal error ratio
B	Low residual errors but high signal error ratio
C	Practically unusable because of the residual errors

Transport protocol classes and additional information on transport layer.
OSI supports five classes of transport protocols (TPs) that use a connection-oriented network service (CONS). In chronological order, these five TPs are

TP0	Simple class
TP1	Basic error-recovery class
TP2	Multiplexing class
TP3	Error-recovery and multiplexing class
TP4	Error-detection and -recovery class

The simple (TP0) class is recommended only with group A network connections. No flow control or connection release is provided. Both flow control and connection release are based on the underlying network service.

The basic error-recovery (TP1) class does resynchronize data flow if an error is detected. No flow control exists. The transport connection can be released on demand; however, the underlying protocol may not.

The multiplexing (TP2) class enforces flow control. This protocol also merges multiple TCs into one network connection (NC). Its underlying NC protocol provides residual error and signaling capabilities; therefore, error recovery is unnecessary.

The error-recovery and multiplexing class (TP3) can merge multiple TCs and perform error recovery as well. It also provides flow control and expedited data transfer.

The error-detection and -recovery class (TP4) provides checksum capabilities, is capable of overcoming network failures, and provides retransmissions as well. It also provides a mechanism for inactivity (timing).

More information on the transport layer can be found in the following specifications:

ISO no.	Document title
8072	*Transport Service Definition*
8072-1	*Connectionless-Mode Transmission*
8073	*Connection-Oriented Transport Protocol Specification*
10736	*Transport-Layer Security Protocol*
10737	*Transport Protocol Management Specification*

Session layer

Functionality. The session layer provides what is commonly known as *session services* (SS). These services are available to the presentation layer. The presentation layer accesses these services at the session service access point (SSAP). Some refer to SSAP as an SS *user,* and some refer to the session and lower layers as the SS *provider.*

The presentation layer uses the connection made by the SSAPs between two session layers. The session layer offers services in the name of service elements. Some of these service elements include

- Connection establishment

- Data transfer

- Release connection

The session layer actually controls communication between applications. It is the session layer that is responsible for establishment of a connection for the application to use. The session layer is also responsible for enforcing protocols between the applications and the reconnection in the attempt of a failure between the two layers. To some degree the session layer is responsible for ensuring that buffers do not become overrun during data transfer.

Addressing. Session addresses have two components: the transport address and a session selector. The transport address includes a transport selector and network address that identifies a location in the network. Selectors identify a particular entity at a given location. A session selector has meaning attached to it because of the particular network address with which it is associated.

A session connection identifier is defined at connection phase. This connection identifier has four parts:

- The calling SS user

- The called SS user

- A common reference

- An additional reference

In addition to the connection identifier, functional units are negotiated at the time of connection establishment. They exist for the duration of the session connection and dictate which session operating primitives are available.

Structure and data flow. Two basic functions of the OSI session layer are to

- Provide structure to the data stream
- Regulate data flow between SS users

Structure can be added by way of synchronization points (also called *sync points*). Sync points can be implemented via major and minor forms. A major sync-point implementation is where a logical connection is broken into units, where identifiable beginnings and endings can be discernible. Thus definition of a major sync point would be the beginning or end of a dialog unit.

Minor sync points can be implemented as well. Minor sync points represent the organization of a dialog where the dialog itself can be represented by one or more smaller units called "activities." This activities structure level can be structured into minor sync points. Thus a dialog can have sync-point functions performed on it at a major or minor level. In either case both sync points have a synchronization-point serial number (SPSN) assigned to it. An additional function includes *resynchronization,* which is used to set the session connection to the previous synchronization point; then the connection state can be reestablished at that point.

Regulation of data flow between the SS users is performed by way of a token implementation. Abstractly defined, a *token* is an attribute of a session connection instance. Tokens may take the form of data, major synchronization, minor activity, or a release token. These tokens are used in dialog control, namely, via transmission characteristics, activity management, governing of sync-point insertions, and determination of a negotiated release.

Types of data transfer.   A data transfer phase can be identified at the session layer. Here four types of data transfer can be realized:

- Capability data transfer
- Expedited data transfer
- Normal data transfer
- Typed data transfer

Capability data transfer permits users to transfer data up to 512 octets while they are not within a data transfer activity. *Expedited* data transfer permits expedited data delivery which, to a degree, frees the session from constraints that the session would otherwise utilize. *Normal* data transfer specifies data transfer over a session connection in half- or full-duplex mode. *Typed* data transfer permits transfer of data over a session connection independent of a token.

Data tokens. ISO specifications use a token in the context of a session. A token is an attribute, assigned dynamically to a session connection. These tokens provide structure to dialogs.

Data tokens are used in management of half-duplex connections. Synchronization (minor) tokens throttle synchronization points. The other type of synchronization token (major) is used to manage initial sync-point settings and activity structure. The release token manages release of connections.

Additional information. Much more information can be obtained about the session layer through the following specifications:

ISO no.	Document title
8326	*Basic Connection-Oriented Session Service Definition*
8327	*Basic Connection-Oriented Session Protocol*
9548	*Connectionless Session Protocol*

Presentation layer

The OSI presentation layer provides selection and transformation of syntax for data exchanged between the application-layer entities above it. Application entities above the presentation layer are considered presentation users. These presentation users access the presentation layer at presentation service access points (PSAPs).

At the presentation layer the structure of information must be defined. This structure is known as *abstract syntax*. This syntax is created via the Abstract Syntax Notation One (ASN.1).

Functions. The following presentation-layer functions can be identified:

- Session establishment
- Data transfer
- Syntax negotiation
- Syntax formatting
- Session termination

Of these functions, a key function is the formatting of syntax. The presentation layer is best described as a boundary.

At the session layer and below data is represented in bit (binary) format. At the application layer, data is presented to users in an understandable format. It is the presentation layer that makes this possible.

At the application layer, in OSI, ASN.1 is used to represent information. By using ASN.1 to specify data, a degree of independence can be achieved. This representation of data is then passed to the presentation layer. The presentation layer takes this ASN.1 representation and converts it into required transfer syntax to move data through the network. The converse is true when the data reaches the target host.

Transfer syntax. Another critical component of the presentation layer is the transfer syntax. This is an encoding method used by the presentation layer to inform presentation entities that the information is being transferred between application entities.

The transfer syntax uses identifiers to specify a data type. A *tag* identifier reveals the tag class and number. A *length* identifier specifies the number of octets in a given field. A *contents field* reflects the actual data value encoded.

The abstract syntax and the transfer syntax define a presentation context. The abstract syntax is conveyed between presentation entities through the provision of the transfer syntax.

Additional information. The presentation layer has numerous ISO specifications that apply. Additional information concerning ASN.1 and the presentation layer can be obtained from the following ISO specifications:

ISO no.	Document title
8822	*Connection-Oriented Presentation Service Definition*
8823	*Connection-Oriented Presentation Protocol Specification*
9576	*Connectionless Presentation Protocol Specification*

Application layer

The OSI application layer has four functional components. One of them is common to all applications (including FTAM, MHS, X.500, and other applications). The following list includes these components, and the remainder of this section explains their functions.

- Association control service element (ACSE)

- Remote-operations service element (ROSE)

- Reliable transfer service element (RTSE)

- Commitment, concurrency, and recovery (CCR) services

Association control service element (ACSE). *Application associations* (sometimes called "associations") are defined as an application-layer specification with the ability to make a presentation-layer connection. Simply put, ACSE is a common denominator between applications such as FTAM, directory services (DS), the message-handling system

(MHS), and the presentation layer. In addition, ACSE is common to all OSI applications. It can best be understood by considering it as an interface for applications to the presentation layer.

ACSE manages application associations. These associations are a necessary common link between two applications. This link includes common ground for communication and also defines the rules for interactions between the two applications. At the time when associations are established an application context and abstract syntax are negotiated.

An application context defines the common environment to be used by the applications. Generally, it consists of application service elements (which is a defined set of capabilities of an application entity). The importance of this lies within any procedures that may need to be defined by means of the protocol used.

Application service element primitives map to an application protocol data unit, which in turn maps to a presentation-layer primitive. The presentation-layer primitive is mapped to a presentation protocol data unit which is, in turn, mapped to a session-layer primitive.

More information about ACSE can be obtained from the following specifications:

ISO no	Document title
8649	Service Definition for the Association Control Service Element
8650	Protocol Specification for the Association Control Service Element
10035	Connectionless ACSE Protocol

Remote-operation service element (ROSE). This service element's protocol has been defined as being that component which makes remote operations between two application entities (AEs) possible. Some of the services it performs are a request to execute an operation, report the results, and report status of the execution.

The functionality of ROSE is such that either participating application can invoke (establish) a communication instance. The ROSE application defines three classes of association. The following includes an example of application entities that can invoke remote operations:

Class 1   The initiating entity

Class 2   The responding entity

Class 3   Either entity

Further classification of ROSE reflects the mode of operation. Two modes apply: (1) asynchronous mode, in which either entity can invoke an operation without waiting for a reply from another entity, and (2) synchronous mode, in which an entity does invoke an operation until acknowledgment is received from the other application entity.

Similar to ACSE, ROSE application protocol data units (APDUs) are used to convey the meaning of ROSE service primitives. An example of an application that can use ROSE is a mail system. In this scenario the concept of peer entities exists: a mail server and a mail user. Here the mail application utilizes ROSE services to aid in the execution of the remote operation.

More information about ROSE can be obtained from the following specifications:

ISO no.	Document title
9072-1	*Remote Operations Part 1: Model and Service Definition*
9072-2	*Protocol Specification*

Reliable transfer service element (RTSE). RTSE uses ACSE services for establishment or relinquishment of an application association. However, when data transfer is to be performed it uses presentation services, bypassing ACSEs.

An example of RTSE use is with the message-handling system (MHS). RTSE regulates communication and transfer between origin and destination MHS systems.

RTSE protocol has built-in mechanisms to perform the following operations:

Open request

Open accept

Open reject

Transfer

Token, please

Abort

In short, RTSE is a method whereby the large messages (or data files) can be moved through a network that is considered unreliable.

More information about RTSE can be obtained from the following specifications:

ISO no.	Document title
9066-1	*Reliable Transfer Part 1: Model and Service Definition*
9066-2	*Protocol Specification*

Commitment, concurrency, and recovery (CCR). CCR, as it is called, is a standard for a two-phase commit protocol between one application

between CCR users. It alone does not provide any transfer abilities, but it does support commitment, rollback, and error-recovery procedures.

Fundamentally, the purpose of CCR is to prevent problems due to lack of control services. For example, in simple protocols a requester can request that an action be performed. The recipient can elect to perform the action or not. The recipient informs the requester whether an action has been performed or sends a negative acknowledgment indicating that the action was not performed. Either way, there is a problem with this scenario.

How are system failures overcome if one occurs in the middle of the request and response process? Second, if multiple systems are used, how is coordination of the systems managed? CCR settles these two issues by having the requester request an ID. When the recipient responds to this request, the action has not yet been taken. However, the recipient confers on an offer to commit back to the requester. The requester can then request the recipient to commit. The action to commit is not performed until the recipient receives the commit request. Then, on execution of the commit request, the recipient can inform the requester of the action performed.

The result of this scenario is communication between the requesting entity and the recipient such that should a failure occur, errors can be averted and synonymous system activity can be maintained.

More information about CCR can be obtained from the following specifications:

ISO no.	Document title
9804	*Service Definition for the Commitment, Concurrency, and Recovery Service Element*
9805	*Protocol Specification for the Commitment, Concurrency, and Recovery Service Element*

18.4   Directory Services (X.500)

The ISO and CCITT's X.500

The X.500 number is a CCITT recommendation for distributed directory service. This X.500 series has been adopted by the ISO for directory services.

Directory service components

Directory services are made possible through a number of components. These components interoperate, thus making the directory services appear to be a single database when in reality it constitutes multiple distributed databases. The following components make up the core of directory services:

- Directory information base (DIB)
- Directory entries (DEs)
- Directory entry attributes (DEAs)
- Directory information tree (DIT)

The DIB consists of directory entries. Each entry in the DIB is information about an object. An *object* is a reference to a person, place, or thing. Objects may also be categorized into classes. Objects in a given class share a commonality. For example, a class might be a country or organization.

Entries consist of one or more attributes. A type and a value are associated with each object attribute. The X.500 standard specifies some of these types and values. Figure 18.5 is an abbreviated example of the X.500 standard specifying types and associated values.

Additional information can be stored in the entries that make up the DIB. Some classes of entries may even have subclasses. Figure 18.6 depicts a more specific example of entries in a DIB.

Geographical	Country Name
	Locality
	State/Province
	Street Address
Telecommunications	Telephone Number
	Telex Number
	Telex Terminal ID
	Fax #
	X.121 Address
	and others
Postal	Address
	Code
	P.O. Box
	Office for Delivery

Figure 18.5 Example of X.500 attribute types.

Entry Identifier	Classes	Attribute Values
PI	Country	US
	Organization	Information World
	Person	Ed

Figure 18.6 Specific DIB entry.

In Fig. 18.6, PI indicates the person and the class lists the country, organization, and person. The attribute values associated with the class identify the country (United States), organization name, and person at the company.

The DIT indicates how the directory is organized. Because the DIT is tree-structured, greater flexibility is provided. Consider an example of how a DIT appears in Fig. 18.7.

DIT structure has inherent flexibility built into it. Implementations that use DIT can accommodate large or small environments.

How the directory services are used

The DS can be used by humans or programs. Typically, DS functions well with the MHS, to be discussed in Sec. 18.5. However, a user is viewed as a directory user agent (DUA), and the directory is made up of one or more directory service agents (DSAs). Directory Access

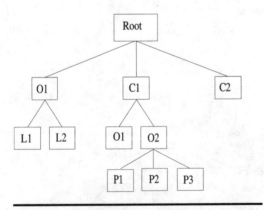

Figure 18.7 Conceptual view of a DIT.

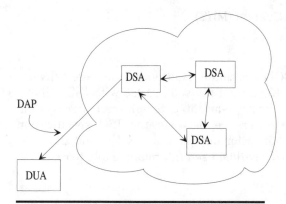

Figure 18.8   Conceptual view of DUAs and DSAs.

Protocol is used between DUAs and DSAs. Figure 18.8 depicts this scenario.

Directory services and security

Since DSs is an application and uses the application and presentation layers in OSI, ASN.1 is used. As a result security is implemented through ASN.1 parameters issued when invoked.

The ISO specification 9594, Part 8 explains the authentication framework for use by other protocols that work with DS. A simple authentication is specified whereby a name and a password must be entered then verified by the recipient. A strong authentication is more robust because it uses cryptography in the exchange of information requiring verification.

How to obtain more information

Additional information can be obtained from the ISO specifications previously listed; they are listed below again for your convenience.

ISO no.	CCITT no.	Document title
9594-1	X.500	*The Directory: Part 1: An Overview of Concepts, Models, and Services*
9594-2	X.501	*Part 2: Information Framework*
9594-3	X.511	*Part 3: Access and System Services Definition*
9594-4	X.518	*Part 4: Procedures for Distributed Operations*
9594-5	X.519	*Part 5: Access and System Protocol Specification*
9594-6	X.520	*Part 6: Selected Attribute Types*
9594-7	X.521	*Part 7: Selected Object Classes*
9594-8	X.509	*Part 8: Authentication Framework*

18.5 Message-Handling System (MHS) and X.400

X.400 and ISO specification

The X.400 number is a CCITT recommendation for message delivery. It was first announced in 1984, then revised in 1988 with ISO collaboration. The ISO had the message-oriented text interchange system (MOTIS). There is little difference between the ISO specification (ISO/IEC 10021) and X.400; so for our purposes X.400 and the message-handling system (MHS) will be used interchangeably here.

MHS components

Message handling is made possible through components implemented according to the architecture. These components include

- User agent
- User
- Message transfer agents
- Message transfer system

User agents (UAs) are applications that generate messages with a format and header field specific to the type of message being transferred. These UAs help users, which may be human or a program to originate or receive messages.

Message transfer agents (MTAs) are programs that store and forward messages between user agents and reside within the *message transfer system* (MTS). This concept of a message store is called *message store* (MS) in MHS. MTAs can connect with other MTAs and multiple UAs within the MTS. Consider Fig. 18.9.

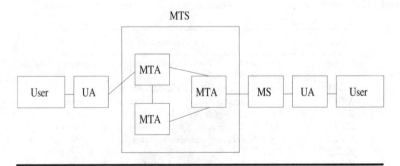

Figure 18.9 Conceptual view of MHS components.

These components make MHS possible. Other functions and components also contribute. The MHS is more than just a system capable of moving text. The actual content of a message can be text, but it can also be voice, fax, video, or coded data; such messages are considered *multimedia messages.*

Message types

Three primary types of messages are used with MHS. The simplest of all MHS messages is the *user message,* which consists of an envelope and contents. The envelope is used to identify the sender and intended receiver of the message. The contents identify the constituent parts by categorizing them as being for either interpersonal use or business reasons.

The *delivery report* message is the second type of MHS message. This is generated by MTS and used to inform a message originator of the outcome of a user message or a probe. A *probe* contains no contents. Its function is to determine whether it is possible to send a message from a user to a destination.

Message information base (MIB)

The MS keeps incoming messages in a MIB. These messages have attributes that characterize the message as to type of content, whether it is new, and who originated it, for example. The MHS MIB is similar to the DIB.

MHS protocol requirements

Message transfer agents use MHS protocols to transfer messages, reports, or probes (probes are similar to messages but are sent first to determine whether a message can be sent from one MHS to another). MHS protocol encompasses approximately 46 arguments, but not all of them are necessary; only 10 are actually required. The arguments and brief descriptions are listed below.

Content: This contains the content of a message.

Content type: This identifies the type of contents the message consists of.

Message identifier: This is the unique identifier that separates the message from other messages in the system.

Originator name: The originator's name for the message.

Originator report request: This indicates a value of either no report, nondelivery, or report. It originates from the originating MTS user.

Originally specified recipient number: This is a value for each receiver.

Originating MTA report request: Indicates the kind of report.

Recipient name: This is an argument containing the name of each recipient.

Responsibility: This indicates who should deliver a message to a MTA or the designated recipient.

Trace information: This is documented information on the route the message takes and any activities occurring along this route.

Other arguments exist. If desired, they can be found in ISO 10021-5 specification.

MHS security

MHS supports different security types. Some of these are

Data confidentiality service: This type of security ensures that the message will not be disclosed to an unauthorized user.

Data integrity service: This type of security protects the contents of a message during transmission so that the data cannot be modified.

Nonrepudiation service: This is a function whereby proof exists that a message originated, was submitted, or was delivered to the intended party.

Message label service: This service can work in conjunction with the security context service to provide a security label for MHS objects.

Origin authentication service: This security service provides a peer-oriented security approach. Proof-of-submission service enables an originator to verify that a message has been received by the MTS. Proof-of-delivery service provides the originator of a message to verify that a message has been delivered to its intended recipient. This security service is usable with messages, probes, and reports.

Security access service: This consists of two components: *peer entity authentication security,* which provides the ability to secure a peer origin and destination that an unauthorized entity does not attempt to use for a previously established connection; and the *security context service,* which is used to limit message scope between MHS objects.

Security management service: This service requires change credential security service. This enables a MHS object to change credentials held by another MHS object. *Register security service* allows security labels to register with a MTA for a specific user.

Theory of operation

MHS can be regarded as an electronic post office. With this in mind the following explanation puts the basic operations of MHS in perspective.

In MHS originators and recipients exist, whereas in today's postal system there are senders and receivers. In any instance a user can be one or the other.

Just as the postal service office offers different services, so does MHS. MHS defines the type of user and what the user wants to mail. For example:

- User types:

  Person

  Company

  Process

- What is mailed:

  Message

  Probe

  Report

As does postal mail, messages in MHS have two parts: the envelope and the contents. An envelope has the required information to route a message to its destination. The contents portion is the main information carried to its destination.

In MHS user agents provide varying services. The basic function of a user agent concerns how a message is transferred. It can be normal or expedited.

MHS service elements provide particular capabilities, features, or functions.

User agents can provide a message-store capability. This is similar to a post office box; messages can be stored securely.

Another similarity between today's post office and MHS is the transfer and delivery of mail. Just as post offices have a controlling entity, regardless of the country, and these post offices exchange mail around the world, MHS has a similar concept. The message transfer agent (MTA) functions similarly to the cooperating entities in the postal services.

Figure 18.10 is a conceptual view of the MHS and its major components.

Figure 18.10 depicts a simple MHS system implementation. When MHS is viewed from a layer perspective, it appears as shown in Fig. 18.11.

The MHS can be divided into sublayers as seen in Fig. 18.11. The user agent layer contains UA functions associated with the format of

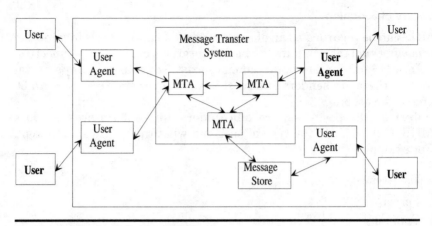

Figure 18.10   Conceptual view of MHS major components.

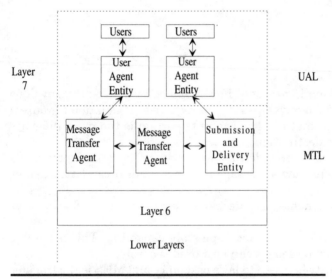

Figure 18.11   A perspective of MHS layers.

messages. This layer is also responsible for transfer and delivery interaction with the MTA. The message transfer layer provides three services to the UA: submission, delivery, and relay.

The UA entity shown in Fig. 18.11 provides message transfer services and interpersonal messaging services. The MTA entity provides layer services for other MTAs. The submission and delivery entity functions with the MTA provide access to the entire MTS.

How to obtain more information

Additional information can be obtained through the following ISO documents.

ISO no.	CCITT no.	Document title
10021-1	X.400	*Part 1: Message-Oriented Text Interchange System (MOTIS) System and Service Overview*
10021-2	X.402	*Part 2: Overall Architecture*
10021-3	X.407	*Part 3: Abstract Service Definition Conventions*
10021-4	X.411	*Part 4: Message Transfer System—Abstract Service Definition and Procedures*
10021-5	X.413	*Part 5: Message Store—Abstract Service Definition*
10021-6	X.419	*Part 6: Protocol Specification*
10021-7	X.420	*Part 7: Interpersonal Messaging System*
10021-8	X.435	*MTS Routing*

18.6 File Transfer and Access Method (FTAM)

Concisely, FTAM permits multiple systems to access parts of files and file systems, move data between systems, remove and add data, and maintain an accurate description of a file, all without requiring knowledge of how filing systems are provided.

Because file types, structure, and access method vary between systems, there is a need to support these various systems and thus provide a common file model and file storage. The latter is called a *virtual filestore.*

Virtual filestore

The basic concept of the virtual filestore is to allow a complete mapping of functions from real filing systems to the common model called the *virtual filestore,* which is an abstract model for describing files and their storage capacity, as well as operations that may have been performed upon them. The purpose is to maintain independent implementation.

A virtual filestore is a collection of many files. In the virtual file store context, each file is considered an entity that consists of the following characteristics:

- File contents
- Unique filename
- Size of file
- History of the file

- Description of the logical structure of the file
- List of actions that can be performed on the file

A virtual filestore can be accessed by a user-developed program or a human user using FTAM service elements. The virtual filestore is also key to providing transparent operations between a filestore and the user. The FTAM virtual filestore is merely an abstraction of a filestore.

Real filestore

A *real filestore* is a group of files including a description of the following:

- Characteristics of each file
- Characteristics of each file's current activity

Theory of operation

FTAM operations are similar to a client/server model of communications. Specifically, in FTAM lingo the term *dialog* is used to refer to the communications between participating entities. A filestore *user* is the one who *initiates* a file transfer, access, or management operation. A filestore user may be a human or a user-developed program. A *responder* is a program that maintains a virtual filestore with virtual files. A FTAM data element is a unit of information transferred between an initiator and a responder. Figure 18.12 depicts this scenario.

Virtual filestore management

A number of management-related operations can be performed on a file within a virtual filestore environment. Some of these operations are

Create	Create a new file; define its attributes
Change	Change a file's attribute values
Close	Close a file, to prevent further operations on the file contents
Delete	Delete a file
Deselect	Prevent further file operations by deselecting a file
Open	Open a file
Read attributes	Read particular values
Select	Select a file by its contents or initiator

FTAM services and functional units

FTAM services are used by an initiator and a responder. FTAM has other services as well. FTAM groups services in what is called a *func-*

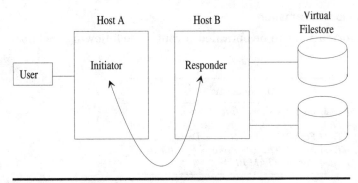

Figure 18.12 Conceptual view of initiator and responder.

tional unit, which is a single unit with a group of services. FTAM functional units and services are as follows:

Functional unit	Services
Enhanced file management	Used to change file attributes
FADU locking	FTAM supports file concurrency and file access data unit (FADU) locking
File access	Used to locate a specific FADU
Grouping	Permits multiple function executions to be carried out in one exchange
Kernel	Supports basic functions for FTAM association establishment and release
Limited file	Aids in file creation, file deletion, and file attribute examination
Read	Supports data transfer to an initiator from a responder
Recovery	Permits an initiator to execute recovery actions when a failure occurs once it is opened
Restart	Permits a restart following a checkpoint
Write	Provides support for data transfer from an initiator to a responder

Synopsis

FTAM's strength is its ability to overcome the internal file structures of end systems. Its ability to support different document types is another strength. Another feature of FTAM is that it allows users to continue the interface to which they are accustomed but to access files on other systems. Consequently, the filestore implementation is practically transparent to the user.

How to obtain more information

Additional information can be obtained through the following ISO documents:

ISO no.	Document title
8571-1	*FTAM Part 1: General Introduction*
8571-2	*FTAM Part 2: The Virtual Filestore Definition*
8571-3	*FTAM Part 3: The File Service Definition*
8571-4	*FTAM Part 4: The File Protocol Specification*
8571-5	*FTAM Part 5: FTAM PICS*

18.7   Virtual Terminal

The concept of a virtual terminal (VT) is a terminal having a generic set of characteristics and able to communicate with a variety of hosts and applications throughout a network. The ISO VT is based on a terminal having such a general set of characteristics rather than specific ones. From the OSI perspective, communication between a terminal and an application is an application-layer function. Consequently, the VT capability is provided through application-layer services.

Basic class virtual terminal

A common communication area (CCA) is shared by display objects, control objects, and device objects. This CCA area is a common area used by the VT user and programs. Two modes of operation are available in the *basic class virtual terminal* (BCVT): asynchronous and synchronous. *Asynchronous mode* allows both users to utilize this CCA, thus communicating at the same time.

The *synchronous mode* allows one user to communicate at a given instance. For example, if a user is typing data on the keyboard, the information from the application results are not displayed on the terminal.

Virtual terminal environment

A *virtual terminal environment* (VTE) is a set of parameters for a VT association. This *profile,* as it is known, is used to define operational characteristics and how data is structured. When a user begins VT operation, a set of default VTE parameters can be used, or the VTE can be negotiated.

The OSI Implementation Workshop defined six profiles that will be mentioned here:

Forms   This defines an application that determines where data input is based. In this sense, local verification at the terminal is performed.

Scroll	Defines a line at a time interaction between terminal and computer application. The definition permits scrolling forward or backward. It also supports the type-ahead function.
S-Mode	This definition is where an entire screen of information can be entered on the terminal and then sent to the application.
TELNET	This definition is based on the TELNET definition of the TCP/IP suite of protocols.
Transparent	This definition calls for bidirectional passing of character sequences at the same time.
X3	This definition is based on the CCITT specification explained in the X.3 and X.29 definition. In essence, it operates on packet assembly and disassembly.

BCVT concepts

In addition to the modes of operation for the BCVT, other concepts apply. They are briefly reviewed here.

Access rules: The notion of access rules is used to aid in the determination of when a VT user can update a display object. These rules include variables for write access and the related ownership.

Break-in: Another concept with VT user is the ability to interrupt another user during data transfer. A *destructive interrupt* is used to prevent a current function in progress. A *nondestructive interrupt* does not cause destructive action against the target VT user; rather, it sends a signal that does not disrupt current operations.

Conceptual communication area: This is a shared area used by display and device objects along with control objects. This service is a common link for entities needing to share parameters.

Control objects: Many different control objects exist in the use of VT operations. Their fundamental purpose is to provide control mechanisms required for the execution of useful VT operations.

Delivery: Delivery of data to the destination can be controlled through the concept and abilities provided via delivery control.

Device objects: These objects are abstract and are used to reflect a real device. The result is that manipulation and association of real devices can be manipulated with greater ease.

Conclusion

The VT is a protocol that a user can use to access applications regardless of the host on which they reside. Practically any terminal user can use VT to access many different hosts. Because of the generic set of capabilities, the VT can use a variety of implementations.

How to obtain more information

Additional information can be obtained through the following ISO documents:

ISO no.	Document title
9040	*Virtual Terminal Service*
9041	*Virtual Terminal Protocol*

18.8   Additional Applications

Other applications exist in addition to those explained previously. For our purposes here, a brief synopsis of some of them is sufficient to convey the functions available in OSI protocols and applications. Some of these applications include

- Common management information protocol

- Remote database access

Common management information protocol

Common management information protocol (CMIP, as it is commonly known) is the OSI method (protocol) for managing OSI networks. OSI networks are managed by considering everything as objects. Information about the state of a managed object is maintained in a management information base (MIB). The concept of a management information tree (MIT) exists with CMIP. This MIT is a directory of structured objects. Since the OSI network has named objects, particular functions can be performed on them.

Event reporting, alarm reporting, log control functions, and other functions can be performed with CMIP. CMIP can also perform get, set, creation, and deletion operations on objects.

Further information on CMIP can be obtained from the following ISO specifications:

ISO no.	Document title
9595	*Common Management Information Service*
9596	*Common Management Information Protocol*

Remote database access

The *remote database access* (RDA) concept is built around the notion of peer entities. This means that a distributed database implementation can be achieved. Its functional structure mirrors the client/server architecture. In this case, a client can initiate RDA services to invoke certain functions on the database server.

Additional information on this topic can be obtained from the following ISO specifications:

ISO no.	Document title
9579-1	RDA *Part 1: General Model, Services, and Protocol*
9579-2	RDA *Part 2: SQL Specifications*

18.9 Summary

The OSI model is a collection of specifications pertaining to each of its layers. Each layer of the OSI model has specifications making implementation flexible to a site's specific requirements. Some applications in use today are FTAM, X.400 or MHS, and X.500 (directory services).

The number of OSI products being produced is increasing. The U.S. government has already made its statement of direction with the GOSIP announcement. It seems currently that OSI products are being designed, introduced to the market, and implemented to various degrees. To what degree remains to be seen in retrospect.

OSI is a complex network protocol. It is vast, and many well-written books have been published on the topic. One foremost book on the topic is Mijendra N. Jain and Ashok K. Agrawala, *Open Systems Interconnection: Its Architecture and Protocols,* ISBN 0-07-032385-2. The book is published by McGraw-Hill, Inc.

OSI is evolving into a mature protocol. It is embraced by corporations such as Digital Equipment (DEC), IBM, and many other respected corporations. Particular implementations offered by many corporations are such that OSI products can coexist with current equipment.

Since work began on OSI protocols and specifications, much momentum has been gained. Its constant penetration into all facets of business is facilitating its growth. This chapter has presented some of the core components of OSI, and further information should be obtained to reveal the depths of what OSI has to offer.

19

Transmission Control Protocol/Internet Protocol (TCP/IP)

Transmission control protocol / Internet protocol (TCP/IP) is an upper-layer network protocol. It is generally referred to as TCP/IP. It is in widespread use around the world today. This chapter presents the core components and issues related to TCP/IP, beginning with a historical perspective.

19.1   A Historical Perspective

A good place to begin is in the late 1960s. An entity in the U.S. government, the Advanced Research Projects Agency (ARPA), was exploring technologies of all sorts. One of those technologies led to a need (desire) to create a network based on packet-switching technology to help them experiment with what they built. It was also seen as a means of using the then current telephone lines to connect scientists and personnel who were in physically different locations to work together in this network.

By late 1969 the necessary components had come together to create the ARPANET. In short order a few individuals had put together a network that was capable of exchanging data. Time passed, and additions and refinements were made to ARPANET.

In the 1970s

In 1971 the Defense Advanced Research Projects Agency (DARPA) succeeded the Advanced Research Project Agency (ARPA). As a result, the

ARPANET came under the control of DARPA. DARPA's forte was concentration on satellite, radio, and packet-switching technology.

During this same time period the ARPANET was using what was called a *network control program* (NCP). This NCP (not to be confused with IBM's NCP) was limited and impacted the ARPANET with regard to capabilities and other requirements. These protocols ARPANET utilized (namely, the NCP) were characteristically slow, and there were periods when the network was not stable. Since ARPANET was now officially under DARPA's umbrella and it was realized that a new approach to ARPANET was needed, a different direction was taken.

By approximately 1974 DARPA sponsored development for a new set of protocols to replace the ones in use at that time. This endeavor led to the development of protocols that were the basis for TCP/IP. The first TCP/IP began to appear during 1974–1975. While these technical matters were in full force another phenomenon was occurring.

In 1975 the U.S. Department of Defense (DoD) put ARPANET under the control of the Defense Communication Agency (DCA); the DCA was responsible for operational aspects of the network. It was then that ARPANET became the foundation for the Defense Data Network (DDN).

Time passed and TCP/IP continued to be enhanced. Many networks emerged working with and connecting to ARPANET with TCP/IP protocols. In 1978 TCP/IP was sufficiently stable for a public demonstration from a mobile location connecting to a remote location. It was a success.

In the 1980s

From 1978 until 1982 TCP/IP gained momentum and was continually refined (however, in a real sense the Internet had existed for years). In 1982 multiple strides were made. First, DoD made a policy statement adopting TCP/IP protocols and making it the overseeing entity for uniting distributed networks. The next year, 1983, DoD formally adopted TCP/IP as the standard for the protocol to use when connecting to the ARPANET.

DoD discontinued support for the NCP and began phasing in TCP/IP. Technically, the term *Internet* was an outgrowth of the TCP/IP protocol suite; namely, IP (Internet Protocol). The term *Internet* has maintained its association with TCP/IP to date.

In the 1990s

The Internet today consists of numerous interconnected networks. The National Research and Education Network (NREN) is a dominant part of the Internet today. Other networks participating in the Internet include the National Science Foundation network (NSF), NASA, Department of Education, and many others including educational institutions.

Commercial, educational, and organizations of all types are connected to the Internet. An industry of service providers for the Internet began and has now mushroomed in growth.

19.2 Forces Contributing to Growth of TCP/IP

Technology

The historical review in Sec. 19.1 sheds some light on technology surrounding TCP/IP and the Internet, but does not explain certain aspects of the Internet that may explain the technological impact it had on TCP/IP.

The Internet (uppercase I) is based on TCP/IP as the U.S. government made it the standard. The Internet is worldwide, and all sorts of entities are connected to it. Knowing this, we can deduce that those entities connected to it are using TCP/IP. This alone counts for a tremendous amount of TCP/IP in the marketplace. And at the current rate, it is increasing rapidly.

The 1980s can be characterized as a decade of rapid technological growth. Many companies capitalized on the U.S. government endorsement of TCP/IP as the standard for the Internet and accordingly began producing products to meet this need.

This influx of TCP/IP products nursed the need for additional products. For example, in the 1980s two technologies dominated: PCs and LANs. With the proliferation of PCs and LANs an entirely new industry began to emerge. These technological forces seemed to propel TCP/IP forward because TCP/IP and PCs made for a good match when implementing LANs. TCP/IP implemented on an individual basis is referred to as an *internet* (lowercase i).

Market forces

A factor that contributed to the growth of TCP/IP in the market was corporate downsizing. This may seem strange, but during the 1980s I witnessed many cases where TCP/IP-based networks grew when others were shrinking. Granted that this was not the only reason for TCP/IP's healthy market share, but it did contribute.

For example, I observed a similar scenario. A corporation (which I will not identify by name) had its corporate offices in the northeastern United States. This corporation had many (over 50) satellite offices around the nation. This corporation needed these satellite offices to have independence for daily operations and at the same time be connected to the corporate data center. It achieved this by implementing TCP/IP-based LANs in its satellite offices, then connecting them to the data center. This example is one of many I have seen.

Availability

TCP/IP could be purchased off the shelf of many computer retailers by the end of the 1980s. This degree of availability says a lot for a product which at the beginning of the decade was not readily available to end users.

Another factor played a role in the availability of TCP/IP. The DoD not only encouraged use of TCP/IP; they funded a company called Bolt, Beranek, and Newman to port TCP/IP to UNIX. In addition, DoD encouraged the University of California, Berkeley (UCB), to include TCP/IP in the BSD UCB UNIX operating system. This meant that by acquiring UCB UNIX, users got TCP/IP free. It was not long before TCP/IP was added to AT&T's UNIX operating system. I suppose this conveys how available it was becoming.

Individual knowledge

By the late 1970s and surely into the 1980s TCP/IP was in most colleges and many educational institutions. Since by the mid-1980s it was shipped free with UCB UNIX and was available there, it became dominant in learning institutions. Then the obvious occurred. Individuals everywhere began graduating from educational institutions, and if their backgrounds included computer science, odds were they had been exposed to TCP/IP.

Granted this premise, consider this. These individuals entered the workplace and began penetrating the technical and managerial departments. When it came to contributing to a decision about a network protocol, which would be the likely choice in many cases?

In the 1980s the marketplace paid a premium for those who understood TCP/IP. Now, in the mid-1990s the market has considerable numbers of individuals who have varying degrees of TCP/IP knowledge.

All these factors weave together to make TCP/IP as dominant as it is today. Surely other factors have contributed as well, and TCP/IP has become a prevalent upper-layer protocol worldwide.

19.3 Layer Analysis

Author's note

In the early days of the Internet the term *gateway* became commonplace. It generally meant a connection from a specific location into the

Internet. This was adequate at the time; however, now confusion abounds with the use of this term.

According to the *American Heritage Dictionary*, the term *gateway* is defined as: "1. An opening, as in a wall or a fence, that may be closed by a gate. 2. A means of access." I believe the original intent of the term's meaning was "a means of access." This is fine, and you are probably wondering why it is even mentioned. Well, today an entire industry called *internetworking* and *integration* has appeared, and with it specialized devices exist. One such device is a gateway.

There is a consensus among integrators and those who integrate heterogeneous networks regarding the meaning of the term *gateway*. It is a device that at a minimum converts upper-layer protocols from one type to another. It can, however, convert all seven layers of protocols.

The purpose of explaining this here is simple. Throughout the presentation on TCP/IP the term *gateway* may appear. The term has such a foothold in the TCP/IP community that it is still used. Ironically, when this term is used in many instances with TCP/IP and the Internet, technically the term should be *router*. More in-depth information on networking devices is presented in Part 4 of this book.

Overview and correlation to the OSI model

TCP/IP is an upper-layer protocol. TCP/IP is implemented in software; however, some specific implementations have abbreviated TCP/IP protocol stacks implemented in firmware. TCP/IP can, however, operate on different hardware and software platforms, and it supports more than one data-link-layer protocol.

The OSI model is a representation of the layers that *should* exist in a network. Figure 19.1 compares TCP/IP to the OSI model.

Notice that TCP/IP has three layers: network, transport, and the upper three layers combined together functioning aggregately as the application layer. TCP/IP is flexible when it comes to the lower two layers. It can be implemented in a variety of ways.

TCP/IP can operate with a number of data-link-layer protocols. Some are listed in Fig. 19.1. The remainder of this section highlights popular components at each layer.

Network-layer components and functions

OSI model layer 3 is the network layer. In TCP/IP it is the lowest layer in the TCP/IP protocol suite. TCP/IP network-layer components include the following:

Internet protocol (IP):   IP has an addressing scheme used to identify the host and network. IP is involved in routing functions.

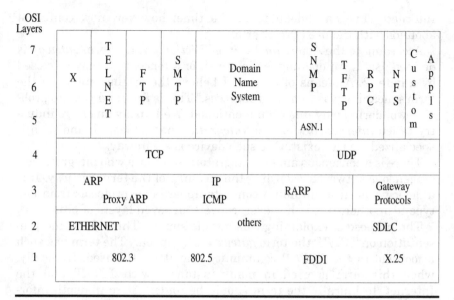

Figure 19.1   Comparison of TCP/IP to OSI layers.

Internet control message protocol (ICMP):   ICMP is a required component in each TCP/IP implementation. It is responsible for sending messages through the network via the IP header.

Address resolution protocol (ARP):   ARP dynamically translates IP addresses into physical (hardware interface card) addresses.

Reverse address resolution protocol (RARP):   RARP requests its host IP address by broadcasting its hardware address. Typically a RARP server is designated and responds.

Routing information protocol (RIP):   RIP is a routing protocol used at the network layer. If implemented, it performs routing of packets in the host in which it resides.

Open shortest path first (OSPF):   This is a routing protocol implemented at the network layer as RIP, but it utilizes knowledge of the internet topology to route messages the quickest route.

Transport-layer components and functions

Layer 4 at the OSI model is the transport layer. In TCP/IP it is the same. Transport-layer components include

TCP: This transport-layer protocol is considered reliable and performs retransmissions if necessary.

UDP: This transport-layer (user datagram) protocol is considered unreliable and does not perform retransmissions; this is left up to the application using its services.

Popular application-layer offerings

Above the transport layer in TCP/IP there are a number of popular applications. These include

X   This is a windowing system that can be implemented in a multivendor environment.

TELNET   This application provides remote log-on services.

file transfer protocol (FTP)   This application provides file transfer capabilities among systems.

simple mail transfer protocol (SMTP)   This application provides electronic mail (E-mail) services for TCP/IP-based users.

domain name service (DNS)   This application is designed to resolve destination addresses in a TCP/IP network. This is an automated method of providing network addresses without having to update host tables manually.

trivial file transfer protocol (TFTP)   This UDP application is used best in initialization of network devices where software must be downloaded to a device. Since TFTP is a simple file transfer protocol, it meets this need well.

simple network management protocol (SNMP)   This is how most TCP/IP networks are managed. SNMP is based on an agent-and-manager arrangement. The agent collects information about a host, and the manager maintains status information about hosts participating with agents.

network file server (NFS)   NFS is an application that causes remote directories to appear to be part of the directory system to the host which the operator is using.

remote procedure call (RPC)   This is an application protocol that enables a routine to be called and executed on a server.

custom applications   Custom applications can be written using UDP as a transport-layer protocol. If this is done, peer communications can be achieved between applications.

19.4   TCP/IP Network Requirements

Before exploring details of TCP/IP, the basic requirements for a TCP/IP network to function should be known. For example, TCP/IP networks require *all* participating hosts to have TCP/IP operating on them, and they must be connected directly or indirectly to a common link. This may require some gateway functionality for some systems, but Fig. 19.2 is an example of a typical TCP/IP network with multivendor computers.

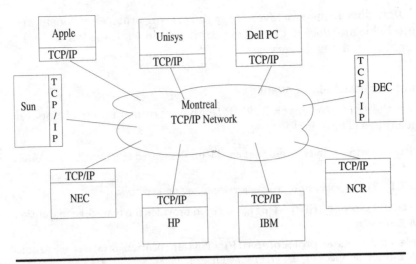

Figure 19.2   TCP/IP networking requirements.

Figure 19.2 includes vendors whose operating systems are different. They also have different hardware platforms. However, if the link to the TCP/IP network is established, the computers shown in Fig. 19.2 can communicate effectively.

With this overview in mind, the remaining portion of this chapter presents detailed information about the TCP/IP protocols and applications.

19.5   Internet Protocol (IP)

IP resides at network layer 3. IP routes packets (units of data) from source to destination. Some individuals refer to a packet in the sense of IP as a datagram. An IP datagram is a basic unit moved through a TCP/IP network.

IP is connectionless. It implements two basic functions: fragmentation and addressing. Fragmentation (and reassembly) is accomplished by a field in the IP header. Fragmentation is required when datagrams need to be smaller to be passed through a small packet-oriented network.

IP header format

The addressing function is also implemented in the IP header. The header includes the source and destination address. The IP header also includes additional information. Figure 19.3 is an example of an IP header.

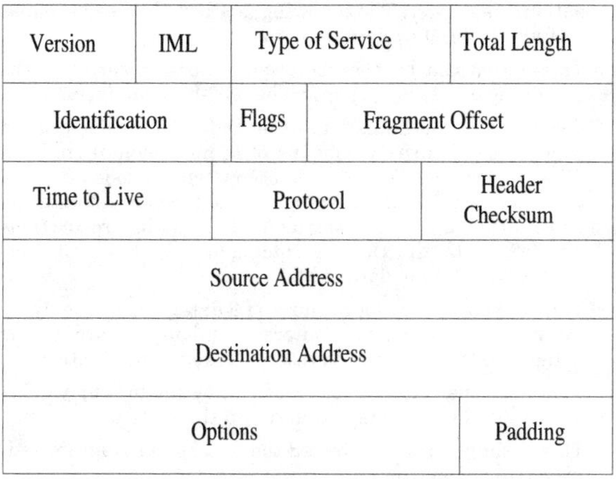

Version	IML	Type of Service	Total Length
Identification		Flags	Fragment Offset
Time to Live		Protocol	Header Checksum
Source Address			
Destination Address			
Options			Padding

Figure 19.3   Conceptual view of an IP header.

The components in the IP header and their meaning are presented in the following list.

VERSION	The version field is used to indicate the format of the IP header.
IHL	IHL stands for internet header length, which is the length of the internet header in 32-bit words and points to the beginning of data.
TYPE OF SERVICE	The type of service field specifies how the datagram is treated during its transmission through the network.
TOTAL LENGTH	This field indicates the total length of the datagram; this includes the IP header and data.
FLAGS	The flag field has 3 bits which are used to indicate whether fragmentation is supported or not to fragment, and indicates more and last fragments.
FRAGMENT OFFSET	This indicates where in the datagram the fragment belongs (assuming that fragmentation has occurred).
TIME TO LIVE	This indicates the maximum time a datagram is permitted to stay in the internet system (whether this is a local internet or the Internet). When the value equals zero, the datagram is destroyed. Time is measured in units per second, and each entity that processes the datagram must decrease the value by one even if the process time is less than 1 s.
PROTOCOL	This field determines whether the data should be sent to TCP or UDP in the next layer in the network.
HEADER CHECKSUM	This is a header checksum only. Some header fields change and the header checksum is recomputed and verified wherever the header is processed.
SOURCE ADDRESS	This is the originator of the datagram. It consists of 32 bits.
DESTINATION ADDRESS	This is target for the header and data. It, too, is 32 bits.
OPTIONS	Options may or may not appear in datagrams. Options must be implemented in IP modules; however, they may not be used in any given transmission. A number of variables in the

options field exist. The following is a list of those variables including a brief explanation.

NO OPTION

This option can be used between options to correlate the beginning of a following option on a 32-bit boundary.

SECURITY

Security is a mechanism used by DoD. It provides hosts a way to use security by means of compartmentation, handling restrictions, and transmission control codes (TCCs). The compartmentation value is used when information transmitted is not compartmented. Handling restrictions are defined by the Defense Intelligence Agency. TCC permits segregation of data traffic.

LOOSE SOURCE and RECORD ROUTE

This provides a way for a source of a datagram to supply routing information for routers to aid in forwarding the datagram. It also serves to record the route information.

STRICT SOURCE and RECORD ROUTE

This option permits the source of a datagram to supply information used by routers and to record the route information.

RECORD ROUTE

This is simply a way to record the route of a datagram as it traverses the network.

STREAM IDENTIFIER

This provides a way for a stream identifier to be carried through networks that do not support this stream concept.

TIMESTAMP

This option includes a pointer, overflow, flag field, and internet address. Simply put, this provides the time and date when a router handles the datagram.

PADDING

The padding is used to ensure that the header ends on the 32-bit boundary.

The next IP

In June 1993 RFC 1475 was published and noted to have the status of a "memo." Succinctly, this request for change (RFC) explores the possibilities for the next generation (version) of the Internet, particularly the IP addressing structure. However, other protocols in the TCP/IP protocol suite are also explored.

In brief, the RFC specifying Internet Version 7 focuses on four major areas:

- IP addressing to 64 bits
- A forward route identifier in each datagram to support increasingly fast data transfer speeds
- Increased speed capabilities for network path delays that TCP operates
- An addition of a numbering layer for administration and more space for subnetting.

Another way of comprehending the 64-bit addressing scheme (if it is adopted) is to attempt to understand it from a practical perspective. A 64-bit addressing scheme would mean that approximately each person on planet Earth would have more than 2 *billion* addresses available for personal use!

Along with the idea of a 64-bit addressing scheme, another component would be added to the current 32-bit structure to accommodate space for network and host addresses. The 64-bit addressing scheme idea includes bits to identify an administrative domain. This domain identifies an administration that may be a service provider or other entity.

Much more information is available about this topic of "the next internet" in RFC 1475. RFC 1475 is considered a *memo,* not a standard. It is approximately 35 pages in length and is a good reference source for this topic.

19.6   Internet Control Message Protocol (ICMP)

ICMP works with IP; it is also located at layer 3 with IP. Since IP is connectionless-oriented, it has no way to relay messages or errors to the originating host. ICMP performs these functions on behalf of IP. ICMP sends status messages and error messages to the sending host.

ICMP message structure

ICMP utilizes IP to carry the ICMP data within it through a network. Just because ICMP uses IP as a vehicle does not make IP reliable; this simply means that IP carries the ICMP message.

The structure of an ICMP message is shown in Fig. 19.4.

The first part of the ICMP message is the TYPE field. This field has a numeric value reflecting its meaning; this field identifies its format as well. The numeric values and their meanings that can appear in the type field are shown in Fig. 19.5.

The next field in the ICMP message is the CODE field. It too has a numeric value assigned to it. These numeric values have an associated meaning as shown in Fig. 19.6.

The CHECKSUM is computed from the ICMP message starting with the ICMP type.

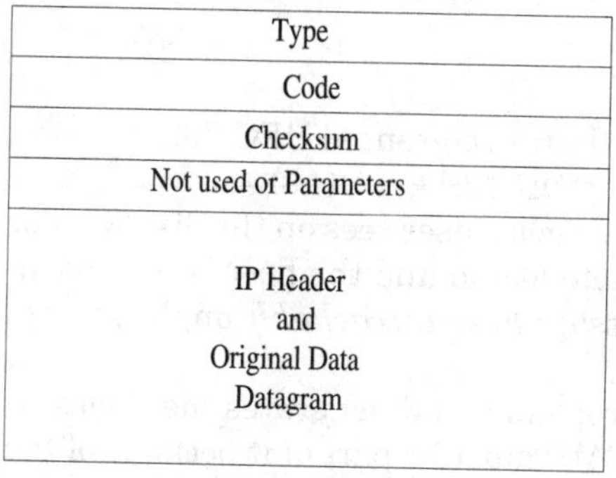

Type
Code
Checksum
Not used or Parameters
IP Header and Original Data Datagram

Figure 19.4   ICMP message format.

Type	Meaning of Message
0	Echo reply
3	Destination unreachable
4	Source quench
5	Redirect
8	Echo request
11	Time exceeded for a Datagram
12	Parameter problem on a Datagram
13	Timestamp request
14	Timestamp reply
17	Address mask request
18	Address mask reply

Figure 19.5 ICMP message types and meanings.

0	Network unreachable
1	Host unreachable
2	Protocol unreachable
3	Port unreachable
4	Fragmentation needed
5	Source route failed
6	Destination Network unknown
7	Destination Host unknown
8	Source Host isolated
9	Administrative restrictions to destination Network. Communication prohibited
10	Communication with destination Host prohibited by Administration
11	Network unreachable for service type
12	Host unreachable for service type

Figure 19.6 ICMP codes and meanings.

The NOT USED field means just that; I referenced RFC 792.

The next field is the IP HEADER AND DATA DATAGRAM.

ICMP is the source for many messages a user sees on the display. For example, if a user attempts a remote log-on and the host is not reachable, then the user will see the message *host unreachable* on the screen. This message comes from ICMP.

ICMP detects errors, reports problems, and generates messages as well. For IP to be implemented, ICMP must be part of it because of the design of IP.

19.7   Address Resolution Protocol (ARP)

Perspective

ARP, as it is commonly known, is located at layer 3 along with IP and ICMP. ARP maps IP addresses to the underlying hardware address. Actually, ARP dynamically binds these addresses.

Since TCP/IP works at layer 3 and above, it must have a mechanism to function with interface boards. When TCP/IP is implemented, it is done so in software. Each host participating on a TCP/IP network must have TCP/IP and have a unique IP address. This IP address is considered a software address since it is implemented at layer 3 in software.

Because any one of many data-link protocols could be used, IP requires a way to correlate the IP address and the data-link address. Data-link addresses are generally considered hardware addresses. For example, if TCP/IP is implemented with ETHERNET, there is a 48-bit ETHERNET address that must be mapped to the 32-bit IP address. Or, if token ring is used, a 12-digit hexadecimal (hex) address is used as the hardware address. Neither of these data-link protocol addresses matches the 32-bit IP address of TCP/IP. This is the reason for ARP.

Theory of operation

Using ETHERNET as a data link protocol, ARP can be explained as follows. Assume five hosts residing on an ETHERNET network. Assume that a user on host A wants to connect to host E. Host A uses ARP to broadcast a packet that includes A's IP and ETHERNET address and host E's IP address.

All five hosts on the network "hear" the ARP broadcast for host E. However, only host E recognizes its IP address inside the ARP request. Figure 19.7 depicts this scenario.

When host E recognizes its hardware address, it replies back to host A with its IP address. Figure 19.8 is an example of this process.

It is obvious that all hosts shown in Fig. 19.7 must examine the ARP request. This is expensive in terms of network utilization. To avoid this, an ARP cache is maintained. This is a list of network hosts' physical and IP addresses. As a result, this curbs the number of ARP packets on the network.

When a host receives an ARP reply, that host keeps the IP address of the other host in this ARP table. Then, when a host wants to communicate with another host on the network, it first examines its ARP cache for the IP address. If the desired IP address is found in the cache, there is no need to perform an ARP broadcast. This communication occurs via hardware communication, for example, with ETHERNET boards communicating with one another.

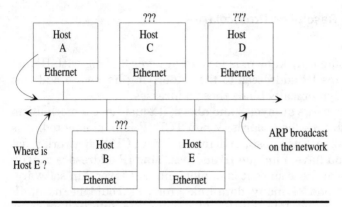

Figure 19.7   ARP request.

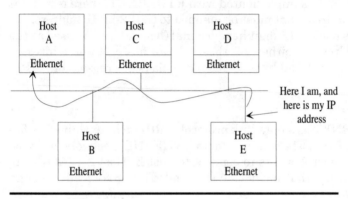

Figure 19.8   ARP response.

ARP message format

Figure 19.9 is an example of ARP message format.

The following lists the fields in the ARP packet and briefly explains their meanings.

HARDWARE TYPE	Indicates the hardware interface type
PROTOCOL TYPE	Specifies the upper-level protocol address that the originator sent
HARDWARE ADDRESS LENGTH	Specifies the length of the bytes in the packet
PROTOCOL ADDRESS LENGTH	Specifies the length in bytes of the high-level protocol
OPERATION CODE	Specifies one of the following: ARP request, ARP response, RARP request, or RARP response
SENDER HARDWARE ADDRESS	If known, it is supplied by the sender
SENDER PROTOCOL ADDRESS	Like the hardware address; it is sent only if known
TARGET HARDWARE ADDRESS	Destination address
TARGET PROTOCOL ADDRESS	Contains the IP address of the destination host

Physical Layer Header
Hardware Type
Protocol Type
Hardware Address Length
Protocol Address Length
Operation Code
Sender Hardware Address
Sender Protocol Address
Target Hardware Address
Target Protocol Address

Figure 19.9 Example of an ARP packet.

Since ARP functions at the lowest layers within a network, the ARP request itself must be encapsulated within the hardware protocol frame because the frame itself is what physically moves through the network at this level. Conceptually, the frame carrying the ARP message and the frame appear as shown in Fig. 19.10.

ARP's dynamic address translation provides a robust method for obtaining an unknown address. The efficiency of ARP is in the utilization of the caching mechanism.

19.8 Reverse Address Resolution Protocol (RARP)

Theory of operation

RARP is the reverse of ARP. It is commonly used where diskless workstations are implemented. When a diskless workstation boots, it knows its hardware address because it is in the interface card that connects it to the network. However, it does not know its IP address.

RARP request and server operation

Devices using RARP require that a RARP server be present on the network to answer RARP requests. A typical RARP request would ask:

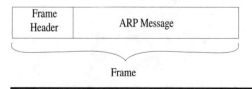

Frame

Figure 19.10 Conceptual view of the frame and the ARP message.

"What is my IP address?" This is broadcast on the network, and a designated RARP server replies by examining the physical address received in the RARP packet, comparing it against its tables of IP addresses, and sending the response back to the requesting host. Figure 19.11 is an example of a RARP broadcast.

Figure 19.11 shows the RARP request going to all hosts on the network. It also shows a RARP server. Notice in Fig. 19.12 that the RARP server answers the RARP request.

For RARP to be used in a network, a RARP server must exist. In most implementations when RARP is used multiple RARP servers are used. One is designated as a primary server and another, as a secondary server.

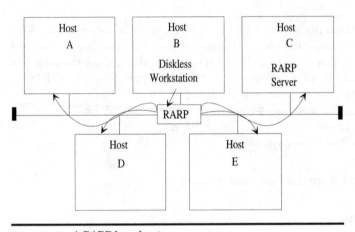

Figure 19.11   A RARP broadcast.

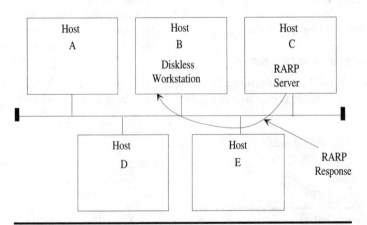

Figure 19.12   A RARP response.

19.9   Router Protocols

Normally this section would be called "gateway protocols," but as mentioned previously, this would be incorrect if these protocols were defined according to functionality. For clarity, this section covers the *gateway* protocols, but here they are called *routers* because that is what they do: *route*.

This section focuses on what are called *interior gateway protocols*. This is defined as routing in an autonomous system. An *autonomous system* is a collection of routers controlled by one administrative authority and uses a common interior gateway protocol. Two popular routing protocols exist in this category: router information protocol (RIP) and open shortest path first (OSPF).

These protocols are used by network devices such as routers, hosts, and other devices normally implemented in TCP/IP software. However, it is feasible to implement these protocols by firmware. This section explains RIP and OSPF.

Router information protocol (RIP)

RIP originated from Xerox's network systems protocol. It was at one time included in the software distribution of TCP/IP with UCB UNIX.

RIP is an example of a de facto protocol. It was implemented before a standard RFC existed. It was part of TCP/IP, it worked, and it was needed; all the ingredients to make a product popular! RFC 1058 brought RIP into a formal standard.

RIP header analysis.   Consider Fig. 19.13, which is an example of the RIP message format.

The following is a brief description of each field in the RIP message format.

COMMAND	Specifies an operation that could be a request or response
VERSION	Identifies the protocol version
ZERO	A blank field
ADDRESS FAMILY IDENTIFIER	Used to identify the protocol family that the message should be interpreted under
ZERO	A blank field
IP ADDRESS	Usually has a default route attached to it
ZERO	A blank field
ZERO	A blank field
DISTANCE TO NET n	A value indicating the distance to the target network.

RIP messages are either conveying or requesting routing information. RIP is based on broadcast technology. Periodically a router (or a

Command
Version
Zero
Address Family Identifier
Zero
IP Address
Zero
Zero
Distance to Net A

Figure 19.13   RIP message format.

designated device) broadcasts the entire RIP routing table throughout the network, either LAN or otherwise. This aspect alone has become a problem in some environments because of the lack of efficiency.

In addition to broadcasts and updates from that process, RIP also gets updates due to changes in the network configuration. These updates are referred to as *responses*.

Another characteristic of RIP is that it relies on other devices (adjacent nodes) for routing information for targets that are more than one "hop" away. RIP also calculates its distances by costs per hop. One hop is defined as a metric. The maximum hops RIP can make along one path is 15.

RIP maintains tables with entries. This table is the one referred to previously that is broadcast throughout the network. Information contained in each entry in this table includes

- The destination IP address
- The number of hops required to reach the destination
- The IP address for the next router in the path
- Any indication of whether the route has recently changed
- Timers along the route

RIP is still used today. Many vendors support it. In certain environments it may be a good *gateway* protocol to use. However, many vendors support OSPF.

Open shortest path first (OSPF)

The philosophy of OSPF differs from RIP. It was recommended as a standard in 1990 and by middle to late 1991 Version 2 OSPF was available. This is an example of its popularity. Some of the tenets OSPF maintains include

- Offers a type of service routing
- Virtual networks can be defined
- Offers route distribution
- Minimized broadcasts
- Supports a method for trusted routers

Other tenets support OSPF and, depending on the vendor, a variety of them may be implemented.

OSPF advertisements.   OSPF uses what is called *advertisements*. These advertisements enable routers to inform other routers about paths. Four distinct types of advertisements are

AUTONOMOUS	Has information of routes in other autonomous systems
NETWORK	Contains a list of routers connected to the network
ROUTER	Contains information about given router interfaces in certain areas
SUMMARY	Maintains route information outside a given area

These advertisements enable a more focused approach to spreading information throughout a network. Besides the advertisements, OSPF uses a number of messages for communication. Some of these messages are

- HELLO
- Database description
- Link state request
- Link state update
- Link state acknowledgment

Two of these are presented and explained in detail to provide insight on the operation of OSPF. Additional information about OSPF is available from RFCs 1245, 1246, and 1247.

OSPF header analysis.   The OSPF packet header appears as shown in Fig. 19.14.
The following OSPF header fields are listed and briefly explained:

VERSION	Indicates the protocol version.
TYPE	This field indicates messages as one of the following: (1) HELLO, (2) database description, (3) link status, (4) link status update, or (5) link status acknowledgment
MESSAGE LENGTH	Indicates the length of the field including the OSPF header
SOURCE GATEWAY IP ADDRESS	Provides the sender's address

Version
Type
Packet Length
Router ID
Area ID
Checksum
Authentication type
Authentication

Figure 19.14   OSPF packet header.

AREA ID	Identifies the area from which the packet was transmitted
CHECKSUM	Performed on the entire packet
AUTHENTICATION TYPE	Identifies the authentication type that will be used
AUTHENTICATION	Includes a value from the authentication type

The HELLO packet includes messages that are periodically sent on each link to establish whether a destination can be reached. The HELLO packet appears as shown in Fig. 19.15.

The following includes a list of fields in the HELLO packet and a brief explanation of each.

OSPF HEADER	Required.
NETWORK MASK	Contains the network mask for the network from where the message originated.
DEAD TIMER	A value (in seconds) indicating that a neighbor is dead and no response is received.
HELLO INTERVAL	This field has a value, in seconds, reflecting the amount of time between a router sending another HELLO packet.
ROUTER PRIORITY	Used if a designated router is used as a backup.
DESIGNATED or BACKUP ROUTER	Identifies the backup router.
NEIGHBOR ROUTER ID	This field, and subsequent ones, indicate the IDs of routers which have recently sent HELLO packets within the network.

The database description packet message includes an OSPF header and fields of information. These fields include information about messages received. They can be broken into smaller units. Information is also provided to indicate whether information is missing. The packet also includes information about the type and ID of the link. A checksum is provided to ensure that corruption has not occurred.

The link state packet header includes an OSPF header and fields that provide information such as router, network, and link station type.

The essence of OSPF is that it reduces traffic overhead in the network because it performs individual updates rather than broadcasts

OSPF Header
Network Mask
Dead Timer
Hello Interval
Router Priority
Designated or
Backup Router
Neighbor 1 IP Address
Neighbor 2 IP Address
Neighbor 3 IPAddress
etc

Figure 19.15 HELLO packet format.

that permeate the entire network. OSPF also provides an ability for authentication. Another strength of OSPF is that it can exchange subnet (work) masks as well as subnet addresses.

19.10   Transmission Control Protocol (TCP)

TCP operates at layer 4 and is a transport protocol. It takes data passed to it from its applications and places it in a send buffer. Then TCP divides the data into what is called a *segment,* which consists of the application data and a TCP packet. This is necessary because data is delivered in datagrams.

Characteristics and functions

TCP treats data in different ways. For example, a user enters a simple command and presses RETURN. TCP uses what it refers to as a *push* function to make this happen. This *push* ability is a function of TCP. On the other hand, if a large amount of data is passed from an application to TCP it segments it and passes it to IP for further processing.

Applications that use TCP pass data to TCP in a stream fashion. This is in contrast to some protocols where applications pass data down the protocol stack in byte fashion.

TCP is a connection-oriented protocol. This means that TCP keeps the state and status of streams passing into and out of it. It is TCP's responsibility to ensure reliable end-to-end service.

Two other major features TCP provides are multiplexing and full-duplex transmission. TCP is capable of performing multiplexing user sessions because of the addressing scheme used in TCP/IP. This addressing scheme is covered later in this chapter in a separate section. TCP also performs a function called *resequencing*. It can perform this function because of sequence numbers used for acknowledgments. This function manages segments if they reach the destination out of order.

TCP header analysis

Many of the aforementioned functions and others can be understood by examining the TCP header in Fig. 19.16.

The following list includes the parts of the TCP segment and a brief description of these parts.

SOURCE PORT	Identifies the upper-layer application using the TCP connection.
DESTINATION PORT	Identifies the upper-layer application using the TCP connection.
SEQUENCE NUMBER	The value in this field identifies the transmitting byte stream. This value is used during connection management operations.
ACKNOWLEDGMENT NUMBER	The value in this field reflects a value acknowledging data previously received.
DATA OFFSET	This value determines where the field of data begins.
RESERVED	This field is reserved, and the bits within it are set to zero.
URGENT	Indicates whether the urgent pointer is used.
ACKNOWLEDGMENT	Indicates whether the acknowledgment field is significant.
PSH	Indicates whether the push function is used.
RESET	Indicates whether the connection should be reset.
SYNCHRONIZE	Used to indicate whether sequence numbers are to be synchronized.
FINISHED	This field is used to indicate that the sender has no more data to send.
WINDOW	This value indicates how much data the receiving host can accept. This value is contingent on the value in the acknowledgment field.
CHECKSUM	This performs a checksum on the 16-bit words in the segment.
URGENT POINTER	This field, if used, indicates that urgent data follows.
OPTIONS	At the current time three basic options are implemented in this field: the end of list, no operation, and maximum segment size.
PADDING	This field is used to ensure that the header length equals 32 bits.
DATA	User data follows in this field.

TCP's reliable data transfer, connection-oriented nature, stream support for applications, multiplexing, full-duplex transmission, push function flow control, and other characteristics make it a popular and *reliable protocol.*

Source Port
Destination Port
Sequence Number
Acknowledgment Number
Data Offset
Reserved
Urgent
Acknowledgment
Push
Reset
Synchronizer
Finished
Window
Checksum
Urgent Pointer
Options
Padding
Data

Figure 19.16   TCP segment.

The TCP protocol is defined in RFC 793. Additional information can be obtained from it if details are required.

19.11   User Datagram Protocol (UDP)

User datagram protocol (UCP) resides at transport layer 4. In many respects it is the opposite of TCP. UDP is connectionless-oriented and unreliable. It does little more than provide a transport-layer protocol for applications that reside above it.

UDP header analysis

The extent of information about UDP is brief compared to that for TCP. An example of the UDP datagram is shown in Fig. 19.17.

The following list includes the components in the UDP datagram and provides a brief description of each.

SOURCE PORT — The value in this field identifies the origin port. (Ports are used in addressing and will be discussed in detail in the section on addressing.)

DESTINATION PORT — This identifies the recipient port for the data.

LENGTH — The value in this field indicates the length of the data sent, including the header.

CHECKSUM — This algorithm computes the pseudo-IP header, the UDP header, and the data.

DATA — The data field is the data passed from applications using UDP.

Source Port
Destination Port
Length
Checksum
Data

Figure 19.17   UDP datagram.

UDP applications

UDP is a useful protocol. In certain situations a custom application may be needed. To accomplish this task, UDP is a good transport protocol to use. Because UDP is unreliable and does not perform retransmissions and other services that TCP offers, the custom applications must perform these functions.

Because of UDP's nuances, this leaves work for application programmers. These necessary operations can be achieved via the application; however, they merely require more work on behalf of the one creating the application.

Messages sent to UDP from applications are forwarded to IP for transmission. Some applications that reside on the UDP protocol pass messages directly to IP and ICMP for transmission.

19.12   TCP/IP Addressing

Addressing in TCP/IP consists of a variety of factors that work together to make TCP/IP a functioning upper-layer network protocol. Some of these factors are

- IP addressing
- Address classifications
- Ports
- Well-known-ports
- Port manipulation
- Sockets
- Hardware addresses

Each of these is presented separately, and a synthesis provided to aid in understanding how they interrelate.

IP addressing

The Internet protocol uses a 32-bit addressing scheme. This addressing is implemented in software; however, in some network devices it is implemented in firmware and/or nonvolatile random-access memory (RAM).

Each host participating in a TCP/IP network is identified via a 32-bit IP address. This is significant because it is different from the host's hardware address (or addresses).

The IP addressing scheme structure appears as shown in Fig. 19.18.

Figure 19.18 shows five classes of IP addresses. The IP addressing scheme is in dotted decimal notation. The address class indicates how many bits are used for a network address and for a host address. Before examining these in detail a word about how these addresses are assigned may be helpful.

As Fig. 19.18 shows, there are multiple classes of addresses. A reasonable question is "Why?" Two implementations of TCP/IP networks are possible: the Internet (big I) and internets (little i).

The Internet is a worldwide network to which thousands of entities are connected. An agency responsible for maintaining Internet addresses assigns IP addresses to entities connecting to the Internet; the entity itself has no authority in the matter. On the other hand, if a TCP/IP network is implemented in a corporation, for example, the IP addressing scheme is left up to the implementers responsible for that corporate network. In other words, it is "locally" administrated. In such cases, the individual(s) involved in the implementation should understand the ramifications of selecting an IP addressing scheme. Multiple

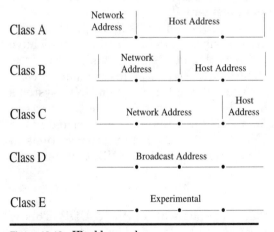

Figure 19.18   IP address scheme.

issues factor (or should factor) into the equation for selecting an IP addressing scheme.

Address classifications

Five classes of addresses were shown in Fig. 19.18. The following list explains the numerical meaning and how this affects hosts implemented with IP addresses.

Address class	Assigned numbers
A	0–127
B	128–191
C	192–223
D	224–239
E	240–255

Class A addresses have fewer bits allocated to the network portion (1 byte) and more bits (3 bytes) dedicated for hosts addressing. In other words, more hosts can be implemented than networks, according to the addressing scheme.

Class B addressing allocates an equal amount of bits for network addressing (2 bytes) and host addressing (2 bytes). This class is popular in locally administered implementations.

Class C addressing allocates more bits (3 bytes) to the network portion and fewer bits (1 byte) to the host portion.

Class D is generally used as a broadcast address. The numerical value in each of the 4 bytes is 255.255.255.255.

Class E networks are used for experimental purposes. This author knows of no class E networks implemented.

Implementing an internet uses these addresses in conjunction with aliases. For example, an address assigned to a host would usually have a name associated with it. If a host had a class B address such as 137.1.0.99, its alias name might be RISC (reduced instruction set computer). This alias and internet address reside in a file on UNIX systems called /etc/hosts. Another file related to this is the /etc/networks file. These two files, in a UNIX environment, are normal in the configuration. Additional information is included in latter sections in this chapter regarding TCP/IP configurations in a UNIX environment.

Ports

Ports constitute the addressable endpoint at TCP and UDP. This is partially how applications on TCP and UDP are addressed.

Well-known-ports. TCP and UDP have popular applications that use them for a transport protocol. Consequently without some standardization of ports and relationships to applications, chaos could exist. As a result, TCP and UDP have applications that are assigned to a well-known-port. Those working in the field of TCP/IP generally know this. It is a standardization to which most adhere. But flexibility does exist.

Port manipulation. Port numbers can be changed, but seldom are. However, there is reason for this capability to exist; this is clarified in Chap. 29.

Nevertheless, TCP and UDP applications can have their ports changed. Some port numbers are available for development of custom applications. During the explanation of UDP the concept of custom applications was presented. This is an example of where the ability to use a "free" port number would be required.

The downside of changing a port number is that if that application using that port is popular, problems could arise in the network from a user perspective.

Sockets

A *socket* is the combination of an IP address and the port number appended to the end. Sockets are used in programming and are rarely of any concern for general users. However, in some instances it is important to understand this socket concept.

Hardware address

TCP/IP operates at layers 3 and above in a network; therefore, it stands to reason an interface of some type is needed for a TCP/IP host to participate in a network. The issue then becomes which lower-layer protocol should be used. If it is ETHERNET, a 48-bit addressing scheme is used; if token ring, a 12-digit hexadecimal address is used, and so forth with other lower-layer protocols—each has its own addressing scheme.

Understanding this addressing scheme is especially important for those who troubleshoot networks. It is also important for those designing networks and implementers who have to make it work.

Synthesis

Understanding the previous information in this section is important for planning a TCP/IP network. The size and purpose of the network and other site-specific parameters should be considered in selecting IP address classes and other issues presented here. Planning, with the technical implications understood in the beginning, can save time and money in the long run.

19.13 Popular TCP Applications

Popular applications that use TCP as a transport-layer protocol are explained in this section. A list of those presented in this section includes

- X window system
- TELNET
- File transfer protocol (FTP)
- Simple mail transfer protocol (SMTP)
- Domain name system (DNS)

X window system

"X," as it is known in the marketplace, is a distributed windowing system. At MIT in the early 1980s developers were looking for a way to develop applications in a distributed computed environment. At that time this was cutting-edge technology. During their work, these MIT researchers realized that a distributed windowing system would meet their needs very well.

After the MIT group met and shared information with individuals at Stanford University who had performed similar work, Stanford University gave MIT a considerable starting point to begin this endeavor. However, the group at Stanford working with this technology had dubbed it "W," for windowing. The individuals at MIT renamed it "X," arguing that it was the next letter in the alphabet. The name stuck.

By the late 1980s X commanded a considerable market share specifically in a UNIX-based environment. One factor for its growth is that it is hardware- and software-independent. Suffice it to say that X is a dominant user interface in the UNIX environment, and it has spread into MS-DOS and VMS environments as well.

X is asynchronous and is based on a client/server model. It can manipulate two-dimensional graphics on a bitmapped display. Before examining some of the operational aspects of X, consider the layers of X and its relationship with the TCP/IP protocol suite shown in Fig. 19.19.

Figure 19.19 shows the TCP/IP protocol suite, but the focus is X. The protocol suite is there to explain the relationship between X and TCP/IP. X is not a transport-layer protocol; however, it uses TCP for a transport protocol.

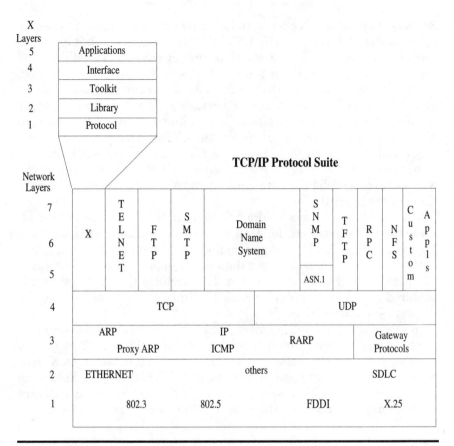

Figure 19.19   Conceptual view of X layers with respect to TCP/IP.

X can be evaluated two ways. From a TCP/IP perspective it makes up layers 5, 6, and 7. However, X itself has five layers. X's layer names and functions include the following:

PROTOCOL	This is the lowest layer in X. It hooks into TCP. This is an actual X protocol component.
LIBRARY	The X library consists of a collection of C language routines based on the X protocol. X library routines perform functions such as responding to pressing of a mouse button.
TOOLKIT	The X toolkit is a higher level of programming tools. Examples of support provided from this layer are the functions they provide in programming-related scrollbar and menu functions.
INTERFACE	The interface is what a user sees. Examples of an interface include Sun Microsystem's OpenLook, Hewlett-Packard's OpenView, Open Software Foundation's (OSF's) Motif, and NeXT's interface, to name only a few.
APPLICATIONS	X applications can be defined as client applications that use X and conform to X programming standards that interact with the X server.

X theory of operation.   X clients and X servers do not function in the way other clients and servers do in the TCP/IP environment. What is considered normal operation of a client is that it *initiates* something and servers *serve or answer* the request of clients. In X this concept is skewed.

An X display manager exists in the X environment. Its basic function is to start and keep the X server operating. The X display manager itself can be started either manually or automatically. With respect to X, the display manager (also referred to as Xdm) is a client application.

An X display server (also known as (Xds) is a go-between hardware component (such as a keyboard or mouse) in X client applications. The Xds operates by catching data entered and directs it to the appropriate X client application.

The correlation between Xdm and Xds can be understood by considering the following scenario. Assume that two windows are active on a physical display. Each window functions as a client application. If this is the case, the idea of directing data to the appropriate X client application takes on a different meaning. This architectural arrangement is required to maintain order because multiple windows may be on the display (say, four or five).

In summary, the X display manager and the X server control the operations on the display, which is what a user actually sees. But, let's go one step further to help clarify how these entities function in an X environment with X client applications. An example of this is the *Xclock,* an *Xterm* which is an emulator, or even a TN3270 emulation software package used to access a 3270 data stream in a SNA environment.

TELNET

TELNET is a TCP application. It provides the ability to perform remote log-ons. TELNET consists of a client and a server. The majority of TCP/IP software implementations have TELNET, simply because it is part of the protocol suite. As stated previously, *clients* initiate something (in this case a remote log-on) and *servers* serve requests of clients. Figure 19.20 shows the TCP/IP protocol suite with TELNET highlighted, showing its client and server.

In Fig. 19.20 TELNET's client and server are shown. This example of TELNET is the same on practically all TCP/IP host implementations if the protocol suite has been developed according to the RFCs. Exceptions do apply. For example, TCP/IP on a DOS-based PC cannot implement a TELNET server because of the architectural constraints of a DOS-based PC; in short, a DOS-based PC cannot truly multitask, and other nuances apply also. Furthermore, this implementation cannot work on some network devices. However, on most host implementations such as

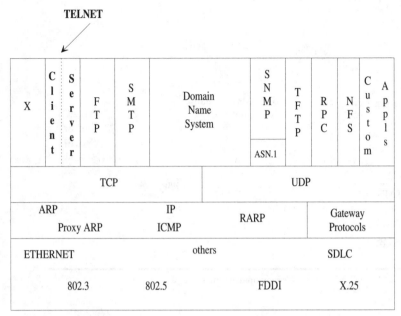

Figure 19.20   Highlighted view of TELNET.

UNIX, VMS, MVS, VM, VSE, and some other operating systems, the TELNET client and server will function.

Figure 19.21 is an example of TELNET client and server interaction on different hosts.

Figure 19.21 shows a RISC/6000 user invoking a TELNET client, native to that machine because it is in the TCP/IP protocol suite. The RISC/6000 user wants to log on to the Sun host. The Sun host has TEL-NET in its TCP/IP protocol suite; consequently, the TELNET server answers the client's request and a logical connection is established between the RISC/6000 and the Sun host. To the RISC/6000 user, they "appear" to be physically connected to the Sun.

This functionality of TELNET works with most major vendors in the marketplace today. The key to understanding the client/server concept is to remember that clients *initiate* and servers *serve* clients' request.

File transfer protocol (FTP)

FTP is a file transfer application that uses TCP for a transport protocol. As does TELNET, FTP has a client and a server; operationally they work the same. What differs between TELNET and FTP is that TEL-NET enables remote log-on whereas FTP permits file transfers.

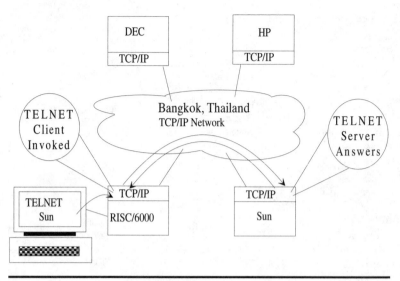

Figure 19.21   TELNET client/server operation.

FTP does not actually transfer a file from one host to another—it copies it. Hence, the original file exists and a copy of it has been put on a different machine. Figure 19.22 depicts this scenario.

Figure 19.22 shows a user on a Sun host performing two steps: (1) the Sun user executes FTP HP and a log-on is established; then (2) the Sun user issues the FTP command GET and designates the filename as FILEABC. The dotted line shows that the file is copied from the HP disk to Sun's disk.

Figure 19.22 also shows a Dell PC, Digital Equipment (DEC), and MVS host. The same operation can be performed on either of these as well. The reason why an FTP can be performed on a PC and not a TEL-NET is because FTP uses two ports to function and is merely request-ing a file transfer—this does not require multitasking on behalf of the host. The DEC host can perform any of the TCP/IP functions as do the Sun or HP.

An interesting twist to the scenario shown in Fig. 19.22 is that a DEC user could TELNET to the HP, and then from the HP would execute a FTP against the Dell PC and move a file *to* or *from* the Dell PC. Such a networking scenario is powerful with regard to what can be accom-plished.

Another twist is with the MVS host shown in Fig. 19.22. In this hypo-thetical scenario a HP user could execute a FTP against the MVS host and *put* the FILEABC into the MVS JES2 subsystem and have it print on the printer attached to the MVS printer.

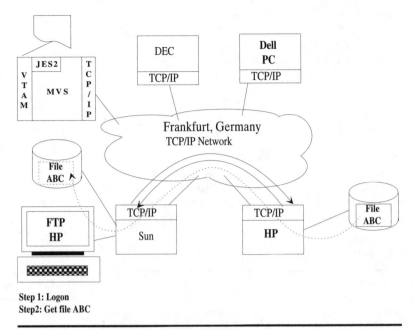

Figure 19.22 FTP client/server operation.

Examples such as this are too numerous to mention here; these are only a *few* examples to convey some simple operations.

Simple mail transfer protocol (SMTP)

SMTP is another TCP application. It does not use a client/server, but the functionality is similar. SMTP utilizes what is called a *user agent* (UA) and a *message transfer agent* (MTA). Figure 19.23 is a simple example of how SMTP operates.

Sending mail is accomplished by invoking a user agent which, in turn, causes an editor to appear on the user's display. After the mail message is created and sent from the user agent, it is transferred to the MTA. The sending MTA is responsible for establishing communication with the MTA on the destination host. Once this is accomplished, the sending MTA sends the message to the receiving MTA, then stores it in the appropriate queue for the user. The recipient of the mail needs only to invoke the user agent on that machine to read the mail.

Domain name system (DNS)

In the early years of the Internet, host files were used to keep track of hosts on the Internet. This meant that when new hosts were added to

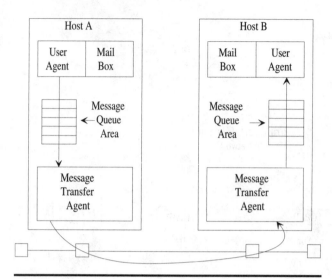

Figure 19.23   SMTP components.

the Internet all participating hosts had to have their hosts file updated. As the Internet grew, this task of updating the hosts file became insurmountable. The DNS emerged from the need to replace such a system.

The philosophy of DNS was to replace the need for FTP to update hosts files throughout the entire network. Thus, the foundation of DNS was built around a distributed database architecture.

DNS structure. DNS is a hierarchical structure that conceptually appears as an upside-down tree. The root is at the top and the layers are below. Figure 19.24 is an example of how DNS is implemented in the Internet.

The legend for the DNS structure in Fig. 19.24 is as follows:

ROOT   The root server contains information about itself and the top-level domains immediately beneath it.

GOV   Refers to government entities.

EDU   Refers to any educational institutions.

ARPA   Refers to any ARPANET (Internet) host ID.

COM   Refers to any commercial organizations.

MIL   Refers to military organizations.

ORG   Serves as a miscellaneous category for those not formally covered.

CON   Refers to countries conforming to ISO standards.

Figure 19.24 shows the Internet implementation of DNS. Three examples are shown to aid in understanding the structure. Notice that IBM is under COM (which is commercial), beneath IBM is Raleigh and Austin, and beneath each of them are research and marketing. The

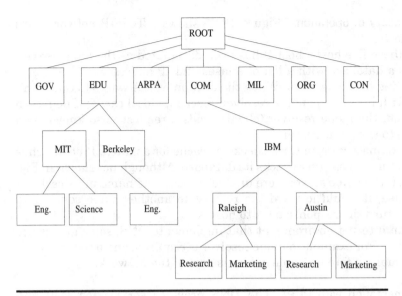

Figure 19.24 DNS structure.

other examples are MIT and Berkeley (UCB). The example with MIT shows two *zones* beneath it: engineering and science. The Berkeley example has one layer beneath it: engineering.

At a local level, such as in a corporation, most sites follow the naming scheme and structure because it is consistent and if a connection to the Internet is made, restructuring of DNS is not necessary.

DNS components. To better understand DNS, knowing the components that make it functional is helpful. These components include

domain The last part in a domain name is considered the domain. For example, in `eng.mit.edu;` `edu` is the domain.

domain name Defined by the DNS as being the sequence of names and domain. For example, a domain name could be `eng.mit.edu.`

label The DNS identifies each part of a domain name as a label. For example, `eng.mit.edu` has three labels: `eng.`, `mit.`, and `edu.`

name server A program operating on a host that translates names to addresses; it does this by mapping domain names to IP addresses. *Also,* name server may be used to refer to a dedicated processor running name server software.

name resolver This is software that functions as a client regarding its interaction with a name server. Sometimes referred to simply as the *client.*

name cache This is storage used by the name resolver to store information frequently used.

zone This is a contiguous part of a domain.

DNS theory of operation.   Figure 19.25 shows a TCP/IP network with five hosts.

Of these five hosts, host B has been designated as the name server. It has a database with a list of aliases and IP addresses of hosts participating in the network. When the user on host A wants to communicate with host C, the name resolver checks its local cache; if no match is found, the name resolver (client) sends a request (also known as a *query*) to the name server.

The name server, in turn, checks its cache for a match. If no match is found, the name server checks its database. Although not shown in Fig. 19.25, if the name server were unable to locate the name in its cache or database, it would forward the request to another name server and then return the response back to host A.

In an internet environment that implements DNS, some givens are assumed. For example, a name resolver is required, and a name server, and usually a foreign name server is part of the network.

Implementation with UDP.   The DNS provides service for TCP and UDP; this is why diagrams have shown DNS residing above part of TCP and part of UDP. It serves the same purpose for both transport-layer protocols.

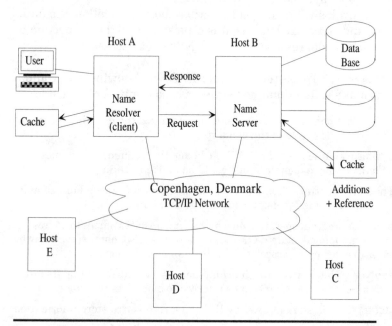

Figure 19.25   Conceptual view of DNS.

Obtaining additional information. Additional information should be consulted on this issue of whether DNS is implemented. The following RFCs are a good starting point.

RFC 882	RFC 1123
RFC 883	RFC 1032
RFC 920	RFC 1033
RFC 973	RFC 1034
RFC 974	RFC 1035
RFC 1034	

19.14 Popular UDP Applications

This section presents popular UDP applications. A list of those covered includes

- Simple network management protocol (SNMP)
- Trivial file transfer protocol (TFTP)
- Network file system (NFS)
- Remote procedure call (RPC)
- Custom applications
- PING and FINGER

Simple network management protocol (SNMP)

SNMP is considered the de facto standard for managing TCP/IP networks as of this writing. SNMP uses agents and application managers (or simply managers). A user agent can reside on any node that supports SNMP, and each agent maintains status information about the node on which it operates. These nodes, which may be a host, gateway, router, or other type of network device, are called *network elements* in SNMP lingo. This term *element* is merely a generic reference to a node.

Normally, multiple *elements* exist in a TCP/IP network, and each has its own agent. Typically, one node is designated as a network management node. Some refer to this node as the *network manager*. This host (network management node) has an application that communicates with each network element to obtain the status of a given element. The network management node and the element communicate via different message types. Some of these messages are as follows:

GET REQUEST	This type of request is used by the network manager to communicate with an element to request a variable or list about that particular network element.

GET RESPONSE	This is a reply to a GET REQUEST, SET REQUEST, and GET NEXT REQUEST.
GET NEXT REQUEST	This request is used to sequentially read information about an element.
SET REQUEST	This request enables variable values to be set in an element.
TRAP	This type of message is designed to report information such as link status, whether a neighbor responds or a message is received, and the status of the element.

Information stored on elements is maintained in a *management information base* (MIB), which is a database containing information about a particular element. Examples of MIB element-specific information include

- Statistical information regarding segments transferred to and from the manager application

- A community name

- Interface type

MIB information structure is defined by the Structure of Management Information (SMI) language. SMI is a language used to define a data structure and methods for identifying an element for the manager application. This information identifies object variables in the MIB. A minimum of object descriptions defined by SMI includes

ACCESS	Object access control is maintained via this description.
DEFINITION	This provides a textual description of an object.
NAMES	This term is also synonymous with object identifiers. This refers to a sequence of integers.
OBJECT DESCRIPTOR	This is a text name ascribed to the object.
OBJECT IDENTIFIER	This is a numeric ID used to identify the object.
STATUS	This describes the level of object support for status.

SNMP implementations use ASN.1 for defining data structures in network elements. Because this language is based on a data type definition, it can be used to define practically any element on a network.

SNMP itself is event-oriented. An event is generated when a change occurs to an element. SNMP operation is such that approximately every 10 to 15 min the manager application communicates with each network element regarding its individual MIB data.

Additional information can be obtained from the following RFCs:

RFC 1155

RFC 1156

RFC 1157

Trivial file transfer protocol (TFTP)

TFTP is an application that uses UDP as a transport mechanism. The program itself is simpler than its counterpart, FTP, which uses TCP as a transport mechanism. TFTP is small enough in size so to be part of read-only memory (ROM) on diskless workstations.

TFTP maximum packet size is 512 bytes. Because of this and the nature of its operation, TFTP is popular with network devices such as routers and bridges. If implemented, TFTP is normally used on initial device boot.

TFTP utilizes no security provision or authentication; however, it does have some basic timing and retransmission capabilities. TFTP uses five basic types of protocol data units (PDUs):

- Acknowledgment

- Data

- Error

- Read request

- Write request

These PDUs are used by TFTP during file transfer. For the first packet TFTP establishes a session with the target TFTP program. It then requests a file transfer between the two. Next it identifies a filename and determines whether a file will be read or written.

These five PDUs represent the operational ability of TFTP. It is straightforward and not as complex as that of FTP.

Additional information can be obtained from the following RFCs:

- RFC 783

- RFC 1068

Remote procedure call (RPC)

RPC is a protocol. Technically speaking, it can operate over TCP or UDP as a transport mechanism. Applications use RPC to call a routine, thus executing like a client, and make calls against servers on remote hosts. This type of application programming is a high-level, peer relationship between an application and a RPC server. Consequently, these applications are portable to the extent that RPC is implemented.

Within RPC is the eXternal Data Representation (XDR) protocol. XDR data description language can be used to define data types when heterogeneous hosts are integrated. Having the capability to overcome the inherent characteristics of different architectures lends to RPC and

XDR a robust solution for distributed application communication. This language permits parameter requests to be made against a file of an unlike type. In short, XDR permits data-type definition in the form of parameters and transmission of these encoded parameters.

XDR provides data transparency by way of encoding (or encapsulating) data at the application layer so that lower layers and lower-layer hardware do not have to perform any conversions. A powerful aspect of XDR is automatic data conversion performed via declaration statements and the XDR compiler. The XDR compiler generates the required XDR calls, thus making the operation less manual in nature. Figure 19.26 is an example of this type of implementation.

RPC implements what is called a *port mapper*. It starts on RPC server initialization. When RPC services start, the operating system assigns a port number to each service. These services inform the port mapper of this port number and its program number, and other information required by the port mapper in order to match a service with a request.

Client applications issue a service request to a port mapper. The port mapper, in turn, identifies the requested service and returns the appropriate parameters to the requesting client application. In other words, the port mapper is similar in function to a manager knowing what services are available and their specific addressable locations.

The port mapper can be used in a broadcast scenario. For example, a requesting RPC call broadcasts a call to all hosts on a network. Applicable port mappers report back to the information sought after by the client; hence the term *remote procedure call* (RPC).

Additional information on RPC and related components can be found in the following RFCs:

RFC 1057

RFC 1014

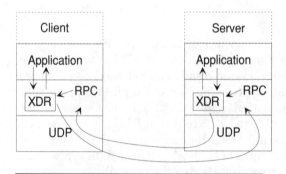

Figure 19.26 Conceptual view of RPC and XDR.

Network file system (NFS)

NFS is a product of Sun Microsystems. It permits users to execute files without knowing the location of these files. They may be local or remote with respect to the user. Users can create, read, or remove a directory. Files themselves can be written to or deleted. NFS provides a distributed file system that permits users to capitalize on access capabilities beyond their local file systems.

NFS uses RPC to make execution of a routine on a remote server possible. Conceptually, NFS, RPC, and UDP (which NFS typically uses) appears as shown in Fig. 19.27.

The idea behind NFS is to have one copy of it on a server that all users on a network can access. As a consequence, software (and updates) can be installed on one server and not on multiple hosts in a networked environment. NFS is based on a client/server model. However, with NFS a single NFS server can function to serve the request of many client requests.

NFS originated in UNIX, where it is implemented in a hierarchical (tree) structure. But NFS can also operate with IBM's VM and MVS operating systems and DEC's VMS operating system.

NFS uses a *mount* protocol, which identifies a file system and remote host to a local user's file system. The NFS mount is known by the port mapper of RPC and thus is capable of being known by requesting client applications.

NFS also uses the NFS protocol; it performs file transfers among systems. NFS uses port number 2049 in many cases; however, this is not a well-known port number (at least at the time of this writing). Consequently, the best approach is to use the NFS port number with the port mapper.

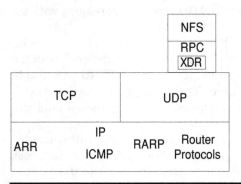

Figure 19.27 Conceptual view of NFS, RPC, and UDP.

In a sense a NFS server operates with little information identified to it. A loose analogy of NFS operation is UDP. UDP assumes that a custom application (or other entity operating above it) will perform requirements such as retransmissions (if required) and other procedures that would otherwise be performed by a connection-oriented transport protocol such as TCP. NFS assumes that required services are implemented in other protocols.

From a user perspective, NFS is transparent. Typical user commands are entered and are then passed to the NFS server, and in most cases a user does not know the physical location of a file in a networked environment.

Additional information about NFS and related components can be obtained through the following RFCs:

RFC 1094

RFC 1014

RFC 1057

Custom applications

Custom applications can be written using UDP as a transport mechanism. One scenario could be where two hosts need peer program communication through a network. Writing a custom application using UDP can achieve this task, as shown in Fig. 19.28.

PING and FINGER

Packet Internet Groper (PING) is actually a protocol that uses UDP as a transport mechanism to achieve its function. It is used to send a message to a host and then waits for that host to respond to the message (if the target host is *alive*). PING uses ICMP echo messages along with the echo reply messages.

PING is a helpful tool on TCP/IP networks that is used to determine whether a device can be addressed. It is also used in a network to determine whether a network itself can be addressed. A PING can also be issued against a remote host *name*. The purpose of this function is name verification; this function is generally used by individuals who troubleshoot TCP/IP networks.

FINGER is a command issued against a host which will cause the target host to return information about users logged onto that host. Some information retrievable via FINGER includes the user name, user interface, and job name that the user is running.

Additional information on FINGER can be obtained in RFC 1288.

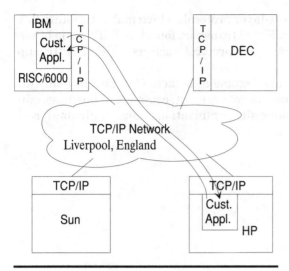

Figure 19.28   Custom applications using UDP.

19.15   Summary

TCP/IP is an upper-layer protocol that has a proven track record. It began around 1975, public demonstration of its capabilities were presented in 1978, and in 1983 the DoD endorsed it as the protocol to use for connection to the Internet.

Many vendors supply TCP/IP products today. TCP/IP can operate on different hardware and software platforms. This flexibility, along with its cost-effective pricing, does put it in a favorable position for those looking for a protocol that provides a variety of services such as

- Remote log-on
- File transfer
- Electronic mail (E-mail)
- A windowing system
- Programmatic interface support
- Network management capabilities
- Distributed processing support

TCP/IP is dominant throughout the global marketplace today; most major vendors around the world support it to varying degrees. Its flexibility with data-link-layer protocols makes it attractive.

TCP/IP has two transport-layer protocols which makes usability flexible; some need the reliability of transport found in TCP, while others need a connectionless transport protocol such as UDP. TCP/IP supports both.

TCP/IP has in many ways become a de facto standard in many different institutions. It is used in government, commercial business, educational institutions, nonprofit organizations, and individual residences as well.

20

NetWare

NetWare is an upper-layer network protocol. In the past decade it has grown into a mature, reputable network protocol capable of working with some of the largest systems available. Examples of how NetWare can operate with such large systems are presented in this chapter, but first a look at its past provides some intriguing and perhaps previously unknown angles on the network protocol.

20.1 Perspective

Novell was organized in 1983 according the 1990 *NetWare Buyer's Guide*. In 11 years the company went from just started (and in many ways a pioneer in the industry) to 1994, when Novell bought the WordPerfect Corporation. Not bad, in anybody's book. During that time frame Novell made strategic moves. Quite simply, Novell kept changing and enhancing its product along with the rest of the industry; however, in many ways Novell charted its own course. For example, Novell was one of the first to provide media independence for its file server. This kind of thinking brought the company from its origins in the PC industry to compatibility with some of the largest systems available in the industry today.

A historical look

Novell has made many milestones in the past 11 years. These milestones had a cumulative effect in bringing the corporation to the prominence it recognizes today. Based on a number of sources, including personal memory, the following recounts some of these early advancements.

1983   The *industry first* file server software is shipped

1983   Media-independent support is announced

1985   Novell supports DOS 3.1 and NetBIOS

1986   Protected mode support for LAN operation is provided

1986   Novell announces a fault-tolerant LAN

1986   Novell announces support for token ring

1988   NetWare for VMS is announced

1988   Novell support for Macintosh is available

1988   Novell supports OS/2

1989   NetWare operating system is announced

1989   Portable NetWare is available

1989   Novell announces 32-bit support

1989   Intel 80386 chip support announced

This list of achievements could continue, but the impact Novell made in the 1980s is clear. Currently, NetWare operates with a program that runs under MVS and VM in the IBM environment: Local Area Network Resource Extension Services (LANRES). More information on LAN-RES is provided later in this chapter. Beyond this, NetWare operates with IBM's SAA, RISC/6000, and AS/400 environments. NetWare is available on many UNIX platforms, also. NetWare has a dominant position in networking protocols.

Forces driving NetWare

Multiple forces have been behind NetWare's growth in the past decade. Clearly, the company made strategic business decisions when necessary and advanced its technology parallel to other technology, but other forces seem to have factored into Novell's success story.

PC growth.   During the 1980s personal computer (PC) growth was unprecedented in the history of computers. Discounting those PCs in the 1970s, which is not a fair comparison to those of the 1980s, anyway, explaining the growth of those PCs in the 1980s is simple; they went from infancy to maturity in about a decade.

So many different vendors entered into the PC market that the list would consume the rest of the space allotted for this chapter. Compaq, Dell, AST, Zero, Packard Bell, HP, Epson, and literally dozens of other vendors entered the market, and their PCs were dubbed *clones*. The term *clone* denotes those systems that copied the IBM PC. Not only did many copy the PC, but some made enhancements to the PC itself. This explosive growth of a new industry split open a new market that pulled in banks, insurance companies, educational institutions, individuals, and companies of all sizes. One example sums up the PC explosion. The IBM PC had 64 kbytes of memory in its introduction; now some IBM PCs can have 64 Mbytes. Kilo means thousand; mega means million! Think about it.

ETHERNET. Another driving force that contributed to the growth of NetWare and its acceptance in the marketplace was growth of the local area network. ETHERNET became a dominant lower-layer protocol in the 1980s. NetWare and ETHERNET were a good match for each other. Together the sum seemed to outweigh the parts.

Since NetWare could operate with ETHERNET, it was practically a shoe-in success. ETHERNET products began appearing in companies everywhere. This was an added plus with the growth of NetWare. Since ETHERNET was growing in notoriety and becoming dominant, it was simple for a customer to see the rationale for using a network protocol such as NetWare.

Token ring. Another factor in the growth and acceptance of NetWare was the impact of token ring. When IBM marshaled its forces behind token ring, the user community knew that it would be a significant protocol in its own right. Novell began supporting token ring as soon as possible. This was a plus for NetWare.

Business community. The 1980s could be considered a decade of downsizing and restructuring from a corporate viewpoint. Because of this, alternatives were required. Many corporations sought to move much data off their traditional mainframes and support departmental LANs. This was additional fuel for the growth of NetWare. NetWare was designed with a solid infrastructure, and that design paved the way for enhancements to be made without recreating the wheel. Consequently, as a product it was solid enough to deliver demands of the largest corporations and scalable enough to be implemented in small companies. These factors seemed to help propel NetWare in a positive manner.

As a spin-off of the downsizing and restructuring a need for distributed file storage emerged. NetWare was designed to accommodate just such a need. The timing of NetWare and the marketplace merged to strengthen a product that continues to grow and adapt itself, now some 12 years later.

20.2 NetWare Protocols: An Orientation

NetWare incorporates multiple protocols as a product. This section explores those protocols to orient the reader for in-depth details later in the chapter.

Read this first

Have you ever seen this title before and deliberately avoided reading it? Probably so; I have done that! It is significant here because of some comments. First, to consider NetWare as simply a PC-oriented network protocol is to totally miss the point. Even to consider the protocol as a success story from an implementation point of view is a mistake.

NetWare is a complex network protocol. It can be implemented with DOS-based systems, Macintosh, AS/400s, UNIX-based systems, DEC-, MVS-, and VM-based systems. Consequently, it is anything but simple. In fact, its installation and configuration vary with each host it operates within. This section, and the remainder of the chapter, explains some of the core components and operations of NetWare.

To this date at least a dozen books have been written on the topic. The complexity of NetWare cannot be understated. However, this complexity brings forth diversity and performance. Not all those systems with which NetWare operates are covered in this chapter; rather, the focus is on the protocol itself, with a detailed explanation of NetWare.

NetWare by layers

NetWare can be evaluated by its layers compared to the OSI model. Figure 20.1 best exemplifies this.

Adapter card. At the lowest layer is support for adapter cards. NetWare supports multiple adapter cards, including

- ETHERNET
- Token ring
- ARCnet
- 802.3
- FDDI

Open data-link interface (ODI). This layer is a specification for the data-link layer providing hardware and thus media protocol independence. Some drawings of the NetWare protocol stack do not show this layer, but it nevertheless exists. The standard was the fruition of collaboration by multiple corporations, including Apple and Novell. Technically, the specification is more than merely a data-link specification; it defines four independent yet cohesive subcomponents. Before examining the details of ODI, consider Fig. 20.2, showing the ODI sublayers.

Figure 20.2 shows the sublayers where ODI operates. ODI is shown compared to the OSI layers, particularly the data-link layer as it is broken into the MAC and LLC sublayers.

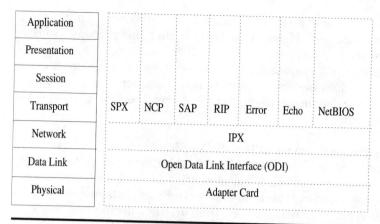

Figure 20.1 OSI and NetWare models compared.

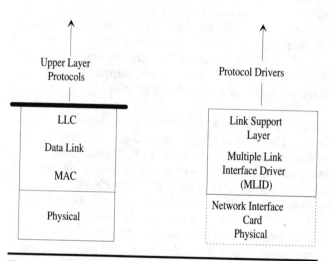

Figure 20.2 Close-up view of ODI.

Internetwork packet exchange (IPX)

IPX operates at layer 3 compared to the OSI model. It is a datagram protocol based on a best-effort delivery system. As Fig. 20.1 depicts, the correlation is directly with the network layer. Its packet delivery has no relationship to other packets. This means that there is no logical order of delivery.

IPX is connectionless-oriented. This means that no acknowledgments are sent from the receiving hosts to the originating host to indicate receipt of packets. Because of this connectionless-oriented nature, acknowledgment and related topics are left to higher-level protocols or programs to perform. Hence, it is obvious that this protocol is faster than those above it as shown in Fig. 20.1.

Even though IPX operates at the network layer 3 with respect to the OSI model, it does perform transport-layer functions. In this sense a one-to-one correlation between NetWare and the OSI model is not strictly accurate.

Sequenced packet exchange (SPX). SPX operates above IPX. It is a connection-oriented protocol. Specialized applications can be built using this as the base protocol. Characteristics of SPX include ensured packet delivery and capability to recover from lost data and errors that may occur in the data being passed from origin to destination. Another characteristic that makes the protocol robust is that it does not acknowledge each and every packet but waits until the maximum number of outstanding packets is reached. In NetWare lingo this is referred to as a *window*. Operationally, SPX performs some functions similar to the transport layer compared to the OSI model, and after it is finished it makes a program call to IPX for packet delivery. In short, SPX and IPX operate together to some degree. A schematic comparison between the OSI and NetWare models is presented in Fig. 20.1.

NetWare core protocol (NCP). NetWare's NCP is a defined protocol which is the procedure that file servers' operating systems utilize to respond to request and accept requests made to it. Functionally, NCP controls client and server operations by defining interactions between them. The NCP provides a service similar to that of SPX in that it performs some packet error checking. It also has built-in session control between entities. The NCP uses a number placed inside the request field of an NCP packet to request a given service. The reason for this is that NetWare services are assigned a number by a NetWare file server. Details of this protocol are Novell proprietary information. Hence, few details can be provided. It does seem, however, that there is general agreement that the NCP is the shell used on workstations.

Service advertising protocol (SAP). Positionally, this protocol resides on top of IPX. It uses IPX to perform its function. SAP does as its name implies; it functions with nodes that provide services to *advertise* available services. Examples of available services include print and file servers. Gateway servers could be included in this as well. Those nodes that provide services broadcast SAP information periodically.

Router information protocol (RIP). RIP is a routing information protocol used on NetWare networks. Functions performed by RIP include location of the best (fastest) route from a workstation to a network. RIP is used by routers to exchange information about routes, respond to requests from routers and workstations, and perform periodic routing table broadcasts among routers.

Error protocol. This protocol is operationally used among peer protocols. Programs that attempt to communicate with a host on a different network use NCP and IPX to attempt to reach that network. If, for some reason, that network is unreachable, an error packet is generated by a router and is sent back to the requesting host regarding the state of the route to the target host. Interestingly, this function is portrayed in Fig. 20.1 above the IPX layer, but functionally it seems to operate at the IPX layer.

Echo. Understanding the echo protocol is similar to understanding PING in TCP/IP. This protocol is used to check the ability to check a path en route to a destination. If the path is functional and the target node is accessible, the echo protocol in the target node is configured such that it literally echoes the packet back to the destination. NetBIOS is also supported providing flexibility to interoperate in that environment.

Review

NetWare is complex. It consists of multiple parts that perform different functions. Some are used in special situations, whereas others are used in most installations. Many variations on the NetWare protocol stack exist. This is a solid representation of those constituent components of NetWare. This section did not address NetWare functionality in different environments because they are discussed later.

20.3 Open Data Interface (ODI) Concepts

ODI is a concept for protocol independence. Its roots are in the philosophy of providing a consistent interface to multiple transport-layer protocols. Hence, network hardware independence can be achieved. When this is achieved, greater implementation and flexibility is realized.

ODI actually consists of three parts, or subcomponents:

- Multiple-protocol interface
- Link support layer
- Multiple-link interface driver

The protocol part of the specification calls for support for a diverse blend of protocols. In fact, if any protocol is coded against this OSI specification, then independence is achievable. These protocol drivers, however, must operate at the network layer and above.

The next part of the specification is the link support layer. The primary purpose of this layer is routing. The routing referred to here is between protocol drivers and multiple-link interface drivers. Figure 20.3 provides a closer view of the link support layer and its interaction with the layers above and below it.

Figure 20.3 shows two interfaces. One provides a connectivity point with the network layer; it is called the *multiple-protocol interface*. As its name implies, the interface is designed to operate with multiple protocols at the network layer and above. This interface was designed for developers creating program code so that they would have a standard interface to program into regardless of the protocol.

The multiple-link interface has the same philosophy behind it as the multiple-protocol interface. This interface was designed as a common ground for data-link-layer protocol developers to have a common standard to code against.

The link support layer includes performing a number of functions, including coordinating numbers assigned to multiple-link interface drivers after these interface drivers have been identified with the link support layer, managing the protocol stack identification assigned to network protocol drivers, managing individual network protocol drivers via their identification number even though frames can be grouped according to MAC frame type, and being capable of manipulating media identification using specific frame formatting.

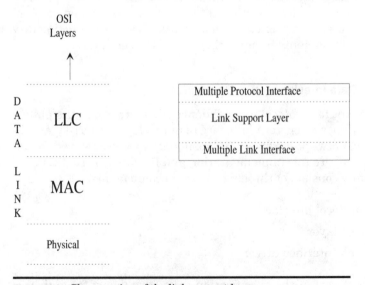

Figure 20.3 Close-up view of the link support layer.

The fundamental purpose of medium and protocol identification is to ensure routing of packets from a given upper-layer protocol stack to the correct lower-layer protocol interface. Basic to this idea is the notion that this is possible without rebooting the system.

The purpose of the *multiple-link interface driver* (MLID) is to pass data to and from the network media. The specification calls for these drivers to be protocol-independent.

Implementing ODI

ODI is implemented differently according to the particular operating system, device driver, network protocol, and NetWare version. The functionality behind it is a threefold concept. Figure 20.4 depicts the first concept behind the implementation.

Figure 20.4 shows multiple upper-layer protocol drivers against one network interface card. This is possible because of the ODI concept. This is an example of one of the possible functions that can be performed.

Figure 20.5 is a different example of ODI functionality.

Figure 20.5 is the converse of Fig. 20.4. In Fig. 20.5 one network protocol driver is used against the link support layer and three multiple-link interface drivers are used. This is an ODI-converse implementation.

Figure 20.6 is another example of how the ODI concept is implemented.

In Fig. 20.6 multiple network protocol drivers are hooked into the link support layer to drive three multiple-link interface drivers. Three is not a magic number with this concept; it could be two or four just as easily.

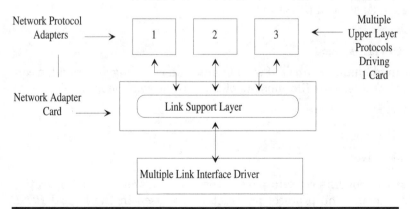

Figure 20.4 Multiple upper-layer protocols with one network protocol adapter card.

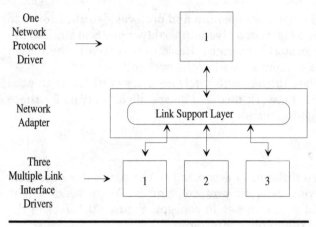

Figure 20.5   ODI-converse implementation.

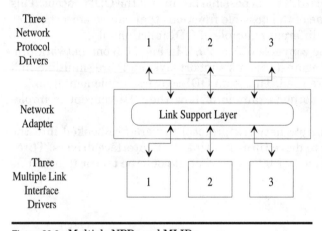

Figure 20.6   Multiple NPDs and MLIDs.

How ODI is managed

The functionality of an ODI implementation is managed through a configuration file. This file consists of three basic components:

- Protocol
- Link support
- Link driver

These components contain parameters that control the ODI operating environment. The protocol parameters are used in the logical *BIND* which is created between the upper-layer network protocol and the MLID. Link support reflects storage used at this part of the layer. The link driver parameters reflect the characteristics of the interface board used. Other configuration parameters may be required by different environments, and because of this variability, the environment manuals should be referenced.

20.4   Internetwork Packet Exchange (IPX)

IPX is used to define addressing schemes used in internetwork and intranetwork environments. NetWare network segments use numbers to identify them primarily for routing purposes. This section explores IPX packet structure and provides additional information concerning IPX function in a NetWare network.

IPX packet structure

IPX packets are carried in the data portion of a MAC frame. Details about MAC frames have been provided previously in this book; the focus here is on the IPX protocol; specifically, its structure and contents. Figure 20.7 illustrates the IPX packet structure and components, and the relationship of IPX to a MAC frame.

Figure 20.7 shows the relationship between a MAC frame and an IPX packet. Inside the IPX packet are field names.

Sockets and ports

Novell refers to destination and source *sockets*. However, practically speaking, these *sockets* are understood better by their function, and that function is a port. Henceforth, when mentioning a Novell reference to a socket, we will use the term *port,* unless the meaning is significantly changed, in which case clarification will be provided. The term *socket* actually refers to a network, host, and port number just as in TCP/IP protocols.

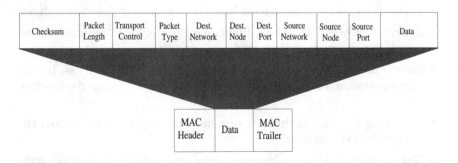

Figure 20.7   IPX packet structure.

IPX field explanation

Each field in the IPX packet has a significant meaning. The following is a brief explanation of those meanings.

- *Checksum:* This field is responsible for performing packet-level checking.

- *Packet length:* This field contains the internetwork packet length. This includes the header and the data section.

- *Transport control:* This field is used by routers between internetworks. It is used primarily by NetWare-based routers.

- *Packet type:* This field indicates the service provided by the packet regardless of whether this service is required or merely offered.

The packet type is indicated by a value. The value indicates the specific service provided. Some of the noted values and corresponding services are

0	Unknown type of packet
1	Routing information packet (RIP)
2	Echo packet
3	Error packet
4	Packet exchange packet
5	Sequenced packet protocol packet
16–31	Designated experimental protocols
17	Network core protocol (NCP)

Other field are

- *Destination network:* This field identifies the target network. Each network in a NetWare networking environment requires unique network numbers.

- *Destination node:* This address identifies nodes on a given network.

- *Destination port:* This address indicates a *process or function* address.

- *Source network:* This field identifies the network (by number) on which the source host is located.

- *Source node:* This address indicates a given node address. In any given instance a host may function as either a source or destination host.

- *Source port:* This is the port number that originally submitted the packet onto the network.

- *Data:* This field includes user data and other information from higher layers.

IPX is the heart of NetWare. All protocols operating above it move down the protocol stack and become enveloped in this packet. Actually, it is similar to TCP/IP in this regard; for example, regardless of whether TCP or UDP protocols are used, either are enclosed into an IP packet.

IPX addressing

IPX uses an addressing scheme similar to that of TCP/IP. A network address is assigned to a NetWare network, node addresses are assigned to each node on a given network, and the network protocol used by nodes on the network have identifiable ports or access points. In TCP/IP the combination of a network, host, and port address create what is called a *socket*. In NetWare, the lingo uses sockets to refer to a parallel concept of a port.

In NetWare the network address is composed of a 4-byte value. Host addresses use 6 bytes for an address. The socket address is a 2-byte address. The socket address reflects that address on which a server will listen and receive requests. The following is an example of some identified sockets:

- File Servers:

  451h NetWare core protocol

- Routers:

  452h service advertising protocol

  453h routing information protocol

- Workstations:

  4000h–6000h: used for the interaction with file servers and other communications

- 455h NetBIOS

- 456 Diagnostics

Additional addressing is used in environments such as with the LANRES product, but these addresses affect that aspect of communications with NetWare which are actually located in a different network. The addressing affects NetWare indirectly.

20.5   Sequenced Packet Exchange (SPX)

SPX is a connection-oriented protocol that applications requiring such services can use to operate in a NetWare network. By default, SPX uses IPX. However, SPX has a completely different set of functions.

SPX packet structure

The SPX packet has its own fields that perform functions differently from IPX, but SPX utilizes IPX as it goes down the protocol stack. Consider Fig. 20.8.

Figure 20.8 shows the SPX datagram behind an IPX header inside a MAC frame. This is how it appears at the data-link layer.

SPX field contents

The following list includes the fields and their purpose in an SPX packet.

- *Connection control:* This field controls bidirectional data flow between connections.

- *Data stream:* This field indicates the type of data found in the packet.

- *Source connection ID:* This field identifies the originating point of this packet. This field is also responsible for multiplexing packets of data as they leave the node if this function is required.

- *Destination connection ID:* This field identifies the target point for the packet. The destination point may perform demultiplexing if required.

- *Sequence number:* This field is responsible for maintaining a packet count on each side of the connection. The sending side maintains a count, and the receiving side maintains a count.

- *Acknowledgment number:* This field performs packet orientation functions. It indicates sequence numbers of the expected SPX packets that should be received.

- *Allocation number:* This is a number that is used to indicate the number of outstanding receive buffers in a given direction at one time.

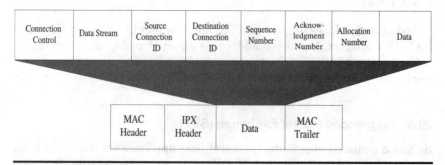

Figure 20.8 SPX packet structure.

- *Data:* This field contains data used by the application requiring the SPX protocol.

This packet, namely, SPX itself, is considered a transport protocol that uses IPX as a delivery service from origin to destination. Packets exchanged between origin and destination SPX points have sequence numbers assigned to them. These numbers can be used to check for out-of-sequence, duplicate, or missing packets.

Not all applications require SPX, but specialized ones such as gateways and applications requiring session-oriented services do.

20.6   NetWare Core Protocol (NCP)

The NCP is similar to a shell. NCP procedures must be followed by a file server's operating system in order to receive and respond to a request sent from a workstation. These protocols define all services that a file server can provide to a workstation and also all the requests a workstation can make against a file server.

Two NCP packets exist according to Novell's *Application Notes* for NetWare. These packets and field explanations include the components described in the following paragraphs.

NCP request packet

The request packet is issued to request services provided by the NCP. Figure 20.9 is an example of this packet.

Request packets are issued by a workstation against a server. In a sense the request and response packets are how workstations and the server exchange information. Details of this are considered proprietary by Novell; hence limited information is available. The details of the fields of both these packets are available, however.

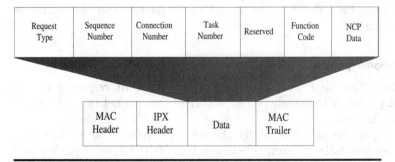

Figure 20.9   NCP request packet.

NCP request packet field contents. Like SPX, NCP is inserted into the IPX packet when it passes down the protocol stack. The field contents and meaning include

- *Request types:* According to Novell, there are seven categories:

  1111   Create a service connection

  2222   File service request

  3333   File service response

  5555   Destroy a service connection

  7777   Packet burst

  8888   A private NCP request issued to a process on the same host

  9999   Indicates that the previous request is still being processed

- *Sequence number:* Once a connection is established and packets begin to flow, packets are issued numbers in sequential order to indicate their sequence. When a server is finished processing these packets, it puts the sequence numbers in the response packet to inform the client that its receipt from the server is correct.

- *Connection number:* This number is used by a file server and clients that connect to it. Each connecting client has a connection number assigned to it and as a result, this number identifies the clients to the server.

- *Task number:* Multitasking hosts can conceivably have multiple tasks operating at one time. Servers use a task number to associate *clients* with opened files so that the server may close these files as the clients finish with them.

- *Function code:* This number identifies the NCP function required.

- *NCP data:* This part of the packet includes data from the workstation.

These fields represent a client's or workstation's requests against a server. One field differs from this type of packet and the response packet format.

NCP response packet

Structure. The NCP response packet is issued in *response* to the request packet it received. The structure of this packet is as shown in Fig. 20.10.

These packets are issued by the server. Notice the similarities, except for one field difference between this NCP response packet and the NCP request packet shown in Fig. 20.9.

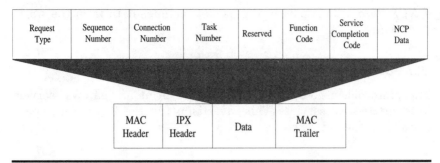

Figure 20.10   NCP response packet.

NCP response field contents

- *Request types:*   According to Novell, there are seven categories:

  1111   Create a service connection

  2222   File service request

  3333   File service response

  5555   Destroy a service connection

  7777   Packet burst

  8888   A private NCP request issued to a process on the same host

  9999   Indicates that the previous request is still being processed

- *Sequence number:*   Once a connection is established and packets begin to flow, packets are issued numbers in sequential order to indicate their sequence. When a server is finished processing these packets, it puts the sequence numbers in the response packet to inform the client that its receipt from the server is correct.

- *Connection number:*   This number is used by a file server and clients that connect to it. Each connecting client has a connection number assigned to it and as a result, this number identifies the clients to the server.

- *Task number:*   Multitasking hosts can conceivably have multiple tasks operating at one time. Servers use a task number to associate *clients* with opened files so that the server may close these files as the clients finish with them.

- *Function code:*   This number identifies the NCP function required.

- *Service completion code:*   This field includes a value indicating whether an error occurred during processing. Any value other than zero indicates that an error of some type occurred; Novell documentation provides additional details on this.

- *NCP data:* This part of the packet includes data from the workstation.

20.7 Service Advertising Protocol (SAP)

This protocol is used by servers such as file, print, and gateway servers to advertise their services. This advertising function is performed periodically on the network.

SAP packet structure

SAP packet structure has considerable internal detail. Examine Fig. 20.11.

Figure 20.11 shows numerous fields, but on close inspection the packet actually contains two groups of the same information. The variables differ because of the information reflected by them.

SAP field contents

Each host that provides a service contains a SAP agent. Each agent acquires information reflected in the fields of this packet and keeps the information in a table.

- *Operation:* This field defines packet operation.

- *Service type:* This field identifies the service type that the server provides.

- *Service name:* This field identifies the server name based on the network number.

- *Network address:* This is the network address on which the server is located.

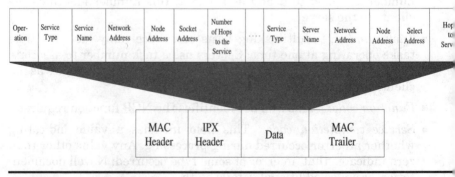

Figure 20.11 SAP packet structure.

- *Node address:* This is the address of the server's location.

- *Socket number:* This number is the process identifier to which packets must be sent.

- *Number of hops to the server:* This indicates the number of hops to the host where the server is located.

The remaining part of the packet may contain additional information about other servers on the network.

Service type

Servers have hexadecimal (hex) numbers identifying the services they provide. These numbers are present in the service field of the SAP packet. Some examples of service types are as follows:

Service type	Value
User	1
User group	2
Print queue	3
File server	4
Job server	5
Gateway	6
Print server	7
Archive server	9
Archive queue	8
Job queue	A
Administration	B
SNA gateway	21
NAS gateway	29
TCP/IP gateway	27
Time synchronization server	2D
Print queue user	53
Network access server	98
Portable NetWare	9E
Wildcard	FFFF
NNS domain	133
NetWare 386 print server	137
NetWare 386	107
Communication execution	130
Advertising print server	47

20.8 Routing Information Protocol (RIP)

RIP is used in a NetWare environment to route packets from one host to another, or from one network to another. Routers that operate with NetWare protocols keep a table with a list of network segments and hosts. Information is exchanged among routers via RIP packets.

RIP packet structure

RIP packet structure appears as shown in Fig. 20.12.

This packet structure is required support for NetWare-based networks. The packet provides information about networks, hosts, and other routers.

RIP field contents

The RIP packets contain few fields, but they may be repeated to meet the needs of a network. The following list explains these fields.

- *Operation:* This is the first field in the packet and indicates whether the packet is a response or a request. Following this field the maximum number of information sets is 50. These information sets include three components: network number, number of hops, and number of ticks.

- *Network number:* This uniquely identifies a network segment.

- *Number of hops:* This refers to the number of routers a packet must pass to reach a given network number.

- *Number of ticks:* This is an estimated reference to the amount of time it takes for a packet to reach a given network number.

When multiple routes to a given network are possible, this information is important. For example, two routes may exist between target and destination networks, but the amount of time between the two routes may vary greatly.

Routing information

Routing information is maintained in tables throughout a network in routers. Information such as the number of network segments a router is aware of, the number of hops to a given network, and the number of

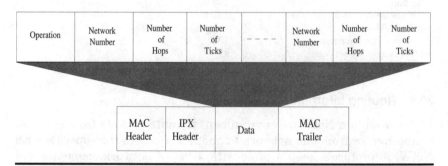

Figure 20.12 RIP packet.

ticks it takes for a packet to reach a given network segment. Other information includes network interface card identifications and the addresses of an additional router, which can be used if the primary route is not possible.

Routing information is spread throughout a network at different times and in different ways. For example, the initial broadcast of routing segments occurs immediately after initialization. Routing can also occur when there is an initial request to receive routing information. From this point in time approximately every 60 seconds periodic updates are performed. Updates can also be performed when routers send broadcasts to other routers connected to other networks, thus notifying other networks of the updates. Broadcasts also occur when a router is powered down; a broadcast is one of its final functions.

20.9   Error, Echo, and NetBIOS Protocols

These three protocols reside on top of IPX and are explained here. These are peer protocols used by NetWare hosts.

Error

Error protocol is used when a destination cannot be reached. For example, when a packet leaves a host and reaches a router and the router does not know the destination, it may generate a network-unreachable message.

Echo

Echo protocol is also a peer protocol. Applications can use this protocol to determine whether a path to a destination is reachable. In this respect the echo protocol is similar to Packet Internet Groper (PING). It works by simply echoing back to the originating point once it reaches the target.

NetBIOS

NetBIOS (derived from *network basic input/output system*) is merely an application programming interface between a network adapter card and applications (programs). Its origins predate LAN operating systems to a significant degree. It is a product of IBM and was introduced with IBM's PC network. NetBIOS is implemented in the Novell environment on top of IPXs much like SPX. In fact, Novell NetBIOS is more of an emulator to the original NetBIOS than a complete clone because its frames are not compatible with IBM's NetBIOS.

NetBIOS was originally, early in the industry, a protocol that vendors could use for application porting. However, a problem with NetBIOS is the lack of continuity between multiple-vendor communication protocols. Ironically, what was meant to be a standard has, at this point in time, become so diverse that a lack of continuity exists among many vendors implementing the protocol.

20.10   System Fault Tolerance (SFT)

System fault tolerance (SFT) emerged in 1987 with NetWare Version 2.1. Included with this was the file server console FCONSOLE, resource accounting, and security.

Version 2.1 at a glance

One of the added features to this version was the FCONSOLE. This gave individuals (namely, system administrators) the ability to monitor a network from any host on the network. This is parallel to the function of NetView in SNA. The resource accounting function was added to so that data could be gathered about users and/or their groups. This function was particularly advantageous because it meant that the protocol was making strides into a well-rounded network protocol. The security introduced with this version improved conditions so that greater control could be obtained over resources and data itself. A number of different levels of security were implemented. For example, file, directory, account, internetwork connections, and groups can all have security through a password system. However, the greatest enhancement was the fault tolerance.

SFT structure and function

System fault tolerance can best be explained by way of example. Consider Fig. 20.13.

Multiple disks exist in Fig. 20.13. In fact, the same data on disk 1 is stored on disk 2. The purpose of disk mirroring is to prevent loss of data in case of possible disk failure. *Disk mirroring* is the ability for either disk to be online at any given instant.

If a problem arises when a disk is being read, SFT detects this and reads the backup disk. After reading the backup disk it marks the primary disk sector as "bad" and then restores information on a primary disk by copying data from the backup disk onto the primary disk. This is referred to as a "hot fix." In short, a hot fix is the capability to detect and correct medium errors while data is being moved.

The notion of *disk duplexing* also exists. This is the duplication of the entire disk, disk channel, power supply, and disk controllers. Disk duplexing is shown in Fig. 20.14.

Primary
Disk

Secondary
Disk

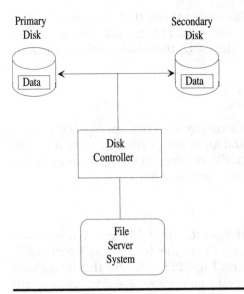

Figure 20.13 SFT example.

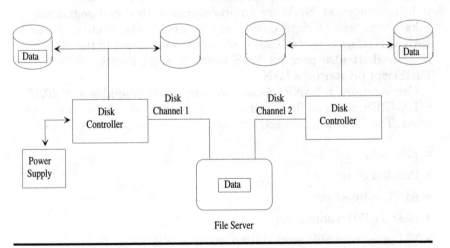

File Server

Figure 20.14 Disk duplexing.

Disk duplexing provides a default advantage; that is, it provides two possible paths for the CPU to access data. This means whichever disk system responds first can serve the CPU; thus disk duplexing is a de facto speed enhancer.

20.11 NetWare Implementations

NetWare can be implemented in a variety of environments. It can operate on multivendor equipment and all in one environment, or it can be used to provide seamless connectivity between heterogeneous systems. Some example implementations are provided here.

LANRES

LANRES is an application that operates in MVS and VM environments. According to IBM, it brings the power behind the S/390 architecture to a NetWare environment. LANRES achieves this by making DASD available to NetWare servers and S/390-based printers available to NetWare clients.

LANRES also permits authorized MVS users to move data to and from a NetWare server. Additionally, NetWare server files and directories can be listed, created, and/or deleted.

LANRES also makes LAN printers available to MVS users. In effect, it brings together NetWare environments with S/390 seamlessly to take advantage of right-positioning of workloads. Another function LANRES offers is centralization of LAN management to the MVS host, if desired. It also permits MVS users to send PostScript files to a PostScript printer on a LAN.

Conceptually, a LANRES environment would resemble Fig. 20.15.

LANRES is versatile because of the connectivity solutions it supports. The following connectivity solutions are supported by LANRES:

- ESCON
- Parallel channels
- APPC connection
- Host TCP/IP connection
- VM programmable workstation services (PWSCS)

The method of connectivity dictates how LANRES is configured on the host. Because of the breadth of support for connectivity solutions, the requirements, installation, and definitions are site-dependent and are directly related to how the product is used. For example, if the prod-

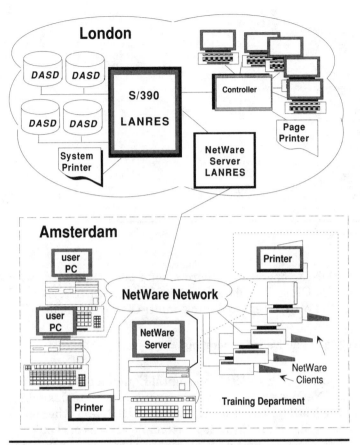

Figure 20.15   Conceptual view of LANRES and NetWare LANs.

uct is used with TCP/IP under MVS, LANRES uses sockets and TCP for connectivity. However, if APPC is used, it connects to APPC MVS via CPI-C, conforming to SAA standards.

NetWare in UNIX

NetWare for UNIX is supported under NetWare Version 3.11 for UNIX. Unlike NetWare operation in a DOS environment, in UNIX NetWare works with the UNIX operating system, file system, memory management schemes, and scheduling resources. In fact, a C programming interface makes NetWare services possible in a UNIX OS environment. NetWare for UNIX, however, is actually C source code NetWare; it has been ported to popular variations of UNIX. Figure 20.16 is an example of this scenario.

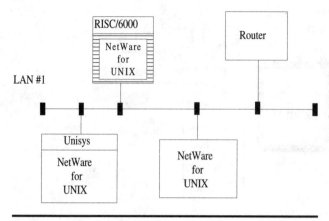

Figure 20.16 NetWare for UNIX.

Multivendor UNIX operating environments are supported with NetWare for UNIX. UNISYS and the RISC/6000 are shown here. Beyond this example is NetWare integrated into a multi-OS environment.

NetWare for VMS and multivendor operating systems

NetWare is supported under VMS from Digital Equipment (DEC). This is another port of NetWare. Such an environment can create flexibility, particularly if the VMS hosts are operating with UNIX and other operating systems as well. Figure 20.17 is an example of this concept.

Figure 20.17 shows VAX and AXP hosts, but UNIX and DOS hosts also operate on the network as well. Because of the independence

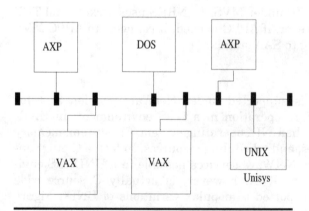

Figure 20.17 NetWare for VMS and multiple operating systems.

achievable through NetWare, communication and file sharing can be achieved among such diverse environments.

20.12 Summary

NetWare is a true success story from a protocol standpoint. It began in the early 1980s with the PC market and grew as the PC architecture grew. The protocol was enhanced to keep pace with network competition and continues to be refined. It is through Version 4.0 as of this writing.

NetWare operates not only on PCs but also on UNIX-based hosts with diverse backgrounds such as IBM's RISC/6000, and UNISYS's version of UNIX. NetWare can also participate with two of IBM's largest operating systems: MVS and VM. By leveraging on the LANRES product, enterprise connectivity and seamlessness can be achieved with NetWare.

Novell introduced its first product in 1983. In practically 12 years it has advanced to an extremely robust network protocol. Many milestones were made in the 1980s. Some of those were mentioned in the first section of this chapter. A number of forces contributed to NetWare growth, including growth in personal computers and ETHERNET, token-ring announcement and growth, and factors in the business community.

One part of Sec. 20.2 is entitled, "Read this first." It explained why NetWare should not be mistaken for a simple network protocol for small environments. NetWare was evaluated by layers compared to the OSI model. The components of NetWare were explained. Details on IPX, SPX, NCP, SAP, RIP, error and echo protocols, and NetBIOS were provided. NetWare now includes support for IP.

The open data interface was explained, and it was shown how upper-layer protocols and applications realize considerable independence from hardware. Implementation and management of ODI were explained.

Field-level details on the following packets were provided:

IPX   SAP
SPX   RIP
NCP

IPX sockets and ports were explained, with a detailed view of the packet type field. IPX addressing was explained, and examples of some identifiable sockets were provided for the reader. SPX was explained and the fields inside its packet described. NCP was discussed, and both the request and response packet were explained at a field level. Delineation between request types was explained. SAP packet structure was presented, and its fields were explained. SAP service types were also listed along with their hex values associated with the service type. RIP was discussed, including its packet structure. RIP routing information was also presented.

Error, Echo, and NetBIOS were discussed. The system fault tolerance (SFT) feature was presented. A brief look a Version 2.1 when SFT was introduced was provided. SFT was discussed along with disk duplexing and multiple-disk implementation. The hot fix was explained.

NetWare implementations were provided to give the reader an insight into the diversity of implementations that can be realized with NetWare. LANRES, NetWare in UNIX, and NetWare for VMS and multivendor operating systems were presented.

21

Digital Network
Architecture (DNA)
Phase V and DECnet/OSI

The Digital Equipment Corporation, better known as DEC in the industry, is another reputable company with powerful and flexible networking architectures. This chapter explores some of DEC's systems and how they are implemented and examines DEC's networking protocols. (A sincere thank you is due to Mr. Larry Walker, Mr. John Adams, and their assistants for sending DEC's most current documentation via air cargo to make this information as current as possible.)

21.1   A Look at the Past

DECnet is the implementation of the architecture and protocols of DNA by DEC, and not other vendors. In other words, it is a genuine DECnet implementation.

DECnet versions

DECnet originated in the mid-1970s. Many use the term *phase* rather than *version* when referring to DECnet upgrades or enhancements to distinguish DECnet from products by other vendors; however, the term *version* is used in this chapter. Version I was used in the mid-1970s. By the late 1970s DECnet had evolved to Version II, and with it came functional enhancements including file transfer capabilities, according to DEC.

In 1980 DECnet Version III was brought to market. A significant enhancement of this version was the remote terminal capability, according to DEC. DECnet Version IV, available in 1982, brought significant improvements as well. It supported an increased number of nodes (which are defined later) and the support for ETHERNET. From 1983 to 1987 Version IV was further enhanced and became the dominant version of DECnet.

In 1987 DEC announced DECnet Version V. This announcement received considerable attention, especially because Version V supported OSI. This announcement sent a clear statement to the marketplace in the direction of DEC's networking architecture. Most refer to this version of DECnet as DECnet/OSI. In 1994 DEC is shipping operating system manuals entitled *OpenVMS,* instead of previous versions called *VMS.* Later in this chapter we will explore in depth the topic of OpenVMS and the changes DEC has made to move toward a considerably more open environment.

Operating systems

DEC's past has not been totally obscured, however; two DEC operating systems are dominant in the workplace even today: VMS (virtual memory systems) and Ultrix. VMS has been DEC's popular operating system for years. Its focus has been on DEC equipment. Users interface with VMS via *Digital Command Language* (DCL), which consists of commands and is considered capable of creating fairly powerful programs.

Ultrix, on the other hand, is Digital's version of UNIX, which can operate on some of DEC's equipment. This makes the VAX line of computers very powerful; they can use VMS, or they can utilize the strength DEC put into ULTRIX.

Hardware

DEC has provided numerous offerings of hardware since its inception in 1957. Some of the first hardware DEC offered were called PDP systems. DEC offers a wide range of scalable hardware to meet customers' needs and provides a migration path for future growth. Historically, some of the hardware offerings, according to DEC, include

- VAX 11-780
- VAX 750
- VAX 730
- VAX 785
- VAX 8600
- VAX 3800 series
- VAX 9000 series
- VAX 6000 series
- VAX 4000 series

DEC has brought to market other systems, but this brief list shows the continual advancement and commitment that DEC has provided with some of the best computing power in the industry. However, the purpose here is not to explore the hardware or software in particular but rather to examine the networking capabilities of DEC's network offerings.

21.2   Orientation to DEC Equipment and Terms

Some readers may be new to DEC equipment, and for that reason we will focus on the types of equipment DEC offers and the names associated with these products. Additionally, some common terms that might be new to the reader are explained later in this chapter according to DEC's most current documentation.

DEC hardware and devices

A good place to start with DEC hardware is probably with the most widespread industry standard: the VT100 terminal.

VT100 terminal. The VT100 terminal is undoubtedly the most widely supported ASCII-based terminal used in environments where emulation is applied. Consider Fig. 21.1.

Figure 21.1 shows a site in metropolitan Kaufman, Texas, where Costello and May, Inc. operates a veterinary office that provides countywide services, which also extends into other states such as South Dakota. The SR Ranch, in Sioux Falls (S. Dak.), is a multi-thousand-acre ranch which includes a main office for personnel maintaining accounts receivables/payables. A large DECnet network is required to maintain operations at this ranch because of its size and the satellite data entry locations on the ranch property.

Figure 21.1 shows a VT100 terminal being used in one location. It provides terminal capabilities locally, and the SNA-to-DECnet gateway provides terminal emulation for the VT100 terminal into the SNA data center in Kaufman (Tex.).

This is just one scenario where a VT100 not only can be used in its native environment but can also be easily integrated into a different environment such as SNA. Figure 21.2 is a simpler example of how VT100 terminal emulation is used.

Figure 21.2 shows Trustmark National Bank headquarters in Jackson (MS), a branch office in Brandon, where Mr. Wiess interacts daily with the main office, and Erin's house. Notice that Erin's house has a PC using software with VT100 terminal emulation to gain access to the SNA data center. Since Erin works at home and creates custom programming for Trustmark, she needs accessibility only to the data center, and the most efficient way to achieve this is for her to have a PC

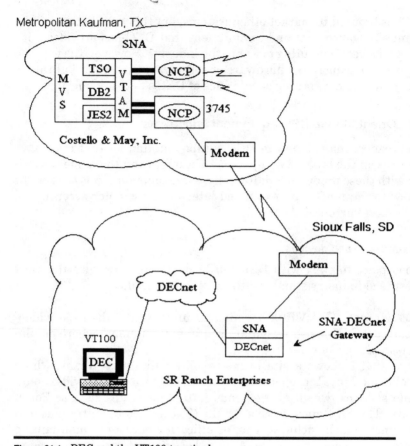

Figure 21.1   DEC and the VT100 terminal.

with a communication package capable of supporting VT100 emulation. This might seem strange, but an efficient way to gain entrance
into an SNA environment, from a terminal perspective, is with VT100
emulation. This assumes that the configuration is set up properly.

The example here where VT100 terminal emulation is used is by far
the most dominant terminal emulation supported in many environments.

Small VAXcluster.   A *VAXcluster* is a collection of equipment interconnected in such a way that the components therein can communicate
with one another and share their resources. Consider Fig. 21.3.

Figure 21.3 shows multiple devices in the VAXcluster. The VAX 6000
series are hosts that have multiple CPUs inside. The *CI* beside each
line to the star coupler indicates a special type of bus. According to
DEC documentation, this is

Computer Interconnect, a high-speed, fault tolerant, dual path bus, which has a bandwidth of 70 Megabits per second [Mbits/s]. With CI, any combination of processor nodes and intelligent I/O subsystems nodes—up to 16 in number—can be loosely coupled in a computer-room environment.

The *star coupler* is a central point where CI cables from different nodes can connect. The star coupler is used for redundancy among paths between devices.

The hierarchical storage controller (HSC) is considered intelligent. According to DEC, it is a self-contained, intelligent, mass storage controller that communicates with VAC processors and implements column shadowing on DEC storage architecture (DSA) disks.

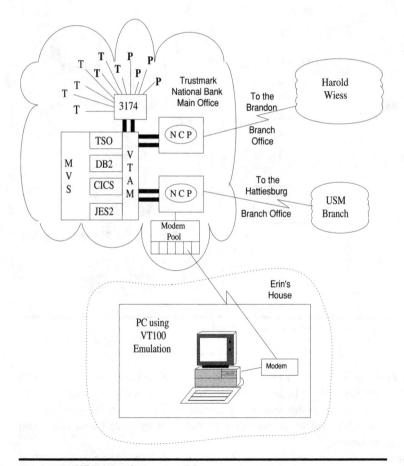

Figure 21.2   VT100 emulation on a PC.

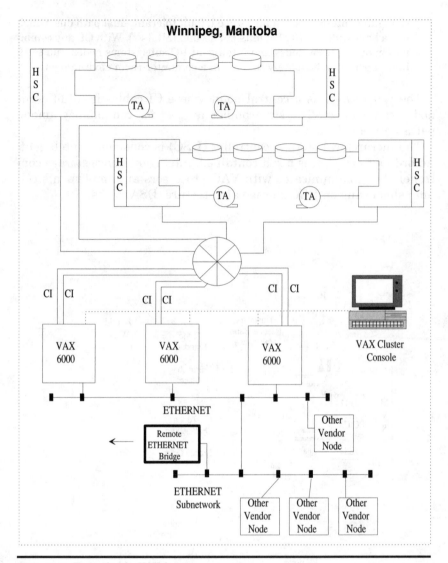

Figure 21.3   Example of a VAXcluster.

The *console* is used to control and issue system commands to the VAXcluster. According to DEC documentation, it is the manual control unit integrated into the central processor. The console enables the operator to start and stop the system, monitor system operation, and run diagnostics.

Other devices are used later in the chapter to convey various concepts and installations, with explanations provided as they are introduced.

Terms

Those new to the DEC environment might find it somewhat difficult to break through the barrier of acronyms that are taken for granted by those who have worked in a DEC environment for some time. Here, some commonly used DEC acronyms, abbreviations, and terms are listed and explained; these are based on the latest original DEC documentation available.

CI   Computer interconnect, a high-speed, fault-tolerant, dual-path bus, which has a bandwidth of 70 Mbits/s. With CI, any combination of processor nodes and intelligent I/O subsystems nodes—up to 16 in number—can be loosely coupled in a computer-room environment.

CS   Refers to a console floppy disk.

DECdts   Digital distributed time service. In DECnet/OSI it is software that synchronizes the clocks in networked systems.

DECnet/OSI   Family of DEC hardware and software products that implement DNA Phase V, which integrates OSI and DNA protocols; compliant with OSI and compatible with DECnet Phase IV and TCP/IP.

DELNI   A DEC local network interconnect; it is a concentrator for up to eight ETHERNET-compatible nonterminal devices.

DEUNA   Network adapter for connecting UNIBUS-based hosts to a CSMA/CD LAN.

DDCMP   Digital data communications message protocol. Byte-oriented, data-link layer, DNA protocol implemented in DECnet software; designed to provide an error-free communications path between adjacent systems. Operates over serial lines, delimits frames by special character, and includes checksums at the link level.

DSA   Digital (DEC) storage architecture. Specifications from DEC governing the design of and interface to mass storage products. DSA defines the functions to be performed by host computers, controllers, and drives, and specifies how they interact to manage mass storage.

DSSI   Digital (DEC) small-systems interconnect. A data bus that uses SCA protocols for direct host-to-storage communications. The DSSI cable can extend to 6 meters and has a peak bandwidth of 4 Mbytes.

DJ   Refers to a removable disk.

DU   Refers to a fixed disk that cannot be removed.

End node   A DEC node that can receive packets addressed to it and send packets to other nodes, but cannot route packets through from other nodes. Also called a *nonrouting node*.

GAP   Gateway access protocol. This is a protocol used between a host DECnet/OSI system and a DECnet/OSI system that is a DTE on a PSDN to provide the X.25 gateway access facility to a user on the host.

Hot standby   A second or backup running computing system that is ready to pick up application processing in the event that the primary computing system fails. That is, the secondary system takes over and continues the processing at the point where the original computing system stopped.

HSC   Hierarchical storage controller. A self-contained, intelligent, mass storage controller that communicates with VAX processors and implements volume shadowing on DSA disks.

ISE   Integrated storage element. This is a disk or tape drive that contains a dedicated controller for the device in a combined package. The ISE contains fully functional mass storage control protocol (MSCP), tape mass storage control protocol (TMSCP) server components, SCS (SNA character string), and port interface logic. ISEs are intelligent multihost controllers, and work in parallel with other ISEs on the interconnect.

Lobe   A set of CPUs connected together by one or more VAXcluster interconnects. A lobe must have its own system disk(s) used by all CPUs in that lobe. A single CPU with a local system disk can be a lobe. Not all CPUs are part of a lobe, however. For example, a set of workstations is not a lobe.

MOP   Maintenance operations protocol. DNA network management protocol used to perform functions such as downline loading, upline dumping, and circuit testing.

NSP   Network services protocol. The DNA protocol that operates in the DNA transport layer. DECnet Phase IV uses NSP. NSP and OSI transport can reside simultaneously on a DECnet/OSI node.

OpenVMS   A group of software programs (or images) that control computing operations. The base operating system is made up of core components and an array of services, routines, utilities, and related software.

Star coupler   A central point where CI cables from different nodes can connect. The star coupler is used for redundancy check among paths between devices.

TT   Reference to a terminal; typically a VT320 type.

VAX PSI   DECnet/OSI for VMS software that implements the CCITT's X.25 and X.29 recommendations; lets users assign and use switched virtual circuits across a packet-switching data network, allowing DECnet/OSI to participate in a packet-switching environment.

VT   Reference to a terminal; typically a VT100 type.

WANDD   DECnet/OSI software that links a hardware device and layered software, for example, VAX packetnet system interface software.

XE   In DECnet environments, a reference to ETHERNET.

21.3   DECnet Layers

As have other protocols in this book, DNA can be compared to the OSI mode. As shown in Fig. 21.4, the correlation between OSI and DNA layers is not one-to-one.

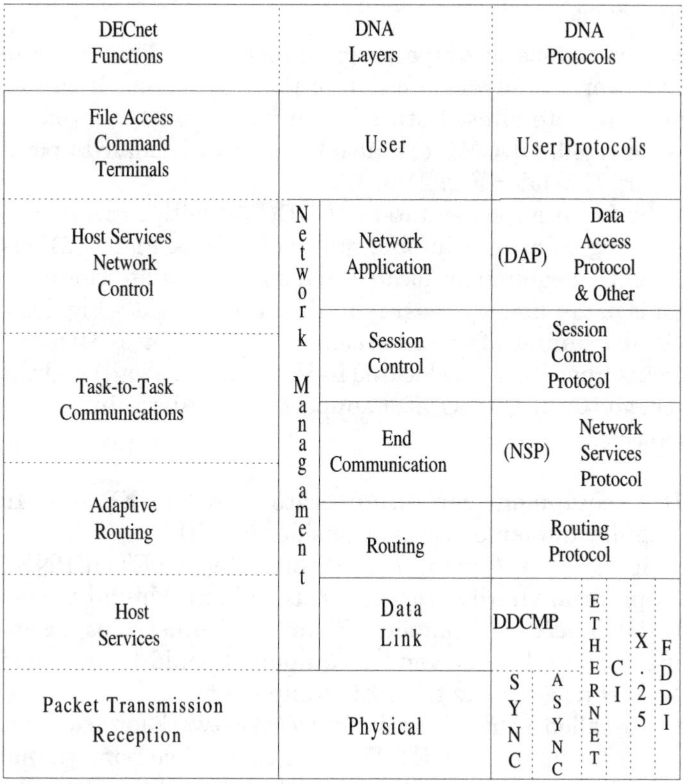

DECnet Functions	DNA Layers	DNA Protocols
File Access Command Terminals	User	User Protocols
Host Services Network Control	N e t w o r k M a n a g e m e n t	Network Application	(DAP)	Data Access Protocol & Other
Task-to-Task Communications	Session Control	Session Control Protocol
	End Communication	(NSP)	Network Services Protocol
Adaptive Routing	Routing	Routing Protocol
Host Services	Data Link	DDCMP	ETHERNET
Packet Transmission Reception	Physical	SYNC	ASYNC	ETHERNET	X.25	FDDI

Figure 21.4   DECnet by layers.

Figure 21.4 has three columns. The DNA protocols, DNA layers, and the DECnet functions. The remainder of this section begins by examining DNA protocols.

Physical-layer protocols

The physical layer is practically the same in all networks. It defines the actual interface where the medium connects to a device. In the DEC environment, multiple physical interfaces are supported. In the case of the DNA protocols, those most popular interfaces associated with data-link-layer protocols are supported. Examples of these interfaces are:

- V.24
- V.25
- V.25bis
- V.35

Data-link-layer protocols

DEC's data communications message protocol (DDCMP). This protocol operates at the lower two layers relative to the OSI model. It can be used locally or in remote sites. Both asynchronous and synchronous protocols are supported. DDCMP can also be used with point-to-point or multipoint lines. Consider Fig. 21.5.

Figure 21.5 shows an implementation of DDCMP with synchronous connections including point-to-point and multipoint lines. Figure 21.5 is also an example of this equipment located in a single location. However, DEC equipment is not limited by geographic distance. Consider Fig. 21.6.

Figure 21.6 is an example of two sites connected together by DDCMP over asynchronous lines. One site located in Honolulu (Hawaii) and the other in San Diego (Calif.). DEC also supports other data-link-layer-equivalent protocols.

ETHERNET. DEC equipment can also operate with ETHERNET. In fact, this is a popular implementation. Consider Fig. 21.7.

In this example where DEC equipment is connected via ETHERNET protocols and appropriate hardware, the functionality is similar to that in any other ETHERNET environment. Figure 21.7 illustrates a common implementation, and other vendor equipment could be installed on this network just as easily as the DEC equipment.

DEC has a device called a *DELNI*, which is a local network interconnect; it is a concentrator for up to eight ETHERNET-compatible nonterminal devices. Figure 21.8 is an example of the implementation of this device in a network similar to that shown in Fig. 21.7. Now consider Fig. 21.8.

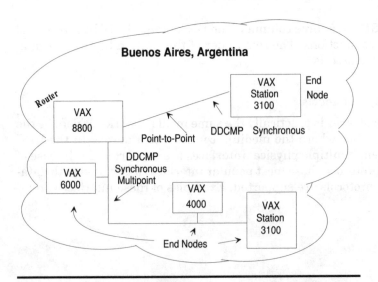

Figure 21.5 Synchronous DDCMP.

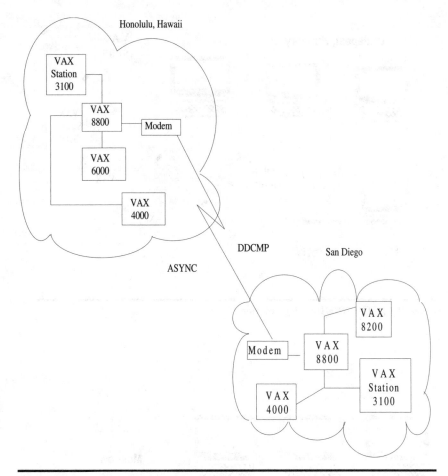

Figure 21.6 Asynchronous DDCMP.

Figure 21.8 is an example of an ETHERNET-based network similar to the one shown in Fig. 21.7. However, in Fig. 21.8 the DELNI is used as a concentrator for three devices as shown here, but the actual DELNI can accommodate more devices than those shown in this example.

CI. The *computer interconnect* (CI) is a high-speed, fault-tolerant, dual-path bus, which has a bandwidth of 70 Mb/s. With CI, any combination of processor nodes and intelligent I/O subsystems nodes—up to 16 in number—can be loosely coupled in a computer-room environment. Figure 21.9 is an example of its usage.

Figure 21.9 shows multiple VAX hosts, disk drives, tape drive, and a star coupler using the CI connection.

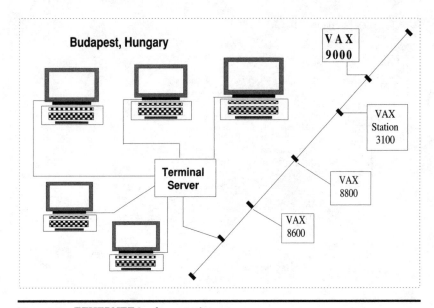

Figure 21.7 ETHERNET implementation.

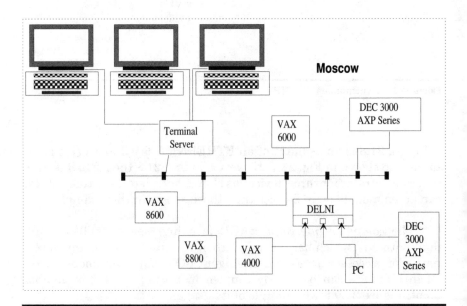

Figure 21.8 ETHERNET and DELNI.

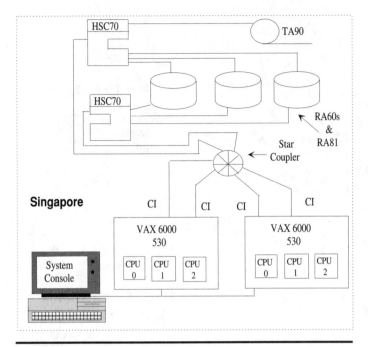

Figure 21.9   Computer interconnect (CI).

X.25.   DEC also supports X.25 connections. Figure 21.10 is an example of multiple sites connected via a X.25 connection. As the figure shows, multiple sites are connected via X.25 by DEC-supported X.25 equipment.

FDDI.   FDDI is also supported by DECnet/OSI. Figure 21.11 is an example of such support.

Figure 21.11 shows multiple devices utilizing the speed of a FDDI ring. In this case, multiple VAXes are used, the AXP series is implemented, and even an ETHERNET is connected to the FDDI backbone via a FDDI-to-ETHERNET bridge.

Routing

Routing occurs at the equivalent of the OSI network layer. Routing is routing regardless of the network protocol; the question is how. Different types of routing are possible in DECnet/OSI because of the routing and nonrouting nodes defined according to DEC specifications. Additional details are provided later concerning routing node types. For our purposes here, the following explanation and example will suf-

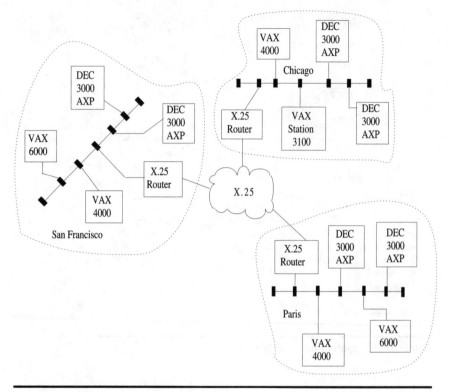

Figure 21.10   X.25 support.

fice. Level 1 (one) routing is performed by a level 1 router. This type of routing is performed within a single area of a network. Level 2 (two) routing is routing within a single area of a network or between two or more networks. Because of the architectural difference between VAX and AXP, differences do exist in what functions routers perform.

DEC does define what is called *full-function* and *end-node routing* for DECnet for OpenVMS environments. Functionality depends on which is selected at the time of purchase; however, changes can be made as growth occurs.

Because the flexibility in the OpenVMS environment is versatile, multiple routing methods can be achieved. For example, FDDI, DDCMP, ETHERNET, and CI routers are offered. They can be implemented in a variety of possible implementations. Figure 21.12 is an example of one of those implementations, according to DEC.

Figure 21.12 shows multiple sites and multiple data-link-layer implementations. Collectively, they create a WAN. Utilization of multiple data-link-layer protocols is easily achieved, and different DEC systems can be used.

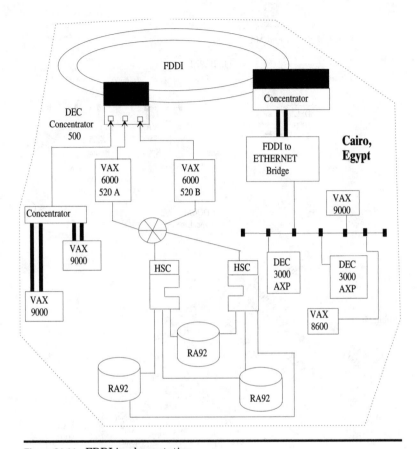

Figure 21.11   FDDI implementation.

Network services protocol (NSP)

NSP is a protocol used for end-to-end communications in DNA to ensure reliable communications between end stations. It operates closely with the session-layer protocol. Fundamental to NSP is the concept of a logical connection between two hosts or nodes.

NSP performs a number of functions, but some are significant and should be mentioned here:

- Establishment and destruction of logical links between systems
- Segmentation of data
- Management of flow control
- Management of error control
- Reassembly of data

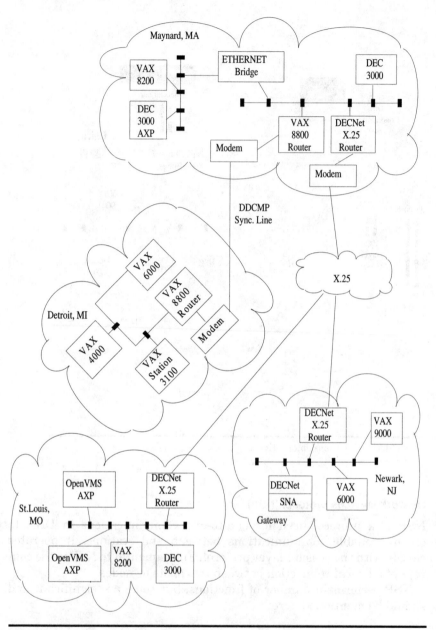

Figure 21.12   DEC router implementation.

Establishment of sessions occurs partially on behalf of the originating NSP module that communicates with a target NSP module. Within the NSP module, a number of *submodules* exist. NSP uses the concept of a port as in TCP/IP ports. In order to achieve NSP functions, databases are maintained by NSP. Some of these databases are maintained for internal functions of the NSP, while others are used for network-wide-related functions.

Session control protocol

The session control protocol defines system-dependent aspects of transport connection communication, according to DEC documentation. The session control protocol includes functions such as name-to-address translation, process addressing, process activation, and some access control. Simply put, session control makes the transport services available to end users of the network. Conceptually, it appears as shown in Fig. 21.13.

Figure 21.13 is a conceptual view of the session control function in a DNA network environment.

Data access protocol (DAP)

DAP is the method DNA uses to communicate across a network to access data. Functionality of DAP is based on a client/server relationship. DAP has been a means of transferring data through a DNA network; however, distributed system services, which perform similar functions, are seemingly gaining ground in what was once a DAP function.

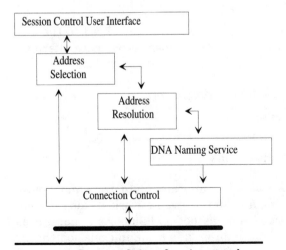

Figure 21.13   Conceptual view of session control.

Additionally, the transition of DNA to OpenVMS and support for multiple OSI protocols is also changing the function of DAP. According to the documentation provided to me by DEC, two DAP gateways are supported.

DAP-FTAM gateway. The DAP-FTAM gateway is basically a server that receives data access protocol (DAP) messages through DECnet software. It uses this information to establish and maintain a connection with a remote FTAM system. The result is support for DAP and now transitional support for FTAM.

The main advantage of the DAP-FTAM gateway is that it simplifies communication capabilities for DECnet nodes because users can communicate with remote OSI systems through the gateway using system commands. For the DCL or ULTRIX shell user, communication with a remote FTAM application is handled the same as any DECnet dialog.

FTAM-FTP gateway (DEC OSF/1). This gateway resides on a DEC OSF/1 system but can be accessed from other OSI systems. According to DEC's latest documentation, FTAM-FTP gateway software provides bidirectional access between OSI systems and internet systems (such as those based on Berkeley 4.2/4.3 BSD TCP/IP implementations); this software permits file exchange between these systems.

OSI end systems and Internet systems that communicate through the FTAM-FTP gateway do not need special software, and remote users do not have to establish accounts on the gateway system. The FTAM installation procedure installs the FTAM-FTP gateway software.

User protocols

User protocols operate at the application layer compared to the OSI network. A variety of functions can be performed by these offerings. Some of them are:

- Retrieving network status information
- Establishing remote log-ons
- Accessing file capability on remote nodes
- Performing task-to-task operations

These functions provide a wealth of information to users and administrators. For example, being able to retrieve network status information on devices and functions such as nodes, links, costs, hops, area information, and the next hop to an area is some information that is possible via the network status information facility.

Establishing remote log-ons is similar to that with the application TELNET native to the TCP/IP stack mentioned in a previous chapter. This provides users on any given host with the capability to log on to and perform functions as if they were physically attached to that system.

Accessing files on remote nodes is another feature for users. This gives users the flexibility to use DCL commands to perform file-related functions. Additionally, this capability provides users with the ability to write high-level language programs and execute them against files on other hosts.

Task-to-Task operations permit programs operating on different systems to communicate with one another. This is true even if these programs are written in different languages. Notice that this concept sounds similar to APPC in the IBM environment. This capability alone is powerful because of the implications behind it. For example, databases can exchange information or online systems can access information in real time.

Much more information could be provided about the DNA protocols and other aspects of Fig. 21.4, but other topics should be covered here.

21.4 OSI Standards Supported by DECnet/OSI for OpenVMS

Now that the DNA layers have been presented and explained, this section lists the OSI standards supported by DECnet/OSI for OpenVMS.

Standards correlation

Layer	Standard	Description
Application	ISO 8571	File transfer, access, and management
	ISO 8649	Service definition—association control
	ISO 8650	Association control service elements
	ISO 9041	Virtual terminal protocol
	ISO 9072	Remote operations service element
Presentation	ISO 8822	Presentation service
	ISO 8823	Presentation (kernel)
Session	ISO 8326	Connection-oriented session service (COSS)
	ISO 8327	Connection-oriented session protocol
Transport	ISO 8072	Service definition
	ISO 8073	Transport definition
Network	ISO 7498	Connectionless mode transmission
	ISO 8208	X.25 packet-level protocol (PLP)
	ISO 8348	Service definition: connection-oriented network service (CONS)

	ISO 8348	OSI addressing format
	ISO 8473	Internetwork protocol (interactive network layer protocol)
	ISO 8473	Connectionless mode network service (CLNS)
	ISO 8878	X.25 to provide CONS
	ISO 9542	ES-IS routing exchange protocol
Data link	ISO 3309, 4335, 7809, 8471, 8885	Point-to-point data links (HDLC)
	ISO 7776	Point-to-point X.25 data links (LAPB, one possible user)
	ISO 8802-1	LAN support (CSMA/CD architecture and maintenance)
	ISO 8802-2	Frame formats for 8802-3 LANs (CSMA/CD logical link control (LLC1)) X.25 logical link control (LLC2)
	ISO 8802-3	LAN support (CSMA/CD)
	ISO 9314	FDDI
Physical	ISO 8802-3	CSMA/CD devices
	ISO 9314	FDDI
	EIA RS-232-C	
	EIA RS-422	
	EIA RS-423	
	CCITT V.35	

DNA Phase V reference model

Another view of DEC's networking model is shown in Fig. 21.14.

Figure 21.14 shows an integrated model of DEC functionality along with OSI-based functions. At all layers within the model OSI support is provided while at the same time maintaining a transition with support for DNA.

Considerably more information is available in DEC's manual entitled *DECnet/OSI for OpenVSM Introduction and Planning,* documentation no. AA-PNHTB-TE. To obtain it, please consult your local DEC representative.

21.5 DECnet Concepts and Topics

Some of these concepts need explanation for those who are not familiar with DECnet or DECnet/OSI. This section focuses on these concepts and topics.

DECnet node types

Through the phases (versions) of DECnet different types of equipment were developed. Some of that equipment is still functioning today in productive environments. Because of the transition into a more open environment, identifying these nodes is important. This is important for purposes of routing configuration and other customization issues.

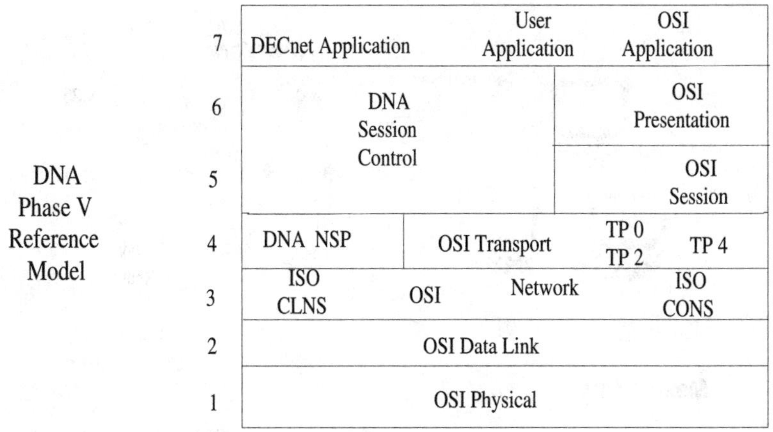

	7	DECnet Application	User Application	OSI Application
	6	DNA Session Control	OSI Presentation	
DNA Phase V Reference Model	5		OSI Session	
	4	DNA NSP	OSI Transport	TP 0 / TP 2 — TP 4
	3	ISO CLNS	OSI Network	ISO CONS
	2	OSI Data Link		
	1	OSI Physical		

Figure 21.14 OSI layers.

Phase II nodes. Phase II nodes can communicate with one another, assuming that a physical connection exists between them. These nodes support point-to-point communications. Additionally, ETHERNET is not supported with these Phase II nodes.

Phase III nodes. These nodes brought adaptive routing capabilities to nodes. The significance of this was that many nodes could participate as a result. Neither ETHERNET nor FDDI was supported with this phase.

Phase IV nodes. This phase of DECnet introduced significant changes from its predecessors. For example, ETHERNET was supported. Area routing is supported with this phase. This allows up to 63 networked areas with each area containing 1023 nodes. This phase also permitted communication with Phase III nodes to a limited degree, but it did not render obsolete the use of Phase III nodes.

DECnet/OSI

This change on behalf of DEC means a breakthrough into an incredibly more open environment. This does not mean that prior phases were not *open,* but it does mean that support for OSI applications such as FTAM, ACSE application-layer functions, OSI addressing, connection-oriented and connectionless-oriented functions provide capabilities not possible before. Also, multiple classes of service are possible and gateway functionality into TCP/IP and other environments is easier to a considerable degree.

Another plus for this phase of DECnet is that multiple data-link support is possible and integration into heterogeneous networks is easier. Consider Fig. 21.15.

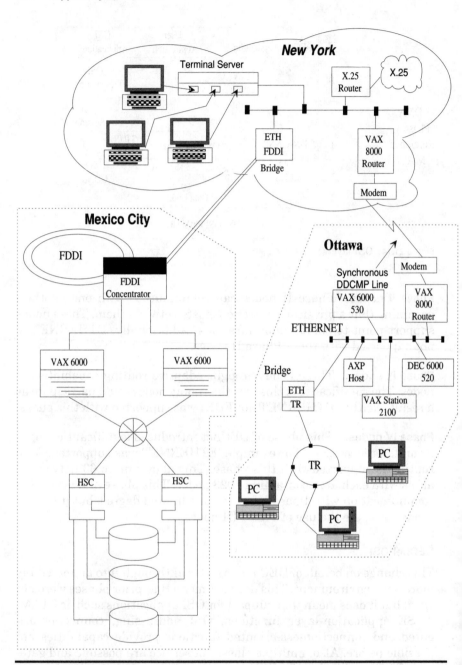

Figure 21.15 Multiple data-link support.

DECnet/OSI and OpenVMS provide more avenues for solutions to diverse environments where multi-vendor equipment exist and multiple network protocols are in use.

VMScluster configurations

VMSclusters are highly integrated organizations of OpenVMS AXP and or VAX systems that communicate over high-speed communication paths. VMScluster configurations provide functions of a single-node system but also provide the ability to share resources such as disk space and CPU power. DEC documentation shows four examples of such systems; they are worthy of consideration here.

CI-based cluster.   This type of cluster is based on CI for communications. It utilizes a star coupler configuration. Consider Fig. 21.16.

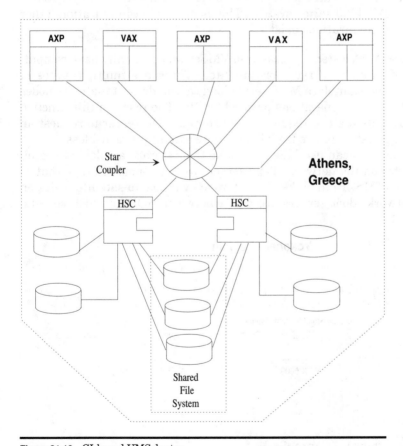

Figure 21.16   CI-based VMScluster.

Figure 21.16 may appear deceptive in that the star coupler seems like a single point for failure. Each star coupler has a redundant backup; hence should one fail, the other would automatically take over.

DSSI-based VMScluster. *Digital small-systems interconnect* (DSSI) is a data bus that uses the systems communication architecture (SCA) protocols for direct host-to-storage communications. DSSI delivers a high bandwidth that can interconnect multiple devices. Consider Fig. 21.17 as an example of this.

This example shows AXP and VAX systems connected together along with disk drives and a tape drive. Each storage device has its own controller and mass storage control protocol (MSCP). These devices are called, in DEC lingo, *integrate storage elements* (ISEs). The nature of these devices makes it possible for them to operate in parallel. Another variation of a DSSI implementation is represented in Fig. 21.18.

Figure 21.18 is an example of utilization of DSSI and ETHERNET with a VAX 4000 series system. This is a sample configuration; other possible configurations exist.

Local area VMScluster. A local area VMScluster generally have computers configured as servers or satellites. There are multiple types of servers. For example, a MOP server loads a boot driver to satellite nodes via maintenance operations protocol (MOP). The result of this function is that a satellite can request a load from a MOP server to request an operating system image load; hence these satellites are diskless.

Disk servers can use MSCP to make their storage available to satellite nodes on the network. Tape servers use tape mass storage control protocol (TMSCP) to make their storage available to satellite nodes on the network. Boot servers are a mix between MOP and disk servers.

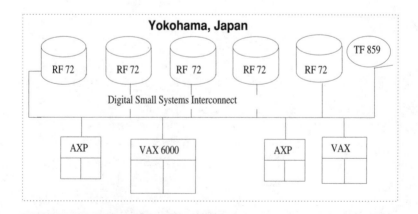

Figure 21.17   DSSI-based VMScluster.

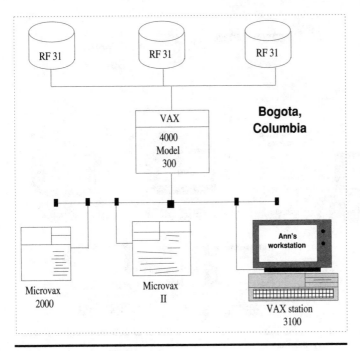

Figure 21.18 DSSI and ETHERNET.

Boot servers make data and applications available for satellite nodes on the network. Figure 21.19 is an example of this local area VMScluster.

Figure 21.19 is an example of a configuration that can maximize resources for the least amount of expense. It is a common implementation.

CI- and ETHERNET-based VMScluster. An ETHERNET- and CI-based system can achieve the best of both implementations because they are integrated. To a considerable degree the example shown in Fig. 21.20 is fault-tolerant.

Other configurations are possible, but these suffice to explain the point behind VMScluster configurations. Other configurations include FDDI implementations and variations thereof.

OpenVMS operating system components

OpenVMS consists of multiple parts. In fact, DEC documentation groups them into three layers: kernel, core services, and utility programs. These components and their subcomponents are examined here.

Kernel. The *kernel* is basically the heart of OpenVMS. It consists of memory management routines, an I/O subsystem, process manage-

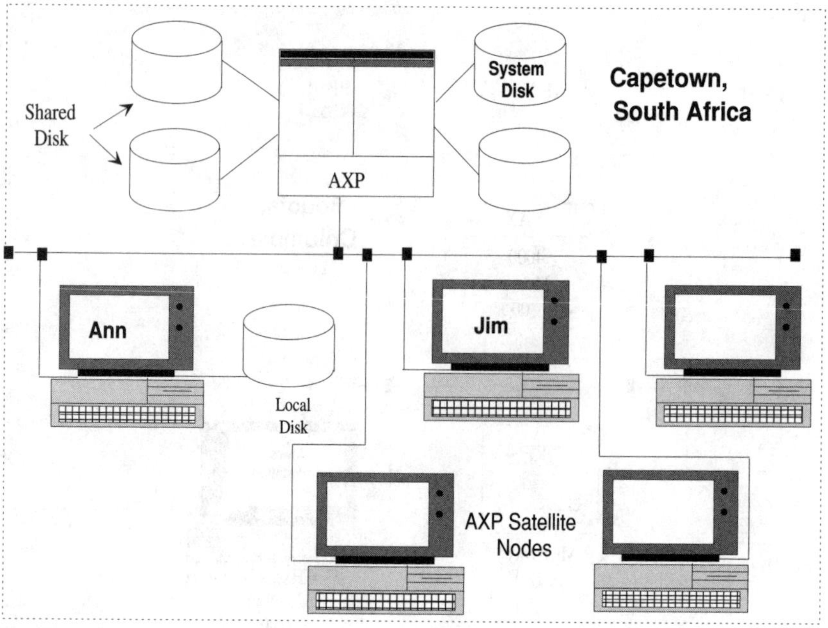

Figure 21.19 Local area VMScluster.

ment capabilities, and a time management system. The memory management subsystem portion of the kernel handles the allocation of memory in the virtual memory environment.

The process and time management functions operate hand in hand. These components schedule and perform process control. A scheduler controls the job starting process on the basis of priority.

The I/O subsystem consists of drivers that operate at the lowest layers in a network (computer). This subsystem is responsible for processing interrupts. The I/O subsystem supports custom device driver interfaces. In turn, device drivers execute I/O instructions to move data to and from devices.

Core services. Core services include user and application interaction within the processing system. These services can be identified; some of them are:

DCL A Command-line interface to the operating system.

POSIX This is optional at this point in time and is the POSIX that is IEEE-compliant. It is a portable operating system with interfaces that support a wide array of functions.

record management subsystem This is a particular way of handling I/O completely independently from devices.

run-time library This includes run-time routines that perform a variety of functions.

Utility functions. OpenVMS is shipped with some utility programs. Those that are oriented toward authorization, AUTOGEN, backup, monitor, mount, and sysman are included. User utilities include mail and help, for example. Text processing utilities include the Digital Standard Editor (EDT) and Extensible VAX Editor (EVE). Program utilities include a linker, debugger, librarian, and File Definition Language (FDL).

Other programs and functions are included in the OpenVMS operating system, such as the Digital distributed transaction manager. This is functional support for distributed transaction programming. Other programs are also available that can operate on the OpenVMS operating system.

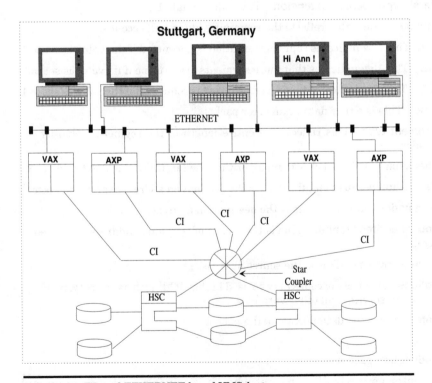

Figure 21.20 CI- and ETHERNET-based VMScluster.

21.6 Phase V Routing

A key area in network configuration and troubleshooting is routing. It is at this layer where devices exchange information throughout a network or multiple networks via messages. Consequently, having more information available about the messages that flow through a given network environment is advantageous. This section provides detailed information on those message packets used in DECnet Phase V routing.

Data and error report format

Data packet headers are used with packets exchanged between different nodes. An error report is sent back to the source node to indicate that a problem was encountered along its path. The data and error packet format appears as shown in Fig. 21.21.

Field meanings:

network-layer protocol identifier   This value equals 129.

length indicator   This value is the length of the header in octets.

version/protocol ID extension   This value equals 1.

lifetime   This value reflects the remaining life in half seconds.

SP   This flag indicates that segmentation is permitted if the value is set to 1.

MS   This flag indicates that more segments are to come if its value is set to 1.

E/R   This flag indicates that an error report is requested if the value is set to 1.

type   This is either data or an error packet.

segment type   This reflects the entire length of the segment including the header.

checksum   This is a checksum algorithm used in ISO 8473, 9542, and 8073.

destination address length   This is the destination address expressed in octets.

destination address   This is the destination address.

source address length   This is the length of the source address expressed in octets.

source address   This is the source address.

options   A variety of codes may be used in this field, such as security, padding, quality of service, and other variables.

data   This is the data portion of the packet.

Redirection packet

A router generates this packet when it receives a packet from an end node. The router then forwards the packet back on the circuit from

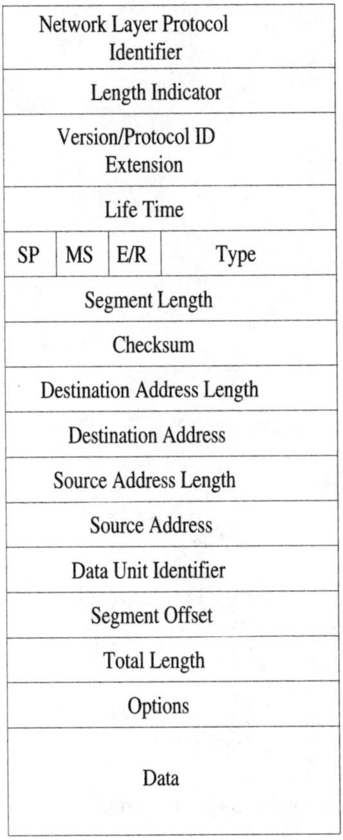

Network Layer Protocol Identifier
Length Indicator
Version/Protocol ID Extension
Life Time

SP	MS	E/R	Type

Segment Length
Checksum
Destination Address Length
Destination Address
Source Address Length
Source Address
Data Unit Identifier
Segment Offset
Total Length
Options
Data

Figure 21.21   Data and error packet format.

which it was received. The purpose of this packet is to inform end nodes of a better path for routing. This packet appears as shown in Fig. 21.22.

Field meanings:

ES-IS   End system-to-intermediate system protocol identifier. This value is 130.

length indicator   This is the header length in octets.

version/protocol ID extension   This value is set to 1.

type   Bits 1–5 are used. Bits 6–8 are reserved.

holding time   This value is measured in microseconds.

checksum   This checksum algorithm is used in ISO 9542 and others.

destination address   This is the routing-layer destination address from the packet which caused the generation of the redirect.

ES-IS Protocol Identifier
Length Indicator
Version/Protocol ID Extension
0

R	R	R	Type

Holding Time
Checksum
Destination Address Length
Destination Address
Subnet Address Length
Subnet Address
Router Address Length
Router Address
Options

Figure 21.22 Redirect packet format.

subnet address This is the data-link-layer address to use for transmitting to the destination address.

router address This is the routing-layer address of the router to which the packet is being redirected. This field is not present if the redirect is directed toward the destination end node; if this is the case, the router address transmit value is 0.

options Data in this field is copied from the original data packet which caused the generation of the redirect.

Other router-generated packets

Endnode hello. The end node hello message is generated by end nodes to inform routers on the network of their existence. It can also be generated by an end node on a broadcast circuit in response to a received data packet which was multicast to ALLENDNODES. Conceptually, this packet appears as shown in Fig. 21.23.

Explanation of fields:

ES-IP protocol identifier The value in this field is set to 130.

length indicator This is the header length in octets.

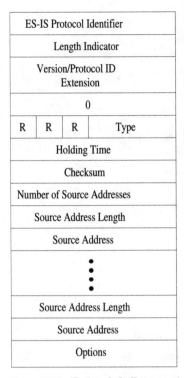

ES-IS Protocol Identifier
Length Indicator
Version/Protocol ID Extension
0
R | R | R | Type
Holding Time
Checksum
Number of Source Addresses
Source Address Length
Source Address
⋮
Source Address Length
Source Address
Options

Figure 21.23   End-node hello.

version/protocol ID extension   This value is set to 1.

type   Bits 1–5 are used, and bits 6–8 are reserved.

holding time   This time is calculated in fractions of seconds.

checksum   This checksum algorithm is used in ISO 9542 and others.

number of source addresses   The number of source addresses specified in this PDU.

source address length   This length is in octets of the following source address.

source address   This is the routing-layer address of the end node generating the hello message.

options   This is a variable-length field and is seldom present on transmission.

Router hello.   The router hello packet is periodically generated by at least one router on the network to inform end nodes of the existence of any router(s). The broadcast is to ALLENDNODES. This packet appears as shown in Fig. 21.24.

```
+-------------------------------------+
|      ES-IS Protocol Identifier      |
+-------------------------------------+
|          Length Indicator           |
+-------------------------------------+
|        Version/Protocol ID          |
|             Extension               |
+-------------------------------------+
|                 0                   |
+-----+-----+-----+-------------------+
|  R  |  R  |  R  |      Type         |
+-----+-----+-----+-------------------+
|            Holding Time             |
+-------------------------------------+
|              Checksum               |
+-------------------------------------+
|        Source Address Length        |
+-------------------------------------+
|           Source Address            |
+-------------------------------------+
|               Options               |
+-------------------------------------+
```

Figure 21.24   Router hello.

Field meanings:

ES-IS protocol identifier   This field value is set to 130.

length indicator   This value expresses the header length in octets.

version/protocol ID extension   This field value is set to 1.

type   Bits 1–5 are used in this field; bits 6–8 are not used.

holding time   This field's value is measured in fractions of a second.

checksum   This checksum algorithm is used in ISO 9542 and others.

source address length   This identifies the length of the source address to follow.

source address   This is the address of the host generating the hello packet.

options   In this field the CODE option can be used and if so, it can be a value of 198 that suggests the ES configuration timer. If it is used, the length value will be used and the value subfield will reflect the value of the timer in seconds.

Level 1 router-to-router hello.   This message is broadcast to DNA routers on broadcast circuits. Its purpose is to automatically discover the identity of any other DNA level 1 routers on the circuit. Figure 21.25 is an example of this type of packet.
    Field meanings:

IS-IS protocol discriminator   If the preset IS-IS format has the value "ISO," then use the IS-IS protocol identifier; if not, then use the DNA private protocol identifier.

length indicator   This value has a header length of 27 octets.

version/protocol ID extension   This field value is set to 1.

reserved   This field is transmitted as zero.

type   Bits 1–5 and 15 are used. Bits 6–8 are reserved, meaning that they are ignored on receipt.

version   This field value is set to 1.

ECO   This field value is set to zero.

user ECO   This field value is set to zero.

reserved/circuit type   The most significant 6 bits are reserved. A 0 is a reserved value; however, if it is specified, the entire packet is ignored. Bits 1 and 2 indicate which level. If bit 1 is used, it indicates level 1 only. If bit 2 is used, it means only level 2 is used. If both are used, this indicates bit 3 is used.

source id   This is a 6-byte node identifier from the transmitting router.

holding timer   The holding timer reflects the time to be used by this router.

segment length   This refers to the entire length of the PDU in octets including the header.

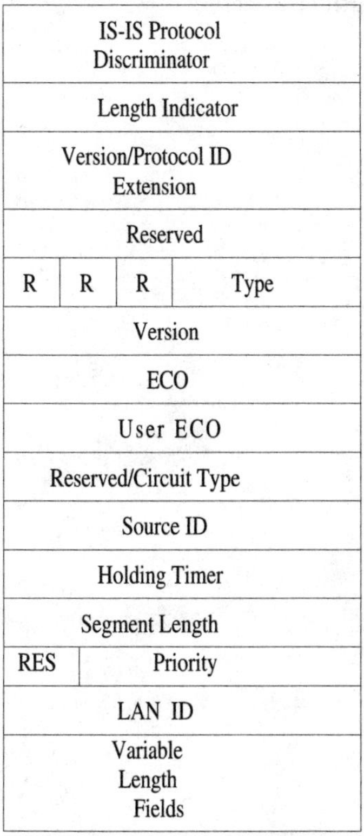

IS-IS Protocol Discriminator
Length Indicator
Version/Protocol ID Extension
Reserved
R R R Type
Version
ECO
User ECO
Reserved/Circuit Type
Source ID
Holding Timer
Segment Length
RES Priority
LAN ID
Variable Length Fields

Figure 21.25   Level 1 router-to-router hello.

reserved/priority Bits 1–7 indicate priority for being LAN level 1-designated router. The higher numbers have a higher priority for being LAN level 1-designated router.

LAN ID This is a 7-byte field. It is composed of the 6-byte unique ID of the LAN level 1-designated router in addition to the low-order octet assigned by LAN level 1-designated routers.

variable-length fields There are actually three subfields, including a code, length, and value. The contents of this field dictate the values input to the subfields.

Level 2 router-to-router hello

This message packet is broadcast to level 2 routers that are on a broadcast circuit. This packet is used by DNA routers to discover other DNA routers on the same circuit. Figure 21.26 is an example of this packet.

Field meanings:

IS-IS protocol discriminator Either ISO or DNA private values are used in this field.

length indicator This value has a header length of 27 octets.

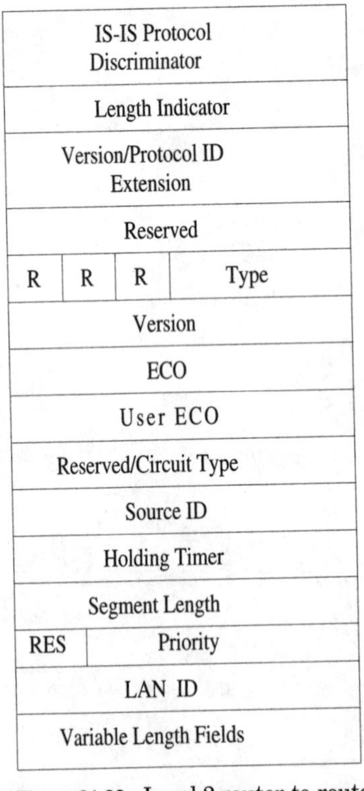

IS-IS Protocol Discriminator
Length Indicator
Version/Protocol ID Extension
Reserved

R	R	R	Type

Version
ECO
User ECO
Reserved/Circuit Type
Source ID
Holding Timer
Segment Length

RES	Priority

LAN ID
Variable Length Fields

Figure 21.26 Level 2 router-to-router hello.

reserved This is transmitted as a value of zero.

type Bits 1–5 and 16 are used. Bits 6–8 are reserved, meaning they are transmitted as zero value.

version This value is transmitted as 1.

ECO This value is transmitted as zero.

user ECO This value is transmitted as zero.

received/circuit type The most significant bits are reserved. A 1 is used to indicate a level 1, and 2 indicates a level 2; a 3 is used to indicate that both levels 1 and 2 are used.

source ID This 6-byte node ID is from the transmitting router.

holding timer This timer is used for the router.

segment length This refers to the entire length of this PDU. It is represented in octets and includes the header.

RES/priority Bits 1–7 indicate priority for being a LAN level 2 router. The higher the number, the higher the priority.

LAN ID This is a 7-byte field composed of the unique 6-byte LAN ID level 2 router.

variable-length fields These are actually three subfields, and including a code, length, and value. These fields are used to reflect those values necessary for routing purposes.

Point-to-point router-to-router hello

This packet is transmitted by DNA routers on *non*broadcast circuits once they have received a *router hello* from the neighbor node. The purpose of this packet is to determine whether the neighbor is a level 1 or level 2 router. Figure 21.27 is an example of this packet.

Field meanings:

IS-IS protocol discriminator Either ISO or DNA private values are used in this field.

length indicator The header length value is 20.

reserved Transmitted as zero on receipt.

type Bits 1–5 and 17 are used. Bits 6–8 are reserved, meaning that they are transmitted as zero value.

version Equals 1.

ECO Equals 0.

user ECO Equals 0.

received/circuit type The most significant 6 bits are reserved. A 1 is used to indicate a level 1, and 2 indicates a level 2; a 3 is used to indicate that both levels 1 and 2 are used.

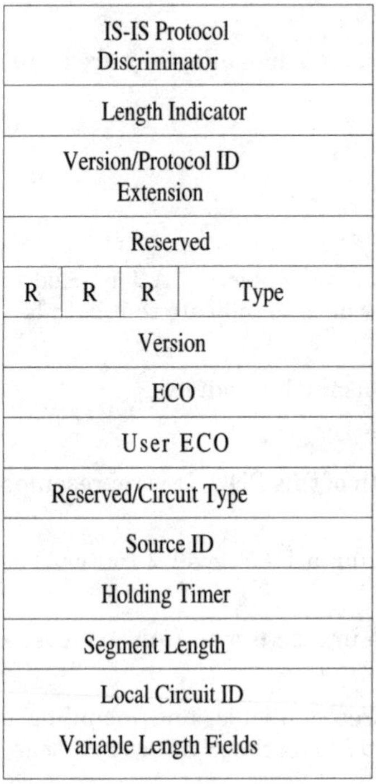

IS-IS Protocol Discriminator
Length Indicator
Version/Protocol ID Extension
Reserved

R	R	R	Type

Version
ECO
User ECO
Reserved/Circuit Type
Source ID
Holding Timer
Segment Length
Local Circuit ID
Variable Length Fields

Figure 21.27   Point-to-point router-to-router hello.

source ID   This 6-byte node ID is from the transmitting router.

holding timer   This timer is used for the router.

segment length   This refers to the entire length of this PDU.

local circuit ID   1 byte is assigned to this circuit when it is created by this router. The actual ID is known to both ends of the link.

variable-length fields   These are actually three subfields, including a code, length, and value. These fields are used to reflect those values necessary for routing purposes.

Link-state packet—level 1.   This type of packet is generated by level 1 and level 2 DNA routers and then propagated through a network. The purpose of this packet is to indicate the state of its adjacencies to its neighbor routers, pseudonodes, and end nodes of the router that originally generated the packet. Figure 21.28 is an example of this.

Field meanings:

IS-IS protocol discriminator   Either ISO or DNA private values are used in this field.

length indicator   Value equals 27.

version/protocol ID extension   The value is equal to 1.

reserved   Transmitted as zero on receipt.

type   Bits 1–5 and 18 are used. Bits 6–8 are reserved, meaning that they are transmitted as a zero value.

version   Equals 1.

ECO   Equals 0.

user ECO   Equals 0.

segment length   This reflects the entire length of the entire packet including the header.

remaining lifetime   This indicates the number of seconds before the link-state packet is considered expired.

source ID   This 8-byte ID is from the originator of the packet.

sequence number   This indicates the number of link-state packets (LSPs).

checksum   The checksum of contents of LSP from the source ID to the end.

R   Bits 5–8 are reserved.

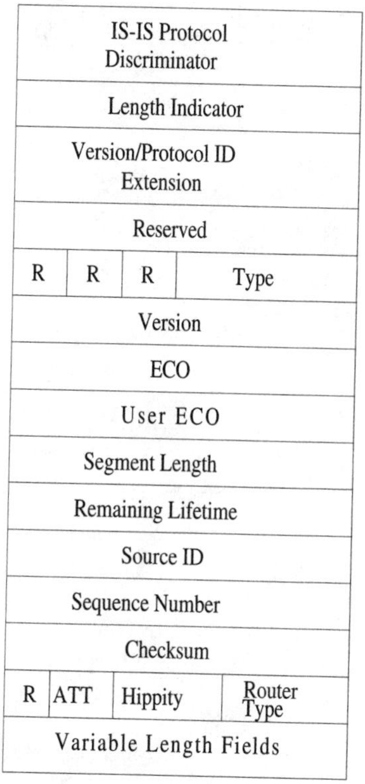

IS-IS Protocol Discriminator			
Length Indicator			
Version/Protocol ID Extension			
Reserved			
R	R	R	Type
Version			
ECO			
User ECO			
Segment Length			
Remaining Lifetime			
Source ID			
Sequence Number			
Checksum			
R	ATT	Hippity	Router Type
Variable Length Fields			

Figure 21.28 Link-state packet—level 1.

ATT   Bit 4 means that the router issuing this packet is attached to other areas. This is computed using a cost metric.

hippity   Bit 3-0 indicates zero hippity costs; 1 indicates a hippity cost.

router type   Bits 1 and 2 indicate the type of router.

variable-length fields   These are actually three subfields, including a code, length, and value. These fields are used to reflect those values necessary for routing purposes.

Link-state packet—level 2.   This type of packet is generated by level 2 DNA routers and is then propagated through a level 2 network. The purpose of this packet is to indicate the state of adjacencies to neighbor routers, pseudonodes, and end nodes of the router that originally generated the packet. Figure 21.29 is an example of this.

Field meanings:

IS-IS protocol discriminator   Either ISO or DNA private values are used in this field.

length indicator   Value equals 27.

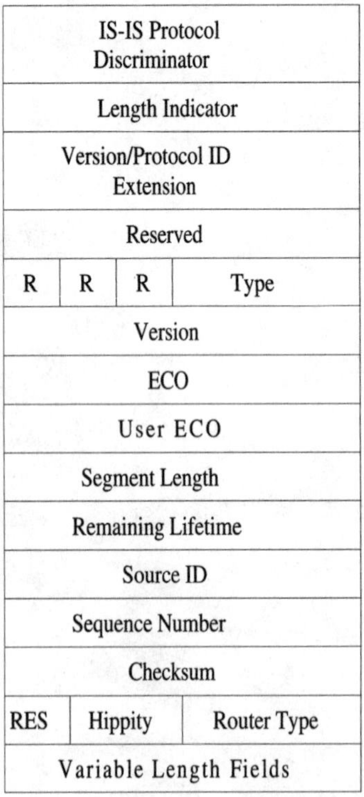

IS-IS Protocol Discriminator			
Length Indicator			
Version/Protocol ID Extension			
Reserved			
R	R	R	Type
Version			
ECO			
User ECO			
Segment Length			
Remaining Lifetime			
Source ID			
Sequence Number			
Checksum			
RES	Hippity	Router Type	
Variable Length Fields			

Figure 21.29   Link-state packet—level 2.

version/protocol ID extension   The value is equal to 1.

reserved   Transmitted as zero on receipt.

type   Bits 1–5 and 20 are used. Bits 6–8 are reserved, meaning that they are transmitted as a zero value.

version   Equals 1.

ECO   Equals 0.

User ECO   Equals 0.

segment length   The entire length of the packet, including the header.

remaining lifetime   This indicates the number of seconds before the LSP is considered expired.

source ID   This 8- byte ID is from the originator of the packet.

sequence number   This indicates the number of LSPs.

checksum   This is the sequence number of the LSP.

R   Bits 5–8 are reserved.

ATT   Bit 4 means that the router issuing this packet is attached to other areas. This is computed using a cost metric.

Hippity   Bit 3-0 indicates zero hippity cost, 1 indicates a hippity cost.

router type   Bits 1 and 2 indicate the type of router.

variable-length fields   These are actually three subfields, including a code, length, and value. These fields are used to reflect those values necessary for routing purposes.

Complete sequence numbers packet—level 1. This message packet is similar to some previous packets, but its meaning is different. This is because it operates with level 1 routers. Figure 21.30 is an example of this packet's field format.

Field meanings:

IS-IS protocol discriminator   Either ISO or DNA private values are used in this field.

length indicator   This field has a value of 33.

version/protocol extension   This field value is 1.

reserved   This is transmitted as a value of zero.

type   Bits 1–5 and 24 are used. Bits 6–8 are reserved, meaning that they are transmitted as zero value.

version   This value is 1.

ECO   This value is transmitted as zero.

user ECO   This value is transmitted as zero.

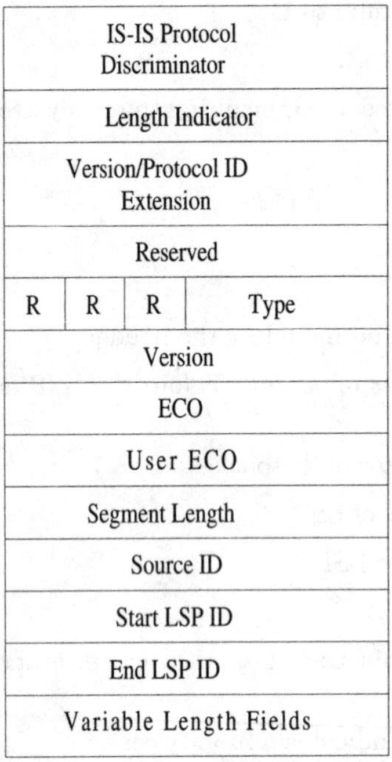

IS-IS Protocol Discriminator
Length Indicator
Version/Protocol ID Extension
Reserved

R	R	R	Type

Version
ECO
User ECO
Segment Length
Source ID
Start LSP ID
End LSP ID
Variable Length Fields

Figure 21.30 Complete sequence numbers packet—level 1.

segment length   This reflects the entire length of the PDU.

source ID   This 7-byte router ID is from the source generating this packet.

start LSP ID   This is an 8-byte ID of the first LSP in that range covered by this complete sequence numbers packet.

end LSP ID   This 8-byte ID is of the last LSP in the range covered by this complete sequence numbers packet.

variable-length fields   These are actually three subfields, including a code, length, and value. These fields are used to reflect those values necessary for routing purposes. In this particular packet their LSP entries may appear multiple times.

Complete sequence numbers packet—level 2.   This message packet is similar to some previous packets, but its meaning is different. This is because it works with level 2 routers. Figure 21.31 is an example of this packet's field format.

Field meanings:

IS-IS protocol discriminator   Either ISO or DNA private values are used in this field.

length indicator   This field has a value of 33.

version/protocol extension   This field value is 1.

reserved   Transmitted as a value of zero.

type   Bits 1–5 and 25 are used. Bits 6–8 are reserved, meaning that they are transmitted as zero value.

version   This value is 1.

ECO   Value equals 0.

user ECO   Value equals 0.

segment length   This reflects the entire length of the PDU.

source ID   This 7-byte router ID is from the source generating this packet.

Start LSP ID   This is an 8-byte ID of the first LSP in that range covered by this complete sequence numbers packet.

end LSP ID   This 8-byte ID is of the last LSP in the range covered by this complete sequence numbers packet.

variable-length fields   These are actually three subfields, including a code, length, and value. These fields are used to reflect those values necessary for routing purposes. In this particular packet their LSP entries may appear multiple times.

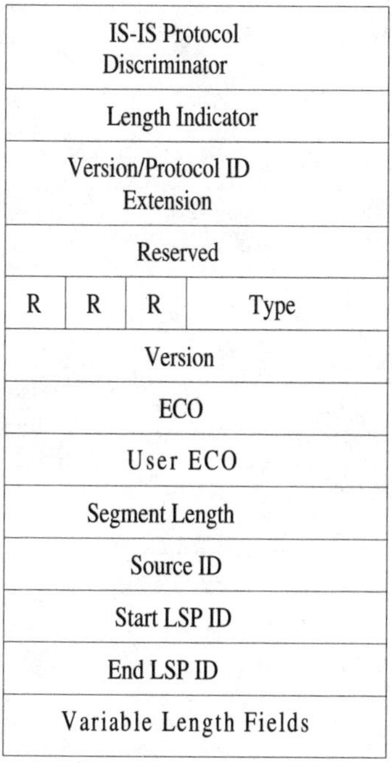

IS-IS Protocol Discriminator
Length Indicator
Version/Protocol ID Extension
Reserved
R	R	R	Type
Version
ECO
User ECO
Segment Length
Source ID
Start LSP ID
End LSP ID
Variable Length Fields

Figure 21.31 Complete sequence numbers packet— level 2.

Partial sequence numbers packet—level 1. This packet works with level 1 routers. Figure 21.32 is an example of this packet's field format. Field meanings:

IS-IS protocol discriminator Either ISO or DNA private values are used in this field.

length indicator This field has a value of 17.

version/protocol ID extension This field value is 1.

reserved Transmitted as a value of zero.

type Bits 1–5 and 26 are used. Bits 6–8 are reserved, i.e., are transmitted as zero value.

version This value is 1.

ECO Value equals 0.

use ECO Value equals 0.

segment length This reflects the entire length of the PDU.

source ID This 7-byte router ID is from the source generating this packet.

variable-length fields These are actually three subfields, including a code, length, and value. These fields are used to reflect those values necessary for

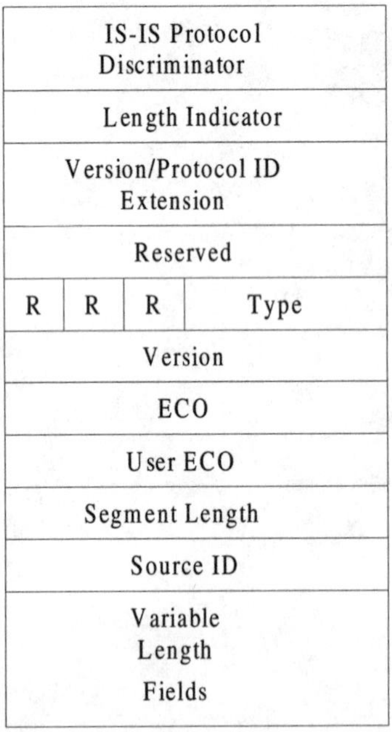

IS-IS Protocol Discriminator			
Length Indicator			
Version/Protocol ID Extension			
Reserved			
R	R	R	Type
Version			
ECO			
User ECO			
Segment Length			
Source ID			
Variable Length Fields			

Figure 21.32 Partial sequence numbers packet—level 1.

routing purposes. In this particular packet their LSP entries may appear multiple times.

Partial sequence numbers packet—level 2. This packet works with level 2 routers. Figure 21.33 is an example of this packet's field format.

Field meanings:

IS-IS protocol discriminator Either ISO or DNA private values are used in this field.

length indicator This field has a value of 17.

version/protocol ID extension This field value is 1.

reserved transmitted as a value of zero.

type Bits 1–5 and 27 are used. Bits 6–8 are reserved. They are transmitted as a zero value.

version This value is 1.

ECO Value equals 0.

user ECO Value equals 0.

segment length This reflects the entire length of the PDU.

source ID This 7-byte router ID is from the source generating this packet.

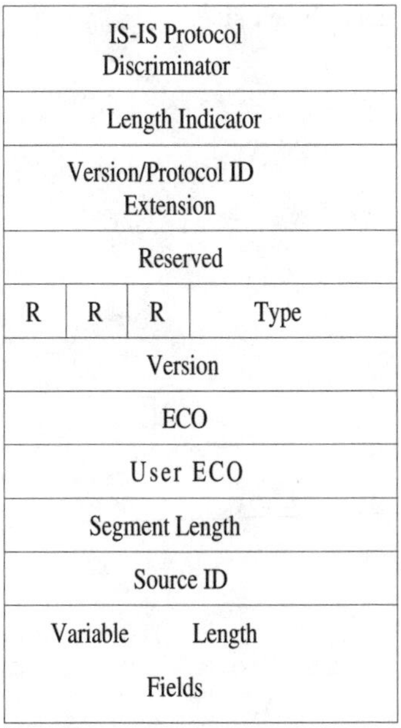

IS-IS Protocol Discriminator			
Length Indicator			
Version/Protocol ID Extension			
Reserved			
R	R	R	Type
Version			
ECO			
User ECO			
Segment Length			
Source ID			
Variable Length			
Fields			

Figure 21.33 Partial sequence numbers packet— level 2.

variable-length fields   These are actually three subfields, including a code, length, and value. These fields are used to reflect those values necessary for routing purposes. In this particular packet their LSP entries may appear multiple times.

XID message.   The exchange identification message fields contain fields similar, but not identical, to those of other packets. Figure 21.34 is an example of this packet.

Field meanings:

DNA private protocol discriminator   This field reflects a DNA private protocol identifier.

length indicator   The header length is 12 bytes.

version/protocol ID extension   This value is 1.

reserved   A zero value is transmitted.

type   Bits 1–5 and 23 are used. Bits 6–8 are reserved. They are transmitted as a zero value.

version   The value is 3.

ECO   The value is zero.

user ECO   The value is zero.

protocol ID   This is the DNA protocol ID for routing.

data-link block size   Reflects the maximum receive block size.

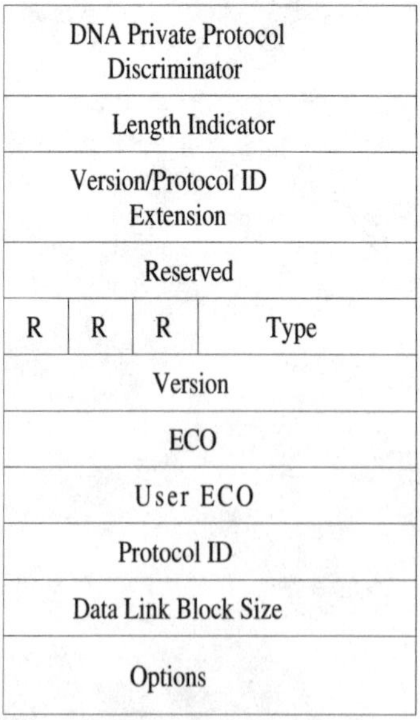

DNA Private Protocol Discriminator			
Length Indicator			
Version/Protocol ID Extension			
Reserved			
R	R	R	Type
Version			
ECO			
User ECO			
Protocol ID			
Data Link Block Size			
Options			

Figure 21.34   Xid message.

options   Three subfields are currently identified: code, length, and value. The code value is 7; the length reflects the total length of the value subfield in octets. The value subfield is the verification data.

Conclusion

This section has examined the message formats for Phase V networking routing-layer protocols. Understanding these message field contents is extremely helpful and hence was the rationale for providing them here. You can obtain additional information by contacting your local DEC representative for the appropriate documentation concerning this matter.

21.7   DDCMP Message Formats

Digital data communications message protocol (DDCMP) is a popular implementation at the data-link layer in a variety of implementations. DDCMP supports asynchronous and synchronous data transfer, as explained previously. This section explores the message formats and their field meanings.

Perspective

DDCMP operates with the philosophy of source and destination. In any given instance a node could be a source or a destination. Messages are exchanged between the nodes, and different types of messages are required to perform necessary functions.

Message types

Numbered data message.   This type of message carries user data from source to destination. Figure 21.35 is an example of its format.

Field meanings:

SOH   This is the numbered data message identifier. It has a hex value of 81.

count   This field is the byte count field which specifies the number of 8-bit bytes that are in the DATA field.

flags   These *link* flags are used to control link ownership and message synchronization.

resp   This field has a response number that is used to acknowledge messages that were received correctly.

SOH	Count	Flags	Resp	NUM	Addr	BLCK1	Data	BLKCK2

Figure 21.35   Numbered data message structure.

num   This is the number used to denote the number of the data message.

addr   This identifies the station address. It designates the tributary stations on multipoint links.

BLKCK1   This is the block check on the numbered message header.

data   This is the numbered message data field. It is transparent to the protocol and is not limited to bit patterns. This field does require an 8-bit byte specification in the count field.

BLKCK2   This is a block check on the data field. A polynomial technique is used to perform this operation.

Control message.   This type of message carries channel, transmission, and status information. Figure 21.36 is an example of a general control message structure.
   Field meanings:

ENQ   This is the unnumbered control message identifier. Its value is hex 5.

type   This indicates the control message type.

subtype   This field provides additional information for those messages which require it.

flags   These *link* flags are used to control link ownership and message synchronization.

RCVR   This field is used to convey information from a sending station to a receiving station. Each control message type has a different use for it.

SNDR   The reverse of the RCVR field. This field is used to pass information from the receiving node to the sending node.

addr   This identifies the station address. It designates the tributary stations on multipoint links.

BLKCK3   This is a block check on the control message. The computation is performed on fields ENQ through addr.

Acknowledgment message.   Acknowledgment messages are used to *acknowledge* the correct number of data messages. Figure 21.37 is an example of this format.
   Field meanings:

ENQ   This identifies the message as a control message.

acktype   This field uses a value of 1.

acksub   This field uses a value of 0.

E N Q	Type	Subtype	Flags	R C V R	S N D R	Addr	BLKCK3

Figure 21.36   Control message structure.

ENQ	Acktype	Acksub	Flags	RESP	Fill	Addr	BLKCK3

Figure 21.37 Acknowledgment message structure.

flags   This field contains the link flags.

RESP   The number here is used to acknowledge that the messages received were correct.

fill   The value of this field is 0.

addr   This is the station address.

BLKCK3   This is a control message block check as with other messages.

Negative acknowledgment message.   This message is used to pass error information between the sending and receiving stations. In this case the reason for the error is included in the subtype field. Figure 21.38 is an example of this message format structure.

ENQ   Identifies the control message.

NAKtype   Negative acknowledgment type.

Reason   Cause of acknowledgment.

Flags   These are link flags.

RESP   The number acknowledging the amount of messages received correct.

Fill   A fill byte.

Addr   The station address.

BLKCK3   A control message block check as with other messages.

Reply to message.   This message is used to request status information from the receiver. Figure 21.39 is an example of this format.
  Field meanings:

ENQ   This identifies the control message.

reptype   This reply message type has a value of 3.

repsub   This reply message subtype has a value of 0.

flags   These are the link flags.

fill   This is a fill byte with a value of 0.

ENQ	NAKtype	Reason	Flags	RESP	Fill	Addr	BLKCK3

Figure 21.38 Negative acknowledgment message structure.

E N Q	Reptype	Repsub	Flags	Fill	Num	Addr	BLKCK3

Figure 21.39   Reply to message number structure.

num   This field contains the number of the last numbered data message sent by the originating node.

addr   This is the station address.

BLKCK3   This is the control message block check.

Start message.   The start message does as its name implies. It is used to perform initial contact and synchronization on a DDCMP link. This type of message is used only during reinitialization or link startup. Figure 21.40 is an example of this message.
   Field meanings:

ENQ   This is the control message identifier.

strttype   The start (strt) message type has a value of 6.

strtsub   The start subtype begins with a value of 0.

flags   This is the link flag. For the start message each flag is set to 1.

fill   This is a fill byte with a value of 0.

fill   This is another fill byte with a value of 0.

addr   This is the station address.

BLKCK3   This is the control message block check.

Start acknowledgment message (STACK).   This STACK message is returned in response to a strt (start) once the station has completed an initialization and has reset its message numbering. Figure 21.41 is an example of this message.
   Field meanings:

ENQ   This is the control message identifier.

stcktype   The STACK message type has a value of 7.

stcksub   The STACK subtype has a value of 0.

flags   These are the link flags. Each flag is set to a value of 1.

fill   This fill byte has a value of 0.

E N Q	Strttype	Strtsub	Flags	Fill	Fill	Addr	BLKCK3

Figure 21.40   Start message structure.

| E N Q | Stcktype | Stcksub | Flags | Fill | Fill | Addr | BLKCK3 |

Figure 21.41   Start acknowledgment message (STACK) structure.

fill   This is another fill byte with a value of 0.

addr   This is the station address.

BLKCK3   This is the control message block check.

Synopsis.   These messages used with DDCMP perform different functions. They utilize station addresses and link control flags for requirements on multipoint connections and half-duplex channels. For further details on this topic, if needed, contact your local DEC representative.

21.8   Summary

Digital Equipment Corporation has been in business for decades and has produced very fine equipment to meet market needs. Five phases of DECnet have been created, refined, and brought to market. The first phase of DECnet was in the 1970s. DECnet/OSI is the current networking environment that DEC supports, and this provides flexibility beyond any prior phases. In addition to supporting OSI and applications such as FTAM, DEC has a series of AXP systems that greatly enhance computing power. VMS, DEC's operating system, has emerged to OpenVMS today. Its flexibility is robust enough to support the underlying hardware.

DEC's systems can operate with Digital Network Architecture (DNA) or TCP/IP, or can utilize the benefits of the OSI additions to their support. This flexibility and the ability to run VMS on different hosts make DEC products open by nature.

De facto terminal emulation for ASCII is the VT100. This support can be found in gateways, as was discussed in Fig. 21.1, as well as communication software packages. In fact, it is a good example of a de facto industry standard. The terminals can be used with the terminal servers DEC provides. This is especially beneficial because the networking architecture is scalable for the needs for any given site.

DEC, like other large corporations with networking architectures, has a vocabulary particular to the equipment it provides. Some of DEC's terminology was provided in this chapter. A considerable number of DEC terms are included in the Glossary at the end of this book.

DECnet can be evaluated by layers as was done in this chapter. DECnet does not correlate exactly to the OSI reference model, but then most network architectures do not. However, DEC's network architecture is defined by layers, and specifications are readily available; some were listed in this chapter. A layer-by-layer breakdown of DECnet with reference to the OSI model was presented, along with a description of the OSI standard and the function(s) it provides in an OpenVMS environment.

DECnet has different concepts and node types, some of which were presented in this chapter. For example, node types, VMScluster configurations, and OpenVMS operating system functions were explained.

An entire section was devoted to Phase V routing. Message formats were presented, and their field terms were explained. Another section was devoted to DDCMP message formats and field meanings. These two aspects of DECnet are important for troubleshooting and systems-level network operations.

22

AppleTalk

Apple Computer Corporation is a success story, American style. The story basically began in *rags* and has continued in *riches*. Steve Jobs and Steve Wozniak, among others, were the driving forces behind Apple Computer in its early years. In many ways it was ahead of its time. For example, it predated IBM's entry into the personal computer market. And, in 1984 the Macintosh was revealed. This was an incredible entry into the market because of its interface. It was not *line-mode*-oriented; it was user-friendly. The software was invoked by selecting the desired icon with a mouse and then pressing a mouse button to invoke a program.

Apple Computer approached technology in an unusual way during the late 1970s to mid-1980s. However, looking back, in retrospect, it *is* not so unusual. It seems that some designs behind Apple computers were viewed from a user's perspective. The interface was different from those of IBM, Compaq, and others that began entering the marketplace, and a different twist on networking was perceived. To a considerable degree networking capabilities were built into some systems.

This chapter explains AppleTalk protocol; its nature differs from that of other protocols. For example, in some ways it is flexible and supports some functions in a plug-n-play fashion when compared to other protocols. It differs from other protocols in that it was not intended to be what DECnet or SNA is. The original design intent of these and most protocols presented in this book are different. Apple's broad design intent was to make a computer user-friendly for nontechnical individuals, and to that end they are very successful.

22.1   AppleTalk Protocols by Layers

Most protocols in this book have been compared and contrasted to the OSI model, and it is appropriate to compare and contrast AppleTalk in like manner. This section focuses on that issue.

AppleTalk is a set of upper-layer protocols. The division separating upper and lower protocols is layer 2. Figure 22.1 is an example of AppleTalk protocols compared to the OSI model.

Physical-layer protocols

Apple networks can use different physical-layer interfaces. Three interfaces are identified in Fig. 22.1: LocalTalk, EtherTalk, and

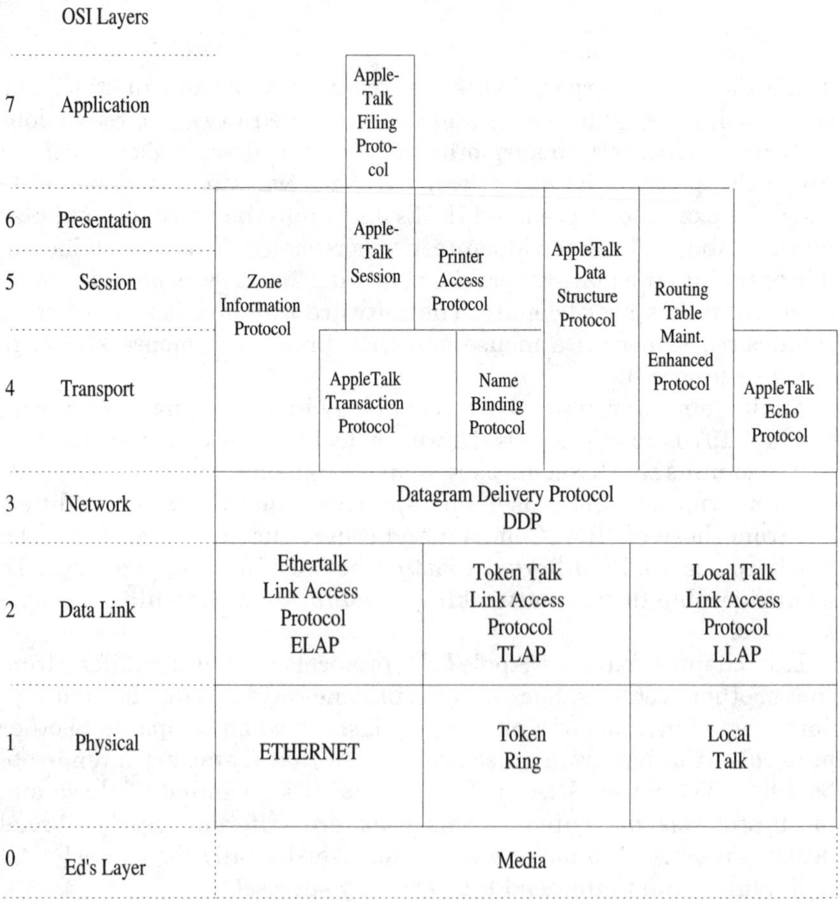

OSI Layers

7	Application	Apple-Talk Filing Protocol					
6	Presentation	Zone Information Protocol	Apple-Talk Session	Printer Access Protocol	AppleTalk Data Structure Protocol	Routing Table Maint. Enhanced Protocol	AppleTalk Echo Protocol
5	Session						
4	Transport		AppleTalk Transaction Protocol	Name Binding Protocol			
3	Network	Datagram Delivery Protocol DDP					
2	Data Link	Ethertalk Link Access Protocol ELAP	Token Talk Link Access Protocol TLAP			Local Talk Link Access Protocol LLAP	
1	Physical	ETHERNET	Token Ring			Local Talk	
0	Ed's Layer	Media					

Figure 22.1   AppleTalk by layers.

TokenTalk. But since the designs of these three systems are different, some delineation is in order here.

LocalTalk. LocalTalk does not use AppleTalk protocol. LocalTalk was designed to be plug-n-play. Macintoshes, Apple IIgs, and the laser writer printer along with other equipment can be connected as shown in Fig. 22.2.

Figure 22.2 shows multiple Macintoshes and other devices. This type of network is simple. The cables literally plug into a device, and a splitter is used to enable the bus structure. Experience and recommendation indicate that 32 devices such as these can be connected and the maximum distance should be 300 meters.

This type of configuration uses *LocalTalk link access protocol* (LLAP), which operates at the equivalent of the data-link layer as Fig. 22.1 depicted. The design behind this protocol was to provide a means for passing packets of data from one node to the other on a network. However, it is a best-effort delivery system. The protocol calls for dynamic address definition, a core function of LocalTalk. This is partially possible because of the transmit and receive hardware in devices that support it. As brief as it may seem, this is LocalTalk networking. However, technically, LocalTalk is considered part of the AppleTalk protocol. LocalTalk uses the LLAP *multiple-protocol package* (MPP) device driver. MPP includes additional drivers, which are discussed later in this chapter.

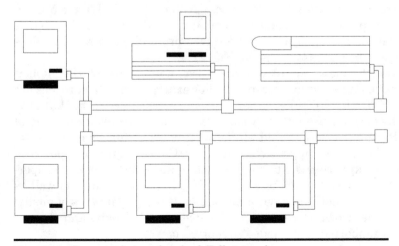

Figure 22.2 Conceptual view of an AppleTalk network.

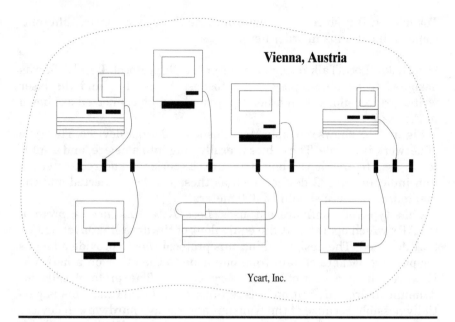

Figure 22.3   View of EtherTalk network.

EtherTalk. *EtherTalk* is an implementation of AppleTalk over ETH-ERNET. The ability to use AppleTalk in this manner is based on the *EtherTalk link access protocol* (ELAP), which operates at the data-link layer. Consider Fig. 22.3 as an example of an EtherTalk network.

Figure 22.3 shows multiple devices using ETHERNET for a data-link-layer protocol, implemented in a bus topology. In this figure AppleTalk is assumed to be upper-layer protocol; however, this does not have to be the case. Consider Fig. 22.4.

Figure 22.4 shows a multivendor environment. However, the implicit may not be immediately perceivable. For example, the Macintoshes are using ETHERNET like other hosts on the network; but the Macintoshes can use either AppleTalk or the Macintoshes version of TCP/IP. Implicit here is that all network hosts at Information World, Inc. use TCP/IP for upper-layer protocols. Consequently, the power of internetworking comes into play. In fact, dual-protocol stacks can operate in this environment successfully. This implementation, as well as others in the EtherTalk environment, can use different types of media; for example, thicknet coax (coaxial cable), twisted pair, and thinnet coax are supported. AUI is also supported now.

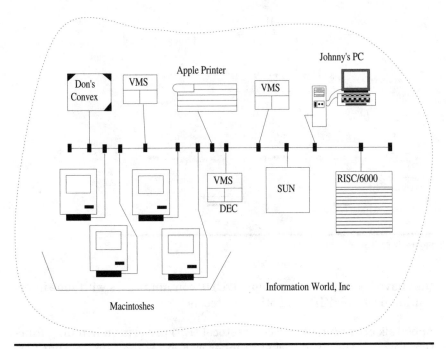

Figure 22.4   EtherTalk and a multivendor environment.

TokenTalk.   TokenTalk uses the IEEE 802.5 specification, which is the token-ring protocol. Figure 22.5 illustrates this type implementation.

This example shows Macintoshes connected to a MAU. Notice this is the same MAU described in the chapter on token ring (Chap. 13). This particular implementation uses either shielded twisted pair or unshielded twisted pair for media. Data rates supported by the MAU are 4 or 16 Mbits/s. This example implicitly conveys that AppleTalk is used as the upper-layer protocol.

Data-link- and upper-layer protocols

Datagram delivery protocol (DDP).   DDP provides node-to-node delivery of datagrams. The datagram includes a network, host, and socket address. DDP performs a number of functions. At startup, DDP acquires an AppleTalk node's address. DDP has an interface to route packets to protocols above and below it. The interface communicates with the socket listener and passes datagrams to the listener on receipt. It is also responsible for closing a socket. The interface performs a function on a dynamic socket when needed and opens sockets

MAU

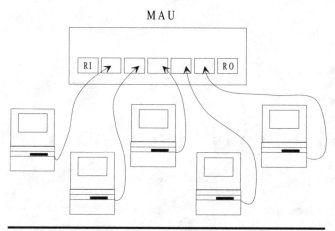

Figure 22.5   TokenTalk.

that have been statically assigned. DDP communicates with zone information protocol (ZIP) and other protocols.

AppleTalk echo protocol.   Echo protocol (EP) performs a listening function through the fixed socket number 4. Echo is the protocol that permits a node to send a packet to another node and, assuming that the destination node receives it, then it returns or *echoes* it back to the originating hosts. In a sense this could be compared to the function of PING in the TCP/IP protocols; granted, it is not a one-to-one correlation, but the functions are similar.

Routing table maintenance protocol.   Routers are used with AppleTalk basically the same way they are used with other networking protocols. Periodically, routers exchange information in their tables. In short, each router has a table of nodes and network addresses. Figure 22.6 is an example of multiple networks connected together via routers using the routing table maintenance protocol.

As Fig. 22.6 shows, routers exchange information to the others connected through the virtual network. The result is that all participants in the AppleTalk network can communicate with one another. For example, a user on Ceil's network can communicate with a user on George's network. All users can communicate with users on Bob's network, for more than one reason.

AppleTalk data stream protocol (ADSP).   ADSP resides on top of DDP and provides connection-oriented service and support for full-duplex data streams between entities on one network or on multiple networks. It is also considered reliable in that it does perform a certain degree of error

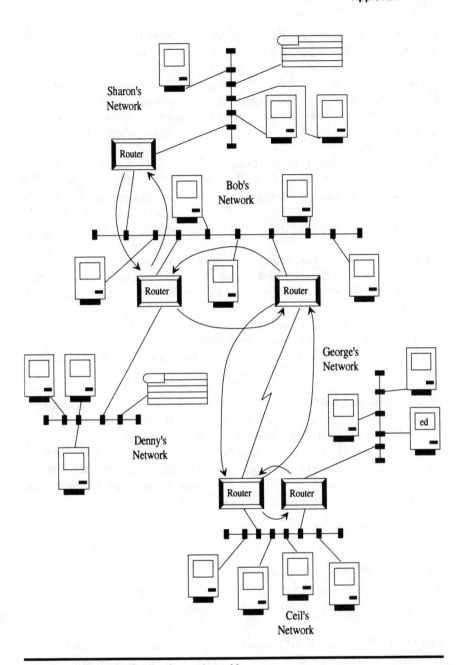

Sharon's
Network

Bob's
Network

George's
Network

Denny's
Network

ed

Ceil's
Network

Figure 22.6 Conceptual view of a routing table.

checking. It also provides flow control. In a sense this is parallel to the functions that TCP provides in a TCP/IP network.

Name-binding protocol (NBP). NBP's primary purpose is to translate aliases into network numbers (addresses). Part of the NBP is the name-binding services. Four services are performed: registration, deletion, confirmation, and lookup. Positionally, it resides on top of DDP.

AppleTalk transaction protocol (ATP). ATP is a protocol that defines communicating socket clients and a socket *server*. The purpose for the protocol, beyond that of having a protocol for socket communication, is that intercommunication is sometimes required to execute a task. In order to do this, a mechanism must be available.

Zone information protocol (ZIP). ZIP maintains tables of routing information between networks and zones. A *zone* is defined as an identifiable set of nodes within an internet. Figure 22.7 is a conceptual example of zones.

In Fig. 22.7 a virtual network is implemented in the Dishman & Perry Corporation. This is achieved by connecting network 1 and network 2 via a router. It is additionally segmented by *zones*. This is advantageous in Apple networks.

AppleTalk session protocol (ASP). ASP operates as a client to AppleTalk transaction protocol. Figure 22.1 shows that it (ASP) resides on top of ATP. ASP aids in session establishment and maintenance between *clients* and *servers*. A *session* is that logical connection between two entities through which data, commands, and so on are passed. ASP is not responsible for execution of commands but is responsible for command delivery. Some highlights of ASP functions are

- Session establishment
- Session termination
- Passing data and commands through the session
- In general, passing information between client and server
- Performing other session-specific functions as well

AppleTalk filing protocol (AFP). AFP is used to manipulate files in the Apple file system. AFP uses calls for intersystem communication, parameters that augment the calls, and the file structure itself. This protocol is responsible for reading, writing, and file searching. This protocol integrates with other components such as the *AppleTalk filing interface* (AFI), which is used to pass calls and sends to a file server.

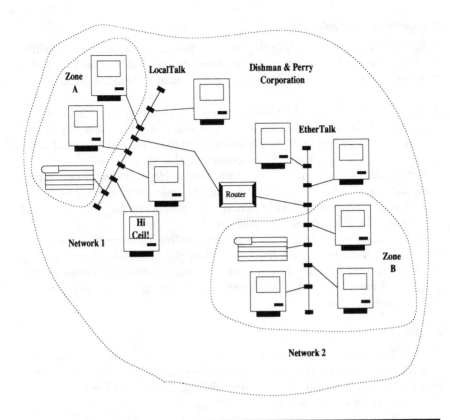

Figure 22.7   Conceptual view of zones.

Perspective

Most protocols just discussed have multiple components. These components interact with one another and with other component protocols defined in Fig. 22.1. Some of these interactions, along with the associated protocols, are discussed in the remainder of this chapter.

22.2   Zone Information Protocol (ZIP)

This protocol is implemented primarily by routers. Some nonrouting devices actually use a portion of ZIP on initialization to acquire information pertaining to their area. ZIP is used in the routing process. AppleTalk routing is achieved through intelligent devices. Interestingly, Apple documentation refers to them as internet routers. This can be somewhat confusing to those who are new to AppleTalk, TCP/IP, and the Internet.

Zone information table (ZIT)

A ZIT maintains network and zone information in a table. It literally maps network addresses to zones and vice versa. Before examining the role this table plays, consider some background information first.

Zone. A *zone* is a logical grouping of devices in an AppleTalk network. This concept is similar to dividing an internet into subnets. By creating a zone, users can be logically associated even if their physical locations are disparate. Figure 22.8 illustrates this idea.

Figure 22.8 shows four physical networks. However, only two zones have been defined. This zoning segments users into *work groups* in such a way that some level of operational control can be achieved. This is an example of logical segmentation.

A *zone name* is used to identify an area defined as a zone to a router. Generally zone naming conventions reflect departmental functionality or other simple means of remembering the association between the purpose of zone users. A *zone list* reflects the names of multiple zones which are available.

ZIP packets

ZIP utilizes two types of packets: query and reply packets. Routers functioning with ZIP use a *zone information socket* (ZIS), which is a statically defined socket in every router. ZIS uses the number 6, and it is this number to which routers respond.

ZIP queries. ZIP queries are simply requests. In plain English, this is a request from a node. The request contains a network number (or numbers if more than one exists). The node's request directed to a router is for the router to respond with a corresponding list of zone names that correlate with the network number(s). Figure 22.9 is an example of a ZIP query packet.

Figure 22.9 shows the ZIP query packet network count. This differs from that of the ZIP response packet.

The ZIP response packet appears as depicted in Fig. 22.10.

One function of a ZIP response is to indicate the number of zones to the requester. The response is actually a listing of network numbers and zone names, but most expressions of this concept typically word it differently; this is what it does.

A number of ZIP packets are used in network operations; however, the focus here is not on providing an exhaustive listing of ZIP packets. Exhaustive listings are best obtained from Apple Computer documentation.

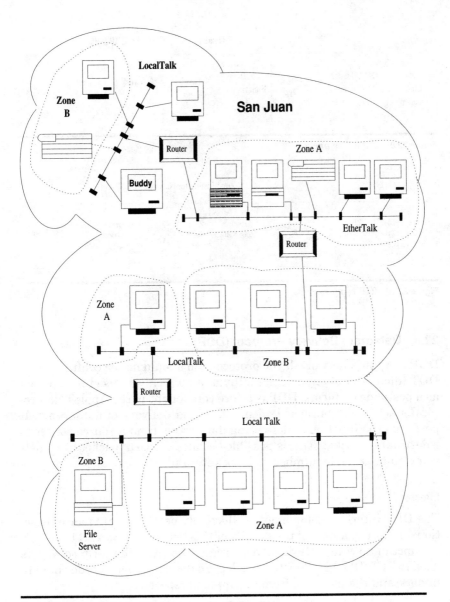

Figure 22.8 Conceptual view of zones.

				ZIP Header		ZIP Data	
Data Link Header	DDP Header Destination Socket #6	DDP Type 6	ZIP Function 1	Network Count	Network 1	Network 2	— —

Figure 22.9 ZIP query packet.

					Zone Listing for Network #1					
Service Socket #6	DDP Type #6	ZIP Function 2 or 8	Network Count	Network 1	Length of Zone Name #1	Zone Name #1	Network 1	Length of Zone Name #2	Zone Name #2	Network 2

Figure 22.10 ZIP reply packet.

22.3 Datagram Delivery Protocol (DDP)

DDP is Apple's network-layer protocol and should not be confused with DDT (dichlorodiphenyltrichloroethylane), which was used in the past as a pesticide on farms. DDP is a core component in the AppleTalk protocol stack. It is concerned with end-to-end delivery of datagrams. It performs the function of delivering datagrams from a source socket to a destination socket. This is possible because the DDP packet has a network number, node number, and socket number.

Interface

The DDP interface components collectively perform a number of functions. Before examining the details of this, consider Fig. 22.11.

Functionality in DDP involves sockets. Because the term *sockets* is used in TCP/IP and UNIX terminology, a description is in order here to understand the meaning from an Apple perspective.

DDP-related terms

listener   Specifically, the term used is *socket listener*. This is simply software that is similar to ports in TCP/IP network protocols. The function is to *listen* for calls.

socket   In Apple lingo this refers to an addressable endpoint in a software process. It could be considered as the function and meaning of a port number in the TCP/IP network protocols. But, according to Apple Computer technical

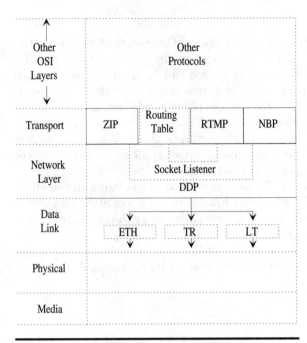

Figure 22.11   Conceptual view of the DDP interface.

personnel, two groups of sockets exist; the statically designated socket and dynamic socket identification.

socket client   Apple uses this term to refer to software in a network node. Simply, it is software. Clarification of the term *network node* is required here because APPN networking protocols use this term. In APPN the network node performs specific functions which are not parallel to the Apple meaning of this term. A network node in Apple is merely a node on the network. The importance of understanding the different connotations of terms cannot be overestimated.

socket number   This is a number that identifies a socket. Again, this concept is parallel to that of a port number in TCP/IP, an example of which could be the TELNET server port number, which *listens* for calls made against it. A maximum of 254 numbers can be used in a given node.

socket table   This is a descriptor that correlates to an active socket listener in a node.

Interface in brief.   The DDP interface interoperates with ZIP, RTMP, and NBP. The interface performs socket-based functions, either opening or closing a socket or sending a datagram. The functional operation of the interface includes four operations: send a datagram, close socket, and open a socket—either dynamically or statically assigned.

DDP packet

The DDP protocol has two types of packets. One is considered short, the other long. Precisely, this refers to the header length. Operationally, the short packet is used when the source and destination sockets have the same network identifier. The converse is true with the long header. Consider Fig. 22.12, which depicts the short header packet.

The short header packet shown in Fig. 22.12 does not include the details of the long header, which contains the full internet socket address of the source and destination nodes, along with the socket number, network number, and node number. Consider Fig. 22.13.

Figure 22.14 contains additional information not found in the short header packet. Significant in this header is the hop count. New packets begin with a hop count of zero, and each router that the packet traverses increments this field by a value of one.

Apple documentation provides bit-level information about these packets, and should additional information be required, Apple documentation should be consulted.

Header

LLAP 1	Most Significant Bits	Data-gram Length	Destination Socket Number	Source Socket Number	DDP Type	Datagram Data

Figure 22.12 Short header packet.

Header

HOP Count	Datagram Length	DDP Checksum	Destination Network Number	Source Network Number	Destination Node ID	Source Node ID	Destination Socket Number	Source Socket Number	DDP Type	Datagram Data

Figure 22.13 Long header packet.

Data Link Header	DDP Header	Source Connection ID	Packet First Byte Sequence	Packet Next Receive Sequence	Packet Receive Window	ADSP Descriptor	ADSP Data

Figure 22.14 ADSP packet structure.

22.4   AppleTalk Data Stream Protocol (ADSP)

This protocol provides full-duplex data transfer support between two nodes. ADSP utilizes DDP to transfer packets from target to destination. This means that the ADSP packets are embedded into the DDP packets and passed from source to destination.

State of a connection

Connections are either open or closed. In an open state, data can be sent and received. When a connection is closed, data can be neither sent nor received. Connections are identified by their network number, node number, and socket number. After a connection is established, it receives a connection identifier.

The notion of a half-open connection exists. For example, if either end of a logical connection loses contact with its matching or partner end, then it is considered a half-open connection. The identification of a half-open connection is derived by a timer that is maintained during the logical connection. Once a certain amount of time passes (which is short), a connection is considered half open.

ADSP packet structure, sequencing information, and attention messages

ADSP packet structure.   ADSP packets include sequencing information along with other header information. Figure 22.14 is an example depicting a DADSP packet.

The ADSP packet structure includes a data-link header, datagram delivery protocol header, AppleTalk data stream packet header, and ASDP data.

Sequencing information.   The packet first-byte and packet next-byte receive sequence fields indicate the sequencing numbers of data packets. The receive window field indicates how much space the sender has to receive data. This information, along with the converse information (for send data) is maintained at each end of a connection. This is generally referred to as the *connection state descriptor.*

Manipulation of this information and sequencing is achieved via queues; both send and receive queues. The execution and exploitation of these resources result in flow control that is managed by ADSP variables which are proportional to data flow.

Attention messages.   *Attention messages* are another ADSP function. These messages provide a function for endpoints in each node to transfer messages to one another outside the normal flow of data between the nodes. Again, this is analogous to another concept in TCP/IP, specifically, with regard to file transfer protocol (FTP).

FTP uses two ports and TCP for a transport protocol. One port FTP uses is to move data from source to destination. The other port is used to pass commands from source to destination. For example, with FTP a file can be copied from one system to another, and without bringing down the logical connection. Consider this explanation as a parallel example.

Once a FTP connection is established between two nodes, a file can be copied in either direction. The key issue here is that this functionality must be performed on the node from which the FTP client was invoked. Assume that a FTP client was invoked from host and the target was host B. A user on host A who invoked the FTP client can copy a field to host B or copy a file from host A to host B. Now, a different scenario exists with the execution of a command. A user on host A can issue commands against host B and perform a variety of functions. In a similar way, endpoints using ADSP communicate outside the normal data stream.

The packet structure for attention messages is different from the previous ADSP packet structures. Consider Fig. 22.15.

This message packet operates differently from other ADSP packets. Specifically, attention messages utilize a packet-oriented number reflecting that the sequence is independent of any data stream packets that may be flowing.

Connection open-packet structure. Once a connection is established, it is considered open. This is significant because when endpoints sense the absence of certain parameters that flow with connection packets, the connection is assumed to be terminated.

Figure 22.16 is an example of the open-connection packet.

Data Link Header	DDP Header	Source Connection ID	Packet Attention Send Sequence	Packet Attention Receive Sequence	Packet Attention Receive Window	ADSP Descriptor	Attention Code	Attention Data

Figure 22.15   Attention message packet structure.

				ADSP Header					
DDP Header	Source Connection ID	Packet First Byte Sequence	Packet Next Receive Sequence	Packet Receive Window	ADSP Descriptor	ADSP Version	Destination Connection ID	Packet Attention Receive Sequence
							Open Connection Parameters	

Figure 22.16   Open-connection message structure.

The open connection packet is generally considered a control packet in a loose sense because it performs an acknowledgment function for the open-connection request control packet.

Apple documentation details the fields in this packet to a bit level. For further information, that documentation should be consulted.

22.5   LAP Manager Topics

The Apple LAP manager performs utility-oriented functions between data-link protocols such as EtherTalk, TokenTalk, and LocalTalk and datagram delivery protocol (DDP).

The transition queue

The *transition queue* manipulates associations between drivers and applications. The function of the LAP manager is to notify the transition queue when certain events occur. Some examples of those events are opening the .MPP driver, closing the .MPP driver, and other events that may merit a position in the queue.

LAP manager protocol

This protocol operates with some of the IEEE definitions such as the 802.1, 802.2, 802.3, and 802.5, for example. Operation of the LAP manager includes handling 802.X packets to the 802.X protocol handlers. These handlers, in turn, perform functions such as attach and detach against specific 802.X protocols.

Focus

The LAP manager performs low-level functions at the data-link layer. Routines that are called are based on assembly language. Apple documentation provides details concerning these topics if they are of interest.

22.6   Apple Network Services

Network services is a nebulous phrase referring to different functions. Generally, it refers to functions that most users perform regularly.

Print

The print services concept is basically the same for all network protocols. Figure 22.17 depicts this concept.

Figure 22.17 shows multiple hosts on a network, a dedicated print server, and a printer. Hosts A and B are sending print files to the printer. In this scenario, these files are intercepted by the print server

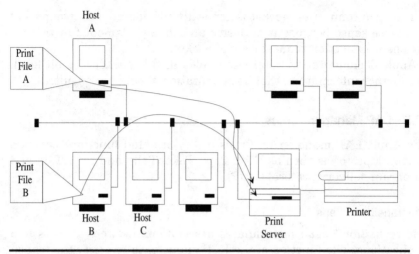

Figure 22.17   AppleTalk print server.

(which is a dedicated host serving as a waiting room for print jobs). The print server queues the jobs and sends them to the printer accordingly.

File

File services are similar to print services in that a host is dedicated to maintaining files that multiple users may need to access. Consider Fig. 22.18.

Figure 22.18 shows a file and print server and multiple individuals working with different files. For example, Ceil is updating the accounts receivable files, Rolando is working with personnel files, and Wanda is busy with organizing checks for payroll distribution.

The power behind such a configuration is that one centralized station can run licensed programs that can offer word processing, database, spreadsheet, and other services. Additionally, having a centralized storage system means that network users can share files with those users who have security authorization. Another advantage with this arrangement is the capability for centralized backup procedures; thus, maintaining corporate records in an off-site environment is much easier to accomplish. The other solution is for each user to have a copy of the programs and perform backups on each workstation.

E-mail

Electronic mail in AppleTalk is similar to E-mail in other network environments. In many ways E-mail operation is contingent on the soft-

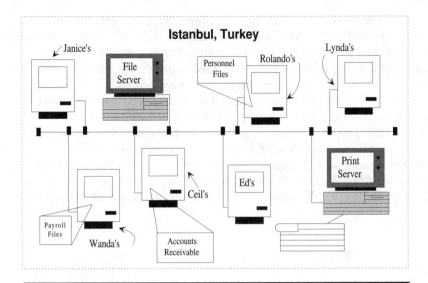

Figure 22.18 Workstation users and the file server.

ware package used. Since a LocalTalk environment supports multiple upper-layer protocols, the functionality of E-mail is dependent on the network protocol implementation.

22.7 Summary

Apple Computer is a success story, *American* style. It truly was a pioneer in the PC industry. The foresightedness of Apple Computer is manifested not only in different ventures in computing but also in networking.

AppleTalk and LocalTalk are different protocols. The former is an upper-layer protocol, and the latter is a lower-layer protocol. AppleTalk does not correlate one-to-one with the OSI model; however, Fig. 22.1 shows the AppleTalk protocol stack. The data-link protocols supported by Apple include token ring, ETHERNET, and what Apple calls LocalTalk.

At the heart of the protocols is the datagram delivery protocol, which operates at the equivalent of the network layer of the OSI model. Other protocols operate above it. Some of those include the routing table maintenance protocol, AppleTalk data stream protocol, name-binding protocol, and AppleTalk transition protocol.

AppleTalk networks can be logically separated by means of zoning. This is a logical partitioning of the network which facilitates management-related functions.

AppleTalk uses the term *socket* differently from other networks, namely, TCP/IP. However, a detailed explanation of the meaning of this term was provided in the section on datagram delivery protocol.

Some of the depth in AppleTalk protocol was provided by the various packets, the details of which were explained. Apple Computer has exhaustive documentation on much of its network protocols; however, some aspects of Apple Computers are considered confidential.

AppleTalk provides network services in the same way as do other protocols examined previously. For example, print, file, and E-mail services are available in an AppleTalk network. Considerably more information on AppleTalk is available, and for those who need additional information, the Apple computer documentation should be consulted.

Network Devices

23

Network Devices

A Practical Perspective

As recently as the late 1970s or early 1980s some of these devices presented here did not exist; or if they did, it was in isolated technological implementations. In fact, the decade of the 1980s may well go down in history as the decade of design and development of internetworking devices. One would only need glance at some of the trade journals and weeklies that began dominating the industry to realize that something was going on; the question was what?

So many specialized devices were designed and brought to market during the 1980s that a nomenclature problem prevailed, and still does to some degree. This part of the book explores and explains the major networking devices that exist today. But before this exploration begins, consider an example.

In the 1970s the term *gateway* was used to mean what the term *routers* means today. Unfortunately, depending on with whom you speak (or what you read), the term *gateway* may have a deceptive or elusive connotation. Nevertheless, separate chapters are devoted to routers and gateways in this part of the book, so it should be possible to discern the two meanings.

23.1 Device Analysis

This chapter presents basic definitions of the devices discussed during the remainder of this part of the book. As in many other chapters in the book, the OSI model is a good reference point for discussion of the operations of these devices. Consider Fig. 23.1.

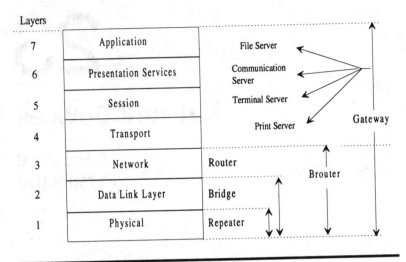

Layers

Figure 23.1   Network devices correlated with OSI layer functions.

Figure 23.1 shows multiple network devices compared to layers of the OSI model. Granted that the OSI model may not fit all protocols needed to perform an analysis, but it does provide a good baseline for showing where certain network devices operate.

Repeaters

Repeaters are devices that operate at the physical layer relative to the OSI model. They are generally considered the simplest network devices. They perform one basic function: repeating a signal. The signal may be either electrical or optical or both.

The basic purpose of a repeater is to extend the distance of something. It may be a network, devices connected together, or even processors—however, the word *repeater* is seldom used to refer to the latter implementation.

Bridges

Bridges operate at layers 1 and 2 relative to the OSI model. A bridge may be described in numerous ways. For example, some bridges operate with *like* lower-layer protocols, while others can work with *unlike* lower-layer protocols. They may also be described by their operations. Some bridges can operate only in a single physical environment, while others can operate in remote environments over switched or leased lines. Bridges can also be characterized as to how they perform their bridge functions. Some may perform source routing, while others may be capable of what is called *learning*.

Routers

Almost no definition is needed here, but for the sake of clarity one is given. Routers route! This may be the most profound statement in the book, but it is true. Routers *can* be characterized as to *what* and *how* they route, however.

First, it is best to note that routers operate at layer 3 relative to the OSI model. This is significant when multiprotocol routers are explained. Routers route data from one location to another. This may be another floor in a high-rise office building, or it may be halfway around the world.

One way of examining routers is by asking which upper-layer protocol it supports. This is what routers do: route upper-layer network protocols. On the other hand, multiprotocol routers *route* multiple upper-layer protocols. This does not mean that they perform any protocol conversion; it merely means they route multiple protocols to the target location.

Brouters

The *brouter* is a hybrid device. It can perform bridge and routing functions. Often the functionality of a brouter is vendor-specific. Depending on the device, some can perform bridging and routing functions at the same time.

Servers

Many types of specialized servers are available today. Interestingly, I have had numerous conversations with individuals who referred to a router as a server. I suppose they have a point in the broadest sense of the word; that is, routers do *serve*—the question is what.

Servers today come in many varieties. For example, there are file, print, terminal, and communication servers, to name only a few. And to be honest, there are probably others on the market that I am not aware of and which fit into a niche outside those described here.

Generally, *servers* are devices that concentrate on performing a single function. For example, the file server is probably the easiest to understand at first glance. Most file servers perform the function of maintaining files and making them available for requesting parties on an as-needed basis. The importance of this principle is that it removes the redundancy of all participating individuals having sizable hard disks. It also serves another function: making files accessible to others who may need access to them if authority is granted.

Gateways

Gateways are the most complex devices in the network device category. *Gateways* perform protocol translation and operate at layer 3 and above relative to the OSI model. Gateways can operate at all layers in a network, meaning that they perform protocol conversion at all seven layers.

The basic purpose of gateways is to interconnect heterogeneous networks. Gateways are required to do this or the functionality thereof. Networks that are not architecturally the same, or architecturally compatible, require protocol conversion at all or some layers. The only question remaining is: "Where is this going to be performed?"

23.2 Protocol-Specific Devices

Some vendors who have proprietary standards require a specific device from themselves or a licensed manufacturer's copy. For example, consider IBM's ESCON lower-layer protocol. It is fiber-based and uses highly specialized equipment. IBM sells repeaters for ESCON, and other reputable companies to whom IBM has agreed to license the technology to create such devices do the same.

The same is true with devices which operate with upper-layer protocols. For example, some gateways can be implemented to operate at OSI-equivalent layers 3 and above in a network. If so, this is done via software. Again, depending on the vendor, contingencies may exist. One thing is for sure, at the time of this writing, and that is that many network devices *are not* plug-n-play.

23.3 Summary

Various types of network devices are available on the market. Most are specialized and perform limited functions. However, some can perform multiple functions, as was explained with the brouter. The names of network devices do connote a specific function, and most credible vendors and others in the industry use them this way; unfortunately, a few do not.

24

Repeaters

Repeaters are the simplest networking devices. Their primary purpose is to regenerate a signal received from input and correct it to its original state for output. In short, repeaters provide signal amplification and also the retiming required to connect the segments.

24.1 Implementations

LANs utilizing ETHERNET as a lower-layer protocol can use a variety of cabling types, including copper-stranded twisted-pair cabling, thicknet coaxial cable, thinnet coaxial cable, and even fiber-optic cabling. By definition thicknet coaxial cable is 75 Ω and generally RG-50, RG-59, or RG-59. Thinnet cable is also 75 Ω but typically RG-6, RG6a, or RG-225. Many ETHERNET implementations use what is considered thicknet coaxial cable for a network backbone.

24.2 Types of Repeaters

Single-port repeater

A single-port repeater operates with actually two segments: (1) the segment it takes a signal from to boost and pass to the next segment and (2) the segment that serves as a multiport repeater. Figure 24.1 is an example of a physical plant with three departments: orders, parts, and shipping. Notice that three repeaters are used to connect the different ETHERNET-based LANs in each department.

In Fig. 24.1 the configuration is such that all hosts in any department can communicate with one another because logically one network, rather than three physical networks, exists. This repeater implementation is straightforward, with one cable segment connected to

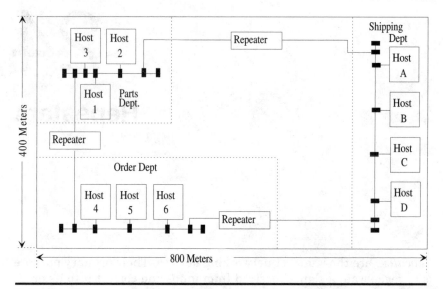

Figure 24.1   Conceptual view of a physical plant layout.

another cable segment. However, there is yet a different type of repeater.

Multiport repeater

A multiport repeater is as its name implies; it has one input and multiple outputs. Consider Fig. 24.2.

Figure 24.2 is a good example of how a multiport repeater can be implemented utilizing two different methods of ETHERNET technology. This is especially beneficial using multiple point-to-point connections. Figure 24.2 shows a 10Base5 (10 Mbits/s, using baseband signaling, with a maximum contiguous length of 500 m) ETHERNET segment connected by a media access unit (MAU). In turn, multiple connections to this repeater are possible via 10Base2 (10 Mbits/s) using baseband signaling with a maximum segment length of 200 m.

Smart repeater

Smart repeaters can perform packet filtering. In reality, this is a hybrid-type device and is very similar to a bridge in functionality. This device operates by capturing packets destined for another segment, but the repeater waits to transmit the packet(s) until the network is not congested. Ironically, this type of repeater can have an inverse effect on the connecting segments by creating a higher risk of collisions.

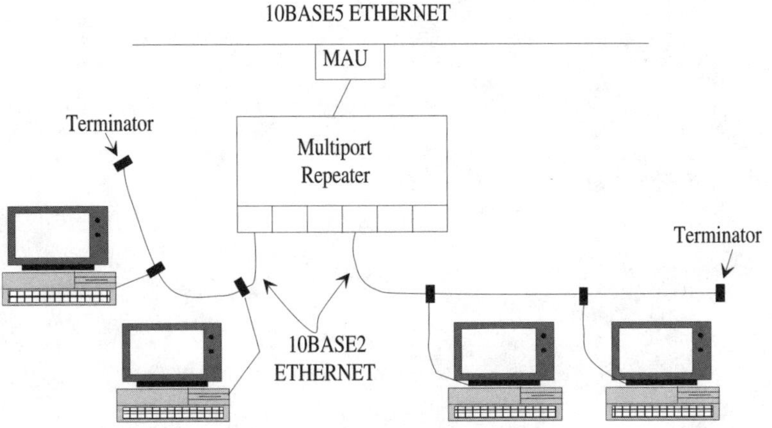

10BASE5 ETHERNET

MAU

Terminator

Multiport
Repeater

Terminator

10BASE2
ETHERNET

Figure 24.2 Conceptual view of a multiport repeater.

Optical repeater

Repeaters that *repeat* optical signals do the same as those which operate with electronic signals; they repeat the signal. IBM has a particular fiber-optic channel extender link, the model 3044. According to the system selection guide, it is used in environments to extend the distances of traditional nonfiber media to that which fiber media can obtain.

Other vendors have repeaters that repeat optic signals over long distances; however, in the traditional sense of copper-stranded cabling, fiber can operate without a repeater in many instances.

24.3 Summary

Repeaters are the simplest of all networking devices. They perform signal amplification and retiming for the segments connected. Repeaters can be used with electrical or optical signals; however, the former are more popular because of the nature of electrical signals.

Single- and multiport repeaters are available. Single-port repeaters are used to connect two segments of cabling. Multiport repeaters take a single segment and provide the ability for multiple segments to connect.

Smart repeaters are really hybrid devices and are not considered repeaters in the classical sense; rather, they fit into devices known as bridges.

25

Bridges

Bridges operate at layers 1 and 2 relative to the OSI model. They work with lower-layer protocols. Bridges are more complex than repeaters, if for no other reason than the fact that they function with two layers of protocols. This chapter focuses on what bridges do and how they do it.

25.1 Functionality within a Network

Bridges can serve multiple functions in a network environment. Some functions bridges perform are vendor-specific; because of the diversity in what a bridge can do, vendors differ in their offerings. However, companies that sell bridges typically offer products that perform the most basic functions. This may sound like a circular statement—contradictory or rhetorical—but there is hope! The point is that all bridges I have worked with have certain commonalities, but some are capable of performing specialized functions.

Conceptual view of operation

As mentioned previously, bridges operate at the physical and data-link layers. Figure 25.1 is an example of where this occurs.

Figure 25.1 is an example of two hosts connected by a bridge. Figure 25.2 is a flowchart of a bridge.

Figure 25.2 shows two physical interfaces, one for host A and the other for host B. There is only one data-link layer because this is where bridging is actually performed.

Functional advantages

Many real-world scenarios have multiple LANs throughout an entity—whether a corporation, government agency, or whatever. When this is

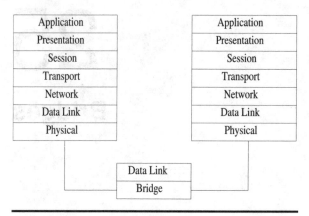

Figure 25.1   Conceptual view of a bridge.

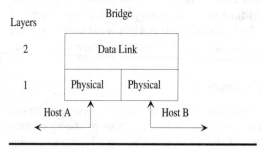

Figure 25.2   Detailed view of a bridge.

the case, it is not uncommon for multiple higher-level protocols to be implemented at higher levels in the networks. Since some upper-layer protocols have limited capability with other networks, a bridge can be used advantageously to connect multiple networks, provided they use the same lower-layer protocol. Another advantage of bridges is that they may be a better choice than a router in certain situations. Bridges are also relatively cheaper and easier to install. Other advantages of bridges include

- Many bridges can connect networks of different speeds.
- They are easily managed.
- Many can be adapted to an environment as it expands.

In all fairness, bridges do have some disadvantages. Common ones include

1. Bridges can be physically installed in such a way that in reality the net result is one logical network; consequently, sometimes troubleshooting can be difficult when problems arise.

2. In some implementations, as in a cascaded topology, a problem can occur with fast protocols because of delay factors.

3. Bridges are transparent to end systems. For this reason, and because of the potential delays encountered relative to the number of bridges, the actual limit could be indirectly imposed on the number of bridges utilized.

4. Bridges could potentially impede the use of some applications over the Internet. An example of this scenario could be multiple copies of an application operating and (unfortunately) using the same naming or addressing scheme.

Practically speaking, bridges are good network devices when used for the right need.

25.2 Theory of Operation

Bridges can be described in different ways. Three methods are discussed in this section. Bridges can perform forwarding, filtering, and learning functions. *Forwarding* is passing a frame toward its ultimate destination. *Filtering* operates by discarding frames before they reach their final destination. *Learning* is a function a bridge performs when it does not receive a positive response in return for comparing a frame to its host's table. The following figures explain these functions.

Forwarding

Forwarding is best explained by examining Fig. 25.3.

Two LANs are shown here: LANs 1 and 2. LAN 1 has three Dell PCs connected. They are known on the LAN as D1, D2, and D3, respectively. A bridge connects both LANs. Two Dell PCs—D4 and D5—are connected to LAN 2. Notice the highlighted table aside from the bridge used to "know" which hosts are located on which LAN. In the bridge's table host D5 is shown to be on LAN 2. The bridge intercepts the frame because it knows its target is on LAN 2. The packet is still broadcast on LAN 1 as well, but it is *spent* timewise in only fractions of a second; therefore, it does not remain on LAN 1.

Filtering

Figure 25.4 shows two LANs, a bridge connecting them, and IBM Personal System/2s (PS/2s) on LAN 1 and two Personal System/Value Point (PS/VP) systems on LAN 2.

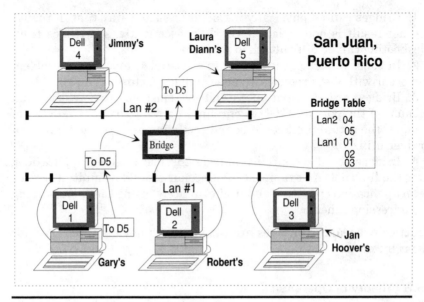

Figure 25.3   Conceptual view of forwarding frames.

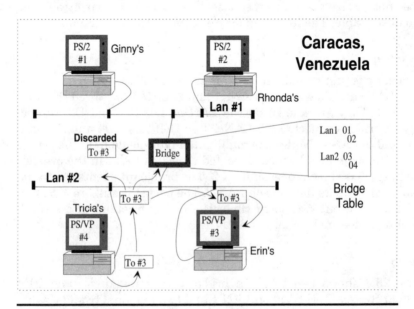

Figure 25.4   Conceptual view of filtering frames.

Figure 25.4 is an example of the filtering function of a bridge. Notice the frame leaving host 4 destined for host 3. The frame is captured by the bridge and compared against the bridge's table. After performing the compare function, the bridge discards the frame. Host 3 then receives the frame destined for it.

Learning

Bridges are said to "learn" about a host (which can be a computer or another device) when a frame is received by a bridge and the bridge does not have the address of that device in its table. If not, the bridge dynamically updates its table and then "knows" about this device; therefore, the bridge is considered to have learned of a device. Consider Fig. 25.5.

In Fig. 25.5 there are two LANs and a bridge connecting them together. Assume that host 3 has recently been added to the LAN. Now assume that host 3 wants to communicate with host 5 on the other LAN. The bridge knew the location of host 5 and now it knows the location of host 3 because it learned its address, both host and network address. Now, any of the hosts on LAN 2 can communicate with any hosts on LAN 1.

25.3 Bridges by Protocol

Bridges can also be characterized by protocol. This is logical because they work with lower-layer protocols. The simplest approach to understanding how bridges work with different protocols is to focus on how the bridge operates with similar and dissimilar protocols.

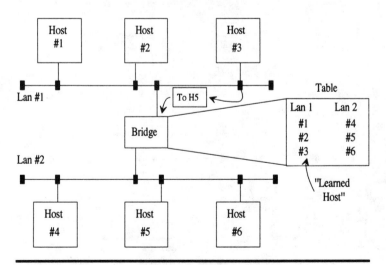

Figure 25.5 Conceptual view of bridge learning.

Like protocols

Different vendors support varying protocols with their bridges. Some of the most popular ones are ETHERNET and token ring.

ETHERNET. ETHERNET is a popular lower-layer protocol and is widely used throughout the marketplace. It is common for LANs to be created in departments, with the existence of multiple disparate LANs becoming apparent over time. Multiple LANs can be connected together by bridges, thereby creating one logical network while still permitting independence of departments. Consider Fig. 25.6.

Figure 25.6 shows three departments that perform distinctly different functions. But these departments need to share files for purposes of creating documents, informing prospective customers of products soon to be released, and general communication. Figure 25.6 depicts such a scenario.

Token ring. Token ring is another popular lower-layer protocol. As with ETHERNET, multiple upper-layer protocols can operate on top of token ring. Figure 25.7 is an example of how multiple token rings can be connected to form a single logical network.

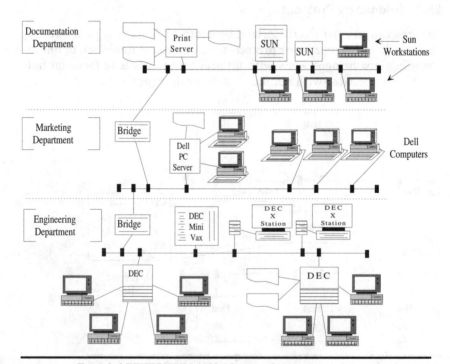

Figure 25.6 Extended ETHERNET LAN bridging.

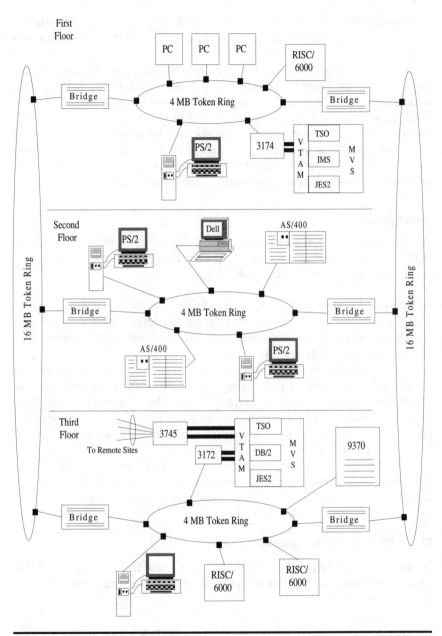

Figure 25.7 Token-ring bridging.

Notice in Fig. 25.7 that both 4- and 16-Mbit/s token-ring speeds are used. Many popular bridge vendors offer solutions that fit such a scenario.

Figure 25.7 shows three floors of a corporation. Each floor is a different department. Floor 1 is the collection department, floor 2 is the billing department, and floor 3 is the central data center for this corporation. Each floor has considerable flexibility. Token ring is considered a self-healing technology and therefore hosts can be inserted and removed from the network at will. Likewise, any given floor can be removed from the 16-Mbit/s corporate backbones. This scenario provides independence and flexibility and is characteristically very dynamic.

Unlike protocols

Bridges can also perform lower-layer protocol conversion. Many reputable vendors have such devices. In fact, these devices have become a commodity in a comparatively short time.

ETHERNET to token ring. An ETHERNET-to-token-ring bridge operates bidirectionally. Users on an ETHERNET network can communicate with users on a token-ring network and vice versa. Figure 25.8 depicts such an environment.

Figure 25.8 shows three ETHERNET-based LANs, a 16-Mbit token-ring backbone, a 4-Mbit token-ring LAN with multiple hosts, and two large SNA environments connected to another 16-Mbit token ring. Although some additional components such as software and possible configuration changes may be required, this configuration enables all hosts to interoperate.

Some bridges support protocols beyond those shown in Sec. 25.4. However, most bridge vendors support these protocols. These hypothetical implementations are examples of real installations.

25.4   Bridges by Geographic Location

Bridges can also be characterized with respect to their support for local and remote operations. Some examples of various bridge implementations are presented here.

Local bridging

Bridges are generally good devices to use in this type of environment. They permit segmentation and connectivity of LANs at the same time. Consider Fig. 25.9.

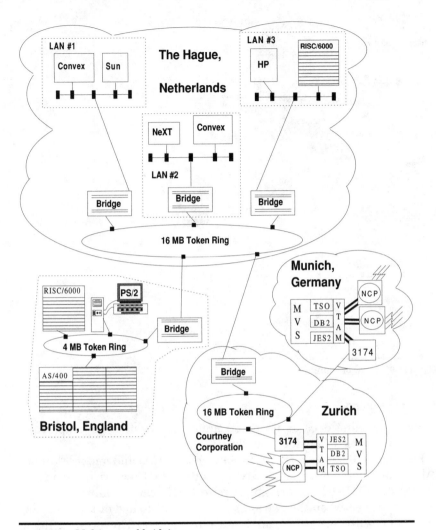

Figure 25.8 Multiprotocol bridging.

Figure 25.9 shows two networks connected by a bridge. A twofold benefit is realized.

First, both networks' hosts can communicate with hosts on the other network; thus enterprisewide connectivity is achieved. Second, isolation of network computing can be maintained because of the way bridges operate. Third, a degree of load balancing can be realized as a result of this scenario.

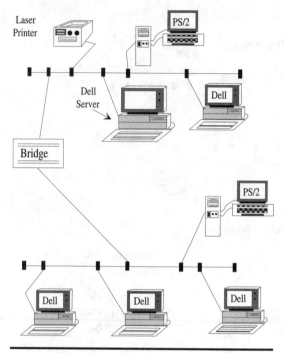

Figure 25.9   Local bridging.

Remote bridging

Remote bridging does as its name implies; it bridges geographically remote networks. Consider Fig. 25.10.

Figure 25.10 shows a site in San Francisco (CA) and one in Florence (MS). Both LANs must be able to communicate. Now, with the advent of remote ETHERNET bridging, these LANs can communicate. The connectivity between the two sites could be a switched or a leased line, generally with speeds of 56 kbits/s or higher. However, using a switched line entails bandwidth on demand, and there are some specific requirements for bridging in this type of environment.

The result is that users on both networks view all hosts as being located on one LAN. In a sense, with bridging between DonSing Enterprises and the Hawthorne & Clark Company (Fig. 25.10), a virtual LAN has been created.

Both local and remote bridging can be performed with token-ring networks. Bridges supporting the respective protocols are required. Depending on the vendor offering, other network protocols may be supported as well.

Some vendors offer redundant line support for remote bridges. Others offer data compression on the fly; depending on the vendor, some algorithms can achieve approximately a 4:1 ratio. This is effective utilization of bandwidth. Some vendors also offer network management support for their devices. These and other aspects of a bridge should be discussed with a vendor whose forte is bridges.

Author's comment

Remote bridging is similar to other aspects of internetworking technology. One question always remains: "Where is the bottleneck?" I am not implying that remote ETHERNET, token ring, or other protocol bridging will result in a bottleneck; however, I do intend to raise this question because in data communications a bottleneck always exists. The question is where and to what degree. It may or may not have anything to do with bridging, routing, or gateways, but it nevertheless is present.

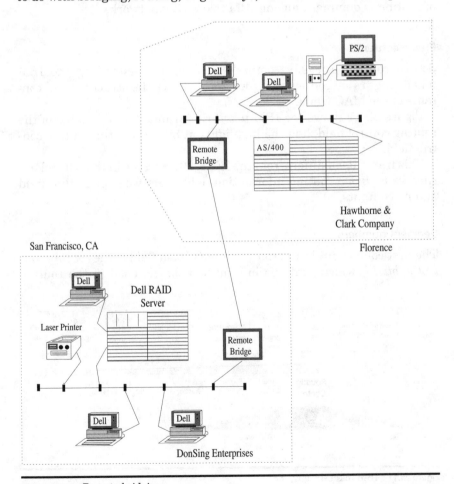

Figure 25.10   Remote bridging.

25.5   Source Routing and Transparent Bridges

How bridges obtain routing information is important. "Bridge routing" is somewhat of a misnomer, however; bridges do not route in the sense that a router does, but they do have to pass frames from their source toward their destination point, wherever that may be. The transparent bridge function was discussed briefly in Sec. 25.2, but additional details are provided here. First, source routing needs explanation.

Source routing

In *source routing,* which is an IBM function, the route to the destination is determined before the data leaves the originating point. This function is *source route bridging* (usually referred to as SRB). This type of routing is dominant among IBM token-ring networks.

Frame contents

Source routing is easily comprehended by understanding the IBM token-ring frame contents. Figure 25.11 shows the structure and contents of the MAC frame.

Figure 25.11 shows the IBM token-ring frame, the highlights of the routing control field, and the highlights at bit level of the routing control field.

This frame itself differs from the IEEE 802.5 frame in that the token-ring frame has a routing information field. Here we explore that field and its contents.

Segment numbers

The first component in the routing information field is the *routing control subfield,* which contains information that is used in the routing

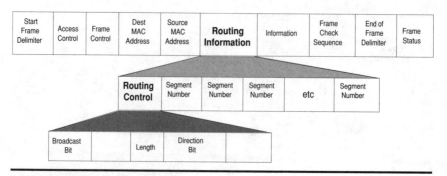

Figure 25.11   IBM token-ring frame.

function and is explained in greater detail shortly. Segment number subfields follow the routing control subfield.

Each segment number reflects two pieces of information: a ring number and a bridge number. Each ring is a LAN, and each LAN has a number associated with each ring. Each bridge used is assigned a number. The segment number is a composite of the ring and bridge numbers. If multiple rings are connected via bridges, then multiple segment subfields exist as shown in Fig. 25.11.

Routing control subfield

The routing control subfield has two significant pieces of information: the broadcast bit and the direction bit. The broadcast bit indicates the type frame: whether a broadcast or a nonbroadcast frame. The direction bit indicates which way the frame is going. It is en route either from original source to destination or vice versa. This is important because the setting of this bit dictates how the segment number bits are interpreted.

Transparent bridges

The working definition of *transparent bridges* is that they learn those hosts reachable according to data link by observing frames as they pass. A transparent bridge is sometimes referred to as a "spanning-tree bridge."

This type of bridge forwards frames, as discussed previously in this chapter. It also maintains and updates a table of MAC addresses of the hosts which are reachable across the link used to attach multiple rings. This type of bridge is also known as a "learning bridge."

A transparent bridge implementation was used in Fig. 25.9, but for the reader's convenience it is shown again in Fig. 25.12.

25.6 Source Routing Theory of Operation

Now that other aspects of source routing have been covered, explanation of IBM source routing operation is provided. Two types of frames can exist in an IBM token-ring network: nonbroadcast and broadcast frames.

Nonbroadcast frame

The term *nonbroadcast* is virtually the same as *multicast*; if there is a difference, it is minimal. What is important is how nonbroadcast frames are handled in a multiring environment. Consider Fig. 25.13.

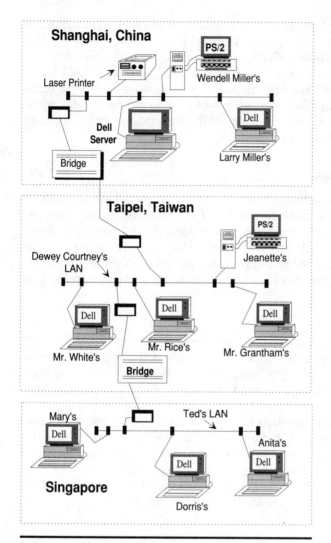

Figure 25.12 Conceptual view of a transparent bridge.

A nonbroadcast frame reveals the significance of the segment numbers discussed previously. Referring to Fig. 25.13, assume that host A on ring 1 is the source host and the destination host is host G on ring 2. Observe that bridge X connects rings 1 and 2 directly. In this case bridge X recognizes its bridge number and ring number. The bridge simply copies the frame from ring 1 onto ring 2. At the same time bridge Z receives the same frame. It examines its bridge number and its ring number; no match is made, so the frame is discarded.

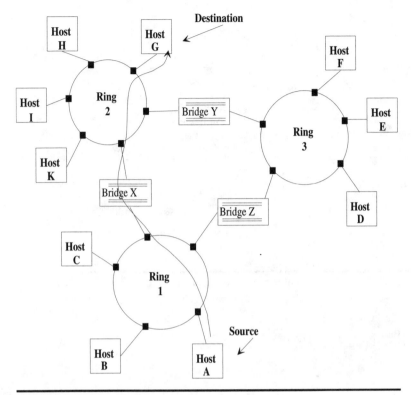

Figure 25.13 Nonbroadcast operation.

Broadcast frame

Source routing is implemented by two types of MAC frames via broadcast addresses, according to IBM: (1) a frame that contains a certain value that is received by all hosts located on the ring and (2) a frame containing the actual broadcast address. Frame type 1 contains a special hexadecimal (hex) value. Figure 25.14 illustrates this concept.

Figure 25.14 illustrates frame type 1 by the box labeled H6 as being received by all hosts on the ring where it originated. The H6 indicates that the frame has reached its destination. However, notice that two rings are connected by a bridge and that the H6 frame is not repeated onto ring 2.

The other type of MAC frame (type 2, defined above), which contains the broadcast address, has a hex value different from that of MAC frame type 1 (shown in Fig. 25.14). The broadcast frame (frame type 2) is sent to all hosts, all rings, and hence all bridges connecting them. Figure 25.15 shows this environment.

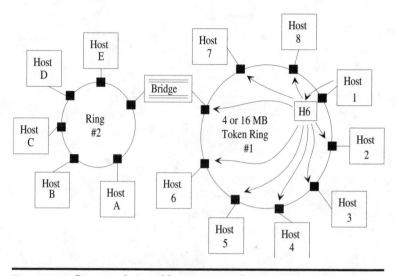

Figure 25.14 Conceptual view of frame contained on one ring.

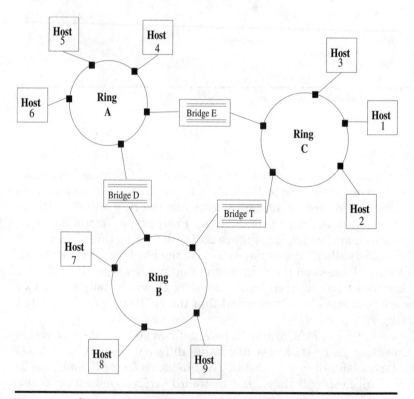

Figure 25.15 Conceptual view of frame contained on multiple rings.

Assume that host 8 on ring B sends a broadcast frame. It is received by all hosts, each ring, and consequently each bridge. Operationally it works like this: Host 8 sends the frame. Then bridge D copies the frame, and adds its bridge number and the number associated with ring C. At the same time the frame reaches bridge T. Bridge T copies the frame and adds its bridge number along with the number for ring A. The frame then passes to bridge E, which copies the frame and adds its bridge number and ring C's number. The process is complete, but a question remains: "Why do this?"

When many hosts exist, isolation of networks must be achieved, but total connectivity between LANs is required; Fig. 25.14 shows how to do this. This arrangement provides multiple paths to any given ring; and, the more rings, the more reason to have redundancies in paths.

Discovering routes

The process of discovering routes is twofold. First, assume that a source wants to communicate with a destination host. First, the source host will attempt to send a frame on the ring on which it (the source host) is located. If no positive response is encountered, then another process is used.

A source host sends a route discovery frame to a destination host whose address is known to the source; the question is how to get there. Assume that multiple rings are attached to the ring where the source host is located. Each bridge attaching these rings copies the frame from the source, inserts its segment information, and then passes the frame on. Once the frames have reached all connected rings and bridges, they begin returning to the source host. If the destination host is located, it inserts routing information into the original route discovery frame and sends it back to the source. Multiple frames may return to the source. When this happens, the source determines the route for additional frames to be sent to the destination host.

25.7 Summary

Bridges operate at the lower two layers relative to the OSI model. They can serve a variety of functions. For example, bridges can be used to merge multiple LANs into one virtual network. Bridges can also perform protocol conversion on LANs at the lower two layers and can convert ETHERNET to token ring and vice versa. Some vendors offer bridges that support FDDI bridging.

Transparent bridges perform three basic functions: forwarding, filtering, and learning. This simply means that, after receiving the frames, a transparent bridge *forwards* the frames to their destinations or *filters* out those frames which are not destined for another ring to which that bridge is attached. Transparent bridges also *learn* of hosts throughout the network(s) in which they operate and store this information in tables, which they in turn use for routing purposes.

Source routing is a function of another type of bridge. IBM uses source routing, and in the MAC frame, a field exists for insertion of routing information. This type of bridge operates by including the destination address into the frame along with the source address. Other functions performed by source routing bridges include different types of frame broadcasting.

Bridges can be used to effectively manage sites where multiple token-ring networks exist and there is also a need for redundancy. Because token-ring technology is considered self-healing, it is easy to remove and add hosts and devices to these networks.

Bridges also offer advantages to ETHERNET-based LANs, particularly where multiple ETHERNET LANs are geographically dispersed. Remote bridges can create a virtual LAN environment where users *think* they are all attached to the same network in the same physical location.

Implementing bridges remotely should be tempered and evaluated in light of the load in the respective sites, the bandwidth available through the link, and other operational considerations.

Many well-known vendors offer good products to achieve a desired result.

Chapter

26

Routers

Routers operate at layer 3 relative to the OSI model. Routers route upper-layer protocols. They do not perform protocol conversion, assuming that they are routers in the classical sense. Routers, like bridges, can be examined from different angles. This chapter explores those angles.

26.1 Routers by Another Name

An explanation is in order concerning the term *router*. This is especially true for those who may be new to networking devices.

The routing function can be defined as moving data from point A to point B, wherever that may be. This concept can be traced back to the 1970s, when networking began to flourish. Those working in the Internet community were involved with networks that were sometimes located in different cities. At any rate, multiple networks existed, and the prospect of connecting to networks and connecting networks together became increasingly attractive.

As an outgrowth of this interest, devices designed to perform this function began to appear on the market. At that time they were generally referred to as *gateways*. Yes, gateway. An obvious question is why. According to the *American Heritage Dictionary, gateway* is defined as: "1. An opening, as in a wall or fence, that may be closed by a gate. 2. A means of access." I believe the informal consensus at that time probably favored the latter definition provided here.

At that time, for a number of reasons, calling devices that performed routing functions *gateways* seemed appropriate. Unfortunately, confusion regarding this term (gateway vs. router) still exists today. Defining a gateway as a device that permitted connectivity of networks *was* reasonable. Those devices did in fact provide such a function. And, beyond this, they routed information to various networks and locations. With this in mind, coupled with the mind-set of the technical community during the 1970s, use of the term *gateway* is plausible. However, technology changes.

Today an entire industry exists around network devices. In fact, a part of this book (Part 4) is devoted to explaining some of these devices. I do not believe that in the 1970s most technically oriented individuals could envision the explosive technological growth of the 1980s with such specialized devices. It may have been conceivable, but not to the detailed level that hindsight provides today. Hence, the dilemma still exists.

The terms *router* and *gateway* are constantly used in various applications, and at best the meaning is skewed. For the record, routers route; period. They may perform other peripheral functions, but the focus of routers is simply to route data.

Gateways, on the other hand, perform protocol conversion between heterogeneous networks at (OSI-equivalent) layer 3 and above at a minimum and can perform protocol conversion at all seven layers between two networks. Gateways perform protocol conversion; routers do not. Interestingly, some gateways can perform routing functions because of their architectural nature, but they do not necessarily have to. In fact, this is a vendor-specific offering.

The purpose of this section is to point out to the reader that discussing internetworking with different people can result in confusion if clarification of terms is not agreed on. Many who have worked for yeas with TCP/IP-, UNIX-, and Internet-related issues still use the term *gateway* to convey the function of a router. I have often wondered what term they use when referring to the device used to integrate heterogeneous networks. This would seem to be a catch 22 situation.

Another comment about routers and terminology is valuable. There are different types of routers (the functions of which are explained in this chapter), and not all routers perform the same functions; many functions are vendor-specific. There are a number of prominent router vendors on the market today selling good equipment. The point is to understand your needs and obtain the appropriate fit to ensure that those needs are met.

26.2   A Perspective on Routers

Various types of routers perform various functions, as mentioned above. Some of these defined or identified router functions tend to overlap in some instances. Routers can be classified in terms of the geographic distance they support (i.e., local or remote), the upper-layer protocols supported, and their interface support.

Router operation is based on tables of potential networks and routes. These tables are utilized to indicate the path to a given network. Router tables do not locate device addresses as do some types of bridges. Functionally, routers exploit the information available to them to determine the most expedient route. Another unique aspect of routers is their ability to receive data addressed to them by hosts or other routers. Route determination is somewhat contingent on the upper-layer protocol. For example, TCP/IP uses routing algorithms that differ from those used for SNA-based networks. In this respect routers are protocol-specific. Some routers are simple and function with one protocol; however, other types of routers can manipulate multiple protocols—hence the term *multiprotocol routers*.

The remainder of this chapter explores various aspects of routers and the specifics of how they operate in certain environments. And similar to bridges, routers have become a basic commodity in internetworking technology.

26.3   Theory of Operation

Routers are protocol-specific. They operate at network layer 3 relative to the OSI model. Consider Fig. 26.1.

Figure 26.1 is an example of two hosts represented by OSI layers, with a router operating between them.

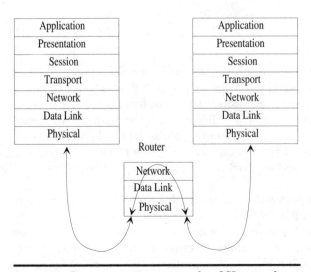

Figure 26.1   Router operation compared to OSI protocol layers.

As Fig. 26.1 shows, the physical and data-link layers are also part of the router, and therefore these aspects must be taken into consideration.

Some large routers (in terms of what they support) offer a wider variety of physical-layer interface support. More than a few reputable router vendors offer fine products that support such interfaces as the following:

- RS-232
- V.35
- AUI
- X.21
- HISSI
- RS-449
- Serial

Lower-layer protocols must be supported as well. These, too, vary according to the vendor and the router. Some examples of popular protocols supported are

- ETHERNET
- Token ring
- SDLC
- HDLC
- FDDI
- X.25

The network layer is where routing actually occurs. If you have read other chapters in this book, particularly those in Part 3, you are aware that upper-layer network protocols operate differently. For example, the way TCP/IP performs routing differs from the way SNA or, say, APPN performs routing. Consequently, these upper-layer protocols drive the router decision to a certain degree. This is so because multiprotocol routers exist and support routing of multiple upper-layer protocols.

When routing is implemented, it usually brings together two or more networks. The consequence of this from the user standpoint is that the routing is transparent except for possible time-zone inconveniences, but those issues are easily overcome. In essence, the use of routers provides an end-to-end solution. The remainder of the chapter focuses on some examples of different upper-layer protocol routing and also the aspects of routing in general.

26.4   Reasons for Routing

Routers can be implemented locally or used in an environment where multiple networks exist in different geographic locations and need the ability to exchange data.

Local implementation

Figure 26.2 depicts a scenario in which routers are used within the same physical facility.

Figure 26.2 shows a single physical site with three distinct ETHER-NET LANs. Notice that a router is the common denominator of these LANs. The result of this configuration is that any of the LANs can communicate directly with any of the hosts shown via the internet protocol. Notice that each LAN host has TCP/IP as its upper-layer protocol. By removing the router, connectivity between all three LANs would not be achieved, or other means would have to be implemented to achieve the same results.

This particular implementation is straightforward, flexible, and relatively inexpensive. Additionally, the router fits into the management method that is common among TCP/IP networks.

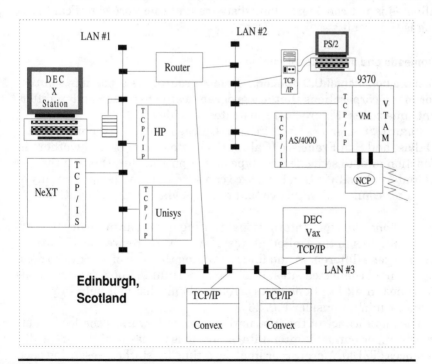

Figure 26.2   Local router implementation.

There are many possible variations of this example because of the flexibility of routers. In a sense they can be customized to meet site-specific needs fairly easily. This is so primarily because routers maintain routing tables within them that can be customized to a diversity of situations.

Metropolitan implementation

The notion of a metropolitan implementation may not be popular, but it is frequently a solution that meets many needs. Consider Fig. 26.3.

Figure 26.3 shows the Walton Insurance Agency headquarters in Dallas (Tex.) where master files, statements, accounts receivables, and other operational functions are performed. However, because the agency prospered, it has offices in Fort Worth, Arlington, and Kaufman (also in Tex.) Each of these three satellite offices is within 35 mi of the headquarters in Dallas.

Because data needs to be sent to and from the satellite office and headquarters this router solution meets the needs. In fact, the router in the Dallas office can route data from Kaufman to the Fort Worth office, if configured appropriately.

Like the example of a local implementation of a router this metropolitan implementation of a router is popular because the link between all sites is not considered long distance and therefore significant saving is the result.

Domestic and international routing

Remote continental U.S. routing.   Remote routing is a popular solution for many corporations that are geographically dispersed and in different time zones. Figure 26.4 illustrates this idea.

Figure 26.4 shows facilities in Chicago (Ill.), Memphis (Tenn.), Dallas, and San Francisco (Calif.). Each site differs from the others in function and, to some degree, type of equipment. But, through the use of routers the site in San Francisco can send data to Memphis or any of the sites connected in this virtual network via the router solution.

International routing.   International routing is similar in theory to routing multisites in the United States. However, there are considerations of time-zone differences and understanding the type of work to be done between all connected sites. It is also important to know the peak traffic times in all locations and correlate them; this will help in performance tuning. Consider Fig. 26.5.

The significance of the international scenario versus the local, metropolitan, or national (interstate) scenario is the issue of time-zone differences (TZDs). Merely coordinating a time for staff representing each site is difficult because of the TZD factor.

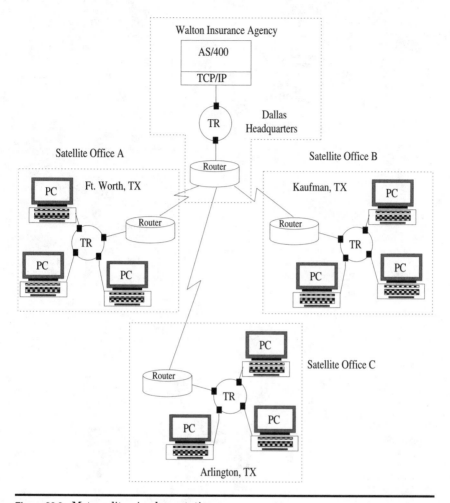

Figure 26.3 Metropolitan implementation.

Another reason for routing is commercial. If a company performs high-volume processing, the best place to do this is where it is most cost-effective while achieving the original goals. With distributed processing and exploitation of network devices such as routers, this can be accomplished.

26.5 Types of Routing

Various methods are used in routing. Technically, routing schemes can be categorically defined. Network protocols of different vendors tend to

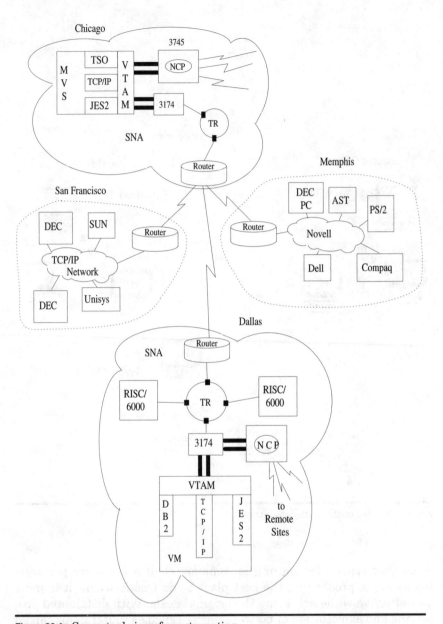

Figure 26.4 Conceptual view of remote routing.

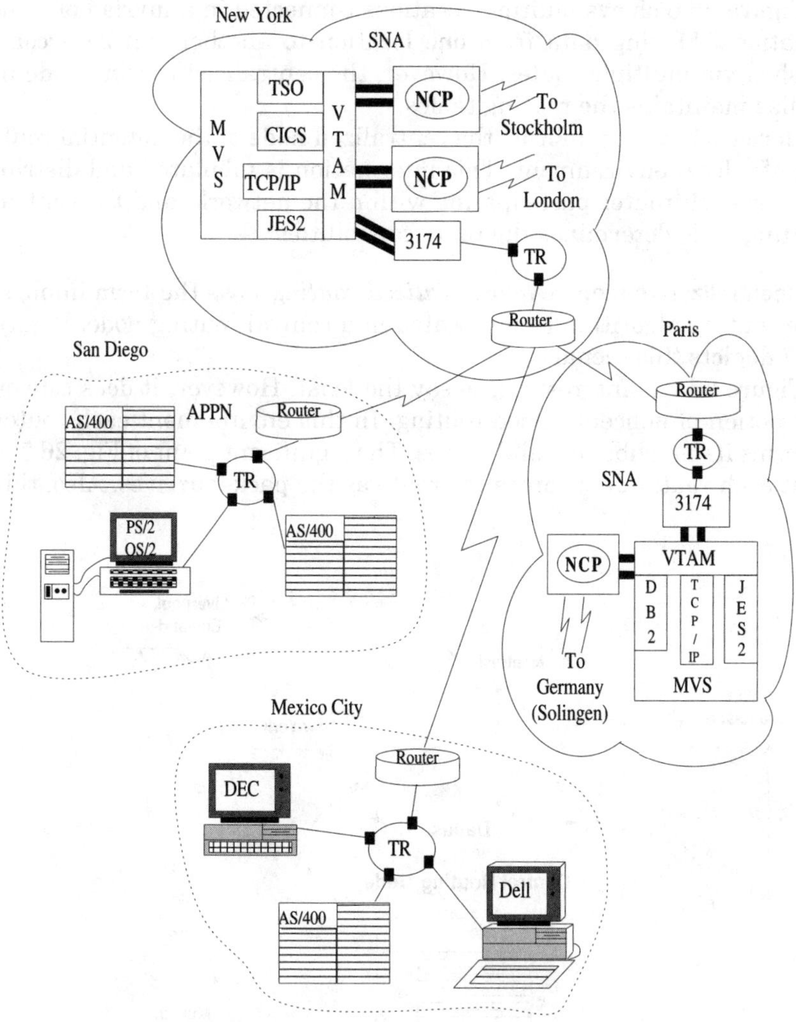

Figure 26.5 International routing.

route differently, but some common threads among the different protocols remain.

Centralized versus noncentralized routing

Centralized routing. In *centralized routing,* or routing from a central base, a central repository of routing information is maintained. Schematically, this appears as shown in Fig. 26.6.

Figure 26.6 shows multiple locations connected in a myriad of combinations. Moving data from one location to another can be accomplished via multiple routes. However, the centralized routing node in Dallas maintains the routing tables.

Here each router informs the centralized node of the potential routing of a local environment. This information is tabulated and distributed to each router participating within the network, and the central routing node determines the route capabilities.

Noncentralized routing. *Noncentralized routing* is as the term implies. The routing algorithm is not located in a central routing node. Figure 26.7 depicts this scenario.

Figure 26.7 is interesting, to say the least! However, it does convey the notion of noncentralized routing. In this environment each router informs its neighbor of valid routes. The significant point of Fig. 26.7 is that each router determines the route as the packet arrives. Also, this

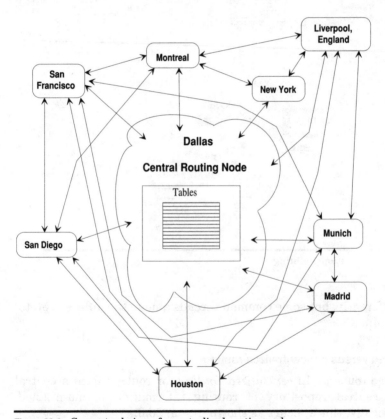

Figure 26.6 Conceptual view of a centralized routing node.

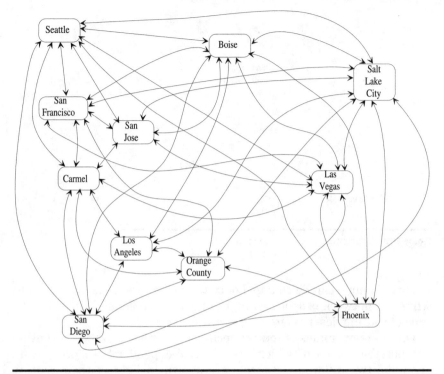

Figure 26.7   Conceptual view of noncentralized routing.

type of implementation is constantly changing as changes are made to individual routes. Each time a change is made, updates occur.

Static routing

In *static routing* tables and routes are created via network management functions. By definition, this means that routing must be performed when the network is nonoperational from a user perspective. The simple way to explain this is that no changes are performed while the network is operational. This means that the routes are therefore *fixed*. Consider Fig. 26.8.

Interpretation of the notion of static routing is easy when considering Fig. 26.8. As the lines drawn between the cities indicate, a routing path is available. Notice paths *not* available. It is not possible to route directly to Toronto (Can.) from Grand Rapids (Mich.) without going through Chicago, and this implies that a routing table for this configuration is predetermined. Also, routing from Detroit (Mich.) to Grand Rapids is not possible without going through Chicago.

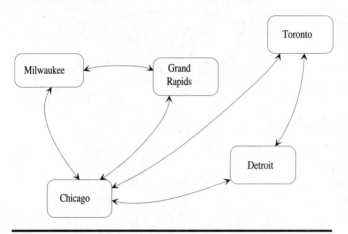

Figure 26.8 Conceptual view of static routing.

This example can be changed once the network is taken off line, but while the network is operational, no changes can be made. Hence, this type of routing deserves its name: static routing.

These are examples of routing technology. Other methods of routing are possible, some of which are protocol-dependent. To provide an example, two different protocols are shown in light of their routing schemes.

SNA routing

To a considerable degree, routing in SNA is performed on the front end (the communications controller). The network control program (NCP) inside has the routes defined and utilizes this software to exploit the hardware and gear involved in the routing process. Consider Fig. 26.9.

Figure 26.9 is an example of predefined routes. Each county or city—Boise (Idaho), Orange County (Calif.), Chicago, and Houston—has its hosts configured to route data through the indicated connections. For example, notice that no direct connection exists between Chicago and Orange County. Does this mean they cannot pass data? Not necessarily; because Boise and Houston are connected to both Chicago and Orange County, they could be configured to route data between those two destinations as well as the ones to which they are physically attached.

In this example, line speed, data compression, type of line (dedicated or switched), and other data communication issues are relatively unimportant because the point in this example is that routing between two

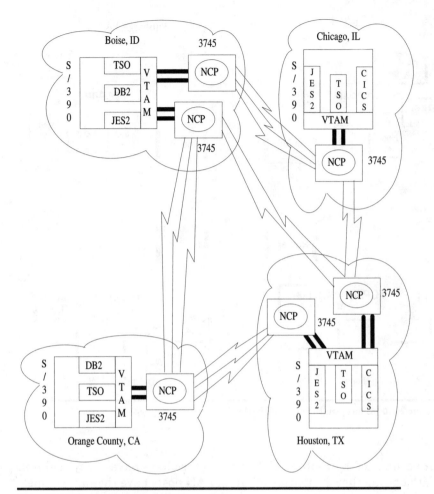

Figure 26.9 Conceptual view of SNA routing.

entities can occur even if these entities are not physically connected. This is a matter of software, and obviously some physical route must exist, but beyond this it is a matter of configuration.

TCP/IP routing

Routing in TCP/IP is performed by IP, as was explained in the chapter dedicated to TCP/IP protocol (Chap. 19). Figure 26.10 is an example of TCP/IP networks and routers connecting them.

Figure 26.10 shows TCP/IP networks in Columbus (Ohio), Pittsburgh (Pa.), Norfolk (Va.), and Nashville (Tenn.). These locations

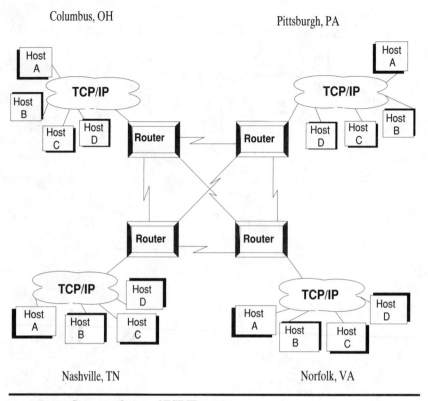

Figure 26.10 Conceptual view of TCP/IP routing.

are connected by routers. Notice that all routers have a physical link with one another. Notice also that all LAN hosts have the same names, and they could be given the same addresses. What differentiates them is the addressing scheme used; therefore, there are no naming or addressing conflicts.

In this example, routing is made possible by the Internet protocol (IP) software in conjunction with one of the routing protocols. The routing protocols could be router information protocol (RIP), open shortest path first (OSPF), or another protocol. Obviously, a routing device is required, but nevertheless it is the combination of IP and the routing protocol that actually effects this process.

Maintaining routing tables is a separate issue here. This could be done via the /etc/networks and /etc/hosts files or the domain name system (DNS). Most likely it will be the latter because of the power with which it operates.

These examples of how different network protocols perform routing suffice to show that in a heterogeneous networking environment routing is not a simple matter. Consider the routing issues involved in the Internet!

26.6 Bandwidth-on-Demand Routing

This topic has moved into the forefront of routing in the past few years. *Bandwidth-on-demand routing* is as its name implies; in plain English, that means the router uses the line when it needs it and does not keep it *dedicated*. Consider the following examples showing the three phases of this routing.

Figure 26.11 shows two networks; one in Monterey (Calif.) and the other in Birmingham (Ala.)

Figure 26.11 shows two multivendor networks with multiple hosts attached to each. It also indicates that no link is established between the two locations. This is important because it means *no* dedicated line exists, which means that some degree of money is involved—typically in the form of savings.

Figure 26.12, on the other hand, shows a different scenario.

Figure 26.12 shows the same environment as depicted in Fig. 26.11, but this time a link is established between the routers. Also, it indicates that host A in Monterey is communicating with host C in Birmingham.

Figure 26.13 shows how the environment would appear once communication is complete.

This example of a bandwidth-on-demand router may well fit the needs of many situations. More than one or two vendors have such

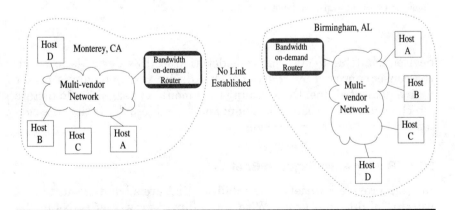

Figure 26.11 Bandwidth-on-demand routing.

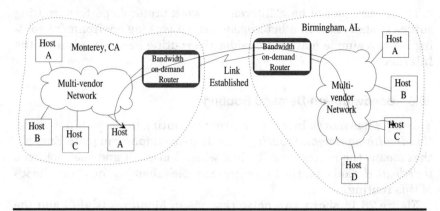

Figure 26.12   Bandwidth-on-demand routing.

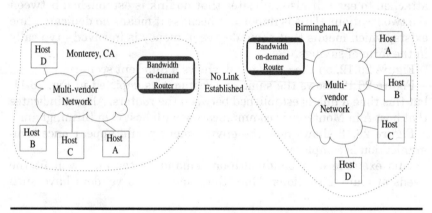

Figure 26.13   Bandwidth-on-demand routing.

devices available today. One consideration for this device is to weigh the consequences of purchasing such a device versus having a router with a dedicated line. Determining the amount of usage should dictate which would be the better solution; and my philosophy is that the *customer* should make the decision.

26.7   Router Advantages in Brief

Routers provide a variety of capabilities that are advantageous to networking of different types. With the influx of growth of LANs in the past decade, routers are sometimes the best solution to meet needs of a given situation.

Segment isolation

A major advantage routers provide is the ability to split a large network into smaller, more easily manageable ones. This is important because the number of LANs and the number of specialized devices attached to them are both increasing. Figure 26.14 is an example of how routers can be used to create a more manageable LAN environment.

Figure 26.14 portrays a single-backbone LAN. It is apparently crowded and difficult to manage. Many different types of traffic pass across a single backbone. This puts a load on the network and consequently causes performance degradation.

Figure 26.15 is an example of the same equipment implemented differently.

In Fig. 26.15 multiple LAN segments have been created and a real sense of load balancing is the result. Notice the three routers that tie the segments together. They are able to control the passing of packets to the correct segment or keep packets on a given segment and thus not impede performance. Another advantage is the ability this configuration provides. With this arrangement a given segment can be removed from other segments if any changes or additional management-related functions that would affect other LAN segments are needed.

Multiprotocol support

Many router vendors offer routers that support multiple lower-layer protocols quite effectively. Since the IEEE 802.X protocols call for delineation of the data-link layer, it is possible to mix and match a variety of lower-layer protocols. Consider Fig. 26.16.

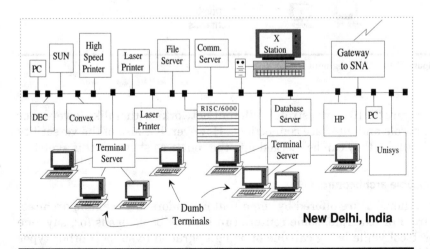

Figure 26.14 Single-backbone LAN.

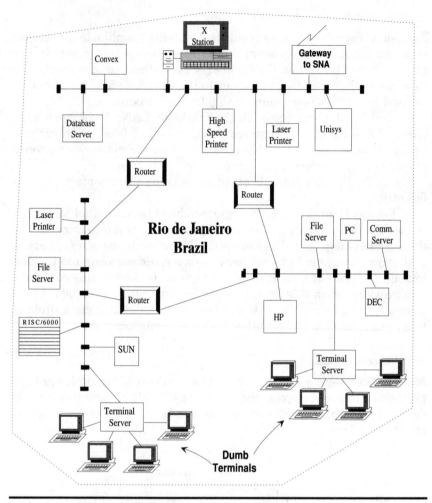

Figure 26.15   Multisegment network.

Figure 26.16 demonstrates different networks with unlike lower-layer protocols and interface connections. However, because of the versatility of routers, they can be merged into the router as shown in this sketch.

Scalable architecture

Another feature offered by some router vendors is a scalable architecture. For example, some routers can provide basic needs for, say, one protocol while being capable of being upgraded to support other types of lower-layer protocols. Network topologies are another factor to con-

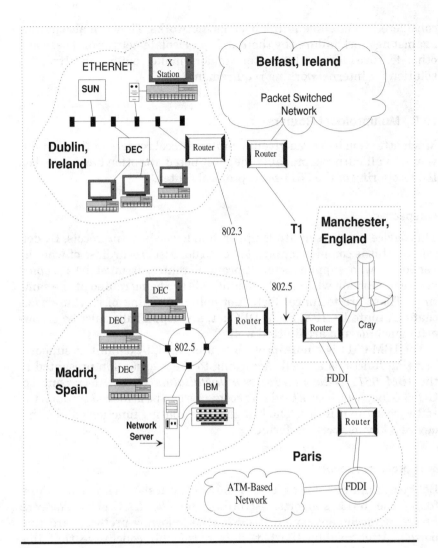

Figure 26.16 Mixed protocol support.

sider in routers. Some router vendors offer routers that have the flexibility to change and expand to interoperate with topologies that change
over time.

Other advantages of routers include price/performance ratio, flexibility to support local and/or remote sites simultaneously, and man-
ageability by network management technologies such as simple network management protocol (SNMP). Routers also provide what is

considered an intelligent link between networks. They can also protect against network failures by their ability to isolate networks. These and other features offered by router vendors make routers an attractive solution for internetworking requirements.

26.8   Multiprotocol Routers

Much attention has been given to multiprotocol routers in the past few years. In all fairness, probably the most noted attention has focused on IBM's offering of the 6611 multiprotocol router.

Perspective

This device supports multiple upper- and lower-layer protocols. By definition, this is complex for one device to do. Also, regardless of who the vendor is who supplies a multiprotocol router, it must be a *robust* device to function when multiple protocols are being routed at the same time. These are technical facts, not opinions. Think of it. One device that can route IP, IPX, SNA, DECnet, and AppleTalk must be a powerful machine and architected well.

The IBM 6611 is a multiprotocol router. And, by default, it supports bridging functions as well. According to the information presented in the *IBM 6611 Network Processor Installation Guide,* document no. GG66-3254, a variety of bridging scenarios can be achieved through the software setup utilities. The bridging and routing functions make this device a key network performer.

6611 protocol support

Before examining some aspects of the 6611, you should know what protocols the 6611 supports. According to *The IBM 6611 Network Processor,* document no. GG24-3870, the following protocols are supported. After reading the list, it is reasonably conclusive that this device must be robust.

Physical-layer protocols

RS-232

RS-422

RS-449

V.24

V.35

V.36

X.21

Physical layers specified by IEEE 802.X

Data-link-layer protocols

SDLC by IBM

ETHERNET V.2 by DIX (DEC, Intel, and Xerox)

X.25 by CCITT

PPP as in RFC 1171 and 1172

MAC and LLC IEEE protocols

Frame relay by CCITT and ANSI

Upper-layer protocols

SNA

TCP/IP

AppleTalk

NetWare

OSI

DECnet

XNS

NetBIOS

Routing-layer protocols

Internet protocol (IP)–TCP/IP

Internetwork datagram protocol (IDP)–XNS

Datagram delivery protocol (DDP)–AppleTalk

Connectionless network protocol (CLNP)–OSI

SNA

DECnet

Hypothetical examples

On the basis of the preceding information, some examples showing what this means is in order. Consider Fig. 26.17.

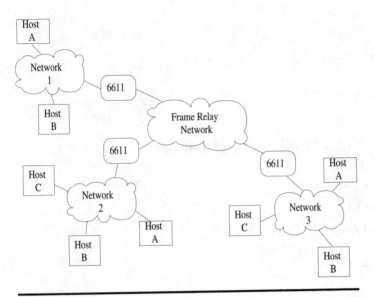

Figure 26.17 Frame relay and the 6611.

Figure 26.17 is an example of three 6611s, three networks, and the backbone frame relay network. Since the 6611 supports this data-layer protocol, connectivity among networks 1, 2, and 3 is possible.

Figure 26.18 is a different example of the 6611 implementation.

In Fig. 26.18 mixed lower-layer protocol networks are supported by the 6611s, thus making data exchange among them possible.

Figure 26.19 portrays yet a different arrangement.

Figure 26.19 is an example of multiple networks using different upper- and lower-layer protocols. This example shows routing capabilities among DECnet networks and also routing among the TCP/IP networks.

Figure 26.20 is an example of two NetWare and TCP/IP networks integrated using the 6611 router.

In this figure the routers are interconnected together, thus rendering the NetWare and TCP/IP networks physically and logically connected. Since NetWare supports TCP/IP, a degree of interoperability can be achieved via this arrangement.

According to the IBM manuals, a variety of implementations can be achieved. The number of permutations with the 6611 network processor is considerably high.

IBM is not the only vendor with a multiprotocol router. Other fine well-established corporations have them as well. The use of the 6611 here as an example is to shed additional light on some skewed interpretations that have evolved concerning it.

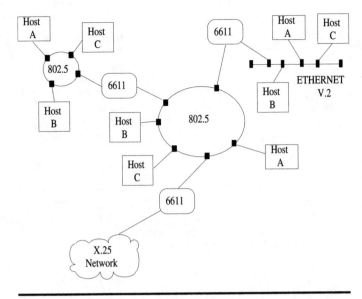

Figure 26.18 The 6611 and mixed protocols.

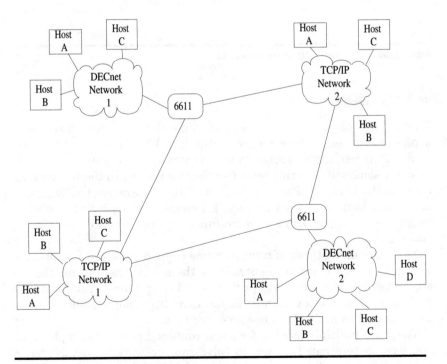

Figure 26.19 DECnet and the 6611.

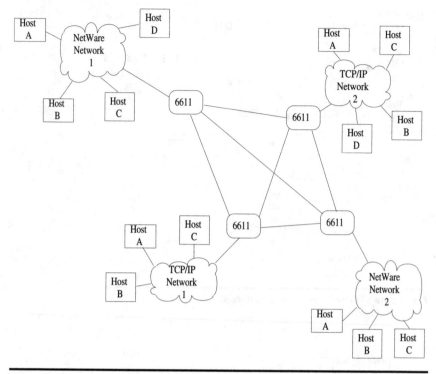

Figure 26.20   NetWare, TCP/IP, and the 6611.

26.9 Summary

Routers are complex devices and in some form or other have been around for at least two decades. In the late 1960s and the 1970s the device that performed router functions was called a gateway, and to this day some still use this term to refer to a router. In the networking community today legitimate devices that perform router functions exist, and they are called routers. Likewise, devices called gateways exist; they, too, perform a specific function, and generally this is not routing.

There are many types of routers. Some support different upper-layer protocols and are protocol-specific in their implementation. Others support different lower-layer protocols and are protocol-specific to that end. Any given router generally supports multiple physical-layer interfaces as well as multiple data-link-layer protocols.

There are multiple reasons for using routers. From a geographic perspective routers can be used to solve internetworking needs. Some routers are focused on what is considered a metropolitan implementation which includes a major facility and satellite offices located within reasonable proximity. Remote routers can be used to solve integration issues of geographically dispersed networks. The same holds true for what is considered internal routing.

Different types of routing are possible. Centralized routing focuses the routing tables within a centralized processing device. Noncentralized routing implements a strategy where routers exchange information about those routers with which they operate. In static routing, routing tables are predetermined; in other words, the tables and paths for routing are determined and configured when the network is nonoperational from a user perspective.

Two examples of different approaches to routing were presented. SNA routes data primarily via its front-end processor (FEP) by way of the network control program, but this appears to be changing. This seems to be the case as indicated by announcements and products that are now capable of performing functions that not too long ago were considered FEP-type functions. Additionally, with the SNA Networking Blueprint, many areas are in a state of flux.

TCP/IP was used as an example of how routing is achieved. The functions and other details of the software components that are involved in this process were explained. The versatility of the TCP/IP protocol stack makes it possible for multiple possibilities to be considered for implementation.

A type of router has arrived on the market in the past few years that differentiates it from some of its predecessors. The bandwidth-on-demand router—which offers an alternative to a dedicated line for two geographically distant sites that need a router for intermittent connectivity; that is, on demand—was also described.

Multiple advantages are realized for those sites where routers are implemented. For example, segment isolation is possible with the implementation of a router. This is significant because troubleshooting and maintenance are easier with a segmented network than with a single-backbone network where crowding exists. Those routers that offer multiprotocol support at the upper and lower layers offer greater flexibility for growing environments. Some routers are scalable, meaning that their processing capabilities can be expanded or upgraded to provide increased support as future needs may dictate.

A multiprotocol router has a tall order to achieve many functions at the same time. As the router example was presented, a router that supports so many different protocols at the physical interface, data link, and upper-layer protocol layers must have vigorous processing capabilities.

Routers may be a part of an integration equation when multiple protocols are used, but they may not meet all the needs possible. As was mentioned early in this chapter, routers route.

Chapter

27

Brouters

Brouters are hybrid devices. This type of network device is offered predominantly as a combination bridge and router. It can perform both functions. A bridge and a router perform dissimilar roles, and some brouters can achieve routing functions via bridge configurations, specifically via filtering. Brouters are capable of handling multiple upper- and lower-layer protocols. In a sense, this sounds like a multi-protocol router. One source I read said in essence that a brouter is the most complex network device to operate. Based on my experience, this is contingent on the complexity of the brouter and networks implemented.

27.1 Perspective

Brouters are the result of integrating bridge and router functions. The issue is how these functions are implemented in a given device and how that device is implemented in a given environment. It is best to begin with a conceptual understanding.

Remote token-ring brouter

Figure 27.1 is an example of a brouter implementation.

Figure 27.1 shows a simple environment consisting of SNA architecture and a blend of APPN. An implicit function shown here is the IBM source routing function. Some brouter vendors supply this as a basic part of their product. Also implicit here is support for dual-ring speeds, both 4- and 16-Mbit.

Token ring A has two brouters attached. Implicit in this drawing is the automatic crossover function in case one brouter fails. When both

575

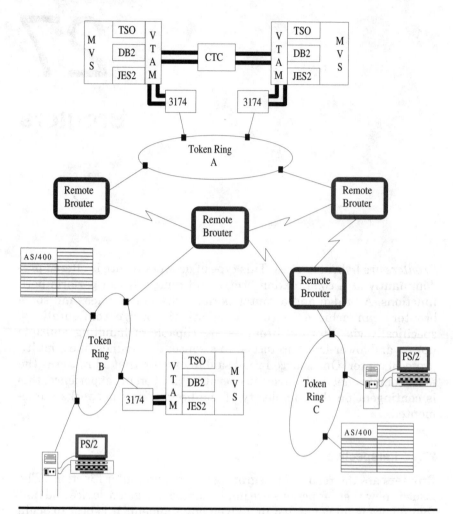

Figure 27.1 Conceptual view of remote brouter implementation.

are operating in normal condition, throughput is enhanced because of dual routes to rings B and C.

Remote ETHERNET brouters

Local area networks. A remote ETHERNET brouter is capable of linking remote ETHERNET networks while providing routing functions at the same time. Consider Fig. 27.2.

Remote ETHERNET brouter functions are similar to the previous token ring example. In Fig. 27.2 host users in San Francisco *think* they

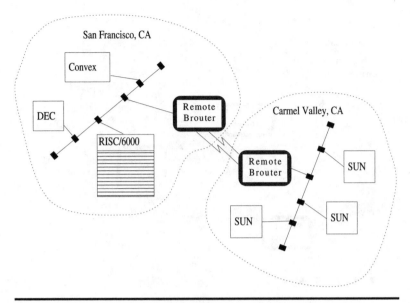

Figure 27.2 Remote ETHERNET brouter.

are directly attached to the hosts on the LAN in Carmel Valley (Calif.). The link is transparent to the user. Increased throughput can be achieved by multiple links between the brouters as shown in the figure.

Wide area networks. It is possible to create what would be considered a virtual WAN with multiple ETHERNET LANs connected via brouters. Figure 27.3 shows an example of this.

Figure 27.3 shows four locations geographically dispersed around the United States. The functionality of brouters here create a virtual ETHERNET WAN. Although not indicated in the figure, the lines connecting the sites could be high-speed leased lines. This depends on the brouter vendor supplying the devices.

Mixed-protocol brouter environment

Most brouters support a variety of protocols. Figure 27.4 is an example of this.

In a mixed-protocol brouter implementation a complex WAN can be successfully achieved. Figure 27.4 is an example of a wide variety of upper- and lower-layer protocols.

The type of environment shown in Fig. 27.4 may not be supported by all brouter providers, but a significant number of them do provide such a solution.

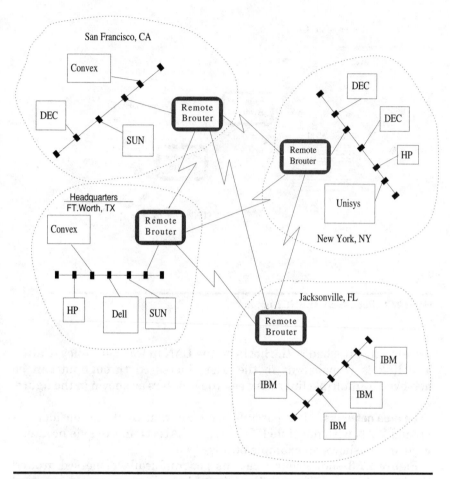

Figure 27.3 Brouters creating a virtual WAN.

The brouter as a hub

Some environments use large brouters as a hub with multiple remote locations. This is similar to the mixed-protocol WAN environment. The difference is from an operational standpoint. In this hub environment there is a centralized operation, whereas in Fig. 27.4 this was not the case.

Consider Fig. 27.5 as an example of a brouter implemented as a backbone hub.

Figure 27.5 shows the brouter operating as a hub. This scenario is not uncommon. In fact, a particular site in central California has an

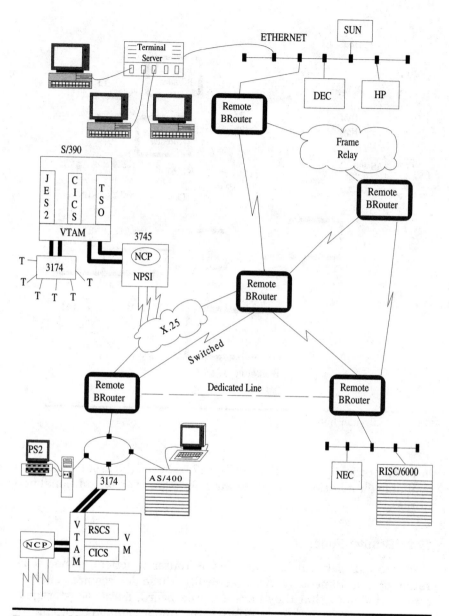

Figure 27.4 Mixed-protocol brouter environment.

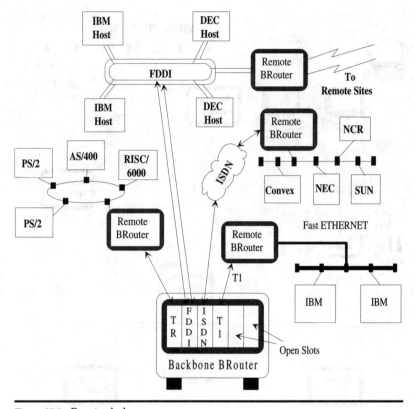

Figure 27.5   Brouter hub.

implementation quite similar. As Fig. 27.5 shows, the brouter has slots available for additional interface cards for expansion of additional networks.

27.2   Brouter Functions

Brouters can perform bridging and/or router functions either separately or simultaneously. Architecturally, these devices are generally constructed such that they have a motherboard, database, storage of some type, vacant slots for interface cards, and a common backplane on which data is passed. These features may vary according to vendor and size of the brouter.

Brouter functions generally vary because different vendors support different protocols and methods of integrating these different protocols. For example, some brouters are limited in terms of interface capacity. Others may be limited in performance.

The brouter as a backbone

A variation on Fig. 27.6, where a brouter is used as a backbone, is Fig.
27.7, which indicates a much simpler implementation.

Figure 27.7 illustrates a common implementation of brouters. It is
conceivable that this type of backbone implementation would be used
in a commercial, institutional, or campus environment.

Path for growth

Because of the brouter architecture that some vendors offer, brouters
that provide a clear migration path for entities may need merely simple bridge-router functions currently, but later may need additional
capabilities. This modular capability is portrayed in Fig. 27.8.

Figure 27.8 shows two ETHERNET-based networks, with a brouter
chosen to meet the connectivity needs between them because of its capability for future expansion. This particular example provides a cost-efficient solution and at the same time a migration path for additional interface boards supporting different network protocols or the same protocol.

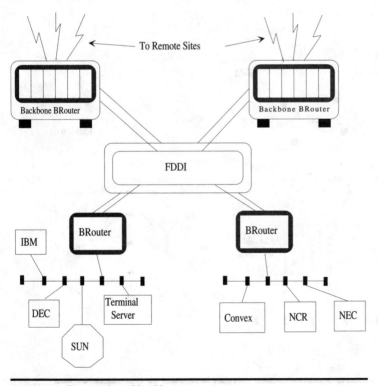

Figure 27.6 Brouter-based backbone.

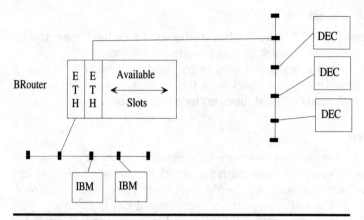

Figure 27.7 Brouter modularity.

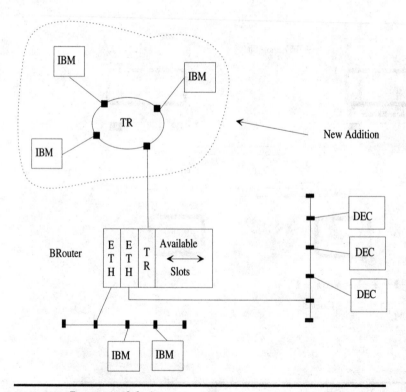

Figure 27.8 Brouter modularity.

Bridging

Brouters can perform bridge functions alone if this function is required. The type of bridge support offered is vendor-dependent; consequently, it is advisable to ask questions concerning how a given brouter handles certain functions as in bridging. For example, IBM bridging calls for source route bridging, and other methods exist. The question is: "Does the offering meet the need at hand?" Such questions should be asked prior to the purchase or negotiation phase, but unfortunately they seldom are.

Routing

Routing has its own issues that deserve attention. For example, with the TCP/IP protocol suite, support for multiple routing protocols is possible. In some instances some vendors support RIP or OSPF, but not both. Beyond this, some vendors may support additional TCP/IP routing protocols. This is another issue to be addressed prior to a purchase. These simple questions will impact installation of such devices; but will the impact be positive or negative.

Brouter management

Managing a brouter should not be overlooked. Two issues should be considered: (1) management of the brouter itself, which introduces three possibilities—local management, remote management, or both types of management—and (2) network management support as a device—for example, how and to what degree a brouter can be managed from TCP/IP, particularly SNMP.

27.3 General Considerations

Included in this section are some baseline considerations for brouters. Because a brouter performs bridge and routing functions, those layers affected by these functions should be examined.

Physical-layer protocols

Some physical-layer interfaces are considered "common," but this is relative. Nevertheless, the following list is an example of dominant physical-layer interfaces to consider:

RS-232	Fiber
V.35	X.21
RS-449	V.24
RS-442	AUI
T1	

Others may be important and should be given appropriate consideration when exploring brouters.

Data-link-layer protocols

Popular among the many data-link-layer protocols in use today are:

SDLC	802.X
Token ring	Frame relay
ETHERNET V 2.0	ISDN
X.25	PPP
FDDI	LAP-8
HDLC	

Upper-layer protocols

Multiple upper-layer protocols are popular, and this listing reflects those covered in this book:

OSI

DECnet

NetWare

SNA

AppleTalk

APPN

TCP/IP

The significance of these lists is to interpret which protocols are presently or potentially needed for a given site. Planning for expansion is always wise when considering integration of devices such as brouters.

27.4   Considerations Prior to Implementation

Sometimes brouters are implemented into sites that are mission-critical. Brouters are complex devices, and thus exploration into what they actually do and do not do is important. Consider total uptime.

Hot backup

To operate mission-critical sites is to operate a 24-h/day, 7-day/week, 365-day/year operation. In short, *no* downtime. This can be interpreted incorrectly and then conveyed incorrectly with respect to some networking devices; brouters are certainly not the only network devices susceptible to this hypothetical situation.

Operation of mission-critical sites is significant for those who require zero (no, none, as in do not even think about it) downtime. The easiest way to explain this situation is by its converse; that is, to achieve total, 100 percent uptime means *no* human intervention. It is that simple. This is a site-specific and product-specific issue. Regardless, it does mean crossover from one system to another within such a short period of time that it is difficult to calculate that brevity. Another way of considering this issue is that any system that requires operation by a human, or even sometimes a program, is not total, 100 percent, completely uninterrupted operational uptime.

Speed

Speed seems to be a consideration for most devices in the internetworking arena. Speed is also a significant factor with brouters. Brouters are valid network devices, but understanding the power of what they can do, interpreting speed capabilities of brouters, and effectively implementing one should involve more than a passing thought. This is especially true if a brouter is to perform as a backbone device and a high volume of hosts and traffic is anticipated.

Management capabilities

Management abilities with brouters are twofold: (1) managing the brouter itself and (2) participation of the brouter in whatever network management is implemented. These are distinct issues and should be dealt with separately.

Compatibility with additional devices

Compatibility is also a consideration for network devices. The rule of thumb should be the more complex the network device, the closer should be examination of potential compatibility with other current devices on a given network and the anticipation of any networking devices that might be purchased in the future.

In particular, if a brouter is integrated into an existing network, the devices already functioning on that network should be considered.

27.5 Summary

Brouters have the ability to perform bridge and router functions. They are complex devices by default. In the right place a brouter may be the precise network device to meet specific needs. Because brouters are complex and can function as a bridge and a router simultaneously, understanding their functions simply involves understanding the functions a bridge and a router perform independently, mentally fusing or merging these two functions together, and then understanding what a vendor offers with the product as a result.

There are various types of brouters, generally differentiated with respect to the various protocol-specific (and hence, usually vendor-specific) functions they perform. Examples of data-link-layer protocols supported include local and remote token ring, local and remote ETHERNET, FDDI, ISDN, X.25, and frame relay, which are presently supported by different vendors today.

The term *multiprotocol router* has emerged in recent years, and practically speaking, most of the multiprotocol routers available today can perform the some functions as those devices called brouters. At the present time it seems that a considerable degree of functions overlap between the two devices. Each device can be implemented to provide the appearance of a virtual WAN. Brouters, however, do have the distinct ability to function as a hub-type device. When implemented this way, all networks or hosts connected to this device must pass data through it.

Brouters can be managed at either the device or network level. Device-level management is what the vendor provides as a mechanism to create or change configurations to enable the device to perform a desired function. Network-level management is where a brouter is managed as another network device. Since this is the case, it is important to understand what type of network-level management support is provided.

As with most network devices, brouters can be evaluated by their speed and ability to perform in a mission-critical environment. Brouter speed is important because of the complexity; not that complexity requires speed, but the degree to which complexity is utilized does. Acknowledging the role a brouter performs in a network environment is important because some environments require total uptime.

Chapter

28

Servers

The term *server* is frequently used generically. However, today there are many highly specialized, vendor-specific types of servers on the market. Even so, the term is still used loosely to refer to devices such as gateways, routers, and other function-specific network devices. This chapter explains some popular servers that perform a specific function.

28.1 Servers: Function and Philosophy

Servers, in the general sense, can be basically any network device. Servers have come of age and are categorized according to functions they perform.

Function

Typically, a server provides specific functions to the whole network or at least a significant portion of it. One point of clarification is in order. Some level of confusion still abounds about the notion of *client/server*. Clients' programs, whether invoked manually or programmatically, always initiate something. Servers, on the other hand, always answer a client's request.

The term *server* is used in a different context. *Servers,* as network devices, do in a sense respond to clients' requests. However, network servers normally provide some functions that separate them from a *server* program. For example, a file server may be a dedicated host with disks dedicated to storing files. These files are used by multiple users throughout the network, and therefore centralized storage of common files eliminates the need for vast amounts of disk space for each user. So, in this sense a server takes on a meaning different from the notion of client/server networking, which connotes peer-oriented communication.

Philosophy

The philosophy behind a server is that it performs some function in a centralized manner rather than having that function distributed throughout a network or tied to particular hosts. The development of servers of all types seems to be the result of networks providing a single commonality among hosts of all types, including PCs. With the proliferation of networks in the 1980s, the notion of delegating functions such as printing to a designated printer seemed to make more sense than each host having a printer when in reality this might not be required.

28.2 File Servers

File servers are devices (hardware and software) that function as centralized network disk storage units which can be accessed by authorized users on a network. These devices can be configured to perform multiple tasks.

File storage

One function a file server can perform is file storage. This is advantageous because one centralized server can be the central repository for all files. Theoretically, a file server would appear on a network as shown in Fig. 28.1.

In Fig. 28.1 multiple PCs are connected to a NetWare network along with a file server that contains the files required by the PC users. In

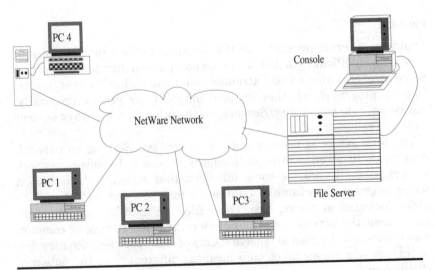

Figure 28.1   Conceptual view of a file server.

such a configuration any PC can access the server to retrieve a file or store a file onto it.

Figure 28.2 depicts a detailed view of the PC users and their operations with the file server.

In Fig. 28.2 PC users 1, 2, and 4 are accessing the file server. All PC users are retrieving different files they are working with on their respective PCs. This is possible because of the multitasking capabilities, storage ability, and configuration of the file server. In this example the file server appears to be a single disk to each user. Individual users *think* they have their own hard disk. In reality the configuration is such that a virtual drive is created and each PC is configured to point to this disk (server).

Program storage

Another function a file server can perform is that of a disk that holds the software package master copy. For example, the Microsoft Word network edition software package can be purchased and loaded onto a server. Multiple users can utilize it and immediate savings can be realized because multiple packages of the software program do not require purchase. Consider Fig. 28.3.

In Fig. 28.3 PC users 2 and 4 are using the Microsoft Word software package based on the file server. Neither user is aware of the other's work because of the nature of the network configuration and how the

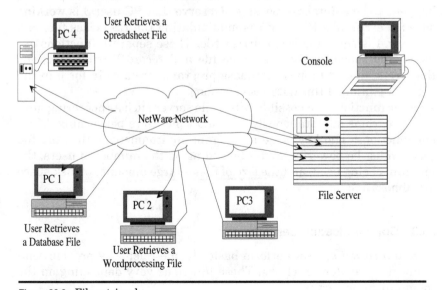

Figure 28.2   File retrieval.

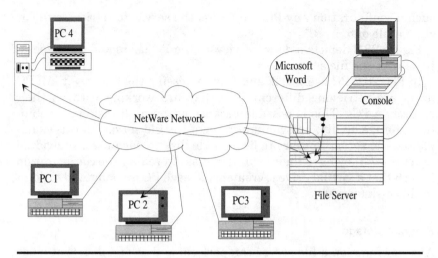

Figure 28.3   File server program storage.

software package operates. This is a straightforward example of how a network can be utilized to save money. Four users are on this NetWare network, and each has access to the Microsoft Word software. One copy of the software, rather than four, was purchased.

Another example of program storage is exemplified in Fig. 28.4.

In Fig. 28.4 a database is loaded onto the file server, and five users are connected to the network. In this example the software package happens to be a database package. Observe that PC user 3 is working on Mr. Perry's file, PC user 1 is manipulating the ESS class file, and PC user 4 is changing Mr. Harris's file. These separate functions are happening concurrently. The same file and record locking functions that are standard to most database programs are also in force in the network edition of this database version.

Other functions are possible with a file server such as operating multiple programs simultaneously with multiple users performing different functions. The key issue in such an environment is that the file server must be powerful, and—depending on the number of users, the programs being used, and the size of files—large amounts of disk space are almost a must.

28.3 Communication Servers

Communication servers perform basically one role; they provide communication-related functions. These functions vary depending on the communication server.

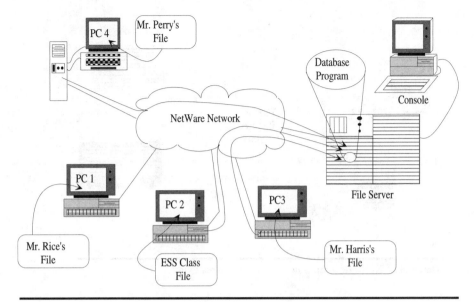

Figure 28.4 Database operations with a file server.

Modem pools

One function a communication server can perform is to house multiple modems, permitting access via various lines and speeds. Consider Fig. 28.5.

Figure 28.5 is a logical view of a communication server and the network with a DEC, Convex, RISC/6000, and two PCs attached. To the users of these hosts, the communication server logically appears as part of the network—as just another resource.

Figure 28.6 is a physical view of the network showing more detail on the line speeds and connections to the hosts themselves.

Notice in Fig. 28.6 that the DEC host has a T1 link, the Unisys host has a switched 19.2-kbit/s line, the Convex host has a 56-kbit leased line, and the two PCs have access to a dual modem with a 9.6-kbit switched line.

Consumer tip

Many vendors are listed in charts and journal listings under the heading of communication servers. Some of the following names and functions are

- Multiprotocol concentrator
- X.25 communications interface

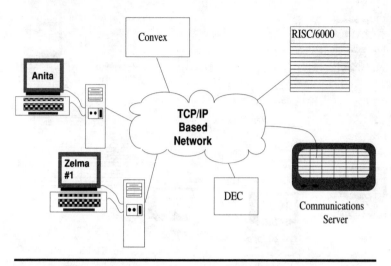

Figure 28.5   Network communication server: logical view.

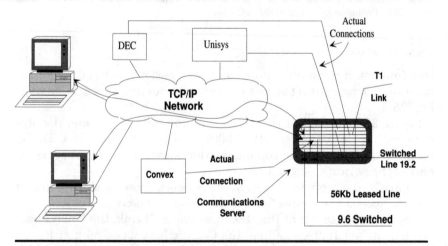

Figure 28.6   Network communication server: physical view.

- Video mux (multiplexer)
- Multiprotocol communication server
- Fax plus
- Remote-access server
- Modem server
- Asynchronous LAN gateway

I believe the point is clear if you read the list. Categorization of a device as a communication server has such broad meanings that research is required to understand what is meant. Some communication servers are highly specialized, so much so that they are vendor-specific and the device may operate only with a given vendor's equipment. This is helpful when reading about *communication servers.*

28.4   Modems

Modems were discussed earlier, but a brief section concerning topics on modems as they relate to servers is in order.

Autobaud

The term *autobaud* is relatively easy to understand because this device does what its name implies. It automatically selects the baud rate based on the speed the destination modem is capable of reaching. Generally, the autobaud process is like bidding. Some modems attempt to move up to the highest speed that either modem will support. However, this is not always the case, and this is the reason for autobaud. If one modem supports a higher baud rate than the other, the higher-baud-rate modem will decelerate to the slower speed if the higher-speed modem can autobaud. This capability, however, can differ with vendors, but the term *autobaud* is generally agreed on. It is preferable to have an autobaud modem in most cases because of their versatility.

Line conditioning

The networks on which modems operate experience interference by default in many instances. As the baud rate increases with higher-speed modems, so does the tendency to incur errors as a result. One problem related to this is *amplitude distortion,* in which the signal strength itself varies independently of noise of the line itself. In order to compensate for this phenomenon, most modems have built-in circuitry for property adjustments so that the signal will not be too disproportional.

The equalizer

This is an important aspect about modems, even though it is not a prevalent topic! Modems operating at baud rates in excess of approximately 2400 baud have built-in equalizers. The purpose of this equalizer is to serve the function of a filter. This filter is special because its phase characteristics and amplitude are inversely proportional to the ones encountered on the lines the modem operates on. Additionally, equalizers perform a function to correct for propagation delay causing distortions.

28.5 Modulation Information

When data is transmitted over a line via a modem, modulation is involved, as was explained earlier in this book. However, additional details are fitting here in view of the subject. *Modulation* is simply the conversion of digital signals into analog signals. Conversely, when this signal arrives at the destination, the modem performs reverse modulation, if you will; it performs *demodulation.*

Additional details of modulation are provided in this section for more in-depth reference purposes.

Amplitude modulation

Amplitude modulation is the use of a single carrier frequency to convert the digital signals to analog. Figure 28.7 is an example of this.

This figure shows the high wave amplitude to indicate a binary one and a low wave amplitude to indicate a binary zero.

Frequency-shift-key modulation

Frequency-shift keying (or FSK, as it is frequently called) uses a constant-amplitude carrier signal along with two additional frequencies to differentiate between a mark and a space. Consider Fig. 28.8.

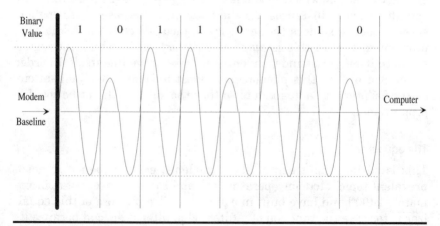

Figure 28.7 Amplitude modulation.

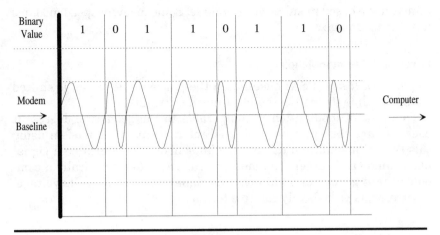

Figure 28.8   Frequency-shift keying modulation.

This type of modulation is not prevalent among higher-speed modems because of its inherent simplicity. Higher-speed baud rates require different modulation techniques.

Differential phase-shift key modulation

This type of modulation technique uses a phase angle comparison of an input signal to the prior di-bit. A *di-bit* is defined as two bit values represented by each phase angle. Consider Fig. 28.9.

Figure 28.9 shows comparison of the wave pattern and the square-wave interpretation thereof. Actually, the modulation technique is a

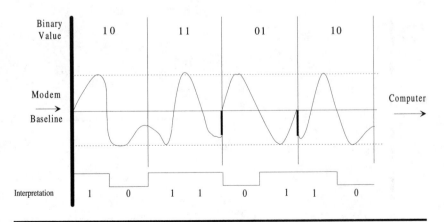

Figure 28.9   Differential phase-shift key modulation.

comparative based modulation technique. Some medium speed modems use this technique.

Phase-shift-key modulation

In *phase-shift key* (PSK) modulation the phase of the signal is shifted at the baseline (transition point). Consider Fig. 28.10.

This type of modulation is comparable to the geometric degree. It uses a phase shift relative to a fixed reference point. A modem using this type of modulation has an oscillator inside to determine a signal phase angle as it enters the modem. (An *oscillator* is basically a component that generates an alternating signal, continuously. Its voltage, or current, is periodically relative to time.)

28.6   Print Server

The term *print server* may be used to refer to one of two possible entities. One is a general reference to a printer attached to a network where all users with network accessibility can use it; in this instance it is generally referred to as a *standalone printer*. According to the second definition of this term, a printer is attached to a network host and its print jobs are queued from that host; thus the printer receives print jobs from the network hosts. In this instance the printer appears to be attached to the network directly when in fact it is attached to a network host. This example is loosely referred to as a network printer. Interestingly, little differentiation exists between the phrases, but the implementation differs radically.

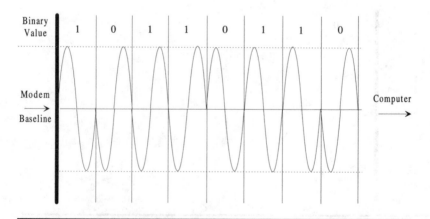

Figure 28.10   Phase-shift key modulation.

Standalone printer

Network printers have existed for some time. The concept is not new; however, network printers have been brought to the forefront of the industry. Figure 28.11 is an example of a network printer.

Figure 28.11 shows multiple hosts attached to a bus-based ETHER-NET. Any of these hosts can use the network printer because it is attached to the network and is addressable.

Figure 28.12 shows the highlight of the ETHERNET network module (NM) attached to the side of a Hewlett-Packard printer. In this example, the NM makes it possible for the printer to be a standalone network device. Notice that the NM attaches one end to the network transceiver and the other to the parallel port on the printer.

In fact, in the arrangement shown in Fig. 28.12 *any* host that can access the network can use this printer to print. This means that utilization of communication servers and other devices make remote-location computing and local network printing possible.

Figure 28.12 is an example of a standalone printer. In most instances this scenario implies small quantities of on-board memory, and thus printer queue bottlenecks are likely. Generally, they are proportional to the number of users attempting to use the printer at any given time.

Network printer

In the *network printer* arrangement, although a printer appears to be attached to the network, it is actually attached to a host, and its jobs are spooled to it via a host on the network. Consider Fig. 28.13.

Notice in Fig. 28.13 that Dave and Tim each have a HP workstation. Even though they do not have a printer attached to their own workstation, they can print on the network printer via the Convex computer to which the printer is physically attached. However, as mentioned pre-

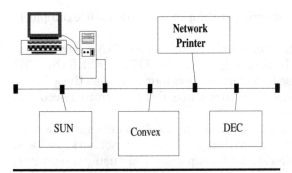

Figure 28.11 Network printer.

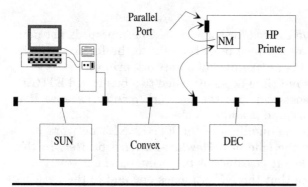

Figure 28.12 Network module highlight.

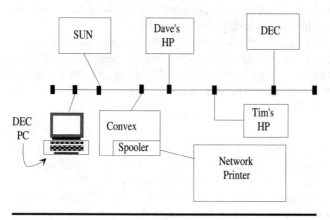

Figure 28.13 Host-queued printing.

viously, Dave and Tim perceive the printer as being attached to the network itself.

Other variations of network printing exist. One such example is shown in Fig. 28.14.

Figure 28.14 shows an example of printing in an SNA environment. Notice that two terminals are connected to the 3174 to which the 3287 printer is attached. These local terminals can print just as those in the New Orleans office and the New York office. But the printing occurs at the headquarters office on Mystic Trail.

Network printing has a variety of meanings. It is not uncommon for the meaning to be associated with a given vendor network protocol. Regardless, the bottom line is that multiple users are using the printer, in contrast to each individual user having a dedicated printer.

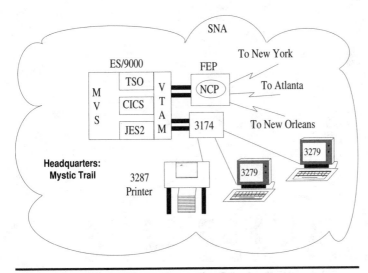

Figure 28.14   SNA printing.

28.7   Terminal Server

Terminal servers have grown from rarely used devices to commodity items in less than a decade. The fundamental philosophy of terminal servers is the implementation of multiple dumb terminals concentrated into one device. Figure 28.15 is an example of this.

Figure 28.15 shows multiple hosts, communications server, terminal server, and six dumb terminals. A "dumb" device is defined technically as one with little intelligence. I suppose this has always been the case, but in networking devices intelligence is measured by levels of degrees.

Figure 28.15 is a TCP/IP-based network. The terminal server has a TCP/IP complement inside in programmable read-only memory (PROM) on the motherboard. Hence, the terminal server has a degree of intelligence. Many vendors market terminal servers, most of which can be configured to accommodate terminals or a modem or two.

The result of a TCP/IP-based terminal server is that dumb terminals can be used. The significance of this is that they cost considerably less than PCs or terminals with a higher level of intelligence. Ironically, what was considered a dumb terminal 6 to 8 years ago is entirely different from what is considered "dumb" at the time of this writing.

There are basically two categories of dumb terminals: those which have the basics such as DCE, DTR, DTE, parity, and other fundamental parameters and those which may have a set of programmable function keys, multiple ports, printer capability, built-in calendars, and so

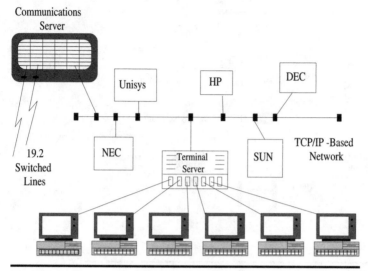

Figure 28.15   Conceptual view of a terminal server.

on. As you may have concluded, the price differs between these two terminal categories as well.

Interestingly, terminal servers are supplied with a variety of ports. Most terminal servers have anywhere from six to nine, and a console port for superuser (maintenance) functions. Terminal servers provide a very cost-effective method of meeting the needs of network users. A dumb terminal–terminal server combination is basically limitless (with the exception of graphic capabilities, and somewhere the ability to perform this function is probably being experimented). Figure 28.16 depicts an operation that is actually a minor task.

In Fig. 28.16 multiple network devices are shown on the TCP/IP-based network. In short, the figure includes a terminal server with two terminals attached, a modem attached to the terminal server, a Convex, DEC, RISC/6000, Unisys, Sun, and IBM hosts. Additionally, the PC is in a location remote from the network. The remote PC user can access the network by dialing the number of the modem indicated in Fig. 28.16.

Five functions are performed sequentially by the "dumb" terminal user (no pun intended), user number 1 (user 1). These five functions are as listed:

A   Function A is the dumb terminal user (user 1) invoking the TELNET client inside the terminal server against the Convex host. The native TELNET server in the TCP/IP stack on the Convex answers the TELNET client request.

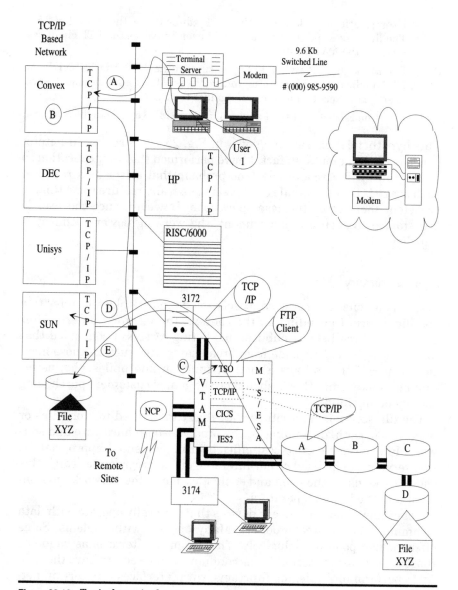

Figure 28.16 Typical terminal server operation.

B   User 1 invokes the TELNET client on the Convex against the 3172 attached to the MVS host. The MVS host has TCP/IP running as an off-load option, but by default a VTAM application. Hence when user 1 invokes the TELNET client against the TELNET server on the MVS host, user 1 sees the VTAM banner screen.

C   User 1 performs a log-on to the ISPF editor under the control of TSO. The line mode option is selected, and user 1 invokes the FTP client resident on the MVS host.

D   Even though TCP and IP are operating on the 3172, the TELNET and FTP applications are running indirectly as VTAM applications. Hence, user 1 can execute a FTP client against the SUN.

E   Function E is where user 1 moves a file from MVS disk D to the Sun host.

This hypothetical operation, shown in Fig. 28.16, is reality in heterogeneous environments. In fact, I have performed this very operation so many times it is uncountable. The problem that arises as a result of such usage is that terminal server users generally require some time to accustom themselves to these operations. However, once the user is accustomed to such an environment, the working power within it is unbelievable.

28.8   Summary

Server is a generic term used loosely in the marketplace to refer to a specific device. Depending on the context, the meaning differs. After examining three lists in publications listing *servers,* I concluded that little continuity exists aside from broad categories such as those mentioned here. Servers have a distinct function and philosophy; ascertaining exactly what this is should be the goal of customers purchasing servers of any kind.

The file server is a specific device that can be used to store files or programs or both, depending on the product—both hardware and software. File servers can be cost-efficient if implemented appropriately. File servers can be used either interactively or programmatically; this choice depends on the need and sophistication of the network environment and the human operators.

Communication servers are devices that generally operate with data communication-related products. Many operate with modems. Some offer modem pooling exclusively. The use of this term is as vague as those of other server terms. Some communication servers have the ability to perform multiplexing functions, with interfaces into certain network environments.

Many communication servers lend themselves to modem support functions. Modems are complex in their own right. The term *autobaud* means a modem capable of regulating operating speed relative to the target modem. Line conditioning is a factor of communications over telephone lines. Most modems have built-in circuitry to compensate for this phenomenon. Equalizers are special filters used to adjust for phase differentiation.

Modulation is a topic in its own right, but is important when discussing modems and communication with modems. Four types of modulation techniques were explained: amplitude, frequency-shift key, differential phase-shift key, and phase-shift key. Modems use a specific modulation technique, and its impact upon modem operation is important.

Print servers are network devices, also. In short, this is a printer that *serves* usually the entire network. Two categories of printers are identifiable: (1) the type with a network interface card attached—which is considered a standalone printer (this type of print server generally has limited memory, even though this is a relative term) and (2) a network printer—this printer configuration is such that it is connected to a host and the host generally has a spooling mechanism with enough memory to accommodate the network users. This is obviously the ideal case; in reality, differences exist.

Terminal servers are network devices that are now at the forefront of the marketplace to a significantly more degree than they were 6 or 8 years ago. Fundamentally, the concept of a terminal server is where multiple dumb terminals are connected to a single device that provides network accessibility. "Dumb" terminals is a technical reference to a terminal that has limited software or firmware intelligence. Some dumb terminals are more intelligent than others.

Terminal servers differ with respect to network protocols, but in principle they achieve the same function; they provide access to a network. Contrasted to a terminal server is a PC or a terminal attached directly to a host. As Fig. 28.16 depicted, the power behind a dumb terminal and a terminal server is typically limited only by the user behind it.

29

Gateways

Gateways are the focus of this chapter. Explanation of the function of this network device is the central purpose. Clarification of the term, implementation of a gateway in different networks, and other topics are covered.

29.1 Perspective

The term *gateway* has been explored prior to this chapter, but here a thorough explanation of the term is provided. The term *gateway* is defined by the *American Heritage Dictionary* as: "1. An opening, as in a wall or fence, that may be closed by a gate. 2. A means of access." Use of this term in the technical community can be traced back to the 1970s easily and possibly even back into the 1960s. Use of the term was prevalent in the Internet community of computer professionals. It also gained technical meaning in its own right in IBM's SNA community.

Use in the Internet community

The Internet community has used the term *gateway* in a more general context. The term was previously used to convey a device that performed functions of what is called a router today. Generally, a gateway conveyed the latter meaning as indicated by the *American Heritage Dictionary*.

Inadvertently, ambiguity of the meaning began to arise from the use of the term in the decade of the 1980s, most probably because of an influx of numerous types of networking devices that were developed and brought to market at that time. As a result, the terms *router* and *gateway* became definable because of the functions they performed. A clear distinction was identifiable between devices that performed router functions and gateway functions. The result was, and is, about the meaning of the term *gateway.*

Gateway operation

By definition, gateways operate at network layer 3 relative to the OSI model and above at a minimum. However, a gateway can operate at all seven layers if necessary. Examine Fig. 29.1.

A gateway performs protocol conversion between unlike networks. The issues are which layers require protocol conversion for integration to be achieved and where the protocol conversion process is performed. Integration of heterogeneous networks is the purpose of a gateway. This is its primary purpose, but it can perform other functions as explained later in this chapter.

Understanding the confusion

Currently there are many good reference sources on networking devices. Many of these sources are *nonbiased,* meaning that they conduct research on products and make this research available to the public through different avenues. Multiple sources I use list devices such as gateways, routers, bridges, and the like by category. The resulting format of such devices typically lists the vendor and the product name and function. In many cases those products listed under gateways are not, in fact, gateways. To be fair, after examining some of the products I am referring to, it seems the validity of listing the product under the

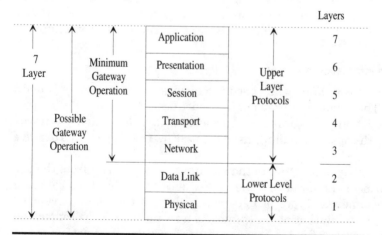

Figure 29.1 Gateway operation correlated to OSI layers.

gateway category is achieved by referring to the product function as performing that role stated in definition 2 in the *American Heritage Dictionary.*

A clear example of this is how a device that permits entry of networks into a X.25 network is categorized as a gateway. X.25 is considered a lower-layer protocol, even though it operates as part of layer 3 when compared to the OSI model functions. The point is that to classify a device that permits entry of networks into a X.25-based network as a *gateway* is semantic hairsplitting. On one hand, this definition could be correct.

For example, consider a token-ring-based network with TCP/IP used as the upper-layer protocol. Now, contemplate the X.25 network using something other than TCP/IP as an upper-layer network protocol. If this is the case, the device that performs this function is in fact a gateway. If, on the other hand, the device does not perform such a function, what is the device called? An interesting question, but nevertheless important.

29.2   Gateways: A Closer Look

Gateways operate with protocols; the question is with which ones and at what layers compared to the OSI model. This section describes gateways and different protocol environments.

Protocol-related gateway types

Internet gateways. The Internet comprises numerous individual networks connected together that collectively make up the Internet. Figure 29.2 is an example of three networks in three different cities that portray this idea.

Figure 29.2 shows each network connected via a router. Traditionally, these routers would have been considered gateways. Indeed, in 1983 the Department of Defense (DoD) announced that anyone desiring to connect to the Internet must use TCP/IP as the protocol. This drew a line in the ground, so to speak, and since that time connecting to the Internet has been a direct TCP/IP-TCP/IP connection or some variation thereof, requiring gateway functionality between the connecting entity and the Internet.

A typical way to connect to a service provider that offers connectivity to the Internet is via a multihome host. Figure 29.3 is an enhanced view of the Dallas site shown in Fig. 29.2.

Figure 29.2 portrayed the Dallas site in such a way that all hosts appeared to be connected to the router and able to connect to the Internet. Figure 29.3 shows reality. All networks and hosts are con-

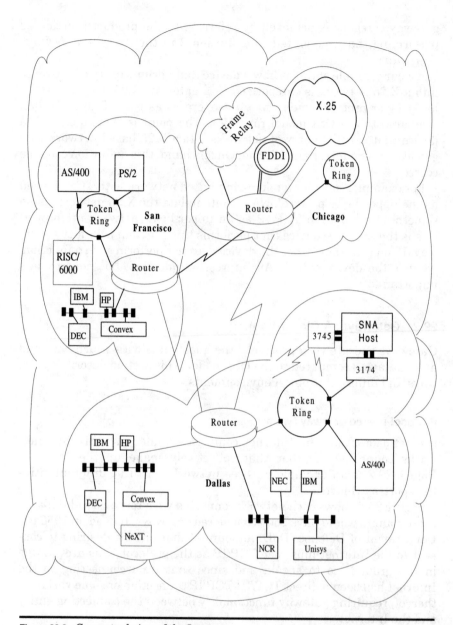

Figure 29.2   Conceptual view of the Internet.

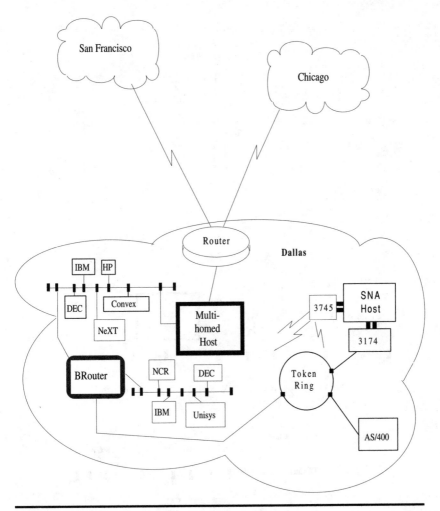

Figure 29.3 Multihome host and the Internet.

nected to an ETHERNET network which has a multihome host. One side of the host attaches to the local ETHERNET and the other to the router, which in turn connects to the San Francisco and Chicago networks.

SNA gateways. SNA gateways function so that connectivity can be achieved between SNA and a different environment. Consider Fig. 29.4.

Figure 29.4 shows a DECnet network connected to a SNA network via an SNA/DECnet gateway. This example shows SNA protocols used in the SNA environment and DECnet protocols used in the DEC envi-

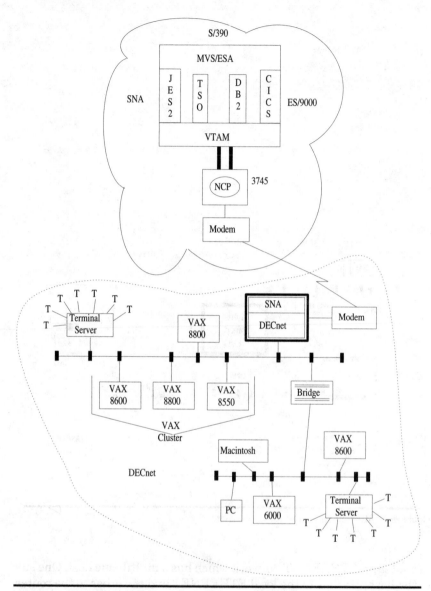

Figure 29.4   SNA/DECnet gateway.

ronment. In this case both upper- and lower-layer protocols have to be converted. In the DECnet ETHERNET is used as a lower-layer protocol. The upper layers are DECnet-oriented. In the SNA environment SDLC is the lower-layer protocol used on the FEP (front-end processor) and SNA is the upper-layer protocol.

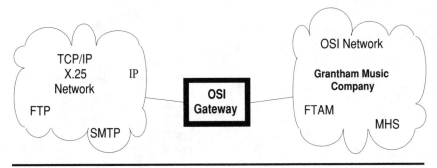

Figure 29.5 OSI-based gateway.

OSI gateways. OSI defines all seven layers. As a protocol, it provides different options at various layers. A chapter (Chap. 18) was devoted to OSI previously in this book. However, many OSI gateways provide access into an OSI-oriented environment where one or more applications are used. Figure 29.5 is an example of an OSI gateway.

It is interesting how OSI gateways are portrayed as being protocol- or application-oriented. In Fig. 29.5 a TCP/IP network is using X.25 and has FTP and SMTP as standard applications. On the other hand, the OSI network has a complete OSI protocol stack implemented, and FTAM along with MHS is available.

This scenario implies the exchange between mail systems (SMTP and MHS) and file transfers between (FTP and FTAM). These applications are distinctly different and require conversion between them.

DECnet gateways. A *DECnet gateway* is a device that permits entry of networks or devices that are using non-DECnet protocols and converts them into DECnet requirements. Figure 29.6 is an example of a scenario such as this.

Figure 29.6 is an example of a sizable DECnet environment and also an APPN network comprising AS/400 hosts that use a 5250 data stream. This is significant because the DECnet network utilizes an ASCII data stream. In this particular instance both the protocol and the data stream must be converted in both directions for successful work to be achieved.

AppleTalk gateways. AppleTalk-based networks use this protocol for the upper layers in the network. Figure 29.7 is a conceptual example of this.

Figure 29.7 is an example of a gateway between an AppleTalk- and TCP/IP-based network. Multiple devices exist on the AppleTalk-based network, and they have interoperability with the hosts on the TCP/IP-based network. The gateway is required because of the difference in upper-layer protocols.

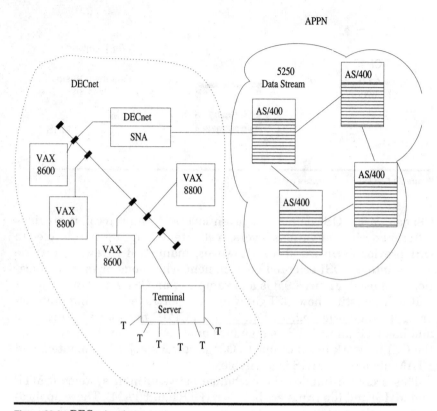

Figure 29.6   DECnet gateway.

Bidirectional functionality

Gateways operate between heterogeneous networks. Some exceptions may apply, but most gateways provide bidirectional communication between networks. The degree to which this is possible is vendor-dependent. Since some gateways perform specific functions, it is difficult to explain the many variations of this concept without performing a vendor analysis, and that is not the purpose of this chapter or the book.

Gateways in light of previous information

Other types of gateways exist; for instance, some provide peer-level communication and are highly specialized. Observation of industry journals provides multiple examples of how the term *gateway* is used. This term has been used to convey communication capabilities with UNIX, X.25, SNA, DECnet, and other applications.

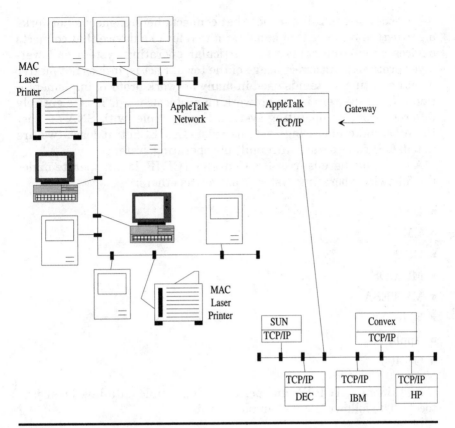

Figure 29.7 AppleTalk gateway.

All usages of a term that has multiple interpretations should be included in the definition of that term. Some vendors use the term *gateway* to define products that give their environment or products the ability to operate with environments such as UNIX, X.25, SNA, or DECnet.

The point here is that UNIX and SNA are not parallel. UNIX is an operating system. SNA is a network protocol. X.25 is considered a data-link-layer protocol, but in reality it operates at part of layer 3, the network layer. If the term *gateway* is used to describe products, or their ability to communicate with a different environment, then a comparative analysis should be performed on equal or parallel terms. To use a familiar phrase, comparing apples to apples should be the order of the day. Explaining the functionality of a device that provides interoperability between devices or networks should be done in context.

For example, to call a device that connects heterogeneous networks a *gateway* and to use that same term to refer to a device that connects devices or environments to a particular operating system or lower-layer protocol is incorrect usage of the term. Fact is, it does not matter what operating system is used in many network protocol implementations. What matters is the networking protocols used. This is entirely different from an operating system. For example, with SNA at least three bonafide operating systems exist. In DECnet at least two are noted. OSI can operate with multiple operating systems.

An example here is in order. Consider TCP/IP. It can operate under the following operating systems, as well as others:

- UNIX
- VMS
- OS/2
- MS-DOS
- MVS/ESA
- VM
- Apple
- OS/400

Additionally, TCP/IP can operate with multiple data-link-layer protocols. The following are examples of this:

- ETHERNET
- FDDI
- X.25
- Token ring
- 802.X series of protocols

These issues should, indeed, be understood and considered prior to purchase of a gateway. Actually, it would be best to have any vendor you are working with explain explicitly *their* gateways' functionality in such a way that *no* ambiguity exists.

29.3 SNA-TCP/IP Gateways

Perspective

A finite number of upper- and lower-layer protocols exist and have been explained in this book. As a result, it is possible to identify popular networking protocols in light of their percentage in the worldwide marketplace.

SNA is probably the most dominant vendor network protocol in the world. TCP/IP is probably the most dominant nonvendor protocol in the world. However, an incredible number of vendors support TCP/IP or offer it as a network protocol. DECnet has a considerable amount of the market share as well. The same can be said about NetWare; it is prevalent also. AppleTalk also has a considerable share in the market. OSI is gaining market share as time passes.

To estimate the percentage that these networking protocols command in terms of market share is difficult because the market has been, and is, in such a state of flux. One thing is certain—these protocols are dominant. In terms of gateways, SNA-based gateways certainly have a considerable share in the market. Of this type gateway, TCP/IP and DECnet are probably the most significant protocols to be integrated into SNA via gateways.

SNA and TCP/IP have received considerable attention in the past 6 to 8 years. It seems TCP/IP has penetrated many institutions and companies that have traditionally been SNA-based. Interestingly, even DEC systems support TCP/IP. This means DEC customers have an option for protocol use; they can use TCP/IP, DECnet, or OSI as it is supported in the DEC environment. In this section, however, the focus is on SNA and TCP/IP gateways.

Multiple types of SNA-TCP/IP gateway implementations exist. Most possibilities are presented here. From the users' perspective, the question is which best meets their needs.

Hardware gateways

Hardware gateways almost always operate at all seven layers within a SNA-TCP/IP network. Figure 29.8 illustrates this type of gateway.

In this scenario, the Brown & Williams Insurance agency has headquarters in Dallas and another major office in Hattiesburg (Miss.). The office in San Diego was recently acquired because of a recommendation from Tricia. Ginny, however, is responsible for all corporate operations including those in Hattiesburg, for which Mrs. Harris is responsible for daily operations.

The final result is a hardware gateway in the San Diego office. This provides interoperability between the San Diego office and both the Dallas and Hattiesburg offices. This arrangement enables users in the Hattiesburg office to use files on hosts in the San Diego office.

This type of gateway is required because the San Diego office has an ETHERNET-based TCP/IP network. In Dallas and Hattiesburg, SNA is the network protocol and native lower-layer protocols are used.

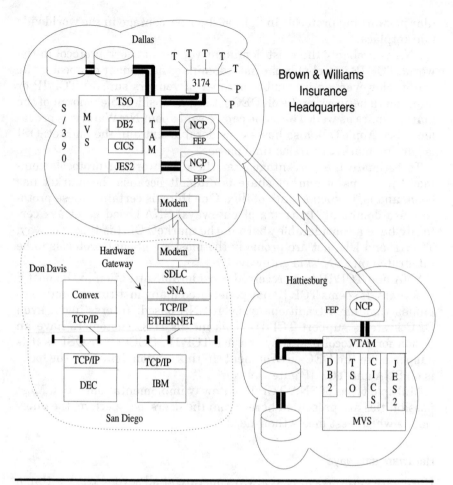

Figure 29.8 Hardware gateway installation.

Consequently, this means that in order for a gateway to integrate San Diego and the SNA offices, ETHERNET and TCP/IP must be converted into SNA and the appropriate lower-layer protocol. The most cost-effective solution in this case is a hardware-based gateway.

In this example both upper- and lower-layer protocol conversion is performed on the hardware gateway. Data translation may or may not be performed on the gateway. Data translation in this case is contingent on how the vendor implements this type of hardware gateway.

This example of an ETHERNET-based TCP/IP network connected to multiple SNA networks is a common scenario in the internetworking community today. A significant number of vendors provide gateways such as this.

TCP/IP on a SNA host

A different SNA-TCP/IP solution is if TCP/IP is loaded onto an SNA host. When this is the case, TCP/IP is GENed as a TCP/IP application. Figure 29.9 is an example of this scenario.

TCP/IP operates as a VTAM application as shown in Fig. 29.9. Just because TCP/IP runs as a VTAM application does not mean that it does not perform protocol conversion. In fact, when SNA and TCP/IP are integrated, protocol conversion and data translation always take place; the only question is where. In this example protocol conversion does, in fact, have to be performed on the host. Data translation may or may not be performed on the host. The location of data translation depends on a number of factors, especially the type of remote log-on and file transfer mechanism that are used.

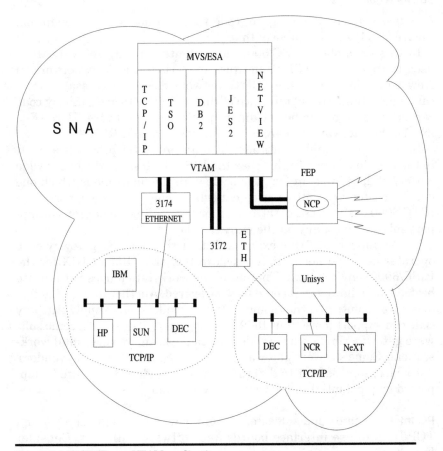

Figure 29.9   TCP/IP as a VTAM application.

Off-load

IBM has what they call an *off-load option,* in which TCP/IP is loaded on a SNA host, but off-loaded onto a 3172 interconnect controller and then only with the customer-requested applications running on the host itself. Figure 29.10 illustrates this scenario.

Figure 29.10 shows TCP/IP on the host, on the 3172, and on disk. So, *where is it?* TCP and IP have been off-loaded onto the 3172 processor. Those applications selected can function where TCP/IP is indicated on the hosts, and the remainder of the TCP/IP suite that is not needed is on hard disk. This type of implementation is versatile. As Fig. 29.10 shows, even a remote TCP/IP network (shown as LAN #1) can be integrated into the TCP/IP network connected to the 3172 processor via a remote ETHERNET bridge.

Software gateways

A *software gateway* is program code that operates on a workstation. Figure 29.11 is an example of this.

In this example a RISC/6000 implements gateway software. It is used to connect the TCP/IP network in Paris to the SNA network in New York. Notice that the TCP/IP network does not necessarily indicate a data link. The significance of this is that software gateway code can operate on a number of different vendors' hardware. Here (Fig. 29.11), the gateway software is operating on an IBM RISC/6000.

In this type of implementation the data-link-layer protocols are left to the host to resolve. In this case the RISC/6000 supports a very wide variety of data links; in fact, the issue with this machine is which one does a customer want to use, not which ones can it support. The RISC/6000 host has one of the most robust data-link-layer protocol support subsystems of any on the market today.

The significance of this example is that the software gateway code operates as any other initiated task in the AIX (IBM's UNIX for the RISC/6000) environment. Therefore, the workstation does not have to be dedicated; however, it could be if so desired. Another interesting fact about this solution is that some software vendors who supply gateway code can support more than 1000 LUs (logical units). In fact, some software gateway vendors have code that can operate on very small workstations. Some software gateway vendors support only certain vendors' hardware; so, it is best to determine which vendors' hardware is supported by a potential software gateway supplier.

PC interface cards and software.   Another way to integrate SNA and TCP/IP is to use interface boards and software with PCs. Consider Fig. 29.12.

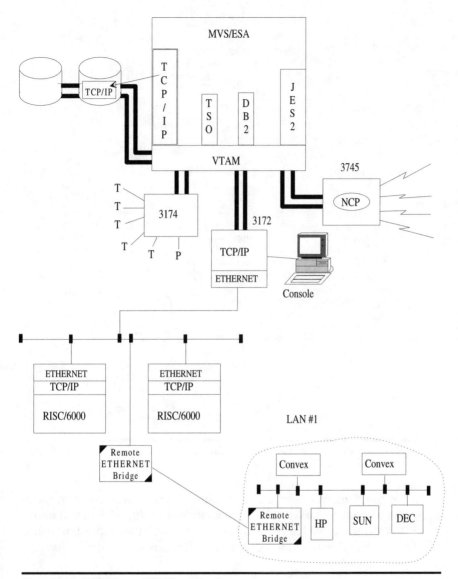

Figure 29.10 TCP/IP as a VTAM application.

Figure 29.12 may be the best solution if this type of configuration applies to the particular situation. However, it is just one of the four previously mentioned. In the example shown in Fig. 29.12 there are multiple PCs located on the site where the S/390 host is located. The PCs have interface boards and software inside, and are attached by coaxial cable to the 3174 establishment controller.

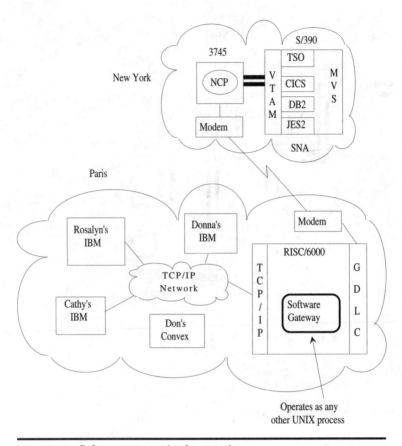

Figure 29.11 Software gateway implementation.

As the previous five examples have shown, there are several ways to integrate SNA and TCP/IP. The last example (Fig. 29.12) depicted multiple PCs with interface boards attached to a local 3174. Indeed, multiple methods of bringing together SNA and TCP/IP are currently available.

29.4 Why Use a Gateway?

To achieve interoperability between SNA and TCP/IP protocol, conversion and data translation must be performed. The only question is where these functions are performed. The previous section provided examples of where protocol conversion was performed: primarily upper-layer protocols. However, it did not indicate where data translation occurs or what user-oriented functions are supported. Some of those aspects are covered in this section.

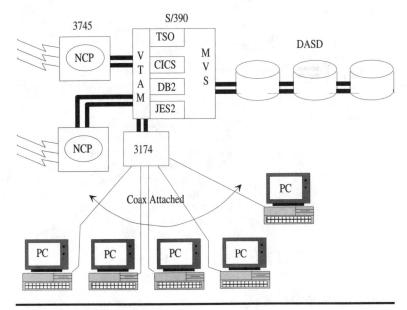

Figure 29.12   Interface boards.

Remote log-on

In SNA and TCP/IP, much of the traffic is considered inbound, that is, to the SNA host. Any log-on from a TCP/IP-based network to a SNA-based network is considered a remote log-on, even if the network is in the same room with the SNA host. When SNA and TCP/IP are integrated via a gateway, the issue of data translation is paramount. In the SNA arena multiple data streams are supported, but the one used interactively is a 3270 data stream. This is significant because 3270 data streams are based on an EBCDIC character set. Most traditional data streams in TCP/IP networks have been based upon the ASCII character set. There are multiple ways to solve this one issue of data translation. Consider Fig. 29.13 as an example.

Figure 29.13 shows a Convex user invoking a TELNET client native to the TCP/IP stack on that host. The user executes the client against the TELNET server on the gateway between the network in Salt Lake City (Utah) and Boise (Idaho). The first part of the connection, labeled *A*, establishes a logical link between the convex host and the gateway. The second portion of the connection, labeled *B*, is established from the gateway to VTAM on the S/390 host. The Convex user *thinks* one logical connection exists between the Convex terminal and VTAM. This type of connection is called a "raw" TELNET because the TELNET client out of the native TCP/IP stack is used to establish a logical con-

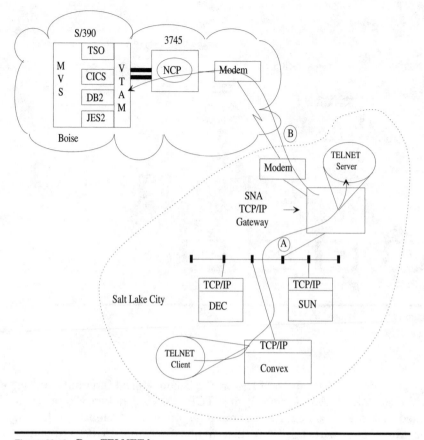

Figure 29.13 Raw TELNET log-on.

nection. In this case data translation is performed on the gateway; this may be either a hardware or software gateway.

Figure 29.14 is another example of how a remote log-on can be achieved.

Figure 29.14 shows a TCP/IP-based network in New York and a SNA network in Ottawa (Can.). Notice the TN3270 client application shown on the IBM host in the New York network. It indicates via logical connection A that it connects to the TN3270 server on the gateway connecting the two networks. The second part of the connection between New York and Ottawa is shown via connection B. In this case data translation occurs on the IBM host.

This is so because TN3270 client applications are designed to output a 3270 data stream. The fact that it is both TCP/IP- and ETHERNET-based is insignificant because data formatting is performed at the presentation layer within a network. By the time any data reaches the physical interface of the host, the data is represented in binary code. Hence, what is shipped across the ETHERNET is ETHERNET frames.

For a TN3270 application to operate, however, one of two conditions must be met. Either a TN3270 server must be present as shown on the gateway in Fig. 29.14, or the SNA-based host must have TCP/IP on it to enable the TELNET server native to that stack to answer the request of the TN3270 client. Figure 29.15 illustrates this.

In Fig. 29.15 a TN3270 client application is present on the Unisys host. A user invokes it against the SNA host attached to the TCP/IP-based ETHERNET network. Notice that the TN3270 client application is executed against the TCP/IP stack running as a VTAM application on the SNA-based host. Why would this be done? When this type of scenario is implemented, it operates this way because within the TCP/IP

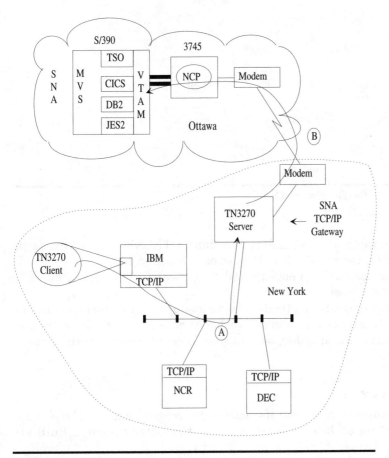

Figure 29.14   TN3270 client and gateway.

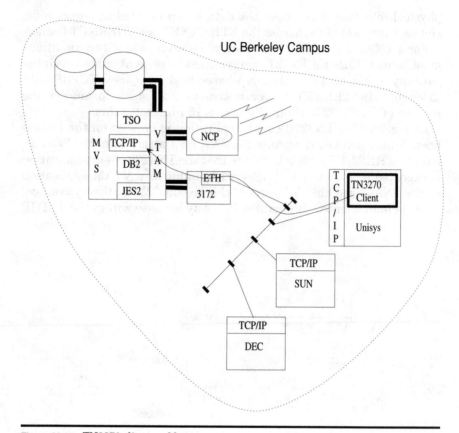

Figure 29.15   TN3270 client and host.

stack running as a VTAM application, a TELNET server exists by default. Consequently, the data stream inbound to the SNA hosts is 3270 (no conversion needed), and this saves CPU cycles—and CPU cycles cost money.

Other proprietary methods exist that support a remote log-on from a TCP/IP-based network and a SNA-based network. Since these are proprietary (vendor-specific), they may or may not coincide with industry standards.

File transfers

File transfers can be interactive or programmatic. The former implies human intervention, whereas the latter does not. Both are discussed here.

Interactive.   There are several ways to perform file transfer between SNA environments and TCP/IP networks. A common way to do this is via FTP. Figure 29.16 is an example of this.

This example shows a SNA network, TCP/IP network, and gateway connecting them together. Notice that the FTP client has been invoked on the Sun (Microsystems) host and is executed against the FTP server native to the TCP/IP stack running as a VTAM application. As this figure shows, file XYZ is written from the DASD in the SNA environment to the Sun disk.

Another method for file transfer between SNA and TCP/IP environments includes support for the IND$FILE program IBM has supported. Figure 29.17 is an example of this.

Figure 29.17 shows a DEC user invoking a program that supports the IND$FILE program by TSO and CICS under VTAM. As the figure shows, file XYZ is moved from DASD on the MVS hosts to a disk connected directly to the DEC host.

Programmatic.   Files can be transferred programmatically. There are several reasons for this capability. Some companies need to download files at night when little or no operations are being performed within the company. Others need programmatic file transfers because of the nature of the interacting entities. Figure 29.18 best depicts this scenario.

Figure 29.18 is an example of two different types of file transfers being performed. On the left portion of the drawing is LU6.2 file transfer using a hardware gateway with proprietary LU6.2 library support on the gateway itself and the host. To the right side of the figure, socket communication is being performed where socket programming with CICS is communicating with a socket-based UDP program on host C in the figure.

Mail

Mail support between SNA and a TCP/IP environment is best depicted by Fig. 29.19.

Figure 29.19 is an example of a SNA network in Oak Harbor (Wash.) and a TCP/IP network in Portland (Oreg.). A SNA-TCP/IP gateway is used to connect the two networks. This particular gateway supports SMTP-to-PROFS (professional office system) mail and vice versa. As a result, those in Oak Harbor can send mail to those users on the network in Portland. All users in their respective networks view mail in the native format that they are accustomed to using. Some vendors have such support for gateways; some do not.

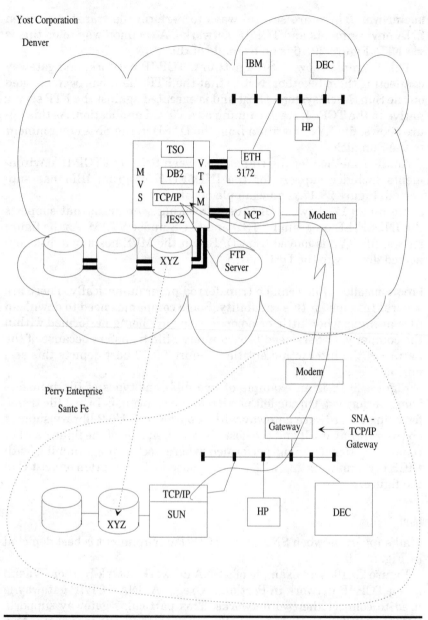

Figure 29.16   FTP client.

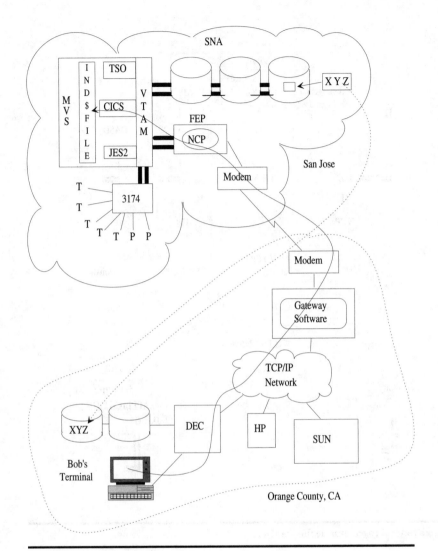

Figure 29.17 IND$FILE file transfer.

Management support

The notion of network management is a topic by itself; however, it is important to know whether the network devices to be integrated into an internetworking arena support the network management used to manage that (those) network(s). SNA and TCP/IP networks connected together via a gateway means that the gateway should support network management for SNA and TCP/IP. In addition, it should have its

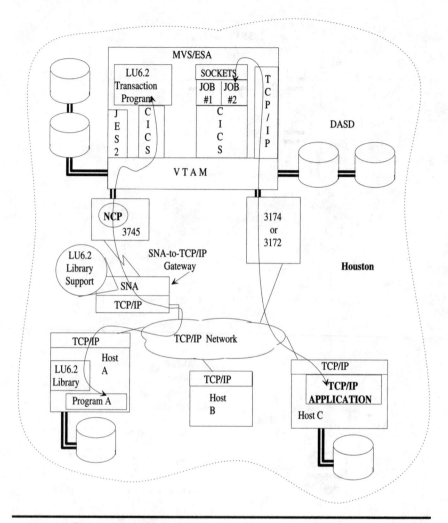

Figure 29.18 Programmatic file transfer.

own management capabilities, meaning that it should be manageable without outside assistance or intervention. Figure 29.20 is an example of network management on both sides of the gateway.

Figure 29.20 shows SNMP as TCP/IP network management. It also shows NetView as SNA network management. No particular management is shown for the gateway itself because this is vendor-specific. However, the gateway should have some functional management capabilities that permit systems personnel access to information from the gateway if needed.

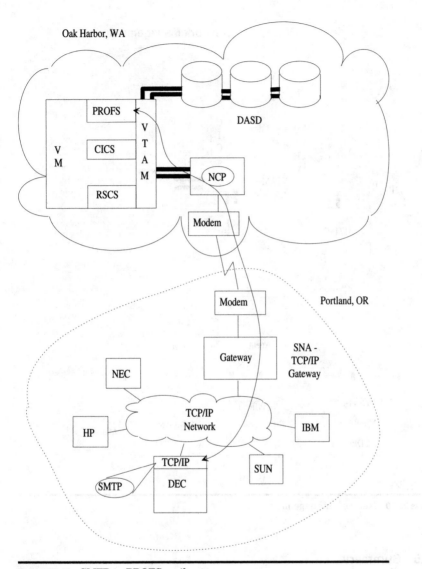

Figure 29.19 SMTP-to-PROFS mail gateway.

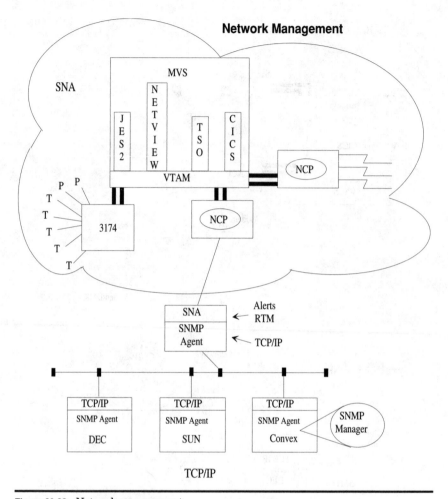

Figure 29.20   Network management.

29.5 Summary

The term *gateway* means different things to different groups of individuals. However, the *American Heritage Dictionary* does give general-usage definitions of the term. Among the Internet community in the 1970s the term acquired a new meaning: routing. The term did connote an entryway, but technically speaking, it was used to convey what the term router means today.

Different types of gateways are marketed today. Gateways to permit connectivity into DECnet, TCP/IP, NetWare, SNA, OSI, and AppleTalk can all be purchased. Gateways perform protocol conversion by default;

this is what makes them gateways. Gateways can operate at layers 3 and above relative to the OSI model, or they can operate at all layers within a network.

SNA- and TCP/IP-based gateways are quite popular today. These two environments include a considerable amount of the internetworking community around the world. Other network protocols have shares in the market as well, but even the Digital Equipment Corporation (DEC) supports TCP/IP for those who want to use it with DEC hosts. DEC also supports OSI.

Multiple types of gateways exist. There are hardware- and software-based gateways. Regarding TCP/IP and SNA, TCP/IP can be loaded as a VTAM application and operated in this way, or it can be off-loaded onto a 3172 interconnect controller. Other possibilities may be available in the near future.

Gateways are used to bring together heterogeneous networks. When this is achieved, multiple user-oriented tasks can be achieved. For example, remote log-ons, file transfers—both interactive and programmatic, and electronic mail can be exchanged; network management can be utilized; and customized offerings may be provided by some vendors.

Gateways are available in a variety of packaging and support a variety of different end-user application and systems-oriented tasks. It is best to check with a specific vendor to determine what is supported by a given gateway.

Glossary

abend   Abnormal end of task.

abstract syntax   Machine-independent types and values, defined using an ASN.1.

Abstract Syntax Notation One   The OSI language used to describe abstract entities or concepts in a machine.

ACB   In VTAM, access (method) control block. In NCP, an adapter control block.

ACB name   The name of a microinstruction. A name typically specified on the VTAM APPL definition statement.

accept   In a VTAM application program, this means to establish a session with a logical unit in response to a CINIT request.

access control entry   Specifies identifiers and access rights to be granted or denied to holders of the identifiers, default protection for directories, or security alarms.

access control list   A list defining the kinds of access granted or denied to users of an object.

access method   A technique for moving data between main storage and input/output devices and for telecommunications.

access method control block (ACB)   A control block that links an application program to VSAM or VTAM.

ACF/VTAM   Advanced communications function for the virtual telecommunications access method.

acknowledgment   A positive response sent indicating successful reception of information.

acquire   In VTAM, to take over resources that were formerly controlled by an access method in another domain.

activate   To initialize a resource.

active   Operational. The state of a resource when it has been activated and is in operation.

active application   An application currently capable of being used by a user.

active window The terminal window where current operations are in the foreground.

ACTLU In SNA, a command used to start a session with a logical unit.

ACTPU In SNA, a command used to start a session with a physical unit.

adapter control block (ACB) (1) In NCP, a control block that contains line control information and the states of I/O operations. (2) A definition in an I/O database describing an adapter or device controller and the I/O interconnect.

adaptive session-level pacing Pacing where session components exchange pacing windows that may vary in size during the course of a session.

address In data communication, this is a designated identifier.

address descriptor record Information that reflects the address of an Apple event.

address mask A bit mask used to select bits from an IP address for subnet addressing.

address resolution Conversion of an IP address into a corresponding physical address, such as ETHERNET or token ring.

address resolution protocol (ARP) A TCP/IP protocol used to dynamically bind a high-level IP address to low-level physical hardware addresses. ARP works across single physical networks and is limited to networks that support hardware broadcast.

address space Addresses used to uniquely identify network-accessible units, sessions, adjacent link stations, and links in a node for each network in which the node participates.

addressing In data communication, the way in which a station selects the station to which it is to send data. An identifiable place.

adjacent control point A control point directly connected to an APPN, LEN, or composite node.

adjacent NCPs Network control programs connected by subarea links with no intervening NCPs.

adjacent nodes Two nodes connected by at least one path that connects no other node.

adjacent SSCP table A table identifying SSCPs that VTAM can enter into a session with.

adjacent subareas Subareas connected by one or more links with no intervening subareas.

Advanced Communications Function (ACF) A group of IBM licensed programs.

Advanced Peer-to-Peer Network (APPN) An upper-layer networking protocol based on peer technology. The fundamental difference between APPN and SNA is that the former is "peer" oriented whereas the latter is hierarchial by design.

Advanced Peer-to-Peer Network (APPN) end node   A node that can register its local LUs with its network node server. It can also have links to multiple nodes, but can have only one CP-CP session with a network node at a time. This type node can attach to subarea SNA as a peripheral node.

Advanced Peer-to-Peer Network (APPN) interchange node   This node can be characterized by its functions, which include controlling network resources, performing CDRM functions in subarea networks, and owning NCPs. This node appears to be an APPN node to an APPN network, as well as a subarea node to a subarea network. It can reside between an APPN network and a subarea network, thus providing integration of the two.

Advanced Peer-to-Peer Network (APPN) LEN node   A node that provides end-user services without explicit, direct use of CP-CP sessions.

Advanced Peer-to-Peer Network (APPN) network node   An APPN network node performs the following functions:

Distributed directory services

Intermediate routing services in an APPN network

Network services for specific end nodes

Intermediate session routing

It is also the management service focal point. The APPN network node cooperates with other network nodes to maintain a network topology database, which is used to select optimal routes for LU-LU sessions based on requested classes of service.

Advanced Program-to-Program Communication (APPC)   A protocol based on T2.1 architecture utilizing LU6.2 to accomplish peer communications. Many references to APPC are simply LU6.2. The basic meaning of APPC is communication between programs, i.e., transaction programs. This type of protocol permits communication between programs written in different languages.

Advanced Research Projects Agency (ARPA)   A U.S. government entity focused upon what would become known as "connectivity." From this agency came the ARPANET—later to be known as the Internet.

alert   In SNA, a message sent to a management subsystem typically using network management vector transport (NMVT) protocol.

alertable   In DECnet architecture, a synchronous alert that delivers data to specific points.

alertsafe   According to DEC (Digital Equipment Corp.) documentation, a routine that can be called without risk of triggering an alert while asynchronous delivery of alerts is enabled.

alias file   A file that contains a pointer to another file, directory, or volume.

alias name   A naming convention sometimes used to refer to a different name which means the same as the alias or that refers to another name which may be used as a pointer.

allocate   A LU6.2 verb that assigns a session to a conversation.

allocation class   According to DEC documentation, this is a unique number between 0 and 255 that the system manager assigns to a pair of hosts and to the dual-path devices that the hosts make available to other nodes in a VMS cluster.

alpha primary bootstrap   According to DEC documentation, the primary bootstrap program that initializes an AXP system with OpenVMS AXP.

alternate route   A route which is not the primary method of moving data from point to point.

ampersand (&)   When used in the command of most UNIX operating systems, it places the task or tasks in background operation.

ancillary control process   A process acting as an interface between software and an I/O driver.

Apple event   According to Apple documentation a high-level function that adheres to the Apple event interprocess messaging protocol.

Apple event handler   According to Apple documentation, this is a defined function that extracts pertinent data from an Apple event.

Apple event interprocess messaging protocol   A standard defined by Apple Computer, Inc., for communication among applications. According to Apple documentation, high-level events that adhere to this protocol are called *Apple events.*

Apple event parameter   According to Apple documentation, this is a keyword specifying a descriptor record containing data which the target application of an Apple event must use.

Apple file exchange   A program that permits file transfer between Apple computers and DOS-based computers.

AppleShare file server   Software that permits users to perform tasks in AppleTalk networks.

AppleShare print server   A Macintosh computer-software combination that manages network printing.

AppleTalk data stream protocol (ADSP)   This is an AppleTalk protocol that provides the capability for maintaining a connection-oriented session between entities on an internet.

AppleTalk echo protocol (AEP)   A response-oriented protocol whereby a response is sent when a packet is received.

AppleTalk filing protocol (AFP)   The protocol used between an application and a file server. AFP is a client of ASP.

AppleTalk Internet router   Router software operating on an Apple computer supporting up to eight AppleTalk networks.

AppleTalk Phase 2   Introduced in June 1989, it extended the capability of AppleTalk Phase I; for example, it supports token ring and provides more efficient routing techniques. It is included in System/7 software.

AppleTalk session protocol (ASP)   An AppleTalk session protocol used to establish and maintain logical connections between a workstation and a server.

AppleTalk transaction protocol (ATP)   An AppleTalk protocol that functions in many ways like a connection-oriented protocol. For example, it orients the packets into the order in which they need to be received, and if any packets have been lost, automatic retransmission of a packet(s) is performed.

AppleTalk transition   A message containing information indicating specific occurrences that relate to the AppleTalk transition queue.

AppleTalk transition queue   An operating system queue.

application layer   According to the ISO OSI model, this is layer 7. It provides application services.

application program   Software that provides functions needed by a user. A user may be defined as a human, API, transaction program, or other logical entity.

application program interface (API)   An addressable point that serves the function of bringing together two or more entities.

application program major node   According to IBM documentation, in VTAM, this is a group of application program minor nodes.

application result handler   A program designed to perform predefined functions of results generated from applications.

application server   A computer used solely to provide processing power for application programs.

application service element (ASE)   A definition explaining the capabilities of an application entity.

application service object   A subobject of an application entity; it may contain ASE elements or service objects as subjects.

application transaction program   The program built around LU6.2 protocols.

apply   A system modification program (SMP) command, used in certain SNA environments, whereby programs and/or program fixes can be added to system libraries.

APPN intermediate routing network   An APPN network consisting of network nodes and their interconnections.

argument list   According to DEC, a vector of entries representing a procedure parameter list and possibly a function value.

argument pointer   According to DEC, general register 12 on VAX systems. By convention, AP contains the address of the base of the argument list for procedures initiated using call instructions.

ARPANET   Served as central backbone for the Internet during the 1970s and some of the 1980s.

ASCII   American Standard Code for Information Interchange. A standard code using a character set based on binary digits (1s and 0s). This character set lists and defines letters, symbols, numbers, and specialized control functions.

assignment statement   In Digital Command Language (DCL), the association of a symbol name to use with a specific character string or numeric value. In

DEC implementations, symbols can define synonyms for system commands or can be used for variables in command procedures.

association control service element   Used to establish and terminate associations between applications.

asynchronous (ASYNC)   In data communication, a term referring to transmission of data without synchronous protocol.

asynchronous system trap   A method of identifying a specific event via a software-simulated interrupt.

asynchronous transfer mode (ATM)   A CCITT standard defining cell relay. This is where data of multiple types of services, such as data, voice, and/or video, is moved in fixed-size cells throughout a network.

attenuation   A term used with fiber optics. It refers to the reduction of signal loss, measured in decibels.

attribute   From a security perspective, it is a identifier or holder of an identifier. When used in a conversation concerning threads, it specifies detailed properties of the objects to be created.

attributes object   A term describing details of the objects to be created.

authentication   A process of establishing identity.

authorized path   In VTAM under MVS, a facility that enables an application program to specify that a data transfer or related operation be carried out in a privileged and more efficient manner.

automatic log-on   In SNA, a process by which VTAM automatically creates a session-initiation request to establish a session between two logical units (LUs). The result of this request is the SNA BIND command from the primary logical unit (PLU) to the secondary logical unit (SLU). Hence, a LU-LU session is established if the BIND image is correct and other parameters are accurate.

automatic record locking   According to DEC documentation, it is a capability provided by OpenVMS record management services that allows a user to lock only one record in a specific shared file at any given time.

A/UX toolbox   According to Apple documentation, this is a library that enables a program running under the A/UX operating system to call Macintosh user interface toolbox and operating system routines.

background process   A term used in most UNIX environments. It means that a process does not require total attention of the system for operation.

backplane interconnect   According to DEC documentation, VAX systems have an internal processor bus that allows I/O device controllers to communicate with main memory and the central processor. These I/O controllers may reside on the same bus as memory and the central processor, or they may be on a separate bus.

BASE disk   In SNA, and specifically the VM operating system, this is the disk containing text decks and macroinstructions for VTAM, NetView, and VM/SNA console support (VSCS).

base priority   Priority a VM system assigns to a process when it is created. Normally, it comes from the authorization file.

basic logical object   An object in a specific logical structure that has no subordinate.

baud   The number of times per second a signal can change on a transmission line.

begin bracket   In SNA, the value of the begin-bracket indicator in the request header (RH). It is the first request in the first chain of a bracket. Its value denotes the start of a bracket.

Berkeley broadcast   A nonstandard IP broadcast address using all zeros instead of all ones in the host portion.

best-effort delivery   A description of network technologies that do not provide reliability at link levels.

Bias   A DC voltage used in electronic circuitry that holds a desired time spacing of transient states from a mark to a space.

Big Endian   A format for storage or transmission of binary data in which the most significant byte comes first.

binary synchronous communication (BSC)   A telecommunication protocol using a standard set of transmission control characters and control character sequences.

BIND   In SNA, this is a request for session activation between logical units.

BIND command   In SNA, this is the command used to start a session and define the characteristics thereof.

BIND image   In SNA, this is a session parameter used to establish and govern the session between logical units. In VTAM, the BIND image is located in the LOGMODE table in the form of entries—one entry per type of image required to define the LU.

BIND pacing   In SNA, BIND pacing can be used to prevent a BIND standoff. It is a technique used by the address space manager (ASM).

bitmap   Generally speaking, an array of data bits used for graphic images.

bitmap device   A device that displays bitmaps.

bitmap font   A bitmap font is created from a matrix of dots.

BIU segment   Basic information unit segment. In SNA, data contained within a path information unit (PIU). It consists of either a request (or response) header (RH) followed by all or part of a request (or response) unit (RU), or only part of an RU.

blocking AST   An asynchronous system trap that can be requested by a process using the lock management system services. A blocking AST is delivered to the requesting process when it is preventing another process from accessing a resource.

boot server   According to DEC documentation, this is the management center for a VMScluster system and its major source provider. The boot server provides disk access and loads satellite nodes downline.

bootstrap block   That part of the index file on a system disk that contains a program which loads the operating system into memory.

boundary function   In SNA, protocol support for attached peripheral nodes.

boundary node (BN)   A boundary node fundamentally performs the function of transforming network addresses to local addresses.

bracket   In SNA, one or more chains of RUs and their responses that are exchanged between session partners.

bracket protocol   A data flow control protocol in which session partners exchange data via IBM's SNA bracket protocol.

broadcast   In TCP/IP, a request for logical connection at layers 1 and 2 in the network. Broadcast technology is exemplified by ETHERNET.

broadcast search   In APPN, the simultaneous search to all network nodes in the form of a request for some type of data.

broadcast storm   A situation in a network using broadcast technology where a considerable number of broadcasts are put on the medium at one time.

browse   With regard to functions that can be performed on an entity, browse merely permits viewing.

Btrieve   According to Novell documentation, this is a complete indexed record management system designed for high-performance data handling.

bucket   According to DEC documentation, this is a storage structure used for building and processing files. A bucket contains one or more records or record cells. Buckets are the unit of contiguous transfer between OpenVMS record management services buffers and the disk.

buffer   A temporary storage area used to hold input or output data.

bundle bit   The FINDER uses information in a bundle bit (BNDL resource) to associate icons with the file.

button   Generally considered to be the button which is pressed on a mouse. Mice may have two or three buttons. Two-button mice are common in non-XWindow environments. Three-button mice are common in XWindow environments and can be used in non-XWindow environments, depending on the mouse vendor.

cache   A high-speed storage buffer.

call   To invoke a program, routine, or subroutine.

call-connected packet   A DTE packet transmission indicating DCE acceptance of the incoming call.

Call progress signal   Communication from the DCE to the calling DTE indicating status of a call.

call request packet   In X.25 communications, a call supervision packet transmitted by a DTE to ask for a call establishment through the network.

calling   The process of transmitting selection signals to establish a connection between data stations.

cancel   To terminate.

carrier   In data communication, a continuous frequency capable of being modulated.

carrier sense   A device (transceiver, interface board, or other entity) capable of detecting a constant frequency.

carrier-sense multiple access with collision detection (CSMA/CD)   A protocol utilizing equipment capable of detecting a carrier which permits multiple access to a common medium. This protocol also has the ability to detect a collision, because this type of technology is broadcast-oriented.

casual connection   In SNA, this type of connection is made in subarea networks with PU T5 nodes attached via a boundary function using low-entry networking capabilities.

catalog   In SNA, a list of pointers to libraries, files, or data sets.

CCW   In SNA, channel command word.

CDRM   In SNA, refers to cross-domain resource manager.

CDRSC   In SNA, refers to cross-domain resource.

central directory server   According to IBM documentation, this is an APPN network node that provides a repository for network resource locations.

central processing unit (CPU)   The circuitry that executes instructions.

channel   Generic term for a path over which data can move. In most IBM documentation, the word *channel* has a specific meaning. For example, IBM supports two types of channels today: parallel and serial.

channel adapter   A device used to attach the communication controller to a host channel.

channel-attached   In SNA terminology, this term connotes a serial or parallel data-link protocol. IBM has other data-link protocols, such as SDLC, ESCON, and token ring.

channel link   A data-link connection between two devices.

character-coded   An SNA term used in VTAM meaning unformatted.

chooser   According to Apple documentation, an accessory that lets a user select shared devices, such as printers and file servers.

CI-only VMScluster configuration   According to DEC documentation, this is a type of VMScluster configuration in which the computer-interconnect (CI) device is used for most interprocessor communication. In these configurations, a node may be a VAX processor or a hierarchical storage controller (HSC).

CINIT A request sent from a SSCP to a primary logical unit requesting that a BIND command be issued.

circuit switching Connectivity on demand between DTEs and DCEs.

cladding The surrounding part of a cable that protects the core optical fibers. This part is located between the fibers and the cable jacket.

class-of-service (COS) In SNA, this term defines explicit routes, virtual routes, and priority, and is used to provide a variety of services within the network.

class-of-service database In APPN, this is a database maintained independently by each network node, and optionally by APPN end nodes.

cleanup A term used in SNA referring to how sessions are terminated between LUs. Specifically, it is a network services request that causes a particular LU-LU session to be ended immediately.

clear-indication packet A call supervision packet that data circuit-terminating equipment (DCE) transmits to inform data terminal equipment (DTE) that a call has been cleared.

clear-request packet A call supervision packet transmitted by a DTE to ask that a call be cleared.

click To press *and release* a mouse button.

client A term used to connote peer technology. Clients are programs used to initiate something. In TCP/IP clients can be found in TELNET and FTP. In the XWindowing system, a client is typically a program that conforms to the X protocol and works in conjunction with a X server.

client application In Apple environments, this is an application using Apple event interprocess messaging protocol to request a service, for example, printing a list of files or spell-checking a list of words.

CLNP Connectionless network protocol. A protocol that does not perform retransmissions or perform error recovery at a transport layer.

clocking The use of clock pulses to control synchronization.

closed user group (CUG) A user group that can communicate with other users in the group, but not with users outside the group.

cluster A network of computers in which only one computer has file system disk drives attached to it.

cluster controller In SNA, this is the precursor to the establishment controller. It is a device to which terminals and printers attach. IBM's 3274 cluster controllers appear as a PU2.0.

CMOT The use of common management information service over TCP/IP. Succinctly, it is the implementation of the OSI network management specification over TCP/IP.

collision An event in which two or more devices simultaneously perform a broadcast on the same medium. This term is used in ETHERNET networks, and also in networks where broadcast technology is implemented.

collision detection   Term used to define a device that can determine when a simultaneous transmission attempt has been made.

command   (1) A request to execute an event. (2) According to DEC, when this is used in reference to Digital Command Language (DCL), it means an instruction, generally an English-type word, entered by the user at a terminal or included in a command procedure.

command facility   A component of IBM's NetView program that is the base for command processors that can monitor, control, automate, and improve the operation of an SNA network.

command-line prompt   Generally, this refers to a place on a terminal where commands can be entered. The direction of these commands is contingent on the system or software from which the command line is generated. According to Hewlett-Packard's documentation, it is that which appears on the screen immediately after log-in. Usually the command-line prompt is either a $ (for Bourne and Korn shells) or a % (for C shells), but it can be modified. A popular modification is to print the current working directory and the history stack number before the $ or %. You can find the command-line prompt by pressing RETURN several times. Every time you press RETURN, the HP-UX operating systems prints the prompt.

command list (CLIST)   In IBM's SNA, this is a list of commands and statements designed to perform a specific function for the user. Examples of CLISTs are NetView, TSO, and REXX.

common management information protocol (CMIP)   The OSI protocol for systems management.

common management information service element.   The application service element responsible for relaying systems management information.

common parent   The lowest-level directory that appears in pathnames of multiple files or directories on the same volume.

communication adapter   Generally agreed to be a device (normally an interface) that provides a common point for a device and the data communication device being used.

communication control unit   A communication device that controls transmission of data over lines in a network.

communication controller   A control unit whose operations are controlled by one or more programs stored and executed in the unit. In most instances it manages lines and routing data through a network.

communication identifier (CID)   A VTAM key for locating the control blocks that represent a session. This key is created during session establishment and is deleted when the session ends.

communication line   Deprecated term for telecommunication line.

communication network management (CNM)   According to IBM documentation as the term applies to SNA, it is the process of designing, installing, operating, and managing distribution of information and control among users of communication systems.

communication network management (CNM) application program According to IBM documentation, it is a VTAM application that issues and receives formatted management service request units for PUs. NetView is an example of a CNM application program.

communication network management (CNM) interface A common point where applications can move data and commands to the access method. This data and these commands are associated with communication system management.

component Generally speaking, this can be hardware or software.

composite LEN node A group of nodes made up of a single type 5 node and subordinate type 4 nodes. To a type 2.1 node, a composite LEN node appears as one LEN node. Examples of a composite LEN node are NCP and VTAM.

composite network node (CNN) A group of nodes made up of a type 5 node and its subordinate type 4 nodes that appear as a single APPN network node to the APPN network.

composite node In IBM networking, specifically SNA with VTAM, a type 5 node and its owned type 4 nodes that collectively appear as a single node to other APPN nodes in an APPN network.

computer interconnect (CI) A fault-tolerant, dual-path bus, with a bandwidth of 70 Mbits/s.

concurrency controls Methods provided by Btrieve to resolve possible conflicts when two applications attempt to update or delete the same records at the same time.

configuration The manner in which the hardware and software of an information processing system are organized and interconnected.

configuration restart In SNA, the VTAM recovery facility that can be used after a failure or deactivation of a major node, VTAM, or the host processor to restore the domain to its status at the time of the failure or deactivation.

configuration services A type of network service in a control point. Configuration services activate, deactivate, and record the status of PUs, links, and link stations.

congestion loss In DEC's DECnet, a condition in which data packets transmitted over a network are lost when the DECnet for the OpenVMS routing layer is unable to buffer them.

connected To have a physical path from one point to another.

connection In data communications, there are two types of connections: physical and logical. A physical connection consists of a tangible path between two or more points. A logical connection is capability to communicate between two or more endpoints.

connection control block A data structure that is used by ADSP to store state information about the connection end.

connection end The combination of an AppleTalk socket and the ADSP information maintained by the socket client.

connection listening socket A socket that accepts ADSP requests to open connections and passes them on to a socket client.

connection network A representation within an APPN network of a shared-access transport facility (SATF), such as a token ring, that allows nodes identifying their connectivity to the SATF by a common virtual routing node to communicate without having individually defined connections to one another.

connection-oriented internetworking A set of subnetworks connected physically and thus rendered capable of connection-oriented network service.

connection-oriented service A type of service offered in some networks. This service has three phases: connection establishment, data transfer, and connection release.

connectivity The notion of device communication interchange, even if such devices are diverse.

connection server A program that accepts an open connection request passed to it by a connection listener and selects a socket to respond to the request.

connectionless internetworking A set of subnetworks connected physically, thus rendered capable of providing connectionless network service.

connectionless service A network service that delivers data or packets as separate pieces. An example of this type of service is TCP/IP's Internet protocol (IP).

console communication service (CCS) A function used in SNA environments that acts as an interface between the control program and the VSCS component of VTAM for VM.

contention A term frequently used in SNA. It has multiple meanings, depending on the context. One example is how a session can be explained. In this case the network-accessible units attempt to initiate the same action at the same time against one another. Another example is contention in LU6.2. Here, it is the attempt to allocate a session by two programs against the other(s) at the same time.

context Information of a process maintained by the process manager.

context dependence When the glyph corresponding to a character can be modified depending on the preceding and following characters.

continue-any mode This specifies whether VTAM is to receive the data in terms of logical records or buffers.

continue-specific mode In IBM's VTAM, the state of a session or APPC conversation permitting its input to satisfy only receive requests issued in specific-mode.

contour A closed loop in a TrueType outline glyph, defined by a group of outline points.

control block A storage area that holds control information.

control logical unit (CLU) A logical unit that resides in a transaction processing facility (TPF) type 2.1 node. It is used to pass private protocol requests between the TPF type 2.1 node and the log-on manager which is a VTAM application program. Communication flow between the CLU and the log-on manager enables a logical unit controlled by VTAM to establish a session with TPF.

control panel A place that allows a user to set or "control" a feature of some sort.

control panels folder In Apple documentation, this is a directory located in the system folder for storing control panels. This allows users to modify their work environment on their Apple or Macintosh computer.

control point (CP) The managing component of a type 2.1 node that manages the resources of that node. In an APPN network node the CP can engage in CP-CP sessions with other APPN nodes.

control program Normally this type of program performs system oriented functions such as scheduling and supervising execution of programs of a computer system. In IBM's Account Interactive eXecutive it is that part of the operating system which determines the order in which basic functions should be performed.

control vector A particular structure that is one of a general class of RU substructures basically characterized by a variable length and having a 1-byte key used as an identifier.

controller A device that coordinates and controls the operation of one or more input/output devices.

controlling application program According to IBM documentation, this is an application program which a secondary logical unit (other than an application program) is automatically put in session with whenever the secondary logical unit is available.

controlling logical unit This can be either an application program or a device-type LU.

conversation Communication between two transaction programs using an LU 6.2 session.

conversational monitoring system (CMS) An IBM virtual machine (VM) operating system facility that provides interactive timesharing, problem solving, and program development capabilities. CMS operates under control of the control program component of a VM system.

converted command An intermediate form of a character-coded command produced by VTAM through use of an unformatted system services definition table. The converted command format is fixed. The unformatted system services (USS) table must be constructed in such a manner that character-coded commands are converted into the predefined, converted command format.

core Apple event According to Apple documentation, an event that nearly all applications can use to communicate.

cost According to DEC documentation, a numeric value assigned to a circuit that exists between two adjacent nodes. In the DECnet for OpenVMS network, data packets are routed on paths with the lowest cost.

CP   Control point or control program. A control point is a part of T2.1 architecture. A control program in IBM terminology is that software that runs on a VM host. The CP communicates with the hardware.

CP capabilities   One definition according to IBM documentation is the level of network services provided by the control point (CP) in an APPN end node or network node. Control program capabilities are exchanged during the activation of CP-CP sessions between nodes.

CP-CP session   A parallel session between two control points, using LU 6.2 protocols and a mode name of CPSVCMG, on which network services requests and replies are exchanged.

CP name   The name of a control point (CP), consisting of a network ID qualifier identifying the network to which the CP's node belongs, and a unique name within the scope of that network ID identifying the CP. Each APPN or LEN end node has one CP name assigned to it at system definition time.

cross-domain   In SNA, a term used to refer to resources in a different domain. Domains, in SNA, relate to ownership.

cross-domain keys   In SNA, this is a pair of cryptographic keys used by a system services control point (SSCP) to encipher the session cryptography key that is sent to another SSCP.

cross-domain link   A subarea link connecting two subareas that are in different domains. A physical link connecting two domains.

cross-domain resource (CDRSC)   A term used in SNA referring to a resource (typically software) that resides in another domain, under the control of a different VTAM.

cross-domain resource manager (CDRM)   A term used in SNA referring to the function in the SSCP which controls initiation and termination of cross-domain sessions.

cross-network LU-LU session   A term used in SNA to refer to a session between logical units (LUs) in different networks.

cross-network session   A session whose path traverses more than one SNA network.

cryptographic   In SNA terminology, it pertains to transformation of data to conceal its meaning.

cryptographic session   In SNA, an LU-LU session in which a function management data (FMD) request may be enciphered before it is transmitted and deciphered after it is received.

CUA   A term used in SNA, particularly in VTAM, to refer to the channel unit address.

currency   The previous, current, and next position of a record in a file. There are two types of currency: logical and physical. The physical previous, current, and next positions form the physical currency. The logical previous, current, and next positions form the logical currency. According to Zac, it means money.

Customer Information Control System (CICS)   An IBM software program that supports real-time transactions between remote users and custom-written transactions. It includes facilities for building, using, and maintaining databases.

cut buffer   A memory area that holds text which has been deleted from a file.

cycle   A complete oscillation of a wave.

CWALL   An NCP threshold of buffer availability, below which the NCP will accept only high-priority path information units (PIUs).

DARPA   Defense Advanced Research Projects Agency, formerly called ARPA. The government agency that funded research and experimentation with the ARPANET.

DASD   Direct-access storage device. This is IBM's terminology for a disk drive in mainframe environments.

data channel   An IBM term used as a synonym for input/output channel.

data circuit   A pair of associated transmit and receive channels providing a means for two-way data communication.

data circuit-terminating equipment (DCE)   In a data station, equipment that provides signal conversion and coding between the data terminal equipment (DTE) and the line.

data communication   Transfer of data among functional units by means of data transmission according to a specific protocol. The transmission, reception, and validation of data.

Data Encryption Standard (DES)   In computer security, it refers to the National Institute of Standards and Technology (NIST) Data Encryption Standard, adopted by the U.S. government as *Federal Information Processing Standard* (FIPS) publication 46, which allows only hardware implementations of the data encryption algorithm.

data flow control (DFC)   In SNA, a request or response unit (RU) category used for requests and responses exchanged between the data flow control layer in one half-session and the data flow control layer in the session partner.

data host node   A term used in SNA. It refers to a host dedicated to processing applications that does not control network resources, except for its channel-attached or communication adapter-attached devices.

data link   In SNA, synonym for link.

data-link control (DLC)   A set of rules used by nodes at layer 2 within a network. The data link is governed by data-link protocols such as ETHERNET or token ring for example.

data-link control (DLC) protocol   Rules used by two nodes at a data link-layer to accomplish an orderly exchange of information. Examples are ETHERNET, channel, FDDI, and token ring.

data-link layer   Layer 2 of the OSI reference model. It synchronizes transmission and handles error correction for a data link.

data-link level   The conceptual level of control logic between high level logic and a data-link protocol that maintains control of the data link.

data network   An arrangement of data circuits and switching facilities for establishing connections between data terminal equipment. A term commonly found in X.25 network implementations.

data packet   In X.25, it refers to a packet used for the transmission of user data on a virtual circuit at the DTE/DCE interface.

data server   According to common definition in Apple documentation, it is an application that acts like an interface between a database extension on a Macintosh computer and a data source, which can be on the Macintosh computer or on a remote host computer. It can be a database server program that can provide an interface to a variety of different databases, or it can be the data source itself, such as a Macintosh application.

data set   A way data, programs, and other representations of information are stored in IBM's MVS operating system environment.

data stream   In SNA, multiple data streams are identified. Generally, this is a continuous stream of data elements being transmitted, in character or binary digit form, using a defined format.

data-switching exchange (DSE)   Equipment at a single location that provides switching functions, such as circuit switching, message switching, and packet switching.

data terminal equipment (DTE)   That part of a data station which constitutes a data source, data link, or both.

data types   In SNA, particularly NetView, these can be alerts, events, or statistics.

data unit   In the OSI environment, the smallest unit of a file content meaningful to an FTAM file action.

database   A collection of data within a given structure.

datagram delivery protocol   In AppleTalk networks, this is a protocol that provides socket-to-socket delivery of data packets.

DCE clear-confirmation packet   A call supervision packet that a DCE transmits to confirm that a call has been cleared.

deactivate   A term frequently used in SNA environments. It means to take a resource out of service.

deallocate   A term used in APPC in which a LU6.2 application program interface (API) terminates a conversation and makes the session free for a future conversation.

decipher   To convert enciphered data in order to restore the original data.

decrypt   In computer security, to decipher or decode. Synonym for decipher.

de facto standard   A standard that is the result of some technology that has been developed and used and has achieved some level of popularity.

default SSCP list   In VTAM this is a list of SSCPs to which a session request can be routed to when a cross-domain resource manager (CDRM) is not specified.

Defense Data Network (DDN)   Used loosely to refer to MILNET, ARPANET, and the TCP/IP protocols they use. More specifically, it is MILNET and associated parts of the connected Internet that connect military installations.

defined context set   A set of presentation contexts negotiated between peer presentation entities.

definite response (DR)   According to IBM documentation, a protocol used in SNA that requests the receiver of the request to return a response unconditionally, whether positive or negative, to that request chain.

definition statement   In IBM's VTAM program, this statement describes an element of the network. In IBM's NCP, it is a type of instruction that defines a resource.

DELTA disk   In SNA, it is the virtual disk in a VM operating system that contains program temporary fixes (PTFs) that have been installed but not yet merged.

de jure standard   A standard set by a body or official consensus.

descent line   An imaginary line usually marking the maximum distance below the baseline of the descenders of glyphs in a particular font.

descriptor   A data buffer parameter passed for an extended-get or extended-step operation.

descriptor type   An identifier for the type of data referred to by the handle in a descriptor record.

desktop folder   In Apple environments, a directory located at the root level of each volume. It is used by the FINDER for storing information about the icons that appear on the desktop area of the screen. The desktop folder is invisible to the user. It is what the user sees on the screen.

destination logical unit (DLU)   The logical unit to which data is to be sent.

device   When the term is used in networking scenarios, it is typically used generically. For example, it could be a modem, host, terminal, or other entity.

dial-in   In most SNA environments, the notion of inbound traffic toward the host.

dial-out   In most networking environments, the notion of outbound capabilities to access resources elsewhere.

Digital Command Language   According to DEC documentation, this is a command interpreter in the operating system. It provides a means of communication between the user and the operating system.

Digital data communication message protocol (DDCMP)   According to DEC documentation, it is the link-level protocol used by DEC in their network products. DDCMP operates over serial lines, delimits frames by a special character, and includes checksums at the link level. It was relevant to TCP/IP because the original NSFNET used DDCMP over its backbone lines.

Digital Network Architecture   According to DEC, a set of protocols governing the format, control, and sequencing of message exchange for all DEC network implementations. The protocols are layered, and they define rules for data exchange from the physical link level up through the user interface level. DNA controls all data that travels throughout a DEC network. DNA also defines standard network management and network generation procedures.

Digital storage architecture   According to DEC documentation, the specifications from DEC governing the design of and interface to mass storage products. DSA defines the functions to be performed by host computers, controllers, and drives, and specifies how they interact to manage mass storage.

digital storage systems interconnect   A data bus that uses the systems communication architecture protocols for direct host-to-storage communications. The DSSI cable can extend to 6 m and has a peak bandwidth of 4 Mbytes.

direct-access storage device (DASD)   A device in which access time is effectively independent of the location of the data.

direct activation   According to IBM documentation, the activation of a resource as a result of an activation command specifically naming the resource.

direct deactivation   According to IBM documentation, the deactivation of a resource as a result of a deactivation command specifically naming the resource.

directed broadcast address   In TCP/IP-based environments, an IP address that specifies all hosts on a specific network. A single copy of a directed broadcast is routed to the specified network where it is broadcast to all machines on that network.

directed locate search   A search request sent to a specific destination node known to contain a resource, such as a logical unit, to verify the continued presence of the resource at the destination node and to obtain the node's connectivity information for route calculation.

directed search   Synonym for directed locate search.

directory   (1) Depending on the environment, this term is used in different ways. For example, in UNIX environments a directory is a listing of files and the files themselves. This definition is generally the case in most environments; however, some vendors contend that a significant difference exists. (2) In IBM's VM/SP environment, a control program (CP) disk file that defines each virtual machine's normal configuration: the user ID, password, normal and maximum allowable virtual storage, CP command privilege classes allowed, dispatching priority, logical editing symbols to be used, account number, and CP options desired. In APPN, this is a database that lists names of resources and records the CP name of the node where each resource is located. Another definition of a directory is the subdivision of a volume, available in the hierarchical file system (HFS). A directory can contain files as well as other directories.

directory access protocol   The protocol used between a directory user agent and a directory system agent.

directory entry   An object in the directory information base to model information. It can be an object entry or an alias entry.

directory information base   A set of directory entries. It contains objects to which the directory provides access and which includes all pieces of information which can be read or manipulated using the directory operations.

directory information shadowing protocol   A protocol used for shadowing between two directory service agents in the directory services standard.

directory information tree   A tree structure of the directory information base.

directory name   Names for directory entries in the directory information base.

directory operational binding management protocol   A protocol used by directory service agents to activate showing agreement. This allows directory service agents to establish, modify, and terminate operational bindings.

directory services (DS)   (1) According to its use in OSI environments, it is an application service element that translates the symbolic names used by application processes into the complete network addresses used. (2) A control point component of an APPN node that maintains knowledge of the location of network resources.

directory service agent   An application entity that offers the directory services.

directory service protocol   The protocol used between two directory system agents.

directory user agent   An application entity that provides the directory services.

disable   In a loose sense, this is similar to deactivate.

disabled   Pertaining to a state of a processing unit that prevents the occurrence of certain types of interruptions.

discarded packet   A piece of data, called a *packet,* that is intentionally destroyed.

disconnection   Termination of a physical connection.

discontiguous shared segment (DCSS)   According to IBM documentation, an area of virtual storage outside the address range of a virtual machine. It can contain read-only data or reentrant code. It connects discontiguous segments to a virtual machine's address space so that programs can be fetched.

discretionary controls   Security controls that are applied at the user's option.

distributed computing environment (DCE)   An Open Software Foundation (OSF) set of standards for distributed computing. Also, DCE is distributed computing in the general sense of the term.

disjoint network   According to IBM documentation, a network of two or more subnetworks with the same network identifiers that are not directly connected, but are indirectly connected, for example, through SNA network interconnection.

disjoint SSCP   According to IBM documentation, an SSCP that does not have a direct SSCP-SSCP session with other SSCPs in its network-ID subnetwork.

disk cache   A part of RAM that acts as an intermediate buffer when data is read from and written to file systems on secondary storage devices.

display   Generally used to refer to a terminal.

display server   As defined in XWindow system arenas, it is the software that controls the communication between client programs and the display, including the keyboard-mouse-screen-combination.

distinguished name   Name of a directory entry.

distributed directory database   According to documentation related to IBM's APPN architecture, it is the complete listing of all resources in the network as maintained in individual directories scattered throughout an APPN network.

distributed network directory   Synonym for distributed directory database.

DNS   Domain name service. In TCP/IP environments, it is a protocol for matching object names and network addresses. It was designed to replace the need to update /etc/hosts files of participating entities throughout a network.

domain   (1) That part of a computer network in which the data-processing resources are under common control. (2) A part of the DNS naming hierarchy. Syntactically, a domain name consists of a sequence of names separated by periods.

domain name system (DNS)   The online distributed database system used to map human-readable machine names into IP addresses. DNS servers throughout the connected Internet implement a hierarchical name space that allows sites freedom in assigning machine names and addresses. DNS also supports separate mappings between mail destinations and IP addresses.

domain operator   According to IBM documentation, a person or program that controls operation of resources controlled by one SSCP.

domain search   In the context of APPN networking, a search initiated by a network node to all its client APPN end nodes when a search request is received by a network node and that network node does not have any entry in its database for the requested resource.

dotted-decimal notation   A phrase typically found in TCP/IP network conversations. Specifically, this refers to the addressing scheme of the Internet protocol (IP). It is the representation of a 32-bit address consisting of four 8-bit numbers written in base 10 with periods separating them.

double-byte character set (DBCS)   A set of characters in which each character is represented by 2 bytes. For example, languages like Japanese, Chinese, and Korean use this method to represent characters.

downline system load   According to DEC documentation, a DECnet for OpenVMS function that permits an unattended target node to receive an operating system file image or terminal server image from another node.

DR   In SNA it can mean dynamic reconfiguration or definite response.

drag   In Apple and XWindow environments, this is to press *and hold down* a mouse button while moving the mouse on the desktop. Typically, dragging is used with menu selecting, moving, and resizing operations.

drain   This term refers to APPC. It means to honor pending allocation requests before deactivating a session with a partner logical unit.

drop cable   In the IBM cabling system, it is a cable that runs from a faceplate to the distribution panel in a wiring closet. In TCP/IP networking environments where thicknet cable and transceivers are used, it is that cable between the devices' network interface card and the transceiver.

drop folder   In Apple environments, a type of folder (holding place) that serves as a private mailbox for individuals. Once someone places a file in a drop folder, only the owner of the drop folder can retrieve it. According to Apple documentation, users can create drop folders by setting the appropriate AppleShare or Macintosh file-sharing access privileges.

DSRLST   Direct search list. In SNA this is a message unit that contains a search request sent throughout subarea networks to obtain information about a network resource such as its name, routing information, and status information.

DTE/DCE interface   The physical interface and link access procedures between data terminal equipment (DTE) and data circuit-terminating equipment (DCE).

dump   A term used frequently in SNA environments. Typically it means to obtain the contents of some aspect of memory. Specifically, it means to record, at a particular instant, the contents of all or part of one storage device in another storage device. One noted professional in the field, namely, Zelma Gandy, defines a dump as "The contents of memory used for debugging." A *core dump* refers to extraction of the contents of main memory. The term *VTAM dump* is used loosely to refer to reading data areas in IBM's VTAM program offering.

duplex   Pertaining to communication in which data can be sent and received at the same time.

dynamic   In a generic sense, this means to do something on the fly. A more specific explanation is to perform an operation that does not require a predetermined or fixed time.

dynamic node ID assignment   According to Apple documentation, the AppleTalk addressing scheme that assigns node IDs dynamically, rather than associating a permanent address with each node. Dynamic node ID assignment facilitates addition and removal of nodes from the network by preventing conflicts between old-node IDs and new-node IDs.

dynamic path update   A generic reference meaning the process of changing network path parameters for sending information without regenerating complete configuration tables.

dynamic reconfiguration data set (DRDS)   According to IBM documentation, this is a term used in VTAM. It refers to a data set used for storing definition data that can be applied to a generated communication controller configuration at the operator's request or can be used to accomplish dynamic reconfiguration of NCP, local SNA, and packet major nodes. This type of reconfiguration data set can be used to dynamically add PUs and LUs, delete PUs and LUs, and move PUs. It is activated with the VARY DREDS operator command.

dynamic window   A window that may change its title or reposition any of the objects within its content area.

EBCDIC   Extended Binary-Coded Decimal Interchange Code. IBM's basic character set used to represent data within the SNA environment. It consists of 8-bit coded characters.

echo   In data communication, a reflected signal on a communications channel.

Electronic Data Interchange (EDI)   A set of standard data formats for electronic information exchange.

element   This term has different meanings in different networking environments. In SNA, the particular resource within a subarea that is identified by an element address.

element address   According to IBM documentation, a value in the element address field of the network address identifying a specific resource within a subarea.

electromagnetic interference (EMI)   A type of noise that is the result of currents induced in electric conductors.

emulation program (EP)   A program that simulates the functions of another program. A generic example could be a 3270 terminal emulation program. Other possibilities also exist.

EN   In IBM's APPN architecture manuals, this refers to an APPN end node.

enable   To make functional. In a loose sense, it means to activate.

enabled   The state of being capable of performing work.

encapsulate   Generally agreed on in the internetworking community to mean surrounding one protocol with another protocol for the purpose of passing the foreign protocol through the native environment.

encipher   According to IBM documentation, this means to scramble data or to convert it to a secret code that masks the meaning of the data to any unauthorized recipient. In VTAM, to convert clear data into enciphered data.

encrypt   Synonym for encipher.

encryption   The process of transforming data into an unintelligible form.

end bracket   A term specifically used in SNA. It is the value of the end bracket indicator in the request header (RH) of the first request of the last chain of the bracket. The value denotes the end of the bracket.

end node   With reference to use in APPN, it is a node that can receive packets addressed to it and send packets to other nodes. It cannot route packets from other nodes.

end-node domain   That area defined by an end-node control point, attached links, and its local LUs.

end user   Defined by IBM documentation as either a program or a human.

end-user verification   LU6.2 identification check of end users by means of identifiers and passwords on the attach function management headers (FMHs).

entry mask   According to DEC documentation, on VAX systems it is a word where the bits represent the registers to be saved when a procedure is called with a CALLS or CALLG instruction, and restored when the procedure executes a RET instruction.

entry point   According to IBM documentation, it is a type 2.0, 2.1, 4, or type 5 node that provides distributed network management support. It sends network management data about itself and the resources it controls to a focal point for centralized processing, and it receives and executes focal-point-initiated commands to manage and control its resources.

ESCON   IBM's fiber channel that followed the channels which used bus and tag cables.

ES-IS routing   An Open Systems term used to refer a routing exchange protocol that provides an automated means for ISs and ESs on a subnetwork to dynamically determine the existence of each other. It also means to permit an IS to inform an ES of a potentially better route toward a destination.

ETHERNET   A data-link-level protocol. It (Version 2.0) was defined by Digital Equipment Corporation, Intel Corporation, and the Xerox Corporation in 1982. It specified a data rate of 10 Mbits/s, a maximum station distance of 2.8 km, a maximum number of stations of 1024, a shielded coaxial cable using baseband signaling, functionality of CSMA/CD, and a best-effort delivery system.

ETHERNET meltdown   A term used where ETHERNET protocol is used as the data-link-layer protocol in a network. It is an event that causes saturation or near-saturation on an ETHERNET data link. This scenario usually results from illegal or misrouted packets and lasts a short time.

EtherTalk   A term used in Apple and ETHERNET environments. It is software that enables AppleTalk protocols to run over industry-standard ETHER-NET technology.

event   This term can mean a predefined occurrence in a given network. It has a specific meaning in SNA, where in NetView it is a record indicating irregularities of operation in physical elements of a network.

event class   According to Apple documentation, it is an attribute that identifies a group of related Apple events. The event class appears in the message field of the Apple event's event record. In conjunction with the event ID attribute, the event class specifies what action an Apple event performs.

event ID   According to Apple documentation, it is an attribute that identifies a particular Apple event within a group of related Apple events. The event class appears in the where field of the Apple event's event record. In conjunction with the event class attribute, the event ID specifies what action an apple event performs.

exception   An abnormal condition relative to what has been predefined as a normal condition.

exception response (ER)   According to IBM documentation, it is a protocol requested in the for-of-response-requested field of a request header that directs the receiver to return a response only if the request is unacceptable as received or cannot be processed.

exception service routine   According to DEC documentation, it is a routine by which VAX and AXP hardware initially pass control to service an exception. An exception service routine passes control to a general exception dispatcher that attempts to locate a condition handler to further service the exception.

exchange identification (XID).   In SNA, this is a specific type of basic link unit used to convey node and link characteristics between adjacent nodes. In the SNA network, XIDs are exchanged between link stations before and during link activation to establish and negotiate link and node characteristics, and after link activation to communicate changes in these characteristics.

EXEC   According to IBM documentation, in a VM operating system, a user-written command file that contains CMS commands, other user-written commands, and execution control statements, such as branches.

executable image   An image that can be run in a process. When run, an executable image is read from a file for execution in a process.

executive   A generic term for the collection of procedures included in the operating system software that provides the operating system's basic control and monitoring functions.

executive mode   According to DEC documentation, this is the second most privileged processor access mode. The OpenVMS record management services and many of the operating system's system service procedures execute in executive mode.

exit   To execute an instruction within a portion of a program in order to terminate the execution of that portion.

exit list (EXLST)   According to IBM documentation, in VTAM it is a control block that contains the addresses of routines that receive control when specified events occur during execution.

exit program   Synonym for exit routine.

exit routine   One of two types of routines: installation exit routes or user exit routes.

expedited flow   According to IBM documentation, this is a data flow designated in the transmission header (TH) used to carry network control, session control, and various data flow control request (or response) units (RUs). Expedited flow is separate from the normal flow which carries primarily end-user data.

explicit route (ER)   According to IBM documentation, in SNA a series of one or more transmission groups that connect two subarea nodes. It is identified by an origin subarea address, a destination subarea address, an explicit route number, and a reverse explicit route number.

explicit route length   The number of transmission groups in an explicit route.

Extended Architecture (XA)   According to IBM documentation, this is an extension of System/370 architecture. It takes advantage of changing features such as the addressability of the hardware architecture.

extended attribute block   According to DEC documentation, it is an OpenVMS record management services user data structure that contains additional file attributes beyond those expressed in the file access block, such as boundary types and file protection information.

Extended Binary-Coded Decimal Interchange Code (EBCDIC)   IBM's coded character set of 256 8-bit characters and control functions.

extended network addressing   In IBM's traditional subarea networking this is the addressing system that splits addresses into an 8-bit subarea and a 15-bit element portion. The subarea portion of the address is used to address host processors or communication controllers. The element portion is used to permit processors or controllers to address resources.

extended recovery facility (XRF)   In SNA this is a facility that provides an alternate subsystem to take over sessions from the failing subsystem.

extended subarea addressing   According to IBM documentation, a network addressing system used in a network with more than 255 subareas.

fan-out box   A term used in local area network circles that refers to a device that functions like a hub. It provides the capability for multiple connections to make a central connection.

FDDI   *see* Fiber Distributed Data Interface (FDDI).

FDM   Frequency-division multiplexing. A technique that provides for division of frequency bandwidth into smaller subbands to provide each user with exclusive use of a subband.

Fiber Distributed Data Interface (FDDI)   An American National Standards Institute (ANSI) standard for a 100-Mbits/s LAN using optical fibers. A datalink protocol compatible to the IEEE 802 specification.

field-formatted   In SNA, this pertains to requests or responses that are encoded into fields, each having a specified format such as binary codes, bit-significant flags, and symbolic names.

file access data unit   A subtree of the hierarchical access structure. It is used to specify a location on a file structure.

file attributes   Terminology whose origins are in OSI which refers to the properties of a file that do not depend on an FTAM dialog.

File Definition Language (FDL)   According to DEC documentation, this is a special-purpose language used to write specifications for data files. These specifications are written in text files called FDL files; they are then used by

OpenVMS record management services utilities and library routines to create the actual data files.

file directory   The OSI equivalent of a directory in a file system.

file filter function   According to Apple documentation, this is a function supplied by your application for determining which files the user can open through a standard file dialog box.

file header   According to DEC documentation, this is a block in the index file describing a file on a Files-11 disk structure. The file header contains information needed by the file system to find and use the file. Some of this information is displayed when the DCL command DIRECTORY is entered. There is at least one file header for every file on the disk.

file ID   In the Apple environment this is an unchanging number assigned by the file manager to identify a file on a volume. When it establishes a file ID, the file manager records the filename and parent directory ID of the file.

file identifier   According to DEC documentation, this is a 6-byte value used to uniquely identify a file on a Files-11 disk volume. The file number, file sequence number, and relative volume number are contained in the file identifier.

file organization   In DEC networks, this is the particular file structure used as the physical arrangement of the data in a file on a mass storage medium. The OpenVMS record management services file organizations are sequential, relative, and indexed.

file server   A generic term used to refer to a computer whose primary task is to control the storage and retrieval of data from hard disks. Any number of other computers can be linked to the file server in order to use it to access data. This means that less storage space is required on the individual computer.

file system specification record   A term used in Apple computer environments. It is a record that identifies a stored file or directory by volume reference number, parent directory ID, and name. The file system specification record is the file identification convention adopted by system software version 7.0.

file transfer, access, and management   An OSI application protocol standard which allows remote files to be transferred, accessed, and managed.

file translator   A generic term referring to a utility program that converts a file from one computer format to another, such as from Macintosh to DOS. Apple file exchange is a file translator that is supplied with Macintosh system software.

Files-11   A DEC term used to refer to the name of the structure used by the RSX-111, IAS, and OpenVMS operating systems.

Files-11 ancillary control process.   A DEC term referring to the interface process that is the file manager for the Files-11 on-disk structure.

Files-11 on-disk structure level 1   According to DEC documentation, this is the original Files-11 structure used by IAS, RSX-11M, and RSX-11S for disk volumes. VAX systems support structure level 1 to ensure compatibility among systems. AXP systems do not support structure level 1.

Files-11 on-disk structure level 2   DEC reference to the second-generation disk file structure supported by the operating system. The Files-11 data structure prepares a volume to receive and store data in a way recognized by the operating system.

filter   A device or program that separates data, signals, or material in accordance with specified criteria.

flow control   In SNA, the process of managing data rate transfer between components within the network. This same concept refers to other networking environments.

focal point   According to IBM documentation, in NetView the focal point domain is the central host domain. It is the central control point for any management services element containing control of the network management data.

font size   The size of the glyphs in a font in points, measured from the baseline of one line of text to the baseline of the next line of single-spaced text.

font style   How a font (character or number) is represented.

foreground process   (1) In the XWindow system, a process that has the terminal window's attention. This is in contrast to a background process. (2) A process currently interacting with the user.

formatted system services.   A term used in IBM's SNA, specifically VTAM. It is a portion of VTAM that provides certain system services as a result of receiving a field-formatted command, such as an initiate or terminate command.

fragment   A term used in TCP/IP network environments. One of the pieces that results when an IP router divides an IP datagram into smaller pieces for transmission across a network. Fragments use the same format as datagrams. Fields in the IP header declare whether a datagram is a fragment, and if so, whether it is the offset of the fragment in the original datagram. IP software at the receiving end must reassemble fragments into complete datagrams.

frame   One definition generally agreed on as being a packet as it is transmitted across a serial line. The term originated from character-oriented protocols. According to the meaning in OSI environments, it is a data structure pertaining to a particular area of data. It also consists of slots that can accept values of specific attributes.

frame relay   A protocol defined by the CCITT and ANSI that identifies how data frames are switched in higher speeds than X.25, but in packet mode.

frame relay frame handler (FRFH)   A term reflecting a router function using the address field in a frame relay frame.

frame relay switching equipment (FRSE)   A term used in frame relay environments. It is a device capable of relaying frames to the next device in a frame relay network en route to a frame relay terminal equipment (FRTE) destination.

frame relay switching equipment (FRSE) subport set   In frame relay technology this is the set of primary and, optionally, substitute subports within an FRSE that represent those used for a given segment set.

frame relay switching equipment (FRSE) support   An agreed-on set of NCP frame relay functions that include the frame relay frame handler (FRFH) functions, defined by American National Standards Institute (ANSI) Standards T1.617 and T1.618.

frame relay terminal equipment (FRTE)   In FR-technology-based networks, it is a device capable of connecting to a frame relay network. An FRTE adds a frame header when sending data to the frame relay network and removes the frame header when receiving data from the frame relay network.

frequency   The rate of signal oscillation, expressed in hertz.

frequency-division multiplexing (FDM)   A method of multiplexing data on a carrier channel based on frequency.

FTAM   A term from the OSI networking protocols. It manipulates file transfer, access, and management.

FTP (file transfer protocol)   A term found in TCP/IP-based networks. It is a program that runs as a TCP application. It does not move a file from one place to another; rather, it copies a source file to a destination file. Consequently, two files exist unless one is deleted. FTP consists of a client and a server. The FTP client is used to invoke the FTP program. The FTP server is used to serve the request of the client. In normal implementations, FTP uses ports 20 and 21.

full-duplex (FDX)   Synonym for duplex.

full-screen mode   Where the contents of an entire terminal screen can be displayed at once.

function management header (FMH)   According to IBM documentation, one or more headers, optionally present in the leading request units (RUs) of an RU chain, that allow one LU to select a transaction program or device at the session partner. It also permits other control functions such as changing the destination or the characteristics of the data during the session, and transmits between session partners status or user information about the destination.

gateway   In internetworking terminology the agreed-on definition is to perform protocol conversion between dissimilar protocols. This may be upper-layer protocol conversion only or may include lower-layer protocol conversion as well; this depends on the vendor offering and the implementation. For example, TCP/IP-to-SNA gateways, DECnet-to-SNA gateways, and AppleTalk-to-TCP/IP gateways exist, to name only a few. In SNA, IBM uses the term in multiple ways. One explanation according to IBM documentation is the combination of machines and programs that provide address translation, name translation, and SSCP rerouting between independent SNA networks. In SNA a gateway consists of one "gateway" NCP and at least one "gateway" VTAM.

gateway-capable host   According to IBM documentation, it is a host node that has a defined NETID and SSCPNAME but does not perform gateway control functions.

gateway NCP   According to IBM documentation, an NCP that performs address translation permitting cross-network session traffic. In this sense the gateway NCP connects two or more independent SNA networks.

gateway VTAM   According to IBM documentation, a SSCP capable of cross-network session initiation, termination, takedown, and session outage notification.

generalized trace facility (GTF)   In SNA, this is an optional program that records significant system events, such as supervisor calls and start I/O operations, for the purpose of problem determination.

generation   In SNA this is the process of assembling and link-editing definition statements so that resources can be identified to all the necessary programs in a network. This is the origin of the term GEN.

generation definition   According to IBM documentation, this is the definition statement of a resource used in generating a program.

generic unbind   Synonym for session deactivation request.

global   Affecting the entire file, the entire system, or the entire image, depending on context.

global symbol   Agreed on as a symbol defined in a module of a program potentially available for reference by another module. The linker resolves (matches references with definitions) global symbols.

global symbol table   In a library, an index of defined global symbols used to access the modules defining the global symbols. The linker also puts global symbol tables into an image.

glyph   A distinct visual representation of a character that a display device, such as a monitor or printer, can display.

gold key   According to DEC documentation, the upper left key on VT100 series terminal keypads. This enables alternate keypad functions.

GOSIP   Government open-systems interconnection profile; a federal information processing standard that specifies a well-defined set of OSI protocols for government communication systems procurement. GOSIP was intended to eliminate the use of TCP/IP protocols on government internets, but clarifications have specified that government agencies can continue to use TCP/IP.

government OSI profiles   Functional standards used by government agencies to procure open-system equipment and software.

graphical data display manager (GDDM)   According to IBM documentation, it is used in the NetView performance monitor (NPM), in conjunction with the presentation graphics feature (PGF) to generate online graphs in the NPM graphic subsystem.

gray region   Typically the term is used in XWindow environments or Apple environments. A gray region defines the desktop, or the display area, of all active devices, excluding the menu bar on the main screen and the rounded corners on the outermost screens. It is the area in which windows can be moved.

group   Generically defined as a set of users in a system.

group control system (GCS)   A VM component that provides multiprogramming and shared memory support.

guest   A term used in the Apple environment meaning a user who is logged on to an AppleShare file server without a registered user name and password. A guest cannot own a directory or folder.

half-duplex (HD, HDX)   Transmission in only one direction at a time.

half-duplex operation   Data-link transmission where data can travel in one direction at a given instance.

half-duplex transmission   Data transmission in either direction, one direction at a time.

half-open connection   A scenario in which one end connection is established but the other end connection is unreachable or has disposed of its connection information.

half-session   According to IBM's documentation, it is a session-layer component consisting of the data flow control and transmission control components. These constitute one end of a session.

hard copy   Generally a printout.

hardware address   Also called a *hard address*. In ETHERNET networks, this is the 48-bit address assigned to the ETHERNET network interface card. In token ring is the 12-digit hex address assigned to the network interface card.

hardware monitor   In SNA, this is a component of NetView. It is called the *network problem determination application*. It is used to identify network problems, such as hardware, software, and microcode.

header   Control information that precedes user data in a frame or datagram that passes through networks. Specifically, this portion of a message contains control information.

heartbeat   Technically, this is known as signal-quality error (SQE). It is a voltage in a receiver that can be sent to a controller (interface board) to inform the controller that collision detection is functional.

help balloon   A term typically used in Apple environments. It is a rounded-type window containing explanatory information for the user.

help panel   This is also called a *help menu*. It is a display of information concerning a particular topic requested.

hertz   Cycles per second.

hierarchical routing   From a TCP/IP perspective, this type of routing is based on a hierarchical addressing scheme. Most TCP/IP routing is based on a two-level hierarchy in which an IP address is divided into a network portion until the datagram reaches a gateway that can deliver it directly. The concept of subnets introduces additional levels of hierarchical routing.

hierarchy   In SNA, this type of networking is considered traditional. Traditional SNA networking requires VTAM for session establishment. The term used in light of networking can be contrasted with peer networking which does not require an intervening component for session establishment.

high-performance option (HPO)   According to IBM documentation, this is an extension of VM/SP. The fundamental purpose of HPO is to provide performance and operation enhancements for large-system environments.

home directory   Generally, this concept exists in all hosts. It is the place where users initiate their operations in any system. For example, in UNIX the .profile file contains the beginning place for a user. This .profile is customized for users and the beginning place or home base can differ. In DEC's

VMS environment, the same is true but different terms are used. So it is with IBM's MVS, VM, and VSE operating systems.

hop In APPN, a portion of a route that has no intermediate nodes. It consists of only a single transmission group connecting adjacent nodes. One definition of a hop is the moving of a packet through a router.

hop count (1) A measure of distance between two points in the Internet. Each hop count corresponds to one router separating a source from a destination (for example, a hop count of 3 indicates that three routers separate a source from a destination). (2) A term generally used in TCP/IP networks. The basic definition is a measure of distance between two points in an internet. A hop count of n means that n routers separate the source and the destination.

host master key In SNA, a deprecated term for mastery cryptography key.

host node According to DEC documentation on DECnet for OpenVMS network, it is a node that provides services for another node. For the VAX packetnet system interface, a node that accesses a packet-switching data network by means of an X.25 multihost connector node. It is also referred to as the node that makes a device available to other nodes in a VMScluster configuration. A host node can be either a processor that adds the device to the mass storage control protocol server database or a hierarchical storage controller server. According to IBM documentation, it is defined as a processor.

host processor A processor that controls all or part of a user application network. Normally, the data communication access method resides on this host.

hpterm According to Hewlett-Packard's documentation, it is a type of terminal window, sometimes called a terminal emulator program, that emulates HP2622 terminals, complete with softkeys. In the HP-UX environment, the hpterm window is the default window for your X environment.

IAB Internet Architecture Board. A group related to TCP/IP protocol. Specifically, it is a group of people who set policy and review standards for TCP/IP and the Internet. The IAB was reorganized in 1989; technically oriented individuals moved to research and engineering subgroups. See **IRTF** and **IETF.**

ICMP Internet Control Message Protocol. Specific to the TCP/IP protocol suite. It is an integral part of the Internet protocol. It handles error and control messages. Routers and hosts use ICMP to send reports of problems about datagrams back to the original source that sent the datagram. ICMP also includes an echo request/reply used to test whether a destination is reachable and responding.

icon A small, graphic representation of an object on the root window. Icons are found in Apple hosts as well as the XWindow system.

icon family In the Apple Computer family of products, it is the set of icons that represent an object, such as an application or document, on the desktop.

idle state A state in which the Macintosh portable computer slows from its normal 16-MHz clock speed to a 1-MHz clock speed. The power manager puts the Macintosh portable in the idle state when the system has been in the active state for 15 s.

IETF Internet Engineering Task Force. A group of people concerned with short-term and medium-term problems with TCP/IP and the connected Internet. IETF is divided into six areas which are further divided into working groups.

image Procedures and data bound together by the linker to form an executable program. This executable program is executed by the process. There are three types of images: executable, sharable, and system.

image mode The default screen mode using multiple image planes for a single screen. The number of image planes determines the variety of colors that are available to the screen.

image name The name of the file in which an image is stored.

image planes The primary display planes on a device that supports two sets of planes. The other set of display planes is known as the *overlay planes*. These two sets of planes are treated as two separate screens in stacked mode and one screen in combined mode.

image privileges The privileges assigned to an image when it is installed.

inactive This term has a variety of meanings depending on the context or environment. It is, however, generally agreed that something is not operational. Pertaining to a node or device not connected or not available for connection to another node or device. In IBM's AIX operating system it pertains to a window that does not have an input focus. In SNA, particularly VTAM, the state of a resource, a major or minor node not activated or for which the VARY INACT command has been issued.

incoming call packet A call supervision packet transmitted by a data circuit-terminating equipment (DCE) to inform a called data terminal equipment (DTE) that another DTE has requested a call.

independent logical unit (ILU) In SNA, a type of LU that does not require VTAM for session establishment after the initial download of parameters.

index A structure that permits retrieval of records in an indexed file by key values.

index file According to DEC documentation, the file on Files-11 volume that contains the access information for all files on the volume and enables the operating system to identify and access the volume.

index file bitmap According to DEC documentation, it is a table in the index file of a Files-11 volume that indicates which file headers are in use.

indexed file organization A DEC-type file organization in which a file contains records and a primary key index used to process the records sequentially by index or randomly by index.

index path According to NetWare documentation, it is a logical ordering of records in a btrieve file based on the values of an index. An index path for each index in a file exists. A file may have up to 24 separate index paths.

indexed sequential file According to DEC documentation, a record file in which each record has one or more data keys embedded in it. Records in the file are individually accessible by specifying a key associated with the record.

indirect activation According to IBM documentation, in VTAM, the activation of a lower-level resource of the resource hierarchy as a result of SCOPE or ISTATUS specifications related to an activation command naming a higher-level resource.

indirect deactivation According to IBM documentation, in VTAM, the deactivation of a lower-level resource of the resource hierarchy as a result of a deactivation command naming a higher-level resource.

information (I) format A format used for information transfer.

information (I) frame A frame in I format used for numbered information transfer.

Information Management System/Virtual Storage (IMS/VS) This is a software subsystem offering by IBM. It is a database/data communication system that can manage complex databases and networks.

Information/Management A feature of the Information/System that provides interactive systems management applications for problem, change, and configuration management.

inhibited According to IBM, it is a logical unit (LU) that has indicated to its system services control point (SSCP) that it is temporarily not ready to establish LU-LU sessions. An initiate request for a session with an inhibited LU will be rejected by the SSCP. The LU can separately indicate whether this applies to its ability to act as a primary logical unit (PLU) or a secondary logical unit (SLU).

initial program load (IPL) An IBM term referring to the initialization procedure that causes an operating system to commence operation. The process by which a configuration image is loaded into storage at the beginning of a workday or after a system malfunction. The process of loading system programs and preparing a system to run jobs.

INITIATE In SNA, it is a network services request sent from a logical unit (LU) to a system services control point (SSCP) requesting that an LU-LU session be established.

Initiating LU (ILU) In SNA, the LU that first requests a session setup. The ILU may be one of the LUs that will participate in the session, or it may be a third-party LU. If it is one of the session participants, the ILU is also called the *origin LU* (OLU).

initiator In OSI, a file service user which requests an FTAM establishment.

inoperative The condition of a resource that has been, but is no longer, active. The resource may have failed, have received an INOP request, or be suspended while a reactivate command is being processed.

input/output channel In a data-processing system, a functional unit that handles transfer of data between internal and peripheral equipment. In a computing system, a functional unit, controlled by a processor, that handles transfer of data between processor storage and local peripheral devices. In IBM terminology, it refers to a specific type of path, either parallel or serial.

installation exit The means (or way by which) an IBM software product may be modified by a customer's system programmers to change or extend the functions of the IBM software product. Such modifications consist of exit routines written to replace one or more existing modules of an IBM software product, or to add one or more modules or subroutines to an IBM software product, for the purpose of modifying or extending the functions of the IBM software product.

installation exit routine   A routine written by a user to take control at an installation exit of an IBM software product.

installationwide exit   Synonym for installation exit.

integrated communication adapter (ICA)   A communication adapter that is an integral part of the host processor.

integrity control   According to Novell documentation, it is the method used to ensure the completeness of files. Specifically, btrieve uses preimaging and NetWare's transaction tracking system to guarantee integrity.

intensive mode recording (IMR)   An NCP function that forces recording of temporary errors for a specified resource.

interactive problem control system (IPCS)   According to IBM documentation, it is a component of VM that permits online problem management, interactive problem diagnosis, online debugging for disk-resident CP abend dumps, problem tracking, and problem reporting.

Interactive System Productivity Facility (ISPF)   An IBM licensed program that serves as a full-screen editor and dialog manager. Used for writing application programs, it provides a means of generating standard screen panels and interactive dialogs between the application programmer and the terminal user.

interapplication communication (IAC)   In Apple terminology, a collection of features, provided by the edition manager, Apple event manager, event manager, and PPC toolbox, that help applications work together.

interchange node   A new type of node supported by VTAM beginning in Version 4 Release 1. Acts as both an APPN network node and a subarea type 5 node to transform APPN protocols to subarea protocols and vice versa.

interconnected networks   According to IBM, SNA networks connected by *gateways,* gateway NCPs.

interface   A shared boundary between two functional units, defined by functional characteristics, signal characteristics, or other characteristics, as appropriate. The concept includes the specification of the connection of two devices having different functions. Hardware, software, or both, that links systems, programs, or devices.

intermediate node   A node located at the end of more than one branch.

intermediate routing node (IRN)   A node containing intermediate routing function.

intermediate session routing (ISR)   A term used in APPN that performs a type of routing function within an APPN network node that provides session-level flow control and outage reporting for all sessions that pass through the node but whose endpoints are elsewhere.

intermediate SSCP   In SNA it is an SSCP along a session initiation path that owns neither of the LUs involved in a cross-network LU-LU session.

International Organization for Standardization (ISO) An organization of national standards-making bodies from various countries established to promote development of standards to facilitate international exchange of goods and services, and develop cooperation in intellectual, scientific, technological, and economic activity.

Internet According to different documents describing the Internet, it is a collection of networks, routers, gateways, and other networking devices that use the TCP/IP protocol suite and function as a single, cooperative virtual network. The Internet provides universal connectivity and three levels of network services: unreliable, connectionless packet delivery; reliable, full-duplex stream delivery; and application-level services such as electronic mail that build on the first two. The Internet reaches many universities, government research labs, and military installations and over a dozen countries.

Internet address According to Apple documentation, it is an AppleTalk address that includes the socket number, node ID, and network number. According to TCP/IP documentation, it refers to the 32-bit address assigned to the host. It is a software address that on local ("little i") internets is locally managed, but on the central ("big I") Internet is dictated to the user (entity desiring access to the Internet).

Internet packet exchange (IPX) A Novell protocol that operates at OSI layer 3. It is used in the NetWare protocols; it is similar to IP in TCP/IP.

Internet protocol (IP) A protocol used to route data from its source to its destination. A part of TCP/IP protocol.

interpersonal messaging system According to multiple explanations in the OSI community, it is a MHS system supporting the communication of interpersonal messages.

interpersonal message According to OSI-related documents, it is defined as a type of message used for human-to-human communication in MHS.

InterPoll According to Apple documentation, it is software from Apple that helps administrators monitor the network and diagnose the source of problems that arise.

interuser communication vehicle (IUCV) According to IBM documentation, it is a VM facility for passing data between virtual machines and VM components.

interpret table In IBM's VTAM, it is an installation-defined correlation list that translates an argument into a string of eight characters. This table can translate log-on data into the name of an application program for which the log-on is intended.

IP Internet protocol. The TCP/IP standard protocol that defines the IP datagram as the unit of information passed across an internet and provides the basis for connectionless, best-effort packet delivery service. IP includes the ICMP control and error message protocol as an integral part. The entire protocol suite is often referred to as TCP/IP because TCP and IP are the two fundamental protocols.

IP address The 32-bit dotted-decimal address assigned to hosts that want to participate in a local TCP/IP internet or the central (connected) Internet. IP addresses are software addresses. Actually, an IP address consists of a network portion and a host portion. The partition makes routing efficient.

IP datagram   A term used with TCP/IP networks. It is a basic unit of information passed across a TCP/IP internet. An IP datagram is to an internet as a hardware packet is to a physical network. It contains a source address and a destination address along with data.

IRSG   Internet Research Steering Group. A committee consisting of the IRTF research group chairpersons plus the IRTF chairperson, who direct and coordinate research related to TCP/IP and the connected Internet.

IRTF   Internet Research Task Force. A group of people working on research problems related to TCP/IP and the connected Internet.

ISDN   Integrated Services Digital Network. A set of standards being developed within ANSI, ISO, and CCITT for the delivery of various services over digital networks.

IS-IS routing   Routing between ISs within a routing domain.

ISODE   A term used in the ISO development environment. A set of public-domain software subroutines that provide an interface between the GOSIP-specified session layer (ISO) and the DoD-specified transport layer (TCP/IP). Allows the development of applications that will execute over both OSI and TCP/IP protocol stacks as a migration path from TCP/IP networks to GOSIP networks.

ISTATUS   According to IBM documentation, in VTAM and NCP, a definition specification method for indicating the initial status of resources.

job   A way by which an accounting unit is assigned to a process and its subprocesses, if any, and all subprocesses that they create. Jobs are classified as batch and interactive. For example, the job controller creates an interactive job to handle a user's requests when the user logs in to the system, and it creates a batch job when the symbiont manager passes a command input file to it.

job control language   A language used in IBM's MVS operating system environment to identify a job to an operating system and to describe the job's requirements.

job controller   The system process that establishes a job's process context, starts a process running the LOGIN image for the job, maintains the accounting record for the job, manages symbionts, and terminates a process and its subprocesses.

job information block   A data structure associated with a job that contains the quotas pooled by all processes in the job.

Katakana   A character set of symbols used in one of the two common Japanese phonetic alphabets, primarily to write foreign words phonetically.

keyboard binding   In an XWindow environment, an association of a special key press with a window manager function. For example, pressing the special keys SHIFT ESC displays the system menu of the active window.

keyboard resources   According to Apple documentation, a category of files that are stored in a resource file by the resource manager and are used by the

Macintosh script management system, including the International Utilities Package.

keyboard script   The script for keyboard input. It determines the character input method and the keyboard mapping, that is, what character codes are produced when a sequence of keys is pressed.

keyword   (1) According to Apple documentation, a four-character code used to uniquely identify the descriptor record for either an attributor or a parameter in an Apple event. In Apple event manager functions, constants are typically used to represent the four-character codes. (2) In programming languages, a lexical unit that, in certain contexts, characterizes some language construct. In some contexts, IF characterizes an if-statement. A keyword normally has the form of an identifier. One of the predefined words of an artificial language. A significant and informative word in a title or document that describes the content of that document. A name or symbol that identifies a parameter.

keyword operand   A term used in the IBM environment, particularly with JCL. It is an operand that consists of a keyword followed by one or more values (such as DSNAME=HELLO).

keyword parameter   A parameter that consists of a keyword followed by one or more values.

LAP manager   According to Apple documentation, it is a set of operating system utilities that provide a standard interface between the AppleTalk protocols and the various link access protocols, such as LocalTalk (LLAP), EtherTalk (ELAP), and TokenTalk (TLAP).

least-weight route   According to IBM's documentation, in APPN, the one route calculated by topology and routing services (TRS) to have the lowest total weight after TRS compares the node characteristics and transmission group (TG) characteristics of each intermediate node and intermediate TG of each possible route for the class-of-service requested, and computes the total combined weight for nodes and TGs in each route. After a least-weight route is calculated between two given nodes, the result may be stored to prevent repetition of this calculation in future route selections.

LEN connection   A link over which LEN protocols are used.

LEN node   A term used in APPN. According to IBM documentation, a node that supports independent LU protocols but does not support CP-CP sessions. It may be a peripheral node attached to a boundary node in a subarea network, an end node attached to an APPN network node in an APPN network, or a peer-connected node directly attached to another LEN node or APPN end node.

level 1 router   According to DEC documentation, a DECnet for OpenVMS node that can send and receive packets, and route packets from one node to another node within a single area.

level 2 router   According to DEC documentation, a DECnet for OpenVMS node that can send and receive packets, and route packets from one node to another within its own area and between areas.

lexical function   According to DEC documentation, a command language construct that the Digital Command Language command interpreter evaluates and substitutes before it parses a command string.

Librarian According to DEC documentation, a program that allows the user to create, update, modify, list, and maintain object library, help library, text library, and assembler macro library files.

limited resource According to IBM documentation, a connection facility that causes a session traversing it to be terminated if no session activity is detected for a specified period of time.

limited-resource session According to IBM documentation, a session that traverses a limited resource link. This session is terminated if no session activity is detected for a specified period of time.

line The portion of a data circuit-external to data circuit-terminating equipment (DCE), that connects the DCE to a data-switching exchange (DSE) that connects a DCE to one or more other DCEs, or that connects a DSE to another DSE.

line control Synonym for data-link-control protocol.

line control discipline Synonym for link protocol.

line discipline Synonym for link protocol.

line group One or more telecommunication lines of the same type that can be activated and deactivated as a unit.

line speed The number of binary digits that can be sent over a telecommunication line in one second (1s), expressed in bits per second (bits/s).

line switching Synonym for circuit switching.

link The combination of the link connection (the transmission medium) and two link stations, one at each end of the link connection.

link-access protocol (1) According to DEC documentation, a set of procedures used for link control on a packet-switching data network; X.25 defines two sets of procedures: (a) **LAP**—the DTE-DCE interface, defined as operating in two-way simultaneous asynchronous response mode with the DTE and DCE containing a primary and secondary function and (b) **LAPB**—the DTE/DCE interface, defined as operating in two-way asynchronous balanced mode; (2) according to Apple documentation, an AppleTalk protocol that controls the access of a node to the network hardware. A link access protocol makes it possible for many nodes to share the same communication hardware.

link-attached Pertaining to devices that are connected to a controlling unit by a data link.

link connection The physical equipment providing two-way communication between one link station and one or more other link stations; for example, a telecommunication line and data circuit-terminating equipment (DCE).

link connection segment A part of the configuration that is located between two resources listed consecutively in the service point command service (SPCS) query link configuration request list.

link level A reference to the physical connection between two nodes and/or the protocols used to govern that connection.

link problem determination aid (LPDA) According to IBM documentation, a series of procedures used to test the status of and to control DCEs, the communication line, and the remote device interface. These procedures, or a subset of them, are implemented by host programs (such as the NetView program and VTAM), communication controller programs (such as NCP), and IBM LPDA DCEs.

link protocol Rules for sending and receiving data over a medium.

link services layer Routes packets between LAN boards with their MLIDs and protocol stacks. The LSL maintains LAN board, protocol stack, and packet buffer information.

link station According to IBM documentation, the hardware and software components within a node representing a connection to an adjacent node over a specific link. In VTAM, a named resource within an APPN or a subarea node that represents the connection to another APPN or subarea node that is attached by an APPN or a subarea link. In the resource hierarchy in a subarea network, the link station is subordinate to the subarea link.

Little Endian A storage format or transmission of binary data in which the least significant byte comes first.

LLC Logical link control. According to OSI documentation, a sublayer in the data-link layer of the OSI model. The LLC provides the basis for an unacknowledged connectionless service or connection-oriented service on the local area network.

local Pertaining to a device accessed directly without use of a telecommunication line.

local access The ability to execute a program on the computer to which you are attached.

local address According to IBM documentation, in SNA, an address used in a peripheral node in place of a network address and transformed to or from a network address by the boundary function in a subarea node.

local area network (LAN) A collection of computers and other related devices connected together on the premises within a limited geographic area.

local area transport According to DEC documentation, this is a communications protocol that the operating system uses within a local area network to communicate with terminal servers.

local area VAXcluster system According to DEC documentation, a type of VAXcluster configuration in which cluster communication is carried out over the ETHERNET by software that emulates certain computer interconnect, or CI, port functions. A VAXcluster node can be a VAX or a micro VAX processor; hierarchical storage controllers (HSCs) are not used.

local client A term used in an XWindow environment. It refers to a program running on your local computer, the same system that is running your X server.

local directory database According to IBM documentation, a set of LUs in a network known at a particular node. The resources included are all those in the node's domain as well as any cache entries.

local management interface (LMI) A set of operational procedures and messages as well as DLCI 1023 are defined in *Frame-Relay Specification with*

Extensions, a document based on proposed T1S1 standards, which are copyrighted by Digital Equipment Corporation (DEC), Northern Telecom, Inc., and StrataCom, Inc. In this context, the term *local management interface* is a deprecated term for link integrity verification tests (LIVT). Current meaning: any frame relay management interface procedures, such as DLCI 1023 or DLCI 0.

local non-SNA major node   According to IBM documentation, in VTAM, a major node whose minor nodes are channel-attached non-SNA terminals.

local SNA major node   According to IBM documentation, in VTAM, a major node whose minor nodes are channel-attached peripheral nodes.

local symbol   According to DEC documentation, it is a symbol meaningful only to the module that defines it. Symbols not identified to a language processor as global symbols are considered to be local symbols. A language processor resolves local symbols. They are not known to the linker and cannot be made available to another object module. They can, however, be passed through the linker to the symbolic debugger.

LocalTalk   According to Apple documentation, it is a type of AppleTalk network that is inexpensive and easy to set up. LocalTalk is commonly used to connect small to medium-sized work groups.

local topology database   According to IBM documentation, a database in an APPN NN or LEN node containing an entry for each transmission group (TG) having at least one end node for an endpoint. In an APPN end node, the database has one entry for each TG connecting to the node. In a network node, the database has an entry for each TG connecting the network node to an end node. Each entry describes the current characteristics of the TG that it represents. A network node has both a local and a network topology database while an end node has only a local topology database.

locate mode   According to DEC documentation, an OpenVMS record management services record access technique in which a program accesses records in a Open VMS RMS I/O buffer area to reduce overhead.

location name   An identifier for the network location of the computer on which a port resides. The PPC toolbox provides the location name. It contains an object string, a type string, and a zone. An application can specify an alias for its location name by modifying its type string.

log-in directory   The default directory a user is assigned to on log-on into a system.

log off   To request that a session be terminated.

log-off   According to IBM documentation, in VTAM, an unformatted session-termination request. In general, to terminate interaction with a system; actually entering a command of some sort to close the connection.

log on   In SNA products, to initiate a session between a application program and a logical unit (LU).

log-on   According to IBM documentation, in VTAM, an unformatted session-initiation request for a session between two logical units. In general, it is used to sign on or to get to the point where work can be done.

log-on manager   A VTAM application program that provides log-on services for the transaction processing facility (TPF).

log-on mode   According to IBM documentation, in VTAM, a subset of session parameters specified in a log-on mode table for communication with a logical unit.

log-on mode table   According to IBM documentation, in VTAM, a set of entries for one or more log-on modes. Each log-on mode is identified by a log-on mode name.

logical channel   In packet mode operation, a sending channel and a receiving channel that together are used to send and receive data over a data link at the same time.

logical channel identifier   A bit string in the header of a packet that associates the packet with a specific switched virtual circuit or permanent virtual circuit.

logical link control (LLC) protocol   In a local area network, the protocol that governs the exchange of transmission frames between data stations independently of how the transmission medium is shared.

logical name   According to DEC documentation, it is a user-specified name for any portion or all of a file specification. For example, the logical name INPUT can be assigned to a terminal device from which a program reads data entered by a user. Logical name assignments are maintained in logical name tables for each process, each group, and the system. Logical names can be assigned translation attributes, such as terminal and concealed.

logical name table   According to DEC documentation, it is a table that contains a set of logical names and their equivalence names for a particular process, a particular group, or the system.

logical record   A group of related fields treated as a unit.

logical unit (LU)   An addressable endpoint.

Logical unit 6.2 (LU6.2)   Those protocols and that type LU which supports advanced program-to-program communication (APPC).

low-entry networking (LEN)   According to IBM documentation, a capability in nodes allowing them to directly attach to one another using peer-to-peer protocols and to support multiple and parallel sessions between logical units. However, LEN does not provide all the capabilities of APPN; for example, it does not provide CP-CP session support.

low-entry networking (LEN) end node   According to IBM documentation, an end node that provides all SNA end-user services, can attach directly to other nodes using peer protocols, and derives network services implicitly from an adjacent network node when attached to an APPN network without a session between its local CP and another CP.

low-entry networking (LEN) node   According to IBM documentation, a node that supports independent LU protocols but does not support CP-CP sessions.

LU group According to IBM documentation, in the NetView performance monitor (NPM), a file containing a list of related or unrelated logical units. The LU group is used to help simplify data collection and analysis.

LU-LU session A logical connection between two logical units in an SNA network that provides communication between two end users.

LU-mode pair According to IBM documentation, in the VTAM implementation of the LU6.2 architecture, the coupling of an LU name entry and a mode name entry. This coupling allows a pool of sessions with the same characteristics to be established.

LU type According to IBM documentation, the classification of an LU in terms of SNA protocols and options it supports for a given session.

LU Type 6.2 (LU6.2) According to IBM documentation, a type of logical unit that supports general communication between programs in a distributed processing environment. LU6.2 is characterized by a peer relationship between transaction programs and efficient utilization of a session for multiple transactions.

LU6.2 session A logical connection utilizing LU6.2 protocols.

macroinstruction According to IBM documentation, an instruction in a source language that is to be replaced by a defined sequence of instructions in the same source language and that may also specify values for parameters in the replaced instructions.

main screen The screen on which a menu bar appears.

maintain system history program (MSHP) According to IBM documentation, it is a program used for automating and controlling various installation, tailoring, and service activities for a VSE system.

maintenance and operator subsystem (MOSS) According to IBM documentation, a subsystem of an IBM communication controller, such as the 3725 or the 3720, that contains a processor and operates independently of the rest of the controller. It loads and supervises the controller, runs problem determination procedures, and assists in maintaining both hardware and software.

major node According to IBM documentation, in VTAM, a set of resources that can be activated and deactivated as a group.

management information base A collection of managed objects. A term used with the concept of SNMP-based network management.

management information tree A tree structure of the management information base.

management services (MS) According to IBM documentation, one type of network service in control points (CPs) and physical units (PUs). Management services are provided to assist in the management of SNA networks, such as problem management, performance and accounting management, configuration management, and change management.

MASSBUS adapter According to DEC documentation, an interface device between the backplane interconnect and the MASSBUS device.

mass storage control protocol   According to DEC documentation, the software protocol used to communicate I/O commands between a VAX processor and DSA-compliant devices on the system.

master file directory (MFD)   According to DEC documentation, the file directory on a disk volume that contains the name of all user file directories on a disk, including its own.

matte   A term used in window-based environments. It is the border located inside the window between the client area and the frame and used to create a three-dimensional effect for the frame and window.

media access control   According to OSI nomenclature, it is a sublayer in the data-link layer which controls access to the physical medium of a network.

medium   A physical carrier of electrons or photons. The medium may be hard, as in a type of cable; or soft, in the sense of microwaves, for example.

medium access control (MAC)   A protocol that comprises the lower part of the second layer in the OSI model.

medium access control (MAC) protocol   Protocol that governs access to the transmission medium to enable the exchange of data between nodes.

medium access control (MAC) sublayer   The MAC sublayer that supports topology-dependent functions and uses services of the physical layer to provide services to the logical link control (LLC) sublayer.

menu   A list of selections from which to make a choice.

MERGE disk   According to IBM documentation, a virtual disk in the VM operating system that contains program temporary fixes (PTFs) after the VMFMERGE EXEC is invoked.

message   Generically, a reference to meaningful data passed from one end user to another. The end user may be a human or a program.

message (MHS)   According to OSI documentation, a structured set of data that is sent from a user agent to one or more recipient user agents.

message block   According to Apple documentation, a byte stream that an open application uses. It is used to send data to and receive data from another open application. The PPC toolbox delivers message blocks to an application in the same sequence in which they were sent.

message store   A term used in a TCP/IP environment. It refers to an entity acting as an intermediary between a user agent and its local message transfer agent.

message transfer agent (MTA)   In TCP/IP, a subpart of the electronic mail component known as simple mail transfer protocol (SMTP). It is an object in the message transfer system. MTAs use a store-and-forward method to relay messages from originator to a recipient. They interact with user agents when a message is submitted, and on delivery.

message unit   According to IBM documentation, in SNA, the unit of data processed by any layer; for example, a basic information unit (BIU), a path information unit (PIU), or a request (or response) unit (RU).

MHS Message handling service. The service provided by the CCITT X.400 series of standards, consisting of a user agent to allow users to create and read electronic mail; a message transfer agent to provide addressing, sending, and receiving services; and a reliable transfer agent to provide routing and delivery services.

migration data host According to IBM documentation, this is a VTAM node support that acts as both an APPN end node and a SNA subarea node.

MILNET Originally this network was part of the ARPANET. In 1984 it was segmented for military installation usage.

minimize To turn a window into an icon.

Minor node According to IBM documentation, in VTAM, this is a uniquely defined resource within a major node.

mixed interconnect VMScluster system According to DEC documentation, any VMScluster system that utilizes more than one interconnect for SCA traffic. Mixed interconnect VMScluster systems provide maximum flexibility in combining CPUs, storage, and workstations into highly available configurations.

MMS Manufacturing messaging service. According to OSI documentation, a messaging service between programmable devices.

mode name In SNA, a name used by the initiator of a session to designate the characteristics desired for the session.

modem Modulator/demodulator. A device that converts digital signals to analog signals and vice versa for the purpose of using computer devices in remote locations.

monitor Generally considered the same as a display screen. It can also mean to watch or observe a task, program execution, or the like.

monitor console routine According to DEC documentation, the command interpreter in an RSX-11 system; also an optional command interpreter in the operating system.

mounting a volume According to DEC documentation, the logical association of a volume with the physical unit on which it is loaded. Loading or placing a magnetic tape or disk pack on a drive and placing the drive on line.

mount verification According to DEC documentation, it is a feature that suspends I/O to and from volumes while they are changing status. Mount verification also ensures that, following a suspension in disk I/O, the volume being accessed is the same as was previously mounted.

move mode According to DEC documentation, it is an OpenVMS record management services record I/O access technique in which a program accesses records in its own working storage area.

MTU Maximum transfer unit. The largest amount of data that can be transferred across a given physical network. For local area networks implementing ETHERNET, the MTU is determined by the network hardware. For long-haul networks that use aerial lines to interconnect packet switches, the MTU is determined by software.

Multicast   A technique that allows copies of a single packet to be passed to a selected subset of all possible destinations. Some hardware supports multicast by allowing a network interface to belong to one or more multicast groups. Broadcast is a special form of multicast in which the subset of machines to receive a copy of a packet consists of the entire set. IP supports an internet multicast facility.

Multicast address   According to Apple documentation, an ETHERNET address for which the node accepts packets just as it does for its permanently assigned ETHERNET hardware address. The low-order bit of the high-order byte is set to 1. Each node can have any number of multicast addresses, and any number of nodes can have the same multicast address. The purpose of a multicast address is to allow a group of ETHERNET nodes to receive the same transmission simultaneously, in a fashion similar to the AppleTalk broadcast service.

Multicasting   A directory service agent uses this mode to chain a request to many other directory service agents.

MultiFinder   According to Apple documentation, prior to Version 7.0 system software, a multitasking operating system for Macintosh computers that enables several applications to be open at the same time. In addition, processes (such as print spooling) can operate in the background so that users can perform one task while the computer performs another.

multihomed host   A TCP/IP host connected to two or more physical networks; thus they have more than one address. They can serve as router-type devices.

multilink transmission group   According to IBM documentation, a transmission group containing two or more links.

multimode   The transmission of multiple modes of light.

multipath channel (MPC)   According to IBM documentation, a channel protocol that uses multiple unidirectional subchannels for VTAM-to-VTAM bidirectional communication.

multiple-domain network   According to IBM documentation, a network with more than one system services control point in traditional subarea SNA. In APPN it is an APPN network with more than one network node.

multiple link interface driver (MLID)   According to Novell documentation, this driver accepts multiple protocol packets. When an MLID device driver receives a packet, the MLID does not interpret the packet; it copies identification information and passes the packet to the link support layer. MLIDs are supplied either by Novell, the network board manufacturer, or a third-party supplier.

multipoint line   A telecommunication line or circuit that connects two or more stations.

Mutex   A term used in DEC network environments. According to DEC documentation, a semaphore is used to control exclusive access to a region of code that can share a data structure or other resource. The mutex semaphore ensures that only one process at a time has write access to the region of code.

MVS/XA   Multiple Virtual Storage/Extended Architecture. An IBM operating system.

name-binding protocol According to Apple documentation, this is the AppleTalk transport-level protocol that translates a character string into a network address.

name-binding protocol (NBP) According to Apple documentation, this is an AppleTalk protocol that maintains a table that contains the internet address and name of each entity in the node that is visible to other entities on the internet (that is, each entity that has registered a name with NBP).

name block According to DEC documentation, it is an OpenVMS record management services user data structure that contains supplementary information used in parsing file specifications.

name resolution The process of locating an entry by sequentially matching each relative distinguished name in a purported name to a vertex of the directory information tree.

name translation According to IBM documentation, it is a SNA network interconnection. It includes the conversion of logical unit names, log-on mode table names, and class-of-service names used in one network to equivalent names for use in another network.

naming context A term used in OSI networks that refers to a substructure of the directory information tree. It starts at a vertex and extends downward to a leaf and/or nonleaf structure.

naming context tree According to popular OSI documentation, a tree structure in which each node represents a naming context.

NCCF Network communications control facility. A part of IBM's NetView. It is the command that starts the NetView command facility. It is a command line in NetView whereby various commands can be offered.

NCP major node According to IBM documentation, it refers to VTAM, where a set of minor nodes represent resources, such as lines and peripheral nodes, controlled by IBM's network control program.

NCP/EP definition facility (NDF) According to IBM documentation, it is a subprogram of the system support program (SSP). It is used to generate a partitioned emulation program (PEP) load module or a load module for a network control program (NCP) or for an emulation program (EP).

negative response (NR) According to IBM documentation, a term used in SNA. It refers to a response indicating that a request did not arrive successfully or was not processed successfully by the receiver.

negotiable BIND According to IBM documentation, a term used in SNA that refers to the capability allowing two half-sessions to negotiate the parameters of a session when the session is being activated.

NetBIOS NetBIOS is the standard interface to networks that is used by IBM PCs and compatibles. With a TCP/IP network, NetBIOS refers to a set of guidelines that describe how to map NetBIOS operations into equivalent TCP/IP operations.

NetView performance monitor (NPM) According to IBM, it is a program that collects, monitors, analyzes, and displays data relevant to the performance of VTAM.

NetView   According to IBM, it is a program used to monitor and manage a network and diagnose network problems.

NetWare Loadable Module (NLM)   According to Novell documentation, it is a program that is part of a file server memory with NetWare. An NLM can be loaded or unloaded while the file server is running, become part of the operating system, and access NetWare directly.

network   A collection of computers and related devices connected together in such a way that collectively they can be more productive than standalone equipment.

network address   In general, each participating entity on a network has an address so that it can be identified when exchanging data. According to IBM documentation, in a subarea network, an address consists of subarea and element fields that identify a link, link station, PU, LU, or SSCP.

network address translation   According to IBM documentation, in a SNA network interconnection, it is the conversion of the network address assigned to a LU in one network into an address in an another network.

network architecture   The logical and physical structure of a computer network.

network connect block   According to DEC documentation, it is a user-generated data structure used in a nontransparent task to identify a remote task and optionally send user data in calls to request, accept, or reject a logical link connection. For the VAX Packetnet System Interface, a block that contains the information necessary to set up an X.25 virtual circuit or to accept or reject a request to set up an X.25 virtual circuit.

network control   The meaning in SNA, according to IBM documentation, is a request or response unit (RU) category used for request and responses exchanged between PUs. The purpose of this is activating and deactivating explicit and virtual routes. The term is used to refer to send load modules to adjust peripheral nodes.

network control program   (1) According to DEC documentation, it is an interactive utility program that allows control and monitor of a network. (2) According to IBM documentation, it is a program that controls the operation of a communication controller.

network layer   According to ISO documentation, it is defined as OSI layer 3. It is responsible for data transfer across the network. It functions independently of the network media and the topology.

network management vector transport (NMVT)   According to IBM documentation, it is a protocol used for management services in an SNA network.

network name   According to IBM documentation, in SNA, it is the symbolic identifier by which end users refer to a network-accessible unit, a link, or a link station within a given network. Another definition by IBM is used in reference to APPN networks, network names are also used for routing purposes. In a multidomain network, the name of the APPL statement defining a VTAM application program. This network name must be unique across domains.

network node server A term defined by IBM documentation as used in reference to an APPN network node that provides network services for its local LUs and client end nodes.

network number According to Apple documentation, it is a 16-bit number that provides a unique identifier for a network in an AppleTalk internet.

network operator A person who performs a variety of functions on a network, some of which are control functions.

network-qualified name A name that uniquely identifies a specific resource within a specific network. It consists of a network identifier and a resource name, each of which is a 1- to 8-byte symbol string.

network range According to Apple documentation, it is a unique range of contiguous network numbers used to identify each ETHERNET and token-ring network on an AppleTalk internet.

network routing facility (NRF) According to IBM documentation, it is an IBM program that resides in an NCP. NRF provides a path for routing messages between terminals and routes messages over this path without going through the host processor.

network services According to IBM documentation, those services within a network-accessible unit that control network operation.

network services header According to IBM documentation, in traditional SNA it is a 3-byte field in a function management data (FMD) request or response unit (RU) that flows in an SSCP-LU, SSCP-PU, or SSCP-SSCP session. This is used primarily to identify the network services category of the request unit (RU).

network services protocol According to DEC documentation, it is a formal set of conventions used in a DECnet for OpenVMS network to perform network management and to exchange messages over logical links.

network terminal option (NTO) According to IBM documentation, it is a program, used in conjunction with NCP, that allows some non-SNA devices to participate in sessions with SNA application programs in the host processor.

network topology database In an APPN network, according to IBM documentation, it is the representation of the current connectivity between the network nodes within an APPN network. It includes entries for all network nodes and the transmission groups interconnecting them and entries for all virtual routing nodes to which network nodes are attached.

NFS Network file system. According to Sun Microsystems, Inc., it is a protocol developed by Sun that uses IP to allow a set of cooperating computers to access each other's file systems as if they were local. NFS conceals differences between local and remote files by placing them in the same name space. Originally designed for UNIX systems, it is now implemented on many other systems, including personal computers like the PC and Apple computers.

NLDM Network logical data manager. According to IBM documentation, it is a subset of NetView. NLDM is a command that starts the NetView sessions monitor. NLDM also identifies various panels and functions as part of the session monitor.

no response   According to IBM documentation, in SNA, it is a protocol requested in the for-of-response-requested field of the request header. It directs the receiver of the request not to return any response, regardless of whether the request is received and processed successfully.

node   (1) Generally, a term used to refer to a computer or related device. In IBM's SNA, certain node types reflect certain functions they can perform. (2) According to DEC documentation, it is an individual computer system in a network that can communicate with other computer systems in the network. A VAXBI interface—such as a central processor, controller, or memory subsystem—that occupies one of 16 logical locations on a VAXBI bus. A VAX processor or HSC that is recognized by system communications services software.

node initialization block (NIB)   According to IBM documentation, in VTAM this is a control block associated with a particular node or session that contains information used by the application program. The information identifies the node or session and indicates how communication requests on a session are to be handled by VTAM.

node name   According to IBM documentation, in VTAM, it is the symbolic name assigned to a specific major or minor node during network definition.

node number   A unique number used to identify each node on a network.

node type   According to IBM documentation, it is a designation of a node according to the protocols it supports and the network-accessible units that it can contain. Five types are defined: 1, 2.0, 2.1, 4, and 5. Within a subarea network, type 1, type 2.0, and type 2.1 nodes are peripheral nodes, while type 4 and type 5 nodes are subarea nodes.

nonclient   A program that is written to run on a terminal and must be fooled by a terminal emulation window into running in the window environment.

noncommand Image   According to DEC documentation, it is a program not associated with a DCL command. To invoke a noncommand image, use the file-name containing the program as the parameter to the RUN command.

nonprivileged   According to DEC documentation, it is an account with no privilege other than TMPMBX and NETMBX and a user identification code greater than the system parameter MAXSYSGROUP. In DECnet for OpenVMS, this term means no privileges in addition to NETMBS, which is the minimal requirement for any network activity.

normal flow   According to IBM documentation referencing SNA, it is a data flow designated in the transmission header (TH) that is used primarily to carry end-user data. It refers to the rate at which requests flow. On normal flow, regulation can be achieved by session-level pacing. Normal and expedited flows move in both the primary-to-secondary and secondary-to-primary directions.

normalize   A term used in windowing environments. It means to change an icon back into its original appearance. The opposite of iconify.

notification   An indication that something in the network requires the operator's attention.

NOTIFY According to IBM documentation, it is a network services request sent by a SSCP to a LU. It is used to inform the LU of the status of a procedure requested by the LU.

NPDA Network problem determination application. According to IBM documentation, it is a part of NetView. It is also a command that starts the NetView hardware monitor. NPDA identifies various panels and functions as part of the hardware monitor.

NPDU Network protocol data unit. In OSI terminology it refers to a packet. A logical block of control symbols and data transmitted by the network-layer protocol.

NSF National Science Foundation. A government agency that has enabled scientists to connect to networks making up the Internet.

NSFNET National Science Foundation NETwork. Reference to a network that spans the USA.

null key According to Novell documentation, this is a key field that allows the value of the field to be a user-defined null character. For this type of key, btrieve does not index a record if the record's key value matches the null value.

NVP Network voice protocol. A TCP/IP protocol for handling voice information.

Object (1) According to Apple documentation, it is the first field in the name of an AppleTalk entity. The object is assigned by the entity itself and can be anything the user or application assigns. (2) According to DEC documentation, it is a passive repository of information to which the system controls access. Access to an object implies access to the information it contains. Examples of protected objects are files, volumes, global sections, and devices. A DECnet for OpenVMS process that receives a logical link request. It performs a specific network function or is a user-defined image for a special-purpose application. A VAX Packetnet System Interface management component that contains records to specify account information for incoming calls and to specify a command procedure that is initiated when the incoming call arrives.

object class According to DEC documentation, on VAX systems, a set of protected objects with common characteristics. For example, all files belong to the FILE class, whereas all devices belong to the DEVICE class.

object entry In Open Systems networking, this refers to a directory entry which is the primary collection of information in the directory information base about an object in the real world, not an alias entry.

object identifier based name Reference to the names that are based on the OBJECT IDENTIFIER type.

object identifier tree In OSI terminology, this term means a tree where edges are labeled with integers.

object identifier type A term used in OSI and other environments that implement ASN.1. It is an ASN.1 type whose values are the pathnames of the nodes of the object identifier tree.

offline Refers to a resource not being available.

online   Refers to a resource being available.

open-application event   According to Apple documentation, it is an Apple event that asks an application to perform the tasks—such as displaying untitled windows—associated with opening itself; one of the four required Apple events.

open-data-link interface   According to Novell documentation, it is a set of specifications defining relationships between one or more protocol stacks, the LSL, and one or more MLIDs. These specifications allow multiple communication protocols such as IPX/SPX, TCP/IP, and AppleTalk to share the same driver and adapter.

open shortest path first (OSPF)   A routing protocol based on the least cost for routing.

operand   In SNA it is an entity on which an operation is performed. That which is operated on. An operand is usually identified by an address part of an instruction.

optional parameter   According to Apple documentation, it is a supplemental parameter in an Apple event used to specify data that the server application should use in addition to the data specified in the direct parameter.

oscillation   The periodic movement between two values.

other-domain resource   According to IBM documentation, it is a representation for a LU that is owned by another domain and is referenced by a symbolic name, which can be qualified by a network identifier.

owner   According to DEC documentation, it is a user with the same user identification code as the protected object. An owner always has control access to the object and can therefore modify the object's security profile. When the system processes an access request from an owner, it considers the access rights in the owner field of a protection code.

pacing   A term used in IBM's SNA referring to a technique by which a receiving component controls the rate of transmission of a sending component to prevent overrun or congestion.

pacing response   According to IBM documentation, in SNA, it is an indicator that signifies the readiness of a receiving component to accept another pacing group. The indicator is carried in a response header for session-level pacing and in a transmission header for virtual route pacing.

pacing window   According to IBM documentation, this refers to the path information units (PIUs) that can be transmitted on a virtual route before a virtual route pacing response is received indicating that the virtual route receiver is ready to receive more PIUs on the route.

packet   A term used generically in many instances. It is a small unit of control information and data that is processed by the network protocol.

packet assembly/disassembly (PAD) facility   This is a term used in packet-switching technology. It is a device at a packet-switching network permitting access from an asynchronous terminal. Terminals connect to a PAD, and a PAD puts the terminal's input data into packets, then takes the terminal's output data out of packets.

page A term used when virtual storage is being discussed. It refers to a fixed-length block that has a virtual address and is transferred as a unit between real storage and secondary storage.

pagelet According to DEC documentation, this is a 512-byte unit of memory in an AXP environment. On AXP systems, certain DCL and utility commands, system services, and system routines accept as input or provide as output memory requirements and quotas in terms of pagelets.

page table base register According to DEC documentation, on AXP systems, the processor register or its equivalent, in a hardware privileged context block that contains the page frame number of the process's first-level page table.

panel According to IBM documentation, it is an arrangement of information that is presented in a window.

parallel links According to IBM documentation, in SNA this is two or more links between adjacent subarea nodes.

parallel sessions According to IBM documentation, it is two or more concurrently active sessions between the same two network-accessible units (NAUs) using different pairs of network addresses or local-form session identifiers. Each session can have independent session parameters.

parallel transmission groups According to IBM documentation, it is multiple transmission groups (TGs) connecting two adjacent nodes.

parameter A generic term used to refer to a given constant value for a specified application and that may denote the application.

parent window In windowing environments, it is a window that causes another window to appear. Specifically, it refers to windows that "own" other windows.

partitioned data set (PDS) According to IBM documentation, this is a type of storage, divided into partitions, called *members*. Each member contains records that are the actual data which is stored.

path In a network, any route between two or more nodes.

path control (PC) According to IBM documentation, it is the function that routes message units between network-accessible units in the network and provides the paths between them. It is depicted in traditional SNA layers. PC converts the BIUs from transmission control (possibly segmenting them) into path information units (PIUs) and exchanges basic transmission units containing one or more PIUs with data-link control.

path information unit (PIU) According to IBM documentation, it is a message unit containing only a transmission header (TH), or a TH followed by a basic information unit (BIU) or a BIU segment.

pending active session According to IBM documentation, in VTAM, it is the state of an LU-LU session recorded by the SSCP when it finds both LUs available and has sent a CINIT request to the primary logical unit (PLU) of the requested session.

performance assist According to DEC documentation, it is the OpenVMS volume shadowing using controller performance assists to improve full-copy and merge operation performance. There are two distinct types of performance assists: the full-copy assist and the minimerge assist.

peripheral host node   According to IBM documentation, it is a type of node defined in SNA terminology. It does not provide SSCP functions and is not aware of the network configuration. The peripheral host node does not provide subarea node services. It has boundary function provided by its adjacent subarea.

peripheral logical unit (PLU)   This term refers to a logical unit in a peripheral node found in SNA networks. It should not be confused with a primary logical unit, also known as a PLU.

peripheral node   According to IBM documentation, it is a node that uses local addresses for routing and is not affected by changes in network addresses. A peripheral node requires boundary function assistance from an adjacent subarea node.

peripheral path control   According to IBM documentation, it is the function in a peripheral node that routes message units between units with local addresses and provides the paths between them.

permanent virtual circuit   A term used in many different types of network environments. Generally, it is a permanent logical association between two DTEs, which is analogous to a leased line. Packets are routed directly by the network from one DTE to the other.

personal computer (PC)   A term becoming more vague with the passage of time. It basically refers to an individual's own computer. By this definition, that means it has its own processor, memory, storage, and display.

phase   The place of a wave in an oscillation cycle.

physical connection   A link that makes transmission of data possible. Generally considered as a tangible link; it may support electron, photon, or other data-type representation transfer.

physical layer   A term used in OSI circles. It refers to the lowest layer defined by the OSI model. However, layer 0 would be the lowest layer in such a model. This layer (layer 0) represents the medium, whether hard or soft.

physical unit (PU)   According to IBM documentation, a component (either software or firmware) that manages and monitors specified resources associated with a node. The type PU, indicated by number, is typically either 5, 4, 2.0, or 2.1. The type PU dictates what supporting services are available.

physical unit (PU) services   According to IBM documentation, that component within a PU that provides configuration services and maintenance services for SSCP-PU sessions.

PING   Packet Internet Groper. A program found in TCP/IP-based networks. It is the name of a program used with TCP/IP networks employed to test reachability of destinations by sending them an ICMP echo request and then waiting for a reply.

pixel   The smallest dot that can be drawn on the screen. Derived from the term *picture element*.

point   A unit of measurement for type.

point-to-point protocol (PPP)   A type of protocol used over asynchronous and synchronous connections for router-to-router or a host-to-network communications.

polling   The process in which data stations are invited, one at a time, to transmit on a multipoint or point-to-point connection.

port   A term used in TCP/IP-based networks. In TCP/IP two transport protocols exist: TCP and UDP. Applications that reside on top of these protocols have a port number assigned to them for addressing purposes. Generally, it is an addressable point.

port name   That which contains a name string, a type string, and a script code. A term used in Apple documentation.

power manager   According to Apple documentation, firmware that provides an interface to the power management hardware in the Macintosh portable computer.

PPDU   Presentation protocol data unit. In OSI terminology it is a term referring to logical blocks of control symbols and data transmitted at the presentation-layer protocol.

presentation layer   According to the OSI model for networks, this is layer 6. Data representation occurs here. Syntax of data such as ASCII or EBCDIC is determined at this layer.

primary application program   According to IBM documentation, in VTAM, this is an application program acting as the primary end of an LU-LU session.

primary end of a session   According to IBM documentation, this is the end of a session that uses primary protocols. The primary end establishes the session. For an LU-LU session, the primary end of the session is the primary logical unit.

PrimaryInit record   According to Apple documentation, it is a data structure in the declaration ROM of a NuBus card that contains initialization code. The Slot Manager executes the code in the PrimaryInit record when it first locates a declaration ROM during system startup.

primary logical unit (PLU)   According to IBM documentation, it is the logical unit that sends the BIND to activate a session with its partner LU.

primary key   According to DEC documentation, the mandatory key within the data records of an indexed file; used by OpenVMS record management services to determine the placement of records within the file and to build the primary index.

PrintMonitor   According to Apple documentation, a background print spooler that is included with the Macintosh MultiFinder.

print server   In networking, this is a term used in a general sense to convey that a computer controls spooling and other printer operations.

private partition   According to IBM documentation, in VSE, this is an allocated amount of memory for the execution of a specific program or application program. Storage in a private partition is not addressable by programs running in other virtual address spaces.

privilege According to DEC documentation, it means protecting the use of certain system functions that can affect system resources and integrity. System managers grant privileges according to users' needs and deny them to users as a means of restricting their access to the system.

process Depending on the context, it can mean an open application or an open-desk accessory.

processor That component which interprets and executes instructions.

processor status According to DEC documentation, on VAX systems, a privileged processor register, known as the *processor status longword,* consisting of a word of privileged processor status and the processor status word itself. The privileged processor status information includes the current interrupt priority level, the previous access mode, the current access mode, the interrupt stack bit, the trace trap pending bit, and the compatibility mode bit.

processor status word According to DEC, on VAX systems, the low-order word of the processor status longword. Processor status information includes the condition codes (carry, overflow, 0, negative), the arithmetic trap enable bits (integer overflow, decimal overflow, floating underflow), and the trace enable bit.

product-set identification (PSID) According to IBM documentation, it is a technique for identifying the hardware and software products that implement a network component.

PROFILE EXEC According to IBM documentation, in VM, a special EXEC procedure with a filename of PROFILE. The procedure is normally executed immediately after CMS is loaded into a virtual machine. It contains CP and CMS commands that are to be issued at the start of every terminal session.

program operator A term used in SNA. According to IBM documentation, it is a VTAM application program that is authorized to issue VTAM operator commands and receive VTAM operator awareness messages.

program temporary fix (PTF) According to IBM documentation, it is a temporary solution or bypass of a problem diagnosed by IBM in a current unaltered release of the program.

protocol An agreed-upon way of doing something.

protocol data unit A general term used to refer to that which is exchanged between peer-layer entities.

PU-PU flow According to IBM documentation, in SNA it is the exchange between physical units (PUs) of network control requests and responses.

proxy ARP In TCP/IP networks, this is a technique where one machine answers ARP requests intended for another by supplying its own physical address.

pulse dispersion The spreading of pulses as they traverse an optical fiber.

queued session According to IBM documentation, in VTAM this pertains to a requested LU-LU session that cannot be started because one of the LUs is not available. If the session-initiation request specifies queuing, the system services control points (SSCPs) record the request and later continue with the session establishment procedure when both LUs become available.

quit application event   According to Apple documentation, this is an Apple event that requests that an application perform the tasks—such as releasing memory, asking the user to save documents, and so on—associated with quitting; one of the four required Apple events. The Finder sends this event to an application immediately after sending it a print documents event or if the user chooses restart or shutdown from the Finder's special menu.

RACF   Resource access control facility. An IBM security program package.

RARP   Reverse address resolution protocol. A TCP/IP protocol for mapping ETHERNET addresses to IP addresses. It is used by diskless workstations who do not know their IP addresses. In essence, it asks "Who am I?" Normally, a response occurs and is cached in the host.

read-only memory (ROM)   Memory in which stored data cannot be modified except under special conditions.

real resource   In VTAM, a resource identified by its real name and its real network identifier.

receive pacing   According to IBM documentation, it is the pacing of message units being received by a component.

record access block (RAB)   According to DEC documentation, it is an OpenVMS record management services (RMS) user control block allocated at either assembly or run time to communicate with RMS. The control block describes the records in a particular file and associates it with a file access block to form a record access stream. A RAB defines the characteristics needed to perform record-related operations, such as update, delete, or get.

record management services (RMS)   According to DEC documentation, it is a set of operating system procedures that is called by programs to process files and records within files. RMS allows programs to issue get and put requests at the record level as well as read and write blocks. RMS is an integral part of the system software; its procedures run in executive mode.

region code   According to Apple documentation, it is a number used to indicate a particular localized version of Macintosh system software.

relative path   The path through a volume's (disk's) hierarchy from one file or directory to another.

release   A distribution of a new product or new function and APAR (authorized program analysis report) fixes for an existing product. Normally, programming support for the prior release is discontinued after some specified period of time following availability of a new release.

remote client   A term used in an XWindow environment. It is an X program running on a remote system, but the output of the program can be viewed locally.

remote operations service element   A term used in open networking environments. It is an application service element that provides the basis for remote requests.

request header (RH)   According to IBM documentation, it is control information that precedes a request unit (RU).

request parameter list (RPL)   According to IBM documentation, it refers to a VTAM control block that contains the parameters necessary for processing a request for data transfer, for establishing or terminating a session, or for some other operation.

request unit (RU)   According to IBM documentation, it is a message unit that contains control information, end-user data, or both.

request (or response) header (RH)   According to IBM documentation, it is control information associated with a particular RU. The RH precedes a RU and specifies the type of RU (request unit or response unit).

request (or response) unit (RU)   A generic term for a request unit or a response unit.

required Apple event   According to Apple documentation, it is one of four core Apple events that the Finder sends to applications. These events are called *open documents, open application, print documents,* and *quit application.* They are a subset of the core Apple events.

reset   Generally, it is a change to the original state of operation.

resource   Generally, a main storage, secondary storage, input/output devices, the processing unit, files, and control or processing programs, or anything else that can be used by a user directly or indirectly.

resource access control facility (RACF)   According to IBM documentation, it is an IBM program that provides for access control by identifying and verifying the users of the system, by authorizing access to protected resources, by logging the detected unauthorized attempts to enter the system, and by logging the detected accesses to protected resources.

resource definition table (RDT)   According to IBM documentation, a VTAM table that describes characteristics of each node available to VTAM and associates each node with a network address. This is a main VTAM network configuration table.

resource hierarchy   According to IBM documentation, a VTAM relationship among network resources in which some resources are subordinate to others as a result of their positions in the network structure and architecture; for example, the logical units (LUs) of a peripheral physical unit (PU) are subordinate to the PU, which, in turn, is subordinate to the link attaching it to its subarea node.

resource registration   According to IBM documentation, the process of identifying names of resources, such as LUs, to a network node server or a central directory server.

resource takeover   According to IBM documentation, it is a VTAM action initiated by a network operator to transfer control of resources from one domain to another without breaking the connections or disrupting existing LU-LU sessions on the connection.

resource types   According to IBM documentation, with reference to NetView, it is a concept describing the organization of panels. Resource types are defined as central processing unit, channel, control unit, and I/O device for one category; and communication controller, adapter, link, cluster controller, and terminal for another category. Resource types are combined with data types and display types to describe display organization.

response   A reply to some occurrence, or the lack thereof.

response header (RH)   According to IBM documentation, a header, optionally followed by a response unit, that indicates whether the response is positive or negative and that may contain a pacing response.

response time   According to IBM documentation, a term used with the product NetView. It refers to the lapsed time between the end of an inquiry or demand on a computer system and the beginning of the response; for example, the length of time between an indication of the end of an inquiry and the display of the first character of the response at a user terminal.

response unit (RU)   According to IBM documentation, it is a message unit that acknowledges a request unit. It may contain prefix information received in a request unit.

restoring   A term used in window-based environments meaning to change a minimized or maximized window back to its regular size.

Restructured Extended Executor (REXX)   According to IBM documentation, a general-purpose, procedural language for end-user personal programming, designed for ease by both casual general users and computer professionals. It is also useful for application macros.

result handler   A routine that the data access manager calls to convert a data item to a character string.

return code   A code used to identify the action or lack thereof of a program execution.

RFC   Request for comments. Proposed and accepted TCP/IP standards.

RLOGIN   Remote LOGIN. A log-on service provided by Berkeley 4BSD UNIX systems that allows users of one machine to connect to other UNIX systems.

RMS-11   According to DEC documentation, a set of routines that are linked with compatibility mode programs and provide similar functional capabilities to OpenVMS RMS. The file organizations and record formats used by RMS-11 are very similar to those of RMS.

root menu   That menu which could be called the main menu. That menu from which other menus originate.

root window   In the XWindowing environment, this is what is presented on the screen once the X graphical user interface is visible to the user. It is the window on which other windows are based.

route selection services (RSS)   According to IBM documentation, it is a subcomponent of the topology and routing services component that determines the preferred route between a specified pair of nodes for a given class of service.

routing   The moving of data through paths in a network.

routing information base   A collection of output from route calculations.

routing table maintenance protocol (RTMP)   According to Apple documentation, a protocol used by routers on an AppleTalk internet to determine how to forward a data packet to the network number to which it is addressed.

RTMP stub   According to Apple documentation, the portion of the routing table maintenance protocol contained in an AppleTalk node other than a router. DDP uses the RTMP stub to determine the network number of the network cable to which the node is connected and to determine the network number and node ID on one router on that network cable.

RTT   Round-trip time. A measure of delay between two hosts.

RU chain   According to IBM documentation, a SNA set of related request/response units (RUs) that are consecutively transmitted on a particular normal or expedited data flow.

RUN disk   According to IBM documentation, this is a virtual disk that contains the VTAM, NetView, and VM/SNA console support (VSCS) load libraries, program temporary fixes (PTFs), and user-written modifications from the disk.

SACK   Selective ACKnowledgment. This is a term used in TCP/IP. It is an acknowledgment mechanism used with sliding-window protocols. This permits a receiver to acknowledge packets received out of order within the current sliding window.

same-domain   According to IBM documentation, this adjective pertains to communication between entities in the same SNA domain.

satellite node   According to DEC documentation, a processor that is part of a local area VMScluster system. A satellite node is booted remotely from the system disk of the boot server in this type of VMScluster system.

screen   In SAA Basic Common User Access architecture, the physical surface of a display device on which information is shown to a user.

screen dump   A screen capture capable of being routed to a file.

secondary end of a session   According to IBM documentation, the end of a session that uses secondary protocols. For a LU-LU session, the secondary end of the session is the secondary logical unit (SLU).

secondaryInit record   According to Apple documentation, a data structure in the declaration ROM of a NuBus card that contains initialization code. The slot manager executes the code in the SecondaryInit record after RAM patches to the operating system have been loaded from disk during system startup.

secondary logical unit (SLU)   According to IBM documentation, the LU that contains the secondary half-session for a particular LU-LU session. A LU may contain secondary and primary half-sessions for different active LU-LU sessions.

segment   According to IBM documentation, with reference to a token-ring network, a section of cable between components or devices. A segment may consist of a single patch cable, several patch cables that are connected, or a combination of building cable and patch cables that are connected. In TCP/IP this is the unit of transfer sent from TCP on one machine to TCP on another.

segmentation   According to IBM documentation, a process by which path control divides basic information units into smaller units, called *BIU segments,* to accommodate smaller buffer sizes in adjacent nodes. Both segmentation and segment assembly are optional path control features. The support for either or both is indicated in the BIND request and response.

semaphore   According to DEC documentation, in a DECnet for OpenVMS network, a common data structure used to control the exchange of signals between concurrent processes.

send pacing   According to IBM documentation, the pacing of message units (in SNA) that a component is sending.

sequential access mode   According to DEC documentation, the retrieval or storage of records where a program reads or writes records one after the other in the order in which they appear, starting and ending at any arbitrary point in the file.

sequential file organization   According to DEC documentation, a file organization in which records appear in the order in which they were originally written. The records can be of fixed or variable length. Sequential file organization permits sequential record access and random access by the record's file address. Sequential file organization with fixed-length records also permits random access by relative record number.

server   An entity that serves the request of a client. This may be in the context of TCP/IP applications with clients and servers, or it could refer to a print or file server.

service access point (SAP)   A logical addressable point.

service primitive   Part of a service element. Four types exist: confirm, indication, request, and response.

session   A logical connection between two addressable endpoints.

session activation request   According to IBM documentation, a request in SNA that activates a session between two network-accessible units and specifies session parameters that control various protocols during session activity; for example, BIND and ACTPU (activate physical unit).

session awareness (SAW) data   According to IBM documentation, this is data collected by the NetView program about a session that includes the session type, the names of session partners, and information about the session activation status. It is collected for LU-LU, SSCP-LU, SSCP-PU, and SSCP-SSCP session and for non-SNA terminals not supported by NTO. It can be displayed in various forms, such as most recent sessions lists.

session connector   According to IBM documentation, a session-layer component in an APPN network node or in a subarea node boundary or gateway function that connects two stages of a session. Session connectors swap addresses from one address space to another for session-level intermediate routing, segment session message units as needed, and (except for gateway function session connectors) adaptively pace the session traffic in each direction.

session control (SC)   According to IBM documentation, one of the following: (1) a component of transmission control (session control is used to purge data flowing in a session after an unrecoverable error occurs, to resynchronize the data flow after such an error, and to perform cryptographic verification) or (2) a request unit (RU) category used for requests and responses exchanged between the session control components of a session and for session activation and deactivation requests and responses.

session control block (SCB)   According to IBM documentation, in NPM, control blocks in common storage area for session collection.

session data   According to IBM documentation, this is data about a session, collected by the NetView program consisting of session awareness data, session trace data, and session response-time data.

session deactivation request   According to IBM documentation, a term used in SNA that refers to a request that deactivates a session between two network-accessible units (NAUs); for example, UNBIND and DACTPU.

session-establishment request   According to IBM documentation, in VTAM, where a request is sent to a LU to establish a session. For the primary logical unit (PLU) of the requested session, the session establishment request is the CINIT sent from the system services control point (SSCP) to the PLU. For the secondary logical unit (SLU) of the requested session, the session establishment request is the BIND sent from the PLU to the SLU.

session ID   According to IBM documentation, a number that uniquely identifies a session.

session initiation request   According to IBM documentation, an initiate or log-on request from a logical unit (LU) to a system services control point (SSCP) that a LU-LU session be activated.

session layer   According to the OSI reference model, this is layer 5. It coordinates the dialog between two communicating application processes.

session-level LU-LU verification   According to IBM documentation, a LU6.2 security service that is used to verify the identity of each logical unit when a session is established.

session-level pacing   According to IBM documentation, this refers to a flow control technique that permits a receiving half-session or session connector to control the data transfer rate (the rate at which it receives request units) on the normal flow. It is used to prevent overloading a receiver with unprocessed requests when the sender can generate requests faster than the receiver can process them.

session limit   According to IBM documentation, a term used to refer to the maximum number of concurrently active LU-LU sessions that a specific LU can support.

session manager (SM)   Typically a third-party product that permits a user on one terminal to log on to multiple applications concurrently.

session monitor   According to IBM documentation, a component of NetView that collects and correlates session-related data and provides online access to this information.

session parameters According to IBM documentation, the parameters that specify or constrain the protocols (such as bracket protocol and pacing) for a session between two network-accessible units.

session partner According to IBM documentation, in SNA it is one of the two network-accessible units (NAUs) having an active session.

session path According to IBM documentation, the half-sessions delimiting a given session and their interconnection (including any intermediate session connectors).

session services According to IBM documentation, it is one of the types of network services in the control point (CP) and in the logical unit (LU). These services provide facilities for a LU or a network operator to request that a control point aid with initiating or terminating sessions between LUs. Assistance with session termination is needed only by SSCP-dependent LUs.

session stage According to IBM documentation, that portion of a session path consisting of two session-layer components that are logically adjacent and their interconnection. An example is the paired session-layer components in adjacent type 2.1 nodes and their interconnection over the link between them.

7.0-compatible According to Apple documentation, this is used to refer to an application that runs without problems in system software Version 7.0.

7.0-dependent According to Apple documentation, this is used to refer to an application that requires the existence of features that are present only in system software Version 7.0.

7.0-friendly According to Apple documentation, this is used to refer to an application that is 7.0-compatible and takes advantage of some of the special features of system software Version 7.0, but is still able to perform all its principal functions when operating in Version 6.0.

shadow resource According to IBM documentation, it is an alternate representation of a network resource that is retained as a definition for possible future use.

sharable image According to DEC documentation, an image that has all its internal references resolved, but must be linked with one or more object modules to produce an executable image. A sharable image cannot be executed. A sharable image file can be used to contain a library of routines.

shared-access transport facility (SATF) A transmission facility, such as a multipoint link connection or a token-ring network, where multiple pairs of nodes can form concurrently active links.

shared image According to DEC documentation, it is an image that is installed so that multiple users in a system can share the memory pages where the image is loaded.

shared partition According to IBM documentation, in VSE this is a partition allocated for a program such as VSE/POWER that provides services for and communicates with programs in other partitions of the system's virtual address spaces. Storage in a shared partition is addressable by programs running concurrently in other partitions.

sibling networks   According to Novell documentation, two or more coequal networks branching off the same node in an internetwork. Workstations on these networks that use NetWare btrieve must have access to a file server loaded with BSERVER.

sift-down effect   According to IBM documentation, the copying of a value from a higher-level resource to a lower-level resource. The sift-down effect applies to many of the keywords and operands in NCP and VTAM definition statements. If an operand is coded on a macroinstruction or generation statement for a higher-level resource, it need not be coded for lower-level resources for which the same value is desired. The value "sifts down," that is, becomes the default for all lower-level resources.

silly-window syndrome   In TCP/IP-based networks, a scenario in which a receiver keeps indicating a small "window" and a sender continues to send small segments to it.

single-byte character set (SBCS)   According to IBM documentation, a character set in which each character is represented by a 1-byte code.

single-console image facility (SCIF)   According to IBM documentation, a VM facility that allows multiple consoles to be controlled from a single virtual machine console.

single-domain network   According to IBM documentation, a network with one SSCP.

single mode   A type of fiber-optic cable containing only one mode.

sleep state   According to Apple documentation, a low-power consumption state of the Macintosh portable computer. In the sleep state, the power manager and the various device drivers shut off power or remove clocks from the computer's various subsystems, including the CPU, RAM, ROM, and I/O ports.

sliding window   A scenario in which a protocol permits the transmitting station to send a stream of bytes before an acknowledgment arrives.

SLIP   Serial-line IP. A protocol designed to run IP protocol over serial lines. An example is using telephone lines.

SMTP   Simple mail transfer protocol. A TCP/IP application that provides electronic mail (E-mail) support. The SMTP protocol specifies how two mail systems interact and the format of control messages.

SNA network   A collection of IBM hardware and software put together in such a way as to form a collective composite greater than its parts. The components composing the network conform to the SNA format and protocol specifications defined by IBM.

SNA network interconnection (SNI)   According to IBM, the connection of two or more independent SNA networks to allow communication between logical units in those networks. The individual SNA networks retain their independence.

SNMP   Simple network monitoring protocol. A de facto industry standard protocol used to manage TCP/IP networks.

socket   (1) A concept from Berkeley 4BSD UNIX that allows an application program to access the TCP/IP protocols. (2) In TCP/IP networks, the internet address of the host and the port number it uses. A TCP/IP application is identified by its socket.

solicited message   According to IBM documentation, a response from VTAM to a command entered by a program operator.

source route   A route determined by the source. TCP/IP implements source routing by using an option field in an IP datagram.

specific mode   According to IBM documentation, in VTAM, the following:

The form of a RECEIVE request that obtains input from one specific session
The form of an accept request that completes the establishment of a session by accepting a specific queued CINIT request

SSCP-dependent LU   A LU requiring assistance from a SSCP to establish a LU-LU session.

SSCP ID   According to IBM documentation, in SNA, it is a number that uniquely identifies a SSCP. The SSCP ID is used in session activation requests sent to physical units (PUs) and other SSCPs.

SSCP-independent LU   According to IBM documentation, a LU that can activate a LU-LU session (i.e., send a BIND request) without assistance from an SSCP. It does not have an SSCP-LU session. Currently, only a LU6.2 can be an independent LU.

SSCP-LU session   According to IBM documentation, in SNA, a session between the SSCP and a logical unit (LU). The session enables the LU to request the SSCP to help initiate LU-LU sessions.

SSCP-PU session   According to IBM documentation, in SNA, a session between a SSCP and a PU. SSCP-PU sessions allow SSCPs to send requests to and receive status information from individual nodes in order to control the network configuration

SSCP rerouting   According to IBM documentation, an SNA network interconnection   A technique used by the gateway system services control point (SSCP) to send session-initiation RUs, by way of a series of SSCP-SSCP sessions, from one SSCP to another, until the owning SSCP is reached.

SSCP-SSCP session   According to IBM documentation, a session between the SSCP in one domain and the SSCP in another domain. This type of session is used to initiate and terminate cross-domain LU-LU sessions.

stack   An area of memory in the application partition that is used to store temporary variables.

start/stop (SS) transmission   Asynchronous transmission in which each signal that represents a character is preceded by a start signal and is followed by a stop signal.

station   An input or output point.

statistic   Significant data about a defined resource.

status   Generally speaking, it is a condition or state of a resource. According to DEC documentation, it is a display type for the NCP commands SHOW and LIST. Status refers to dynamic information about a component that is kept in either the volatile (temporary) or permanent database.

status monitor   According to IBM documentation, it is a component of the NetView program that collects and summarizes information on the status of resources defined in a VTAM domain.

stream   In IBM's SNA, a structured protocol: for example, a 3270 data stream, a GDS data stream, or LU6.2 data stream. According to DEC documentation, it is an access window to a file associated with a record control block, supporting record operation requests. Generally, it is a full-duplex connection between a user's task and a device.

subarea   According to IBM documentation, a portion of the SNA network consisting of a subarea node, attached peripheral nodes, and associated resources.

subarea address   According to IBM documentation, a value in the subarea field of a network address that identifies a particular subarea.

subarea host node   According to IBM documentation, a node that provides both subarea function and an application program interface (API) for running application programs. It provides SSCP functions and subarea node services, and is aware of the network configuration.

subarea link   According to IBM documentation, a link that connects two subarea nodes.

subarea LU   According to IBM documentation, a logical unit that resides in a subarea node.

subarea network   According to IBM documentation, interconnected subareas, their directly attached peripheral nodes, and the transmission groups that connect them.

subarea node (SN)   According to IBM documentation, a node that uses network addresses for routing and maintains routing tables that reflect the configuration of the network. Subarea nodes can provide gateway function to connect multiple subarea networks, intermediate routing functions, and boundary function support peripheral nodes. Type 4 and type 5 nodes are subarea nodes.

subarea path control   According to IBM documentation, the function in a subarea node that routes message units between network-accessible units (NAUs) and provides the paths between them.

subdirectory   According to DEC documentation, it is a directory file, cataloged in a higher-level directory, that lists additional files belonging to the owner of the directory.

subsystem   A secondary or subordinate software system.

summary   According to DEC documentation, the default display type for the NCP commands SHOW and LIST. A summary includes the most useful information for a component, selected from the status and characteristics information.

supervisor   According to IBM documentation, that part of a control program that coordinates the use of resources and maintains the flow of processing unit operations.

supervisor call (SVC)   According to IBM documentation, a request that serves as the interface into operating system functions, such as allocating storage. The SVC protects the operating system from inappropriate user entry. All operating system requests must be handled by SVCs.

switched connection   A data-link connection that functions like a telephone.

switched line   A line where the connection is established by dialing.

switched major node   According to IBM documentation, in VTAM, a major node whose minor nodes are physical units and logical units attached by switched SDLC links.

switched network   A network that establishes connections by a dialing function.

switched network backup   According to IBM documentation, an optional facility that allows a user to specify, for certain types of physical units (PUs), a switched line to be used as an alternate path if the primary line becomes unavailable or unusable.

switched virtual circuit   A temporary logical association between two DTEs connected to a packet-switching data network.

symbiont   According to DEC documentation, a process that transfers record-oriented data to or from a device. For example, an input symbiont transfers data from card readers to disks. An output symbiont transfers data from disks to line printers.

symbiont manager   According to DEC documentation, the function that maintains spool queues and dynamically creates symbiont processes to perform the necessary I/O operations.

symbol   According to DEC documentation, an entity that, when defined, will represent a particular function or entity (e.g., a command string, directory name, or filename) in a particular context.

symbol table   According to DEC documentation, that portion of an executable image that contains the definition of global symbols used by the debugger for images linked with the DEBUG qualifier. A table in which the Digital Command Language places local symbols. DCL maintains a local symbol table for each command level.

synchronous backplane interconnect   According to DEC documentation, that part of the hardware that interconnects the VAX processor, memory controllers, MASSBUS adapters, and the UNIBUS adapter.

synchronization point   According to IBM documentation, it is an intermediate point or endpoint during processing of a transaction at which an update or modification to one or more of the transaction's protected resources is logically complete and error-free.

sync-point services (SPS)   According to IBM documentation, it is the component of the sync-point manager that is responsible for coordinating the managers of protected resources during sync-point processing. SPS coordinates two-phase commit protocols, resync protocols, and logging.

System Communications Services   According to DEC documentation, a protocol responsible for the formation and breaking of intersystem process connections and for flow control of message traffic over those connections. System services such as the VMScluster connection manager and the mass storage control protocol (MSCP) disk server communicate with this protocol.

system control block   According to DEC documentation, on VAX systems, the data structure in system space that contains all the interrupt and exception vectors known to the system.

system definition   According to IBM documentation, the process, completed before a system is put into use, by which desired functions and operations of the system are selected from various available options.

system disk   According to DEC documentation, the disk that contains the operating system. In a VMScluster environment, a system disk is set up so that most of the files can be shared by several processors. In addition, each processor has its own directory on the system disk that contains its page, swap, and dump files.

system file   According to Apple documentation, a file, located in the system folder, that contains the basic system software plus some system resources, such as font and sound resources.

system generation   Synonym for system definition.

system GETVIS area   According to IBM documentation, a storage space that is available for dynamic allocation to VSE's system control programs or other application programs.

system image   According to DEC documentation, the image read into memory from disk when the system is started up.

system management facility (SMF).   According to IBM documentation, a feature of MVS that collects and records a variety of system and job-related information.

system menu   In a windowing environment, particularly the X windowing environment, the menu that displays when you press the system MENU button on the window manager window frame. Every window has a system menu that enables you to control the size, shape, and position of the window.

system modification program (SMP)   According to IBM documentation, a program used to install software changes on MVS systems.

system services control point (SSCP)   According to IBM documentation, a component within a subarea network for managing the configuration, coordinating network operator and problem determination requests, and providing directory services and other session services for end users of the network. Multiple SSCPs, cooperating as peers with one another, can divide the network into domains of control, with each SSCP having a hierarchical control relationship with the physical units and logical units within its own domain.

system services control point (SSCP) domain According to IBM documentation, the system services control point, the physical units (PUs), the logical units (LUs), the links, the link stations, and all the resources that the SSCP has the ability to control by means of activation and deactivation requests.

System Support Programs (SSP) According to IBM documentation, IBM licensed program, made up of a collection of utilities and small programs, that supports the operation of the NCP.

Systems Network Architecture (SNA) IBM's description of the logical structure, formats, protocols, and operational sequences for their network protocol called SNA.

takeover According to IBM documentation, the process by which the failing active subsystem is released from its extended recovery facility (XRF) sessions with terminal users and is replaced by an alternate subsystem.

task specifier According to DEC documentation, it is information provided to DECnet for OpenVMS software that enables it to complete a logical link connection to a remote task. This information includes the name of the remote node on which the target task runs and the name of the task itself.

TCP Transmission control protocol. The TCP/IP standard transport-level protocol that provides the reliable, full-duplex, stream service on which many application protocols depend. It is connection-oriented in that before transmitting data, participants must establish a connection.

telecommunications access method (TCAM) According to IBM documentation, the access method prior to VTAM.

TELNET The TCP/IP TCP standard protocol for remote terminal service.

10Base T An ETHERNET implementation using 10 Mbits/s with baseband signaling over twisted-pair cabling.

terminal Generally agreed on as a point of entry with a display and keyboard.

terminal access facility (TAF) According to IBM documentation, in the NetView program, a facility that allows a network operator to control a number of subsystems. In a full-screen or operator control session, operators can control any combination of such subsystems simultaneously.

terminal-based program In the X windowing environment, a program (nonclient) written to be run on a terminal (not in a window). Terminal-based programs must be fooled by terminal-emulation clients to run on the XWindow system.

terminal emulator A term generally used to refer to a program that performs some type of simulation; typically this simulation is a type of terminal/protocol.

terminal server   A network device that is used to connect "dumb" terminals to a network medium. Consequently, these terminals have virtual terminal access to hosts and devices located on a network.

terminal type   The type of terminal attached to your computer. UNIX uses the terminal type to set the TERM environment variable so that it can communicate with the terminal correctly. In the SNA environment, the terminal type is required in order to know how to configure the system so that it can function.

TERMINATE   According to IBM documentation, in SNA it is a request unit that is sent by a logical unit (LU) to its system services control point (SSCP) to cause the SSCP to start a procedure to end one or more designated LU-LU sessions.

term0   According to Hewlett-Packard documentation, a level 0 terminal is a reference standard that defines basic terminal functions.

TFTP   Trivial file transfer protocol. A TCP/IP UDP standard protocol for file transfer that uses UDP as a transport mechanism. TFTP depends only on UDP, so it can be used on machines such as diskless workstations.

TG weight   According to IBM documentation, a quantitative measure of how well the values of a transmission group's characteristics satisfy the criteria specified by the class-of-service definition, as computed during route selection for a session.

thread   According to DEC documentation, a single, sequential flow of control within a program. It is the active execution of a designated routine, including any nested routine invocations. A single thread has a single point of execution within it. A thread can be executed in parallel with other threads.

threshold   Generally regarded as a percentage value set for a resource.

tile   In the X windowing environment it is a rectangular area used to cover a surface with a pattern or visual texture.

time-division multiplexing (TDM)   A technique used to multiplex data on a channel by a timesharing of the channel.

time-domain reflectometer (TDR)   A device used to troubleshoot networks. It sends signals through a network medium to check for continuity.

time sharing option/extension (TSO/E)   According to IBM, this is the base for all TSO enhancements.

timeout   An event that occurs at the end of a predetermined period of time.

title bar   A term used in the XWindow environment. It is the rectangular area between the top of the window and the window frame. The title bar contains the title of the window object, for example, Xclock for clocks.

TN3270   A client program that uses TELNET protocol but produces an EBCDIC 3270 data stream. The program is normally found as a TN3270 client application that provides access into a 3270-based environment.

token   The symbol of authority passed successively from one data station to another to indicate which station is temporarily in control of the transmission medium.

token ring A network with a ring topology that passes tokens from one attaching device to another.

token-ring interface coupler (TIC) Interface board used to connect a device such as a 3720, 3725, or 3745 communication controller to a token-ring network.

token-ring network A ring network that allows unidirectional data transmission between data stations by a token-passing procedure.

topology and routing services (TRS) According to IBM documentation, it is an APPN control point component that manages the topology database, computes routes, and provides a route selection control vector (RSCV) that specifies the best route through the network for a given session according to its requested class-of-service.

trace A record of events captured and used to troubleshoot hardware and/or software.

transaction According to Apple documentation, a sequence of Apple events sent back and forth between a client and a server application, beginning with the client's initial request for a service.

transaction processing facility (TPF) A software system designed to support real-time applications.

transaction program According to IBM, a program that conforms to LU6.2 protocols.

transceiver A device that connects a host's cable from the interface board to the main cable of the network.

trap An event used in SNMP-managed networks to send data to the network manager. A trap is sent from a SNMP agent.

trash folder According to Apple documentation, a directory at the root level of a volume for storing files that the user has moved to the trash icon. After opening the trash icon, the user sees the collection of all items that the user has moved to the trash icon—that is, the union of appropriate trash directories from all mounted volumes. A Macintosh setup to share files among users in a network environment maintains separate trash subdirectories for remote users within its shared, network trash directory. The Finder for system software Version 7.0 empties a trash directory only when the user of that directory chooses the empty-trash command.

translated code According to DEC documentation, it is the native AXP object code in a translated image. Translated code includes:

AXP code that reproduces the behavior of equivalent VAX code in the original image

Calls to the translated image environment

translated image According to DEC documentation, it is an AXP executable or sharable image created by translating the object code of a VAX image. The translated image, which is functionally equivalent to the VAX image from which it was translated, includes both translated code and the original image.

translated image environment (TIE)   According to DEC documentation, it is a native AXP sharable image that supports the execution of translated images. The TIE processes all interactions with the native AXP system and provides an environment similar to VAX for the translated image by managing the VAX state; by emulating VAX features such as exception processing, AST delivery, and complex VAX instructions; and by interpreting untranslated VAX instructions.

translation   According to DEC documentation, it is the process of converting a VAX binary image to an AXP image that runs with the assistance of the TIE on an AXP system. Translation is a static process which converts as much VAX code as possible to native Alpha AXP instructions. The TIE interprets any untranslated VAX code at run time.

translation table   A table used to replace one or more characters with alternative characters.

transmission group (TG)   According to IBM, it is a group of links between adjacent subarea nodes, appearing as a single logical link for routing of messages.

transmission header (TH)   According to SNA, it is control information, optionally followed by a basic information unit, created and used by path control to route message units and to control their flow within the network.

transmission priority   According to IBM documentation, a rank assigned to a message unit that determines its precedence for being selected by the path control component in each node along a route for forwarding to the next node in the route.

transport layer   According to the OSI model, it is the layer that provides an end-to-end service to its users.

transport network   According to IBM documentation, it is that part of an SNA network that includes the data-link control and path control layer.

TSO/E   Time Sharing Option/Extension. According to IBM documentation, it is a program that provides enhancements to MVS/XA users.

TTL   Time to live. A technique used in best-effort delivery systems to avoid endlessly looping packets. For example, each packet has a "time" associated with its lifetime.

type   According to Apple documentation, it is the second field in the name of an AppleTalk entity. The type is assigned by the entity itself and can be anything the user or the application assigns.

type 2.1 end node   According to IBM documentation, a type 2.1 node that provides full SNA end-user services, but no intermediate routing or network services to any other node; it is configured only as an endpoint in a network.

type 2.1 network   According to IBM documentation, a collection of interconnected type 2.1 network nodes and type 2.1 end nodes. A type 2.1 network may consist of nodes of just one type, namely, all network nodes or all end nodes; a pair of directly attached end nodes is the simplest case of a type 2.1 network.

type 2.1 node   A node that conforms to IBM's type 2.1 architecture.

type 5 node   According to IBM documentation, a node that can be any one of the following:

Advanced Peer-to-Peer Networking (APPN) end node
Advanced Peer-to-Peer Networking (APPN) network node
Interchange node
Low-entry networking (LEN) node
Migration data host
Subarea node

It is also the node that traditionally has the SSCP.

UDP   User datagram protocol. A TCP/IP standard protocol that is in contrast to TCP. UDP is connectionless and unreliable.

UNBIND   According to IBM documentation, it is a request to deactivate a session between two logical units (LUs).

unformatted   According to IBM documentation, pertaining to commands (such as LOGON or LOGOFF) entered by an end user and sent by a logical unit in character form.

unformatted system services (USS)   According to IBM documentation, in SNA products, a system services control point (SSCP) facility that translates a character-coded request, such as a log-on or log-off request, into a field-formatted request for processing by formatted system services and that translates field-formatted replies and responses into character-coded requests for processing by a logical unit.

unit control block   According to DEC documentation, structure in the I/O database that describes the characteristics of and current activity on a device unit. The unit control block also holds the fork block for its unit's device driver; the fork block is a critical part of a driver fork process. The UCB also provides a dynamic storage area for the driver.

Universal Symbol   According to DEC documentation, it is a global symbol in a sharable image that can be used by modules linked with that sharable image. Universal symbols are typically a subset of all the global symbols in a sharable image. When creating a sharable image, the linker ensures that universal symbols remain available for reference after symbols have been resolved.

unsolicited message   According to IBM documentation, it is a message from VTAM to a program operator that is unrelated to any command entered by the program operator.

upline dump   According to DEC documentation, in DECnet for OpenVMS it is a function that allows an adjacent node to dump its memory to a file on a system.

user exit   According to IBM documentation, it is a point in an IBM-supplied program at which a user exit routine may be given control.

user exit routine   According to IBM documentation, a user-written routine that receives control at predefined user exit points. User exit routines can be written in assembly language or a high-level language.

user file directory   According to DEC documentation, a file that briefly cata-logs a set of files stored on disk or tape. The directory includes the name, type, and version number of each file in the set. It also contains a unique number that identifies that file's actual location and points to a list of its file attributes.

user privileges   According to DEC documentation, those privileges granted to a user by the system manager.

UUCP   UNIX-to-UNIX copy program. An application program that allows one UNIX system to copy files to or from another UNIX system.

VAXBI   According to DEC documentation, it is the part of the VAX 8200, VAX 8250, VAX 8300, VAX 8350 hardware that connects I/O adapters with memory controllers and the processor. In VAX 8530, VAX 8550, VAX 8700, or VAX 8800 systems, or VAX 6200 and VAX 6300 systems, the part of the hardware that con-nects I/O adapters with the bus that interfaces with the processor and memory.

VAXcluster configuration   According to DEC documentation, it is a highly integrated organization of OpenVMS systems that communicate over a high-speed communication path. VAXcluster configurations have all the functions of single-node systems, plus the ability to share CPU resources, queues, and disk storage. Like a single-node system, the VAXcluster configuration provides a single security and management environment. Member nodes can share the same operating environment or serve specialized needs.

VAX environment software translator   According to DEC documentation, it is a software migration tool that translates VAX executable and sharable images into translated images that run on AXP systems. VEST is part of the DECmigrate tool set.

VAX vector instruction emulation facility (VVIEF)   According to DEC docu-mentation, it is a standard feature of the operating system that allows vector-ized applications to be written and debugged in a VAX system in which vector processors are not available. VVIEF emulates the VAX vector processing envi-ronment, including the nonprivileged VAX vector instructions and the vector system services. Use of VVIEF is restricted to user mode code.

vector   According to DEC documentation, it is a storage location that contains the starting address of a procedure to be executed when a given interrupt or exception occurs.

vector present system   According to DEC documentation, it is a VAX system that, in its hardware implementation, complies with the VAX vector architec-ture, and incorporates one or more optional vector processors.

virtual disk   According to the IBM corporation, in VM, a physical disk storage device, or a logical subdivision of a physical disk storage device, that has its own address, consecutive storage space for data, and index or description of stored data so that the data can be accessed.

virtual filestore   A concept in OSI that refers to the OSI abstraction of a col-lection of files, directories, and/or references.

virtual machine (VM)   According to IBM documentation, in VM, a functional equivalent of a computing system. On the 370 feature of VM, a virtual machine operates in System/370 mode. On the ESA feature of VM, a virtual machine operates in System/370, 370-XA, ESA/370, or ESA/390 mode. Each virtual machine is controlled by an operating system. VM controls the concurrent execution of multiple virtual machines on an actual processor complex.

virtual machine group According to IBM documentation, in the group control system (GCS), two or more virtual machines associated with each other through the same named system.

Virtual Machine/Enterprise Systems Architecture (VM/ESA) According to IBM documentation, an IBM program that manages the resources of a single computer so that multiple computing systems appear to exist. Each virtual machine is the functional equivalent of a real machine.

Virtual Machine/Extended Architecture (VM/XA) According to IBM documentation, it is an operating system that facilitates conversion to MVS/XA by allowing several operating systems (one or more systems) to run simultaneously on a single 370-XA processor.

Virtual Machine/System Product (VM/SP) According to IBM documentation, a program that manages the resources of a single computer so that multiple computing systems appear to exist.

Virtual Machine/System Product High Performance Option (VM/SP HPO) According to IBM documentation, it is a program that can be installed and executed in conjunction with VM/SP to extend the capabilities of VM/SP with programming enhancements, support for microcode assists, and additional functions.

virtual route (VR) According to IBM documentation, in SNA it is either of the following:

A logical connection between two subarea nodes that is physically realized through an explicit route

A logical connection that is contained wholly within a subarea node for intra-node sessions

virtual route (VR) pacing According to IBM documentation, in SNA, a flow control technique used by the virtual route control component of path control at each end of a virtual route to control the rate at which path information units (PIUs) flow over the virtual route.

virtual routing node According to IBM documentation, it is a representation of a node's connectivity to a connection network defined on a shared-access transport facility, such as a token ring.

virtual storage The appearance of a type of storage which is realized by processor speed, main memory, DASD, and specialized programs. Virtual storage does not exist in reality; it is a concept.

virtual storage access method (VSAM) According to IBM documentation, it is an access method of direct or sequential processing of fixed- and variable-length records on direct-access devices.

Virtual Storage Extended (VSE) According to IBM documentation, it is a program whose full name is the Virtual Storage Extended/Advanced Function. It is a software operating system controlling the execution of programs.

virtual telecommunications access method (VTAM) According to IBM documentation, it is a program that controls communication and the flow of data in an SNA network. It provides single-domain, multiple-domain, and interconnected network capability.

VM/SNA console support (VSCS) According to IBM documentation, it is a VTAM component for the VM environment that provides Systems Network Architecture (SNA) support. It allows SNA terminals to be virtual machine consoles.

VM/370 control program (CP) According to IBM documentation, that component of VM/370 that manages the resources of a single computer with the result of multiple computing systems appearing to exist. Each virtual machine is the functional equivalent of an IBM System/370 computing system.

VMScluster configuration According to DEC documentation, a highly integrated organization of OpenVMS AXP systems, or a combination of AXP or VAX systems, communicating over a high-speed communications path. VMScluster configurations have all the functions of single-node systems, plus the ability to share CPU resources, queues, and disk storage. Like a single-node system, the VMScluster configuration provides a single security and management environment. Member nodes can share the same operating environment or serve specialized needs.

VSE/Advanced Functions According to IBM documentation, it is the basic operating system support needed for a VSE-controlled installation.

VTAM application program According to IBM documentation, a program that has opened an access (method) control block (ACB) to identify itself to VTAM and that can therefore issue VTAM macroinstructions.

VTAM common network services (VCNS) According to IBM documentation, it is VTAM's support for shared physical connectivity between Systems Network Architecture (SNA) networks and certain non-SNA networks.

VTAM definition According to IBM documentation, it is the process of defining the user application network to VTAM and modifying IBM-defined characteristics to suit the needs of the user.

VTAM definition library According to IBM documentation, it is the operating system files or data sets that contain the definition statements and start options filed during VTAM definition.

VTAM internal trace (VIT) According to IBM documentation, it is a trace used in VTAM to collect data on channel I/O, use of locks, and storage management services.

VTAM operator According to IBM documentation, it is a person or program authorized to issue VTAM operator commands.

VTAM operator command According to IBM documentation, it is a command used to monitor or control a VTAM domain.

waveform The representation of a disturbance as a function as it occurs in time and its relationship to space.

wavelength Defined as the distance an electromagentic wave can travel in the amount of time it takes to oscillate through a complete cycle.

well-known-port   A term used with TCP/IP networks. In TCP/IP, applications and programs that reside on top of TCP and UDP, respectively, have a designated port assigned to them. This agreed-on port is known as a well-known-port.

window   A term used with environments such as XWindows. Generally, the term is used in contrast with line or full-screen mode.

window-based program   A program written for use with a windowing system; for example, this could refer to a XWindow environment or the MS Windows environment. Opposite from a window-based program is a terminal-based program.

window decoration   In the XWindow environment it is the frame and window control buttons that surround windows managed by the window manager.

window manager   A program in the X windowing system that controls size, placement, and operation of windows on the root window. The window manager includes the functional window frames that surround each window object as well as a menu for the root window.

Xenix   A version of UNIX that can run on a PC.

X.21   A CCITT standard defining logical link control and media access control in X.25 networks.

X.25   A CCITT standard for packet-switched network-layer services.

X.400   A CCITT-ISO combination of standards for providing electronic mail (E-mail) services.

X.500   A CCITT-ISO combination of standards for providing directory services.

X application   An application program that conforms to X protocol standards.

X library   This is a collection of C language routines based on the X protocol.

X protocol   A protocol that uses TCP as a transport mechanism. It supports asynchronous, event-driven distributed window environments; this can be across heterogeneous platforms.

X terminal   A terminal and machine specifically designed to run an X server. In this type of environment, X clients are run on remote systems.

X toolkit   A collection of high-level programs based on programming from the X library.

X-Window System   A software system developed at MIT whose original design intent was to provide distributed computing support for the development of programs. It supports two-dimensional bitmapped graphics.

ZAP disk   According to IBM documentation, it is the virtual disk in the VM operating system that contains the user-written modifications to VTAM code.

zone   (1) According to DEC documentation, it is a section of a fully configured VAXft fault-tolerant computing system that contains a minimum of a CPU module, memory module, I/O module, and associated devices. A VAXft system consists of two such zones with synchronized processor operations. If one zone fails, processing continues uninterrupted through automatic failover to the other zone. (2) In AppleTalk, a logical grouping of devices in an AppleTalk internet that makes it easier for users to locate network services. The network administrator defines zones during the router setup process. (3) According to Apple documentation, this is a logical grouping of a subset of the nodes on an internet. The zone is the third field in the name of an AppleTalk entity.

zone information protocol   An AppleTalk protocol that maintains a table in each router, called the *zone information table,* that lists the relationship between zone names and networks.

zone name   According to Apple documentation, it is a name defined for each zone in an AppleTalk internet. A LocalTalk network can have just one zone name. ETHERNET and token-ring networks can have multiple zone names, called a *zone list.*

zone of authority   A term used with the domain name system to refer to the group of names authorized by a given name server.

Acronyms and Abbreviations

3270	Reference to a 3270 data stream (arrangement of data)
3770	Reference to remote job entry
370/XA	370/eXtended Architecture
5250	Reference to a 5250 data stream (arrangement of data)
AAA	Autonomous administrative area
AAI	Administration authority identifier
AAL	ATM adaptation layer
AARP	AppleTalk address resolution protocol
AC	Access control
ACB	Application control block; access (method) control block
ACCS	Automated calling card service
ACD	Automatic call distribution
ACDF	Access control decision function
ACE	Access control (list) entry; asynchronous communication element
ACF	Access control field; advanced communications function
ACIA	Access control inner areas; asynchronous communication interface adapter
ACID	Automicity, consistency, isolation, and durability
ACK	Positive acknowledgment
ACL	Access control list
ACP	Ancillary control process
ACS	Access control store
ACSA	Access control specific area
ACSE	Association control service element
ACSP	Access control specific point
ACTLU	Activate logical unit
ACTPU	Activate physical unit

ACU	Autocalling unit
AD	Addendum document to an OSI standard
ADMD	Administrative management domain
ADP	Adapter control block; AppleTalk data stream protocol
ADPCM	Adaptive differential pulse code modulation
ADSP	AppleTalk data stream protocol
AE	Application entity
AEI	Application entity invocation
AEP	AppleTalk echo protocol
AET	Application entity title
AF	Auxiliary facility
AFI	Authority and format identifier; AppleTalk filing interface
AFP	AppleTalk filing protocol
AID	Attention identifier
AIFF	Audio interchange file format
AIX	Advanced (also Account) Interactive eXecutive
ALS	Application-layer structure
ALU	Application-layer user; arithmetic logic unit
AM	Amplitude modulation
AMI	Alternating mark inversion
ANI	Automatic number identification
ANS	American National Standard
ANSI	American National Standards Institute
AP	Application process; argument pointer
APAR	Authorized program analysis report
APB	Alpha primary bootstrap
APD	Avalanche photodiode
APDU	Application protocol data unit
API	Application program (also programming) interface
APLI	ACSE/Presentation Library Interface
APP	Applications portability profile
APPC	Advanced Program-to-Program Communication
APPL	Application program
APPN	Advanced Peer-to-Peer networking
APT	Application Process Title
ARF	Automatic reconfiguration facility
ARI	Address-recognized indicator

ARP	Address resolution protocol
ARPA	Advanced Research Projects Agency
ARQ	Automatic repeat request
ARS	Automatic route selection
AS/400	Application system/400
ASC	Accredited Standard Committee
ASCII	American Standard Code for Information Interchange
ASDC	Abstract service definition convention
ASE	Application service element
ASM	Address space manager
ASN	Abstract syntax notation
ASN.1	Abstract Syntax Notation One
ASO	Application service object
ASP	Abstract service primitive; AppleTalk session protocol; attached support processor
AST	Asynchronous system trap
ASTLVL	Asynchronous system trap level
ASTSR	Asynchronous system trap summary register
AT	Advanced Technology (IBM Computers)
ATM	Asynchronous transfer mode; abstract text method; automatic teller machine
ATP	AppleTalk transaction protocol
ATS	Abstract test suite
AU	Access unit
AUI	Attachment unit interface
AVA	Attribute value assertion
AVS	APPC/VM VTAM support
AXP	A DEC hardware and operating system architecture
B-ISDN	Broadband ISDN
B8ZS	Bipolar 8-zeros substitution
BACM	Basic access control model
BAS	Basic activity subset
BASIC	Beginner's All-purpose Instruction Code
BB	Begin bracket
BC	Begin chain
BCC	Block-check character
BCN	Backward congestion notification
BCS	Basic combined subset

BCVT	Basic class virtual terminal
BECN	Backward explicit congestion notification
Bellcore	Bell Communications Research, Inc.
BER	Box event records; bit error rate
BF	Boundary function
BIS	Bracket initiation stopped
bits/s	bits per second
BISYNC	Binary synchronous (IBM protocol)
BIU	Basic information unit
BLU	Basic link unit
BMS	Basic mapping support
BMU	Basic measurement unit
BN	Backward notification; boundary node
BNN	Boundary network node
BOC	Bell Operating Company
BOM	Beginning of message
BRI	Basic rate interface
BSC	Binary synchronous communication
BSD	Berkeley standard distribution
BSS	Basic synchronization subset
BTAM	Basic telecommunications access method
BTU	Basic transmission unit
CA	Channel adapter (also attachment); certification authority
CAD	Computer-aided design
CAE	Common applications environment
CAF	Channel auxiliary facility
CAI	Computer-assisted instruction
CAR	Car area network
CASE	Common application service element
CATV	Community antenna television
CBEMA	Computer & Business Equipment Manufacturers Association
CC	Chain command
CCA	Conceptual (also "common") communication area
CCB	Connection control block; channel control block
CCIS	Common channel interoffice signaling
CCITT	Consultative Committee in International Telegraphy and Telephony
CCO	Context control object

CCR	Commitment, concurrency, and recovery
CCS	Common communications support; common channel signaling; console communication service
CCU	Central control unit; communications control unit
CCW	Channel command word
CD	Countdown counter; chain data; committee draft
CDDI	Copper-stranded distributed data interface
CDF	Configuration data flow
CDI	Change direction indicator
CDRM	Cross-domain resource manager
CDRSC	Cross-domain resource
CDS	Conceptual data storage (also store); central directory server
CEBI	Conditional end bracket indicator
CEI	Connection endpoint identifier
CEN/ELEC	Committee European de Normalization Electrotechnique
CEP	Connection endpoint
CEPT	Conference of European Postal and Telecommunications Administrations
CF	Control function
CFGR	Configuration
CGM	Computer Graphics Metafile
CHPID	Channel path identifier (ID)
CHILL	CCITT High-Level Language
CI	Computer interconnect
CICS	Customer Information Control System; customer information communication subsystem
CID	Command (also connection) identifier
CIGOS	Canadian Interest Group on Open Systems
CIM	Computer-integrated manufacturing
CIR	Commitment information rate
CLAW	Common link access to workstation
CLI	Connectionless internetworking
CLIST	Command list
CLNP	Connectionless network protocol
CLNS	Connectionless network service
CLP	Cell loss priority
CLSDST	Close destination
CLTP	Connectionless transport protocol
CLTS	Connectionless transport service

CLU	Control logical unit
CMC	Communication management configurations
CMIP	Common management information protocol
CMIS	Common management information service
CMISE	Common management information service element
CMOL	CMIP over logical link control
CMOT	CMIP over TCP/IP
CMS	Conversational monitoring system
CMT	Connection management
CN	Composite node
CNM	Communication network management
CNMA	Communication network for manufacturing applications
CNMI	Communication network management interface
CNN	Composite network node
CNOS	Change number of sessions
CNT	Communications name table
CO	Central office
COCF	Connection-oriented convergence function
CODEC	Coder/decoder
COI	Connection-oriented internetworking
COM	Continuation-of-message DMPDU
CONF	Confirm
CONS	Connection-oriented network service
COS	Class-of-service; Corporation for Open Systems
COSM	Class-of-service manager
COSS	Connection-oriented session service
COTP	Connection-oriented transport protocol
COTS	Connection-oriented transport service
CP	Control point; control program
CPCB	Control program (also point) control block
CPCS	Common part convergence sublayer
CPE	Customer premises equipment
CPF	Control program facility
CPI	Common programming interface
CPI-C	Common programming interface with C language
CPMS	Control point management services
CPU	Central processing unit
CR	Command response

CRC	Cyclical redundancy check
CRT	Cathode-ray tube
CRV	Call reference value
CS	Circuit switching; convergence sublayer; configuration services; console
CS-MUX	Circuit-switching multiplexer
CSA	Common service (also storage) area
CSALimit	Common service area (buffer use) limit
CSMA/CA	Carrier-sense multiple access with collision avoidance
CSMA/CD	Carrier-sense multiple access with collision detection
CSP	Communications scanner processor
CSS	Control, signaling, and status store
CSU	Channel service unit
CTC	Channel-to-channel
CTCA	Channel-to-channel adapter
CTCP	Communication and transport control program
CTS	Clear-to-send; common transport semantics
CUA	Channel unit address; common user access
CUG	Cluster user group
CUT	Control unit terminal
CVT	Communications vector table
DA	Destination address
DACD	Directory access control domain
DACTPU	Deactivate physical unit
DAD	Draft addendum
DAF	Framework for distributed applications; destination address field
DAP	Directory (also data) access protocol
DARPA	Defense Advanced Research Projects Agency
DAS	Dual-attachment station; dynamically assigned sockets; dual-address space
DASD	Direct-access storage device
DAT	Dynamic address translation
dB	Decibel
DBCS	Double-byte character set
DBK	Definition block
DC	Data chaining
DCA	Document-content architecture; Defense Communication Agency
DCC	Data Country Code

DCE	Data communications equipment; distributed computing environment; data circuit-terminating equipment
DCL	Digital Command Language (DEC)
DCLI	Data-link connection identifier
DCS	Defined context set
DCSS	Discontiguous shared segment
DDB	Directory database
DDCMP	Digital's (DEC's) data communications message protocol
DDDB	Distributed DDB
DDM	Distributed data management
DDN	Defense Data Network
DDName	Data definition name
DDP	Datagram delivery protocol
DDS	Digital data service
DE	Discard eligibility; directory entry
DEA	Directory entry attribute
DEC	Digital Equipment Corporation
DECdts	DEC distributed time service
DECNET	Digital equipment (DEC) network architecture
DELNI	DEC local network interconnect
DES	Data Encryption Standard
DEUNA	Digital Ethernet Unibus Network Adapter
DEV	Device address field
DFC	Data flow control
DFI	DSP format identifier
DFT	Distributed function terminal
DH	DMPDU header
DIA	Document Interchange Architecture
DIB	Directory information base
DIS	Draft International Standard
DISC	Disconnect
DISP	Draft International Standardized Profile; directory information shadowing protocol
DIT	Directory information tree
DIU	Distribution interchange unit
DIX	DEC, Intel, and Xerox
DL	Distribution list
DLC	Data-link control (also connection)

DLCEP	Data-link connection endpoint
DLCI	Data-link connection identifier
DLPDU	Data-link protocol data unit
DLS	Data-link service
DLSAP	Data-link service access point
DLSDU	Data-link service data unit
DLU	Dependent (also destination) logical unit
DLUR	Dependent logical unit requestor
DLUS	Dependent logical unit server
DM	Disconnected mode
DMA	Direct memory access
DMD	Directory management domain
DMI	Digital multiplexed interface; definition of management information
DMO	Domain management organization
DMPDU	Derived MAC protocol data unit
DMUX	Double multiplexer
DN	Distinguished name
DNA	Digital Network Architecture
DNHR	Dynamic nonhierarchical routing
DNS	Domain name service (also system)
DoD	U.S. Department of Defense
DOP	Directory operational binding management protocol
DOS	Disk operating system
DP	Draft proposal
DPG	Dedicated packet group
DPI	Dots per inch
DQDB	Distributed queue dual bus
DR	Definite response; dynamic reconfiguration
DRDA	Distributed Relational Database Architecture
DRDS	Dynamic reconfiguration data set
DRSLST	Direct search list
DS	Directory service(s); desired state
DS-n	Digital signaling level n
DSA	Directory service agent; Digital (DEC) storage architecture
DSAP	Destination service access point
DSD	Data structure definition
DSE	DSA-specific entry; data-switching exchange

DSL	Digital subscriber line
DSname	Data-set name
DSP	Directory service protocol; domain-specific part
DSS 1	Digital subscriber signaling system No. 1
DSSI	Digital (DEC) small systems (also storage systems) interconnect
DSTINIT	Data services task initialization
DSU	Digital services unit
DSUN	Distribution services unit name
DT	DMPDU trailer
DTE	Data terminal equipment
DTMF	Dual-tone multifrequency
DTR	Data terminal ready
DU	Data unit
DUA	Directory user agent
DVT	Destination vector table
E-mail	Electronic mail
EAB	Extended addressing bit
EAS	Extended area service
EB	End bracket
EBCDIC	Extended Binary-Coded Decimal Interchange Code
ECC	Enhanced error checking and correction
ECH	Echo canceller with hybrid
ECMA	European Computer Manufacturers' Association
ECO	Echo control object
ECSA	Exchange Carriers Standards Association
ED	End delimiter
EDI	Electronic Data Interchange
EDIFACT	EDI for administration, commerce and transport
EDIM	EDI message
EDIME	EDI messaging environment
EDIMS	EDI messaging system
EDI-MS	EDI message store
EDIN	EDI notification
EDI-UA	EDI-user agent
EEI	External environment interface
EGP	Exterior gateway protocol
EIA	Electronic Industries Association

EIT	Encoded information type
ELAP	EtherTalk LAP
EMA	Enterprise management architecture
EMI	Electromagnetic interference
EN	End node
ENA	Extended network addressing
EOM	End-of-Message DMPDU
EOT	End of transmission
EP	Emulation program; echo protocol
ER	Explicit route; exception response
EREP	Environmental recording editing and printing
ERP	Error-recovery procedure
ES	End system
ESA	Enterprise system architecture; Enhanced Subarea Addressing
ESCON	Enterprise System Connection
ESF	Extended superframe format
ESH	End system hello
ES-IS	End system to intermediate system
ESS	Electronic switching system
ESTELLE	Extended State Transition Language
ETB	End-of-text block
ETR	Early token release
ETX	End of text
EVE	Extensible VAX Editor
EXLST	Exit list
EXT	External trace (file)
EWOS	European Workshop on Open Systems
FADU	File access data unit
FAS	Frame alignment sequence
FC	Frame-check; frame control (field)
FCC	Federal Communications Commission
FCI	Frame-copied indicator
FCS	Frame-check sequence
FDCO	Field definition control object
FDDI	Fiber Distributed Data Interface
FDDI-FO	FDDI follow-on
FDL	File Definition Language (DEC)
FDM	Frequency-division multiplexing

FDR	Field definition record
FDT	Formal description technique
FDX	Full-duplex
FEC	Field entry condition
FECN	Forward explicit congestion notification
FEE	Field entry event
FEI	Field entry instruction
FEICO	Field entry instruction control object
FEIR	Field entry instruction record
FEP	Front-end processor
FEPCO	Field entry pilot control object
FEPR	Field entry pilot record
FER	Field entry reaction
FFOL	FDDI follow-on LAN
FH	Frame handler
FID	Format identification
FIFO	First in, first out
FIPS	*Federal Information Processing Standard*
FM	Function management; frequency modulation
FMD	Function management data
FMH	Function management header
FN	Forward notification
FOD	Office Document Format
FQPCID	Fully qualified procedure correlation identifier
FRAD	Frame relay access device
FRFH	Frame relay frame handler
FRMR	Frame reject
FRSE	Frame relay switching equipment
FRTE	Frame relay terminal equipment
FS	Frame status field
FSG	SGML interchange format
FSK	Frequency-shift keying
FSM	Finite-state machine
F^t	Foot; fault tolerant (e.g., VAXF^t)
FTAM	File transfer and access management
FTP	File transfer protocol in TCP/IP
FX	Foreign exchange service
GAP	Gateway access protocol

Gbits	Gigabits
Gbits/s	Gigabits per second
Gbyte	Gigabyte
GCS	Group control system
GDDM	Graphical data display manager
GDMO	Guidelines for the definition of managed objects
GDS	General data stream
GEN	Generation
GFC	Generic flow control
GFI	General format indicator
GOSIP	Government OSI profile
GSA	General Services Administration
GTF	Generalized trace facility
GUI	Graphical user interface
GWNCP	Gateway NCP
GWSSCP	Gateway SSCP
H-MUX	Hybrid multiplexer
HAN	House area network
HASP	Houston Automatic Spooling Priority
HCD	Hardware configuration definition
HCS	Header check sequence
HDB3	High-density bipolar—3 zeros
HDLC	High-level data-link control
HDX	Half-duplex (also HD)
HEC	Header error correction
hex	Hexadecimal
HFS	Hierarchical file system
HI-SAP	Hybrid isochronous–MAC service access point
HMI	Human-machine interface
HMP	Host monitoring protocol
HOB	Head of bus
HP	Hewlett-Packard
HP-SAP	Hybrid packet–MAC service access point
HRC	Hybrid ring control
HS	Half session
HSC	Hierarchical storage controller
HSLN	High-speed local network
Hz	Hertz (cycles per second)

IAB	Internet Architecture Board
IADCS	Interactivity defined context set
IAN	Integrated analog network
IAP	Inner administrative point
IAS	Interactive Application System
IBM	International Business Machines Corporation
IC	Interexchange carrier
ICA	Integrated communication adapter
ICCF	Interactive computing and control facility
ICD	International code designator
ICF	Isochronous convergence function
ICI	Interface control information
ICMP	Internet control message protocol
ICP	Interconnect control program
ICV	Integrity check value
ID	Identifier or identification
IDA	Indirect data addressing
IDI	Initial domain identifier
IDN	Integrated digital network; interface definition notation
IDP	Initial domain part; internetwork datagram packet (protocol)
IDU	Interface data unit
IEC	Interexchange carrier; International Electrotechnical Commission
IEEE	Institute of Electrical and Electronic Engineers
IETF	Internet Engineering Task Force
IHL	Internet header length
IIA	Information interchange architecture
ILD	Injection laser diode
ILU	Independent (also initiating) logical unit
IMAC	Isochronous media access control
IMIL	International Managed Information Library
IML	Initial microcode load
IMPDU	Initial MAC protocol data unit
IMR	Intensive mode recording
IMS	Information Management System
IMS/VS	Information Management System/Virtual Storage
IN	Intelligent network; interchange node
IND	Indication

INN	Intermediate network node
INTAP	Interoperability Technology Association for Information Processing
I/O	Input/output
IOC	Input/output control
IOCDS	Input/output configuration data set
IOCP	Input/output control (also channel or configuration) program
IONL	Internal organization of network layer
IOPD	Input/output problem determination
IP	Internet protocol
IPC	Interprocess communication
IPCS	Interactive problem control system
IPDS	Intelligent printer data stream
IPI	Initial protocol identifier
IPICS	ISP implementation conformance statement
IPL	Initial program load(er)
IPM	Interpersonal message
IPM-UA	Interpersonal messaging user agent
IPMS	Interpersonal messaging system
IPN	Interpersonal notification
IPR	Isolated pacing response
IPX	Internetwork packet exchange
IR	Internet router
IRN	Intermediate routing node
IRSG	Internet Research Steering Group
IRTF	Internet Research Task Force
IS	International Standard
ISAM	Index-sequential access method
ISC	Intersystem communications in CICS
ISCF	Intersystem control facility
ISDN	Integrated Services Digital Network
ISE	Integrate storage element
ISH	Intermediate system hello
IS-IS	Intermediate system-to-intermediate system
ISO	International Standards Organization
ISODE	ISO development environment
ISP	International Standard Profile
ISPBX	Integrated Services Private Branch Exchange

ISPF	Interactive System Productivity Facility
ISPSN	Initial synchronization point serial number
ISR	Intermediate session routing
ISSI	Interswitching system interface
ISUP	ISDN user part
IT	Information technology
ITC	Independent telephone company
ITU	International Telecommunication Union
IUCV	Interuser communication vehicle
IUT	Implementation under test
IVDT	Integrated voice/data terminal
IWU	Interworking unit
IXC	Interexchange carrier
JCL	Job Control Language
JES2, JES3	Job Entry Subsystem 2, 3
JTC	Joint Technical Committee
JTM	Job transfer and manipulation
kbits	Kilobits
kbits/s	Kilobits per second
kbyte	Kilobyte
kHz	Kilohertz
km	Kilometer
LAB	Latency adjustment buffer; line attachment base
LAN	Local area network
LANRES	Local Area Network Resource Extension Services
LANSUP	LAN adapter NDIS support
LAP	Link-access procedure (also protocol)
LAPB	Link-access procedure balanced
LAPD	Link-access procedures on the D channel
LAPM	Link-access procedure for modems
LAPS	LAN adapter and protocol support
LATA	Local access and transport area
LCF	Log control function
LCN	Logical channel number
LDDB	Local directory database
LE	Local exchange
LEC	Local exchange carrier
LED	Light-emitting diode

LEN	Low-entry networking
LFSID	Local form session identifier
LH	Link header
LI	Length indicator
LIB	Line interface base
LIC	Line interface coupler
LIDB	Line information database
LIVT	Link integrity verification test
LLAP	LocalTalk link access protocol
LLC	Logical link control
LL2	Link level 2
LME	Layer management entity
LMI	Layer (also local) management interface
LOCKD	Lock manager daemon
LOTOS	Language of Temporal Ordering Specifications
LPAR	Logical partitioned (mode)
LPD	Line printer daemon
LPDA	Link problem determination application
LPR	Line printer
LRC	Longitudinal redundancy check
LS	Link station
LSE	Local system environment
LSL	Link support layer
LSP	Link-state packet
LSS	Low-speed scanner
LT	Local termination
LU	Logical unit
m	Meters
MAC	Media (also medium) access control
MACE	Macintosh audio compression and expansion
MACF	Multiple association control function
MAN	Metropolitan area network
MAP	Manufacturing automation protocol
MAU	Media access unit; multistation access unit
MBA	MASSBUS adapter
Mbits	Megabits
Mbits/s	Megabits per second
Mbyte	Megabyte

Mbytes/s	Megabytes per second
MBZ	Must be zero
MCF	MAC convergence function
MCI	Microwave Communications, Inc.
MCP	MAC convergence protocol
MCR	Monitor console routine
MD	Management domain
MFA	Management functional areas
MFD	Master file directory
MFJ	Modified final judgment
MFS	Message formatting services in IMS
MH	Message-handling (package)
MHS	Message-handling service (also system)
MHz	Megahertz
mi	Mile
MIB	Management (also message) information base
MIC	Media interface connector
MID	Message identifier
MIM	Management information model
min	Minute
MIN	Multiple interaction negotiation
MIPS	Million instructions per second
MIS	Management information system
MIT	Management information tree; Massachusetts Institute of Technology
MLID	Multiple-link interface driver
MMF	Multimode fiber
MMS	Manufacturing message specification (also messaging service)
MNP	Microcom networking protocol
MOP	Maintenance operations protocol
MOSS	Maintenance and operator subsystem
MOT	Means of testing
MOTIS	Message-oriented text interchange system
MPAF	Midpage allocation field
MPC	Multipath channel
MPG	Multiple preferred quests
MPP	Multiple-protocol package
MQ	Message queue

MRO	Multiregion operation in CICS
ms	Millisecond
MS	Management services; message store
MSCP	Mass storage control protocol
MSG	Console message
MSHP	Maintain system history program
MSN	Multiple systems networking
MSNF	Multiple systems networking facility
MSS	MAN switching system
MST	Multiplexed slotted and token ring
MSU	Management services unit
MTA	Message transfer agent
MTACP	Magnetic tape ancillary control process
MTBF	Mean time between failures
MTP	Message transfer part
MTPN	Multiprotocol Transport Networking
MTU	Maximum transfer unit
MTS	Message transfer system
MTSE	Message transfer service element
μm	Micrometer
MVC	Multicast virtual circuit
MVI	Major vector identifier
MVL	Major vector length
MVS	Multiple virtual systems
MVS/370	Multiple Virtual Storage/370 (IBM)
MVS/XA	Multiple Virtual Storage/Extended Architecture (IBM)
MVT	Multiprogramming with a variable number of tasks
NAK	Negative acknowledgment in BSC
NAU	Network-addressable (also -accessible) unit
NAUN	Nearest active upstream neighbor
NBP	Name-binding protocol
NC	Network connection; numerical controller
NCB	Node control block
NCCF	Network communications control facility
NCEP	Network connection endpoint
NCP	Network control program; network (also NetWare) core protocol
NCS	Network computing system
NCTE	Network channel-terminating equipment

NDF	NCP/EP definition facility
NDIS	Network driver interface specification
NetBIOS	Network basic input/output system
NETID	Network ID
NFS	Network file system (also server)
NIB	Node identification (also initialization) block
NIC	Network interface card
NIF	Network information file
NIS	Names information socket
NIST	National Institute of Standards and Technology
NIUF	North American ISDN Users' Forum
NJE	Network job entry
NLM	NetWare loadable module
NLDM	Network logical data manager
nm	Nanometer
NM	Network management (also model)
NMP	Network management process
NMVT	Network management vector transport
NN	Network node
NNI	Network node interface
NNT	NetView-NetView task
NOF	Node operator facility
NPA	Numbering plan area
NPCI	Network protocol control information
NPDA	Network problem determination application
NPDU	Network protocol data unit
NPM	NetView performance monitor
NPSI	Network packet-switching interface
NR	Number of receives; negative response
NREN	National Research and Education Network
NRF	Network routing facility
NRN	Nonreceipt notification
NRZ	Non-return-to-zero
NRZI	Non-return-to-zero inverted
ns	Nanosecond
NS	Network service; number of sends
NSAP	Network service access point

NSDU	Network service data unit
NSF	Network search function; National Science Foundation
NSFNET	National Science Foundation Network
NSP	Network services protocol
NTO	Network terminal option
NVLAP	National Voluntary Accreditation Program
NVP	Network voice protocol
OAF	Origination address field
OAM	Operations, administration, and maintenance
OAM&P	Operations, administration, maintenance, and provisioning
OC-n	Optical carrier level n
OCA	Open communication architectures
OCC	Other common carrier
ODA	Office (also open-) document architecture
ODI	Open data-link interface
ODIF	Office document interchange format
ODINSUP	ODI NSIS support
ODP	Open distributed processing
OIT	Object identifier tree
OIW	OSI Implementation Workshop
OLRT	Online real time
OLU	Origin(ating) logical unit
OM	Object management
Ω	Ohm
ONA	Open-network architecture
OPNDST	Open destination
O/R	Originator/recipient
OS/400	Operating System/400 for the AS/400 Computer
OS	Operating System
OSAK	OSI application kernel
OSE	Open-systems environment
OSF	Open Software Foundation
OSI	Open-systems interconnection
OSI/CS	OSI communications subsystem
OSIE	Open-system interconnection environment
OSILL	Open-systems interconnection lower layers
OSIUL	Open-systems interconnection upper layers

OSNS	Open Systems Network Services
OSPF	Open shortest path first
P-MAC	Packet-switched media access control
PA	Prearbitrated
PABX	Private automatic branch exchange
PAD	Packet assembler/disassembler
PAF	Prearbitrated function
PAI	Protocol address information
PANS	Pretty amazing new stuff
PAP	Printer access protocol
PARC	Palo Alto (Calif.) Research Center
PBX	Private branch exchange
PC	Path control; personal computer
PCCU	Physical communications control unit
PCEP	Presentation connection endpoint
PCI	Protocol (also program) control information; program-controlled interruption; presentation context identifier
PCM	Pulse-code modulation
PCO	Points of control and observation
PCTR	Protocol conformance test report
PDAD	Proposed draft addendum
PDAU	Physical delivery access unit
PDC	Packet data channel
PDF	Program development facility
pDISP	Proposed Draft International Standard Profile
PDN	Public data network
PDP	Programmable data processor
PDS	Partitioned data set
PDU	Protocol data unit
PDV	Presentation data value
PELS	Picture elements
PEP	Partitioned emulation program
PER	Program event recording
PETS	Parameterized executable test suite
PGF	Presentation graphics feature
PH	Packet handler (or packet handling)
PhC	Physical-layer connection
PhCEP	Physical connection endpoint

PhL	Physical layer
PhPDU	Physical-layer protocol data unit
PhS	Physical-layer service
Ph-SAP	Physical-layer SAP
PhSAP	Physical-layer service access point
PhSDU	Physical-layer service data unit
PHY	Physical layer
PICS	Protocol information conformance statement
PIN	Positive-intrinsic negative photodiode
PING	Packet Internet Groper
PIP	Program initialization parameters
PIU	Path information unit
PIXIT	Protocol Implementation eXtra Information for Testing
PKCS	Public key cryptosystems
PLC	Programmable logic controller
PLCP	Physical-layer convergence protocol
PLP	Packet-layer (also level) protocol
PLS	Primary link station
PLU	Primary (also peripheral) logical unit
PM	Protocol machine; phase modulation
PMD	Physical-(layer)-medium-dependent
POI	Program operator interface
POP	Point of presence
POSI	Promoting Conference for OSI
POSIX	Portable Operating System Interface
POTS	Plain Old Telephone Service
POWER	Priority output writers and execution processors and input readers
PPDU	Presentation protocol data unit
PPO	Primary program operator
PPP	Point-to-point protocol
PPSDN	Public packet-switched data network
PRI	Primary rate interface
PRMD	Private management domain
PROFS	Professional office system
PROM	Programmable read-only memory
PR/SM	Processor resource/systems manager
PS	Presentation services

PSAP	Public safety answering point; physical service access point
PSC	Public Service Commission
PSDN	Packet-switched data network
PSI	Packet switched interface
PSID	Product-set identification
PSK	Phase-shift keying
PSN	Packet-switched network
PSPDN	Packet-switched public data network
PSTN	Public switched telephone network
PSW	Program status word
PS/2	Personal System/2 (IBM)
PS/VP	Personal System/Value Point (IBM)
PTF	Program temporary fix
PTLXAU	Public Telex Access Unit
PTN	Public telephone network
PIT	Post, Telegraph, and Telephone
PU	Physical unit
PUC	Public utility commission
PUCP	Physical unit control point
PUMS	Physical unit management services
PUT	Program update tape
PVC	Private (also permanent) virtual circuit
PVN	Private virtual network
PWSS	Programmable workstation service
P1	Protocol 1 (message transfer protocol/MHS/X.400)
P2	Protocol 2 (interpersonal messaging MHS/X.400)
P3	Protocol 3 (submission and delivery protocol/MHS/X.400)
P5	Protocol 5 (teletext access protocol)
P7	Protocol 7 (message store access protocol in X.400)
QA	Queued arbitrated
QAF	Queued arbitrated function
QC	Quiesce complete
QEC	Quiesce at end of chain
QMF	Query management facility
QOS	Quality of service
QPSX	Queued packet and synchronous switch
QUIPU	X.500 Conformant Directory Services in ISODE
RAB	Record access block

RAM	Random-access memory
RARP	Reverse address resolution protocol
RBOC	Regional Bell Operating Company
RD	Routing domain; route redirection; request a disconnect
RACF	Resource access control facility
RDA	Relative distinguished names; remote database access
RDI	Restricted digital information
RDN	Relative distinguished name
RDT	Resource definition table
RECFMS	Record formatted maintenance statistics
REJ	Reject
REQ	Request
RESP	Response
RESYNC	Resynchronization
REXX	Restructured Extended Executor (IBM)
RFC	Request for comment
RFP	Request for proposal
RFQ	Request for price quotation
RH	Request (or response) header
RI	Ring in
RIB	Routing information base
RIF	Routing information field
RIM	Request initialization mode
RIP	Router information protocol
RISC	Reduced instruction-set computer
RJE	Remote job entry
RM	Reference model; resource manager
RMS	Record management services (for OpenVMS)
RMT	Ring management
RN	Receipt notification
RNAA	Request network address assignment
RNR	Receive not ready
RO	Ring out
RODM	Resource object data manager
ROSE	Remote-operations service element
RPC	Remote procedure call; remote procedure call in OSF/DCE
RPL	Request parameter list
RPOA	Recognized private operating agency

RQ	Request counter
RR	Receive ready
RS	Relay system
RSCS	Remote spooling communication system
RSCV	Route selection control vector
RSF	Remote support facility
RSP	Response
RSS	Route selection services
RTM	Response-time monitor
RTMP	Routing table maintenance protocol
RTR	Ready to receive
RTS	Request to send
RTSE	Reliable transfer service element
RTT	Round-trip time
RU	Request (or response) unit
S/390	IBM's System/390 Hardware Architecture
s	Second
SA	Source address (field); subarea; sequenced application
SAA	System applications architecture; specific administrative area
SAB	Subnetwork-access boundary
SABM	Set asynchronous balanced mode
SACK	Selective ACKnowledgment
SACF	Single association control function
SAF	SACF auxiliary facility
SALI	Source address length indicator
SAMBE	Set asynchronous mode balanced extended
SAO	Single association object
SAP	Service access point; service advertising protocol
SAPI	Service access point identifier
SAR	Segmentation and reassembly
SAS	Single-attachment station; statically assigned sockets
SASE	Specific application service element
SATF	Shared-access transfer facility
SATS	Selected abstract test suite
SAW	Session awareness data
SBA	Set buffer address
SBCS	Single-byte character set
SBI	Stop bracket initiation

SC	Session connection (also connector); subcommittee; session control
SCA	Systems communication architecture
SCB	Session control block
SCC	Specialized common carrier
SCCP	Signaling connection control point
SCE	System control element
SCEP	Session connection endpoint
SCIF	Single-console image facility
SCM	Session control (also connection) manager
SCP	Service control point
SCS	System conformance statement; SNA character string
SCTR	System conformance test report
SD	Start delimiter
SDDI	Shielded distributed data interface
SDH	Synchronous digital hierarchy
SDIF	Standard document interchange format
SDL	System description language
SDLC	Synchronous data-link control
SDN	Software-defined network
SDSE	Shadowed DSA entries
SDT	Start data traffic
SDU	Service data unit
SE	Session entity
SFD	Start frame delimiter
SFS	Shared file system
SFT	System fault tolerance
SG	Study group
SGFS	Special Group on Functional Standardization
SGML	Standard Generalized Markup Language
SIA	Stable implementation agreements
SIM	Set initialization mode
SIO	Start input/output
SIP	SMDS interface protocol
SLI	Suppress-length indication
SLIP	Serial-line Internet protocol
SLU	Secondary logical unit
SM	Session manager

SMAE	System management application entity
SMASE	System management application service element
SMB	Server message block
SMDR	Station message detail recording
SMDS	Switched multimegabit data service
SMF	Single-mode fiber; system management facility
SMFA	Systems management functional area
SMI	Structure of the OSI management information service; Structure of Management Information (language)
SMIB	Stored message information base
SMP	System modification program
SMS	Service management system; storage management subsystem
SMT	Station management standard
SMTP	Simple mail transfer protocol
SN	Subarea node
SNA	System Network Architecture
SNAcF	Subnetwork access function
SNAcP	Subnetwork access protocol
SNADS	SNA distribution services
SNARE	Subnetwork address routing entity
SNCP	Single-node control point
SNDCP	Subnetwork-dependent convergence protocol
SNI	Subscriber-network interface; SNA network interconnection (also interface)
SNICP	Subnetwork-independent convergence protocol
SNMP	Simple network management protocol
SNPA	Subnetwork point of attachment
SNRM	Set normal response mode
SON	Sent (or send) outside the node
SONET	Synchronous optical network
SP	Signaling point; system performance
SPAG	Standards promotion and applications group
SPC	Signaling point code
SPCS	Service point command service
SPDU	Session protocol data unit
SPE	Synchronous payload envelope
SPI	Subsequent protocol identifier
SPM	FDDI-to-SONET physical-layer mapping standard

SPSN	Synchronization-point serial number
SPX	Sequenced packet exchange
SQE	Signal-quality error
SQL	Structured Query Language
SRB	Source route bridging
SRH	SNARE request hello
SRM	System resource manager
SS	Switching system; start/stop (transmission); session services
SS6	Signaling system 6
SS7	Signaling system 7
SSA	Subschema specific area
SSAP	Source (also session) service access point
SSCP	System service control point
SSCS	Service-specific convergence sublayer
SSDU	Session service data unit
SSM	Single-segment message DMPDU
SSP	System Support Program
ST	Sequenced terminal
STACK	Start acknowledgment (message)
STM	Synchronous transfer mode; station management
STM-n	Synchronous transport module level n
STP	Shielded twisted pair; service transaction program (in LU6.2); signal transfer point
STS-n	Synchronous transport signal level n
STX	Start of text
SUT	System under test
SVA	Shared virtual area
SVC	Switched virtual circuit
SVI	Subvector identifier
SVL	Subvector length
SVP	Subvector parameters
SYN	Synchronous character in IBM's bisynchronous protocol
SYNC	Synchronization
T	Transport
TA	Terminal adapter
TAF	Terminal access facility
TAG	Technology advisory group
TAP	Trace analysis program

Tbyte	Terabyte
TC	Transport connection or technical committee
TCAM	Telecommunications access method
TCB	Task control block
TCC	Transmission control code
TCEP	Transport connection endpoint
TCM	Time-compression multiplexing
TCP	Transmission control protocol
TCP/IP	Transmission control protocol/Internet protocol
TCT	Terminal control table in CICS
TDM	Time-division multiplexing; topology database manager
TDR	Time-domain reflectometer
TE	Terminal equipment
TELNET	Remote logon in TCP/IP
TEP	Transport endpoint
TFTP	Trivial file transfer protocol
TG	Transmission group
TH	Transmission header
THT	Token holding timer
TIC	Token-ring interface coupler
TI RPC	Transport-independent RPC
TIE	Translated image environment
TLAP	TokenTalk LAP
TLI	Transport-layer interface
TLMAU	Telematic access unit
TLV	Type, length, and value
TLXAU	Telex access unit
TMP	Text management protocol
TMS	Time-multiplexed switching
TMSCP	Tape mass storage control protocol
TOP	Technical and office protocol
TP	Transaction program (also processing); transport protocol
TP 0	TP class 0—simple
TP 1	TP class 1—basic error recovery
TP 2	TP class 2—multiplexing
TP 3	TP class 3—error recovery and multiplexing
TP 4	TP class 4—error detection and recovery
TPDU	Transport protocol data unit

TPF	Transaction processing facility
TP-PMD	Twisted-pair PMD
TPS	Two-processor switch
TPSP	Transaction processing service provider
TPSU	Transaction processing service user
TPSUI	TPSU invocation
TR	Technical report; token ring
TRA	Token-ring adapter
TRS	Topology and routing services
TRSS	Token-ring subsystem
TRT	Token rotation timer
TS	Transaction services; transport service
TSAF	Transparent services access facility
TSAP	Transport service access point
TSC	Transmission subsystem controller
TSCF	Target system control facility
TSDU	Transport service data unit
TSI	Time-slot interchange
TSO/E	Time Sharing Option/Extension
TSR	Terminate-and-stay-resident program
TSS	Transmission subsystem
TTCN	Tree and tabular combined notation
TTP	Timed token protocol; transport test platform
TTL	Time to live
TTRT	Target token rotation time
TTY	Teletype
TUP	Telephone user part
TVX	Valid transmission timer (FDDI)
TWX	Teletypewriter exchange service
TZD	Time-zone difference
UA	Unnumbered acknowledgment; user agent; unsequenced application
UART	Universal asynchronous receiver/transmitter
UCB	Unit control block; University of California, Berkeley
UCW	Unit control word
UDI	Unrestricted digital information
UDP	User datagram protocol
UI	Unnumbered information

UNI	User network interface
UOW	Unit of work
UP	Unnumbered poll
USART	Universal synchronous/asynchronous receiver/transmitter
User-ASE	User application service element
USS	Unformatted system services
UT	Unsequenced terminal
UTC	Coordinated Universal Time
UTP	Unshielded twisted pair
UUCP	UNIX-to-UNIX copy program
VAC	Value-added carrier
VAN	Value-added network
VAS	Value-added service
VAX	Virtual Address eXtended (operating system; DEC)
VAXBI	VAX bus interface
VCC	Virtual channel connection
VCI	Virtual channel identifier/indicator (DQDB)
VCNS	VTAM common network services
VDI	Video display terminal
VEST	VAX environment software translator
VIT	VTAM internal trace
VLF	Virtual look-aside facility
VLSI	Very large-scale integration
VM	Virtual machine
VMD	Virtual manufacturing device
VM/ESA	Virtual Machine/Enterprise Systems Architecture
VMDBK	Virtual machine definition block
VMS	Virtual memory system
VM/SP	Virtual machine/system product
VM/SP HPO	Virtual Machine/System Product High Performance Option
VM/XA	Virtual Machine/Extended Architecture
VPC	Virtual path connection
VPI	Virtual path indicator
VR	Virtual route
VRPWS	Virtual route pacing window size
VS	Virtual storage
VSAM	Virtual storage access method
VSCS	VM/SNA console support

VSE	Virtual Storage Extended
VSE/ESA	Virtual Storage Extended/Enterprise Systems Architecture
VT	Virtual terminal
VTAM	Virtual telecommunications access method
VTE	Virtual terminal environment
VTP	Virtual terminal protocol
VTPM	Virtual terminal protocol machine
VTSE	Virtual terminal service element
VVIEF	VAX vector instruction emulation facility
WACA	Write access connection acceptor
WACI	Write access connection initiator
WAN	Wide area network
WANDD	Wide area network device driver
WAVAR	Write access variable
WBC	Wideband channel
WD	Working document
WG	Working group
WP	Working party
X	The XWindow System
XA	Extended Architecture (IBM)
XAPIA	X.400 API Association
XCF	Cross-system coupling facility
XDF	Extended distance facility
Xdm	X display manager
XDS	X/Open Directory Services API
Xds	X display server
XI	SNA X.25 interface
XID	Exchange identification
XNS	Xerox Network Standard
XRF	Extended recovery facility
XTI	X/Open Transport Interface
ZIP	Zone information protocol
ZIS	Zone information socket
ZIT	Zone information table

Trademarks

Trademarks and their proprietors are listed alphabetically as follows (RTM = registered trademark, TM = trademark; DEC = Digital Equipment Corp., IBM = International Business Machines Corp.; MIT = Massachusetts Institute of Technology).

ACF/VTAM: RTM of IBM

ACMS: TM of DEC

AIX: TM of IBM

AIXwindows: TM of IBM

ALL-IN-1: TM of DEC

Alpha AXP: TM of DEC

APDA: RTM of Claris Corp.

Apollo: RTM of Apollo Computer, Inc.

Apple and Apple logo: RTMs of Apple Computer, Inc.

AppleColor: RTM of Claris Corp.

Apple Desktop Bus: RTM of Claris Corp.

AppleShare: RTM of Apple Computer, Inc.

AppleTalk: RTM of Apple Computer, Inc.

Apple IIGS: RTM of Apple Computer, Inc.

AS/400: RTM of IBM

A/UX: RTM of Apple Computer, Inc.

AXP and AXP logo: TMs of DEC

Bookreader: TM of DEC

CDA: TM of DEC

CDD: TM of DEC

CDD/REpository: TM of DEC

CI COHESION: TM of DEC

CICS: RTM of IBM

CICS/ESA: RTM of IBM

CICS/MVS: RTM of IBM

Cisco: RTM of Cisco Systems, Inc.

DATABASE 2, DB2: RTMs of IBM

DEC: TM of DEC

DEC ACCESSWORKS: TM of DEC

DEC GKS: TM of DEC

DEC MAILworks: TM of DEC

DEC PHIGS: TM of DEC

DEC Rdb for Open VMS: TM of DEC

DEC RTR: TM of DEC

DEC VTX: TM of DEC

DEC VUIT: TM of DEC

DECalert: TM of DEC

DECamds: TM of DEC

DECdecision: TM of DEC

DECdesign: TM of DEC

DECdtm: TM of DEC

DECforms: TM of DEC

DECmcc: TM of DEC

DECmessageQ: TM of DEC

DECnet: TM of DEC

DECNIS: TM of DEC

DECperformance Solution: TM of DEC

DECplan: TM of DEC

DECprint: TM of DEC

DECquery: TM of DEC

DECram: TM of DEC

DECscheduler: TM of DEC

DECserver: TM of DEC

DECset: TM of DEC

DECtalk: TM of DEC

DECterm: TM of DEC

DECthrads: TM of DEC

DECtp: TM of DEC

DECtrace: TM of DEC

DECwindows: TM of DEC

DECwrite: TM of DEC

Digital and DIGITAL logo: TMs of DEC

DNA: TM of DEC

EDT: TM of DEC

80386, 80386SX, 80486, 80486SX: TMs of Intel Corp.

EtherCard PLUS: TM of Western Digital Corp.

Etherlink: TM of 3Com Corp.

ETHERNET: RTM of Xerox Corp.

eXcursion: TM of DEC

Finder: RTM of Claris Corp.

HP: TM of Hewlett-Packard Co.

HSC: TM of DEC

HyperCard: RTM of Apple Computer, Inc.

HYPERchannel: RTM of Network Systems Corp.

IBM: RTM of IBM

ImageWrite: RTM of Apple Computer, Inc.

IMS: RTM of IBM

Information Warehouse: RTM of IBM

Intel: RTM of Intel Corp.

Internetwork PacketeXchange: RTM of Novell, Inc.

IPX: RTM of Novell, Inc.

KanjiTalk: RTM of Claris Corp.

Kerberos: RTM of MIT

LaserWriter: RTM of Apple Computer, Inc.

Lat: TM of DEC

LattisNet: RTM of SynOptics Communications, Inc.

LinkWorks: TM of DEC

Lisa: RTM of Apple Computer, Inc.

MacApp: RTM of Apple Computer, Inc.

MacDraw: RTM of Claris Corp.

Macintosh: RTM of Apple Computer, Inc.

MacPaint: RTM of Claris Corp.

MacWorks: RTM of Apple Computer, Inc.

MacWrite: RTM of Claris Corp.

Madge: TM of Madge Networks Ltd.

Microsoft: TM of Microsoft Corp.

Microsoft C: RTM of Microsoft Corp.

Microsoft Windows: RTM of Microsoft Corp.

MPW: RTM of Claris Corp.

MSCP: TM of DEC

MS-DOS: RTM of Microsoft Corp.

Multifinder: RTM of Claris Corp.

MVS, MVS/ESA, MVS/XA: RTMs of IBM

NAP: RTM of Automated Network Management, Inc.

NCP: RTM of IBM

NCS: RTM of Apollo Computer, Inc.

NETMAP: RTM of SynOptics Communications, Inc.

NetView: RTM of IBM

NetWare: RTM of Novell, Inc.

Network Computing System: RTM of Apollo Computer, Inc.

Network File System: RTM of Sun Microsystems, Inc.

NFS: RTM of Sun Microsystems, Inc.

Novell: RTM of Novell, Inc.

NuBus: TM of Texas Instruments

OpenVMS: TM of DEC

OSF, OSF/Motif: RTMs of Open Software Foundation, Inc.

OS/2: RTM of IBM

PATHWORKS: TM of DEC

PC-AT: TM of IBM

PC-NFS: RTM of Sun Microsystems, Inc.

POLYCENTER: TM of DEC

Portmapper: RTM of Sun Microsystems, Inc.

POSIX: TM of Institute of Electrical and Electronic Engineers

PostScript: RTM of Adobe Systems, Inc.

Proprinter: TM of IBM

P2/2: TM of IBM

Reliable Transaction router: TM of DEC

RISC System//6000 (RISC/6000): TM of IBM

RS6000: RTM of IBM

RT: TM of IBM

rtVAX: TM of DEC

SANE: RTM of Apple Computer, Inc.

SDLC: RTM of IBM

SNA: RTM of IBM

Sun: RTM of Sun Microsystems, Inc.

SunOS: RTM of Sun Microsystems, Inc.

Switcher: TM of Apple Computer, Inc.

TMSCP: TM of DEC

Trellis: TM of DEC

TURBOchannel: TM of DEC

ULTRIX: TM of DEC

UNIX: licensed by and RTM of Unix System Laboratories, Inc. (UNIX was originally developed in the 1970s by AT&T and Bell Laboratories)

VAX: TM of DEC

VAX Ada: TM of DEC

VAX APL: TM of DEC

VAX BASIC: TM of DEC

VAX BLISS-32: TM of DEC

VAX C: TM of DEC

VAX COBOL: TM of DEC

VAX DATATRIEVE: TM of DEC

VAX DBMS: TM of DEC

VAX DIBOL: TM of DEC

VAX DOCUMENT: TM of DEC

VAX DSM: TM of DEC

VAX FORTRAN: TM of DEC

VAX LISP: TM of DEC

VAX MACRO: TM of DEC

VAX Notes: TM of DEC

VAX OPSS: TM of DEC

VAX Pascal: TM of DEC

VAX RALLY: TM of DEC

VAX RMS: TM of DEC

VAX SCAN: TM of DEC

VAX SQL: TM of DEC

VAX TEAMDATA: TM of DEC

VAXcluster: TM of DEC

VAXELN: TM of DEC

VAXft: TM of DEC

VAXmail: TM of DEC

VAXshare: TM of DEC

VAXsimPLUS: TM of DEC

VAXstation: TM of DEC

VIDA: TM of DEC

VMS: TM of DEC

VTAM: RTM of IBM

VT100, VT220. VT330: TMs of DEC

Windows: TM of Microsoft, Inc.

Word for Windows: TM of Microsoft, Inc.

WPS: TM of DEC

WPS-PLUS: TM of DEC

Xerox: TM of Xerox Corp.

XNS: TM of Xerox Corp.

X/Open: RTM of X.Open Co. Ltd.

XUI: TM of DEC

XWindow: RTM of MIT

X-Windows: TM of MIT

Bibliography

Abbatiello, Judy, and Ray Sarch, eds., 1987, *Telec Communications & Data Communications Factbook,* New York, N.Y.: Data Communications; Ramsey, N.J.: CCMI/McGraw-Hill.

Apple Computer, 1991, *Planning and Managing AppleTalk Networks,* Menlo Park, Calif.: Addison-Wesley.

Apple Computer, 1992, *Technical Introduction to the Macintosh Family Second Edition,* Menlo Park, Calif.: Addison-Wesley.

Ashley, Ruth, and Judi N. Fernandez, 1984, *Job Control Language,* Wiley.

Aspray, William, 1990, *John Von Neumann and The Origins of Modern Computing,* Cambridge, Mass.: MIT Press.

ATM Forum, The, 1993, *ATM User-Network Interface Specification,* Englewood Cliffs, N.J.: Prentice-Hall.

Bach, Maurice, J., 1986, *The Design of the UNIX Operating System,* Englewood Cliffs, N.J.: Prentice-Hall.

Baggott, Jim, 1992, *The Meaning of Quantum Theory,* New York, N.Y.: Oxford University Press.

Bashe, Charles J., Lyle R. Johnson, John H. Palmer, and Emerson W. Pugh, 1986, *IBM's Early Computers,* Cambridge, Mass.: MIT Press.

Berson, Alex, 1990, *APPC Introduction to LU6.2,* New York, N.Y.: McGraw-Hill.

Black, Uyless, 1989, *Data Networks Concepts, Theory, and Practice,* Englewood Cliffs, N.J.: Prentice-Hall.

Black, Uyless, 1991, *The V Series Recommendations Protocols for Data Communications Over the Telephone Network,* New York, N.Y.: McGraw-Hill.

Black, Uyless, 1991, *The X Series Recommendations Protocols for Data Communications Networks,* New York, N.Y.: McGraw-Hill.

Black, Uyless, 1992, *TCP/IP and Related Protocols,* New York, N.Y.: McGraw-Hill.

Blyth, W. John, and Mary M. Blyth, 1990, *Telecommunications: Concepts, Development, and Management,* Mission Hills, Calif.: Glencoe/McGraw-Hill.

Bohl, Marilyn, 1971, *Information Processing,* 3d ed., Chicago, Ill.: Science Research Associates.

Bradbeer, Robin, Peter De Bono, and Peter Laurie, 1982, *The Beginner's Guide to Computers,* Menlo Park, Calif.: Addison-Wesley.

Brookshear, J. Glenn, 1988, *Computer Science: An Overview,* Menlo Park, Calif.: Benjamin/Cummings.

Bryant, David, 1971, *Physics,* Great Britain: Hodder and Stoughton.

Campbell, Joe, 1984, *The RS-232 Solution,* Alameda, Calif.: SYBEX.

Campbell, Joe, 1987, *C Programmer's Guide to Serial Communications,* Carmel, Ind.: Howard W. Sams.

Chorafas, Dimitris N., 1989, *Local Area Network Reference,* New York, N.Y.: McGraw-Hill.

Comer, Douglas, 1988, *Internetworking with TCP/IP Principles, Protocols, and Architecture,* Englewood Cliffs, N.J.: Prentice-Hall.

Comer, Douglas E., 1991, *Internetworking with TCP/IP, Vol. 1, Principles, Protocols, and Architecture,* Englewood Cliffs, N.J.: Prentice-Hall.

Comer, Douglas, E., and David L. Stevens, 1991, *Internetworking with TCP/IP, Vol. II, Design, Implementation, and Internals,* Englewood Cliffs, N.J.: Prentice-Hall.

Dayton, Robert L., 1991, *Telecommunications: The Transmission of Information,* New York, N.Y.: McGraw-Hill.

Dern, Daniel P., 1994, *The Internet Guide for New Users,* New York, N.Y.: McGraw-Hill.

Digital Equipment, Corp., 1991, *DECnet Digital Network Architecture (Phase V): Network Routing Layer Functional Specification,* EK-DNA03-FS-001, Maynard, Mass.: Digital Equipment Corp.

Digital Equipment Corp., 1993, *DECnet/OSI for OpenVMS: Introduction and Planning,* AA-PNHTB-TE, Maynard, Mass.: Digital Equipment Corp.

Digital Equipment Corp., 1993, *OpenVMS DCL Dictionary: A-M,* AA-PV5LA-TK, Maynard, Mass.: Digital Equipment Corp.

Digital Equipment Corp., 1993, *OpenVMS DCL Dictionary: N-Z,* AA-PV5LA-TK, Maynard, Mass.: Digital Equipment Corp.

Digital Equipment Corp., 1993, *OpenVMS Glossary,* AA-PV5UA-TK, Maynard, Mass.: Digital Equipment Corp.

Digital Equipment Corp., 1993, *OpenVMS Software Overview,* AA-PVXHA-TE, Maynard, Mass.: Digital Equipment Corp.

Edmunds, John J., 1992, *SAA/LU 6.2 Distributed Networks and Applications,* New York, N.Y.: McGraw-Hill.

Feit, Sidnie, 1993, *TCP/IP Architecture, Protocols, and Implementation,* New York, N.Y.: McGraw-Hill.

Forney, James S., 1989, *MS-DOS Beyond 640K Working with Extended and Expanded Memory,* Blue Ridge Summit, Pa.: Windcrest Books.

Forney, James S., 1992, *DOS Beyond 640K,* 2d ed., Blue Ridge Summit, Pa.: Windcrest/McGraw-Hill.

Fortier, Paul J., 1989, *Handbook of LAN Technology,* New York, N.Y.: Intertext Publications/Multiscience Press.

Gasman, Lawrence, 1994, *Broadband Networking,* New York, N.Y.: Van Nostrand Reinhold.

Graubart-Cervone, H. Frank, 1994, *VSE/ESA JCL Utilities, Power, and VSAM,* New York, N.Y., McGraw-Hill.

Groff, James R., and Paul N. Weinbert, 1983, *Understanding UNIX: A Conceptual Guide,* Carmel, Ind.: Que Corp.

Hecht, Jeff, 1990, *Understanding Fiber Optics,* Carmel, Ind.: Howard W. Sams.

Hewlett-Packard Co., 1991, *Using the X Window System,* B1171-90037, Ft. Collins, Colo.: Hewlett-Packard.

Hewlett-Packard Co., 1992, *HP OpenView SNMP Agent Administrator's Reference,* J2322-90002, Ft. Collins, Colo.: Hewlett-Packard.

Hewlett-Packard Co., 1992, *HP OpenView SNMP Management Platform Administrator's Reference,* J2313-90001, Ft. Collins, Colo.: Hewlett-Packard.

Hewlett-Packard Co., 1992, *HP OpenView Windows User's Guide,* J2316-90000, Ft. Collins, Colo.: Hewlett-Packard.

Hewlett-Packard Co., 1992, *Using HP-UX: Hp 9000 Workstations,* B2910-90001, Ft. Collins, Colo.: Hewlett-Packard.

IBM Corp., 1980, *IBM Virtual Machine Facility: Terminal User's Guide,* GC20-1810-9, Poughkeepsie, N.Y.: IBM.

IBM Corp., 1983, *IBM System/370 Extended Architecture: Principles of Operation,* SA22-7085-0, Research Triangle Park, N.C.: IBM.

IBM Corp., 1985, *IBM 3270 Information Display System: 3274 Control Unit Description and Programmer's Guide,* GA23-0061-2, Research Triangle Park, N.C.: IBM.

IBM Corp., 1986, *JES3 Introduction,* GC23-0039-2, Poughkeepsie, N.Y.: IBM.

IBM Corp., 1987, *IBM System/370: Principles of Operation,* GA22-7000-10, Poughkeepsie, N.Y.: IBM.

IBM Corp., 1988, *IBM Enterprise Systems Architecture/370: Principles of Operation,* SA22-7200-0, Poughkeepsie, N.Y.: IBM.

IBM Corp., 1988, *3270 Information Display System: Introduction,* GA27-2739-22, Research Triangle Park, N.C.: IBM.

IBM Corp., 1989, *MVS/ESA Operations: System Commands Reference Summary*, GX22-0013-1, Poughkeepsie, N.Y.: IBM.

IBM Corp., 1990, *Enterprise System/9000 Models 120, 130, 150, and 170: Introducing the System*, GA24-4186-00, Endicott, N.Y.: IBM.

IBM Corp., 1990, *Enterprise Systems Architecture/390: Principles of Operation*, SA22-7201-00, Poughkeepsie, N.Y.: IBM.

IBM Corp., 1990, *IBM 3172 Interconnect Controller: Presentation Guide*, White Plains, N.Y.: IBM.

IBM Corp., 1990, *IBM VSE/ESA: System Control Statements*, SC33-6513-00, Mechanicsburg, Pa.: IBM.

IBM Corp., 1990, *IBM VSE/POWER: Networking*, SC33-6573-00, Mechanicsburg, Pa.: IBM.

IBM Corp., 1990, *MVS/ESA SP Version 4 Technical Presentation Guide*, GG24-3594-00, Poughkeepsie, N.Y.: IBM.

IBM Corp., 1990, *Virtual Machine/Enterprise Systems Architecture*, GC24-5441, Endicott, N.Y.: IBM.

IBM Corp., 1990, *VM/ESA and Related Products: Overview*, GG24-3610-00, Poughkeepsie, N.Y.: IBM.

IBM Corp., 1991, *Dictionary of Computing*, SC20-1699-8, Poughkeepsie, N.Y.: IBM.

IBM Corp., 1991, *Enterprise Systems Architecture/390 ESCON I/O Interface: Physical Layer*, SA23-0394-00, Kingston, N.Y.: IBM.

IBM Corp., 1991, *Enterprise Systems Connection*, GA23-0383-01, Kingston, N.Y.: IBM.

IBM Corp., 1991, *Enterprise Systems Connection: ESCON I/O Interface*, SA22-7202-01, Poughkeepsie, N.Y.: IBM.

IBM Corp., 1991, *Enterprise Systems Connection Manager*, GC23-0422-01, Kingston, N.Y.: IBM.

IBM Corp., 1991, *Installation Guidelines for the IBM Token-Ring Network Products*, GG24-3291-02, Research Triangle Park, N.C.: IBM.

IBM Corp., 1991 *NetView: NetView Graphic Monitor Facility Operation*, SC31-6099-1, Research Triangle Park, N.C.: IBM.

IBM Corp., 1991, *Systems Network Architecture: Concepts and Products*, GC30-3072-4, Research Triangle Park, N.C.: IBM.

IBM Corp., 1991, *Systems Network Architecture: Technical Overview*, GC30-3073-3, Research Triangle Park, N.C.: IBM.

IBM Corp., 1991, *Systems Network Architecture: Type 2.1 Node Reference*, V. 1, SC20-3422-2, Research Triangle Park, N.C.: IBM.

IBM Corp., 1991, *3174 Establishment Controller: Functional Description*, GA23-0218-08, Research Triangle Park, N.C.: IBM.

IBM Corp., 1991, *Virtual Machine/Enterprise System Architecture: General Information*, GC24-5550-02, Endicott, N.Y.: IBM.

IBM Corp., 1992, *APPN Architecture and Product Implementations Tutorial*, GG24-3669-01, Research Triangle Park, N.C.: IBM.

IBM Corp., 1992, *ES/9000 Multi-Image Processing Volume 1: Presentation and Solutions Guidelines*, GG24-3920-00, Poughkeepsie, N.Y.: IBM.

IBM Corp., 1992, *High Speed Networking Technology: An Introductory Survey*, GG24-3816-00, Raleigh, N.C.: IBM.

IBM Corp., 1992, *IBM Networking Systems: Planning and Reference*, SC31-6191-00, Research Triangle Park, N.C.: IBM.

IBM Corp., 1992, *The IBM 6611 Network Processor*, GG24-3870-00, Raleigh, N.C.: IBM.

IBM Corp., 1992, *MVS/ESA: General Information for MVS/ESA System Product Version 4*, GC28-1600-04, Poughkeepsie, N.Y.: IBM.

IBM Corp., 1992, *MVS/ESA and Data in Memory: Performance Studies*, GG24-3698-00, Poughkeepsie, N.Y.: IBM.

IBM Corp., 1992, *Sockets Interface for CICS—Using TCP/IP Version 2 Release 2 for MVS: User's Guide*, GC31-7015-00, Research Triangle Park, N.C.: IBM.

IBM Corp., 1992, *Synchronous Data Link Control: Concepts*, GA27-3093-04, Research Triangle Park, N.C.: IBM.

IBM Corp., 1992, *TCP/IP Version 2 Release 2.1 for MVS: Offload of TCP/IP Processing,* SA31-7033-00, Research Triangle Park, N.C.: IBM.

IBM Corp., 1992, *TCP/IP Version 2 Release 2.1 for MVS: Planning and Customization,* SC31-6085-02, Research Triangle Park, N.C.: IBM.

IBM Corp., 1992, *3172 Interconnect Controller: Operator's Guide,* GA27-3970-00, Research Triangle Park, N.C.: IBM.

IBM Corp., 1992, *3172 Interconnect Controller: Planning Guide,* GA27-3867-05, Research Triangle Park, N.C.: IBM.

IBM Corp., 1992, *3270 Information Display System: Data Stream Programmer's Reference,* GA23-0059-07, Research Triangle Park, N.C.: IBM.

IBM Corp., 1992, *VM/ESA: CMS Primer,* SC24-5458-02, Endicott, N.Y.: IBM.

IBM Corp., 1992, *VM/ESA Release 2 Overview,* GG24-3860-00, Poughkeepsie, N.Y.: IBM.

IBM Corp., 1992, *VSE/ESA Version 1.3: An Introduction Presentation Foil Master,* GG24-4008-00, Raleigh, N.C.: IBM.

IBM Corp., 1993, *The Host as a Data Server Using LANRES and Novell NetWare,* GG24-4069-00, Poughkeepsie, N.Y.: IBM.

IBM Corp., 1993, *IBM Network Products Implementation Guide,* GG24-3649-01, Raleigh, N.C.: IBM.

IBM Corp., 1993, *IBM VSE/Interactive Computing and Control Facility: Primer,* SC33-6561-01, Charlotte, N.C.: IBM.

IBM Corp., 1993, *LAN File Services/ESA: MVS Guide and Reference,* SH24-5265-00, Endicott, N.Y.: IBM.

IBM Corp., 1993, *LAN File Services/ESA: VM Guide and Reference,* SH24-5264-00, Endicott, N.Y.: IBM.

IBM Corp., 1993, *LAN Resource Extension and Services/VM: Guide and Reference,* SC24-5622-01, Endicott, N.Y.: IBM.

IBM Corp., 1993, *MVS/ESA: JES2 Command Reference Summary,* GX22-0017-03, Poughkeepsie, N.Y.: IBM.

IBM Corp., 1993, *MVS/ESA JES2 Commands,* GC23-0084-04, Poughkeepsie, N.Y.: IBM.

IBM Corp., 1993, *MVS/ESA: System Commands,* GC28-1626-05, Poughkeepsie, N.Y.: IBM.

IBM Corp., 1993, *NetView: Installation and Administration,* SC31-7084-00, Research Triangle Park, N.C.: IBM.

IBM Corp., 1993, *NetView: Command Quick Reference,* SX75-0090-00, Research Triangle Park, N.C.: IBM.

IBM Corp., 1993, *System Information Architecture: Formats,* GA27-3136, Research Triangle Park, N.C.: IBM.

IBM Corp., 1993, *System Network Architecture: Architecture Reference,* Version 2, SC30-3422-03, Research Triangle Park, N.C.: IBM.

IBM Corp., 1993, *Virtual Machine/Enterprise Systems Architecture,* SC24-5460-03, Endicott, N.Y.: IBM.

IBM Corp., 1993, *VM/ESA: CMS Command Reference,* SC24-5461-03, Endicott, N.Y.: IBM.

IBM Corp., 1993, *VM/ESA: CP Command and Utility Reference,* SC24-5519-03, Endicott, N.Y.: IBM.

IBM Corp., 1993, *VTAM: Operation,* SC31-6420-00, Research Triangle Park, N.C.: IBM.

IBM Corp., 1993, *VTAM: Resource Definition Reference Version 4 Release 1 for MVS/ESA,* SC31-6427-00, Research Triangle Park, N.C.: IBM.

IBM Corp., 1994, *LAN Resource Extension and Services/VM: General Information,* GC24-5618-03, Endicott, N.Y.: IBM.

IBM Corp., 1994, *LAN Resources Extension and Services/MVS: General Information,* GC24-5625-03, Endicott, N.Y.: IBM.

IBM Corp., 1994, *LAN Resource Extension and Services/MVS: Guide and Reference,* SC24-5623-02, Endicott, N.Y.: IBM.

Jain, Bijendra N., and Ashok K. Agrawala, 1993, *Open Systems Interconnection*, New York, N.Y.: McGraw-Hill.

Kessler, Gary C., 1990, *ISDN*, New York, N.Y.: McGraw-Hill.

Kessler, Gary C., and David A. Train, 1992, *Metropolitan Area Networks Concepts, Standards, and Services*, New York, N.Y.: McGraw-Hill.

Killen, Michael, 1992, *SAA and UNIX IBM's Open Systems Strategy*, New York, N.Y.: McGraw-Hill.

Killen, Michael, 1992, *SAA Managing Distributed Data*, New York, N.Y.: McGraw-Hill.

Kochan, Stephen G., and Patrick H. Wood, 1984, *Exploring the UNIX System*. Indianapolis, Ind.: Hayden Books.

Madron, Thomas W., 1988, *Local Area Networks: The Next Generation*, New York, N.Y.: Wiley.

Martin, James, 1989, *Local Area Networks Architectures and Implementations*, Englewood Cliffs, N.J.: Prentice-Hall.

McClain, Gary R., 1991, *Open Systems Interconnection Handbook*, New York, N.Y.: Intertext Publications/Multiscience Press.

Meijer, Anton, 1987, *Systems Network Architecture a Tutorial*, London, U.K.: Pitman; New York, N.Y.: Wiley.

Merrow, Bill, 1994, *VSE/ESA Concepts and Facilities*, New York, N.Y.: McGraw-Hill.

Merrow, Bill, 1993, *VSE/ESA Performance Management and Fine Tuning*, New York, N.Y.: McGraw-Hill.

Nash, Stephen G., Editor, 1990, *A History of Scientific Computing*, New York, N.Y.: ACM Press.

Naugle, Matthew G., 1991, *Local Area Networking*, New York, N.Y.: McGraw-Hill.

Naugle, Matthew, 1994, *Network Protocol Handbook*, New York, N.Y.: McGraw-Hill.

Nemzow, Martin A. W., 1992, *The Ethernet Management Guide Keeping the Link*, 2d ed., New York, N.Y.: McGraw-Hill.

O'Dell, Peter, 1989, *The Computer Networking Book*, Chapel Hill, N.C.: Ventana Press.

Parker, Sybil P., 1984, *McGraw-Hill Dictionary of Science and Engineering*, New York, N.Y.: McGraw-Hill.

Pugh, Emerson W., 1984, *Memories That Shaped an Industry*, Cambridge, Mass.: MIT Press.

Pugh, Emerson W., Lyle R. Johnson, and John H. Palmer, *IBM's 360 and Early 370 Systems*, Cambridge, Mass.: MIT Press.

Ranade, Jay, and George C. Sackett, 1989, *Introduction to SNA Networking Using VTAM/NCP*, New York, N.Y.: McGraw-Hill.

Rose, Marshall T., 1990, *The Open Book: A Practical Perspective on OSI*, Englewood Cliffs, N.J.: Prentice-Hall.

Rose, Marshall T., 1991, *The Simple Book: An Introduction to Management of TCP/IP-Based Internets*, Englewood Cliffs, N.J.: Prentice-Hall.

Samson, Stephen L., 1990, *MVS Performance Management*, New York, N.Y.: McGraw-Hill.

Savit, Jeffrey, 1993, *VM/CMS Concepts and Facilities*, New York, N.Y.: McGraw-Hill.

Schatt, Stan, 1990, *Understanding Local Area Networks*, 2d ed., Carmel, Ind.: Howard W. Sams.

Schlar, Serman K., 1990, *Inside X.25: A Manager's Guide*, New York, N.Y.: McGraw-Hill.

Seyer, Martin D., 1991, *RS-232 Made Easy: Connecting Computers, Printers, Terminals, and Modems*, Englewood Cliffs, N.J.: Prentice-Hall.

Sidhu, Gursharan S., Richard F. Andrews, and Alan B. Oppenheimer, 1990, *Inside AppleTalk*, 2d ed., Menlo Park, Calif.: Addison-Wesley.

Spohn, Darren L., 1993, *Data Network Design*, New York, N.Y.: McGraw-Hill.

Stallings, William, 1987, *Handbook of Computer-Communications Standards*, Vol. 1, New York, N.Y.: Macmillan.

Stallings, William, 1987, *Handbook of Computer-Communications Standards*, Vol. 2, New York, N.Y.: Macmillan.

Stallings, William, 1988, *Handbook of Computer-Communications Standards*, Vol. 3, New York, N. Y.: Macmillan.

Stallings, William, 1989, *ISDN: An Introduction,* New York, N.Y.: Macmillan.
Stamper, David A., 1986, *Business Data Communications,* Menlo Park, Calif.: Benjamin/Cummings.
Tang, Adrian, and Sophia Scoggins, 1992, *Open Networking with OSI,* Englewood Cliffs, N.J.: Prentice-Hall.
Umar, Amjad, 1993, *Distributed Computing: A Practical Synthesis,* Englewood Cliffs, N.J.: Prentice-Hall.
White, Gene, 1992, *Internetworking and Addressing,* New York, N.Y.: McGraw-Hill.
Zwass, Vladimir, 1981, *Introduction to Computer Science,* New York, N.Y.: Barnes & Noble Books.

Index

Note: For significant developments arranged by dates, *see* Chronology in this index. Numbers are alphabetized under spelled-out form.

ABOUT THE AUTHOR

D Edgar (Ed) Taylor is an independent consultant and lecturer, who speaks frequently at major trade shows. He has contributed in the integration of heterogeneous networks for entities such as Orange County, California, Childrens Hospital, Ohio, Ore-Ida Foods, and others. Mr. Taylor is also the author of *Integrating TCP/IP Into SNA, Demystifying TCP/IP*, and *Demystifying SNA*, and forthcoming books entitled *Multiplatform Network Management* and *Internetworking Command Reference* (McGraw-Hill).